VEHICLE ROUTING

MOS-SIAM Series on Optimization

This series is published jointly by the Mathematical Optimization Society and the Society for Industrial and Applied Mathematics. It includes research monographs, books on applications, textbooks at all levels, and tutorials. Besides being of high scientific quality, books in the series must advance the understanding and practice of optimization. They must also be written clearly and at an appropriate level for the intended audience.

Editor-in-Chief

Katya Scheinberg
Lehigh University

Editorial Board

Santanu S. Dey, *Georgia Institute of Technology*
Maryam Fazel, *University of Washington*
Andrea Lodi, *University of Bologna*
Arkadi Nemirovski, *Georgia Institute of Technology*
Stefan Ulbrich, *Technische Universität Darmstadt*
Luis Nunes Vicente, *University of Coimbra*
David Williamson, *Cornell University*
Stephen J. Wright, *University of Wisconsin*

Series Volumes

VEHICLE ROUTING
Problems, Methods, and Applications
Second Edition

Edited by

Paolo Toth
DEI, University of Bologna
Bologna, Italy

Daniele Vigo
DEI, University of Bologna
Bologna, Italy

Society for Industrial and Applied Mathematics
Philadelphia

Mathematical Optimization Society
Philadelphia

Library of Congress Cataloging-in-Publication Data
Vehicle routing problem.
  Vehicle routing : problems, methods, and applications / edited by Paolo Toth, University of Bologna, Bologna, Italy, Daniele Vigo, University of Bologna, Bologna, Italy. – Second edition.
      pages cm. – (MOS-SIAM series on optimization)
  Revision of: The vehicle routing problem. ©2002.
  Includes bibliographical references and index.
  ISBN 978-1-611973-58-7
  1. Transportation problems (Programming) I. Toth, Paolo, editor. II. Vigo, Daniele, editor. III. Title.
  QA402.6.V44 2014
  388.3'10285–dc23
                              2014029491

 is a registered trademark.    Mathematical Optimization Society is a registered trademark.

List of Contributors

Claudia Archetti
Dipartimento Metodi Quantitativi, Università
di Brescia, Italy,
archetti@eco.unibs.it

Maria Battarra
Department of Mathematical Sciences,
University of Southampton, UK,
m.battarra@soton.ac.uk

Tolga Bektaş
Southampton Management School, University
of Southampton, UK,
t.bektas@soton.ac.uk

Olli Bräysy
VU University of Amsterdam,
The Netherlands,
o.m.p.braysy@vu.nl

Marielle Christiansen
Department of Industrial Economics and
Technology Management, Norwegian
University of Science and Technology of
Trondheim, Norway,
mc@iot.ntnu.no

Jean-François Cordeau
HEC Montréal, Québec, Canada,
jean-francois.cordeau@hec.ca

Guy Desaulniers
Department of Mathematical and Industrial
Engineering, École Polytechnique de Montréal,
Québec, Canada,
guy.desaulniers@polymtl.ca

Karl F. Doerner
Department of Business Administration,
University of Vienna, Austria,
karl.doerner@univie.ac.at

Richard Eglese
Department of Management Science, Lancaster
University Management School, UK,
R.Eglese@lancaster.ac.uk

Kjetil Fagerholt
Department of Industrial Economics and
Technology Management, Norwegian
University of Science and Technology of
Trondheim, Norway,
kjetil.fagerholt@iot.ntnu.no

Michel Gendreau
Department of Mathematical and Industrial
Engineering, École Polytechnique de Montréal,
Québec, Canada,
michel.gendreau@polymtl.ca

Bruce L. Golden
Robert H. Smith School of Business,
University of Maryland, MD, USA,
bgolden@rhsmith.umd.edu

Geir Hasle
SINTEF ICT, Norway,
geir.hasle@sintef.no

Manuel Iori
Dipartimento di Scienze e Metodi
dell'Ingegneria, Università degli Studi di
Modena e Reggio Emilia, Italy,
manuel.iori@unimore.it

Stefan Irnich
Chair of Logistics Management, Gutenberg
School of Management and Economics,
Johannes Gutenberg University Mainz,
Germany,
irnich@uni-mainz.de

Ola Jabali
Department of Logistics and Operations Management, HEC Montréal and CIRRELT, Québec, Canada,
ola.jabali@hec.ca

Attila A. Kovacs
Department of Business Administration, University of Vienna, Austria,
attila.kovacs@univie.ac.at

Gilbert Laporte
HEC Montréal, Québec, Canada,
gilbert.laporte@cirrelt.ca

Oli B. G. Madsen
Department of Transport, Technical University of Denmark, Kongens Lingby, Denmark,
ogm@transport.dtu.dk

Marcus Poggi
Departamento de Informática, Pontifícia Universidade Católica de Rio de Janeiro, Brazil,
poggi@inf.puc-rio.br

Walter Rei
Dŕpartement de management et technologie, Université du Québec a Montréal, Québec, Canada,
rei.walter@uqam.ca

Panagiotis P. Repoussis
Stevens Institute of Technology, Hoboken, NJ, USA,
panagiotis.repoussis@stevens.edu

Stefan Ropke
Department of Management, Engineering, Technical University of Denmark, Kongens Lyngby, Denmark,
ropke@dtu.dk

Juan-José Salazar-González
Departamento de Estadística, Investigación Operativa y Computación, Universidad de La Laguna, Tenerife, Spain,
jjsalaza@ull.es

Michael Schneider
Logistikplanung und Informationssysteme, Technische Universität Darmstadt, Germany,
schneider@bwl.tu-darmstadt.de

Frédéric Semet
Ecole Centrale de Lille, Villeneuve d'Ascq Cedex, France,
frederic.semet@ec-lille.fr

M. Grazia Speranza
Dipartimento di Economia e Management, Università degli Studi di Brescia, Italy,
speranza@eco.unibs.it

Christos D. Tarantilis
Department of Management Science and Technology, Athens University of Economics and Business, Greece,
tarantil@aueb.gr

Paolo Toth
Department of Electrical, Electronic, and Information Engineering "G. Marconi", Università di Bologna, Italy,
paolo.toth@unibo.it

Eduardo Uchoa
Departamento de Engenharia de Produção, Universidade Federal Fluminensc, Niterói, Rio de Janeiro, Brazil,
uchoa@producao.uff.br

Thibaut Vidal
Laboratory for Information and Decision Systems, Massachusetts Institute of Technology, Cambridge, MA, USA,
vidalt@mit.edu

Daniele Vigo
Department of Electrical, Electronic, and Information Engineering "G. Marconi", Università di Bologna, Italy,
daniele.vigo@unibo.it

Edward A. Wasil
Kogod School of Business, American University, Washington, DC, USA,
ewasil@american.edu

Contents

List of Figures

List of Tables

Preface to the Second Edition

The projects of reediting the Toth and Vigo book on vehicle routing and of editing a book on arc routing germinated during the ROUTE Conference in Sitges, Spain, in June 2011. The first edition of the vehicle routing book had been highly successful, and it was then felt that the evolution of the field over the past 10 years justified a significantly revamped reedition. This led Corberán and Laporte (while exploring the cellars of the Codorníu Winery during the conference excursion) to think up a proposal for a similar arc routing book that would be produced in parallel with the second edition of the vehicle routing book, with a similar structure and the same format. Again, the last major edited book on arc routing had been published more than 10 years before and the field had evolved considerably since then. Both proposals were presented to SIAM in the summer and were accepted. Today we are proud to offer to the research community two up-to-date collections of scientific contributions written by specialists in various areas of vehicle routing and arc routing. The two books are entitled "Vehicle Routing: Problems, Methods, and Applications, Second Edition", Paolo Toth and Daniele Vigo, editors, and "Arc Routing: Problems, Methods, and Applications", Ángel Corberán and Gilbert Laporte, editors, both published by SIAM.

The vehicle routing book contains 15 chapters. A few of these are amalgamations or significantly revised versions of chapters published in the first edition, while most of the others are entirely new. The first chapter offers an overview of the field of the Vehicle Routing Problem (VRP) and its main variants. The remainder of the book is made up of three parts: the capacitated VRP, important variants of the VRP, and applications. The first part contains two chapters on classical and new exact algorithms, as well as a chapter on heuristics. The second part surveys several variants: the VRP with time windows, pickup-and-delivery problems for goods or people transportation, stochastic VRPs, and miscellaneous variants. The third part is devoted to applications and covers the VRP with profits, real-time and dynamic VRPs, software and emerging technologies, ship routing, VRP applications in disaster relief, as well as green vehicle routing.

The arc routing book is new and contains 16 chapters. It opens with a chapter on historical perspectives, followed by three main parts: arc routing problems with a single vehicle, arc routing problems with several vehicles, and applications. The first part starts with a chapter on complexity, which is followed by four chapters on the Chinese Postman Problem and on the Rural Postman Problem. The second part contains four chapters on the Capacitated Arc Routing Problem and two on arc routing problems with min-max and profit maximization objectives. The last part covers some of the most important arc routing applications, including meter reading, salt spreading, snow removal, garbage collection, and newspaper delivery.

We thank all authors for the quality of their contributions, as well as all referees who carefully reviewed the chapters, and Claudio Gambella for his help in editing the final manuscript of the VRP book. Thanks are also due to Dr. Thomas Liebling, Ms. Elizabeth Greenspan, Ms. Ann Manning Allen, and Ms. Sara J. Murphy from SIAM for their support and encouragement.

Ángel Corberán, Universitat de Valencia
Gilbert Laporte, HEC Montréal
Paolo Toth, Università di Bologna
Daniele Vigo, Università di Bologna

May 2014

Preface to the First Edition

The Vehicle Routing Problem (VRP) calls for the determination of the optimal set of routes to be performed by a fleet of vehicles to serve a given set of customers, and it is one of the most important, and studied, combinatorial optimization problems.

More than 40 years have elapsed since Dantzig and Ramser introduced the problem in 1959. They described a real-world application concerning the delivery of gasoline to service stations and proposed the first mathematical programming formulation and algorithmic approach. A few years later, in 1964, Clarke and Wright proposed an effective greedy heuristic that improved on the Dantzig–Ramser approach. Following these two seminal papers, hundreds of models and algorithms were proposed for the optimal and approximate solution of the different versions of the VRP. Dozens of packages for the solution of various real-world VRPs are now available on the market. This interest in VRP is motivated by both its practical relevance and its considerable difficulty: the largest VRP instances that can be consistently solved by the most effective exact algorithms proposed so far contain about 50 customers, whereas larger instances may be solved to optimality only in particular cases.

This book covers the state of the art of both exact and heuristic methods developed in the last decades for the VRP and some of its main variants. Moreover, a considerable part of the book is devoted to the discussion of practical issues.

The realization of this project would have been impossible for us alone to accomplish. We thus involved an enthusiastic group of very well known experts, whose contributions form a large part of the recent history of the VRP (as well as that of Mathematical Programming and Combinatorial Optimization). As editors, we constantly devoted our efforts to reducing as much as possible the overlap between chapters and to preserving coherence and ensuring uniformity of the notation and terminology.

Although focused on a specific family of problems, this book offers a complete overview of the effective use of the most important techniques proposed for the solution of hard combinatorial problems. We, however, assume that readers have a basic knowledge of the main methods for the solution of combinatorial optimization problems (complexity theory, branch-and-bound, branch-and-cut, relaxations, heuristics, metaheuristics, local search, etc.).

The book is divided into three parts, preceded by an introductory chapter in which we present an overview of the VRP family, define the most important variants of the problem, and introduce the main mathematical models. The first part covers the basic and extensively studied version of the VRP, known as capacitated VRP. Three chapters examine the main exact approaches (branch-and-bound, branch-and-cut, and set-covering-based methods), while two other chapters review traditional heuristic approaches and metaheuristics, respectively. For all methods extensive computational results are analyzed. The second part covers three main variants of the VRP: the VRP with time windows, the VRP with backhauls, and the VRP with pickup and delivery. In each chapter, both exact and heuristic methods are examined. Finally, in the third part, the issues arising in real-world VRP applications, as the presence of dynamic and stochastic components, are discussed by analyzing relevant case studies and presenting software packages.

We warmly thank all the people who contributed to this project, which occupied a considerable amount of the past 3 years: our coauthors, whose competent, patient, and collaborative activity made possible the completion of this volume; the referees whose comments greatly improved the overall presentation; Peter Hammer, editor-in-chief of SIAM Monographs on Discrete Mathematics and Applications, who since the very beginning encouraged us and followed all the steps of the project; and Vickie Kearn, Deborah Poulson, Lou Primus, Sara Triller, Marianne Will, Donna Witzleben, Sam Young, and all the people of SIAM who greatly helped us in the preparation of the overall manuscript.

Paolo Toth
Daniele Vigo
Bologna, December 2000

Chapter 1

The Family of Vehicle Routing Problems

Stefan Irnich
Paolo Toth
Daniele Vigo

1.1 ▪ Introduction

A generic verbal definition of the *family of vehicle routing problems* can be the following:

> **Given:** A set of *transportation requests* and a *fleet of vehicles*.

The problem is then to find a plan for the following:

> **Task:** Determine a set of *vehicle routes* to *perform* all (or some) transportation requests with the given vehicle fleet *at minimum cost*; in particular, decide which *vehicle handles which requests in which sequence* so that all *vehicle routes* can be *feasibly* executed.

In this type of problem, subsumed under the term *Vehicle Routing Problem* (VRP), the transportation requests to be served are generally concentrated in specific points of a road network as opposed to the *Arc Routing Problems* (ARP; see the companion book by Corberán and Laporte [35]), where the requests are dispersed along the arcs, i.e., street segments of the underlying road network. The following sections will shed light on the basic VRP components, which are the *transportation requests* and how they can be performed, the *fleet* of vehicles, the related *costs* and *profits* (if relevant), and the *feasibility of routes*.

Before going into the details, however, we discuss the economic relevance of computer-supported vehicle routing. Indeed, the large number of real-world applications have widely shown that the use of computerized solution methods for the solution of VRP, both at the planning and the operational levels, yields substantial savings in the global transportation costs. The success of the utilization of optimization techniques is due not only to the power of the current computer systems and to the full integration of the information systems into the operations and commercial processes, but it can also be attributed to the development of rigorous mathematical models, which are able to take into account almost all the characteristics of the VRP arising in real-world applications. Furthermore, the corresponding algorithms and their computer implementations (*software tools*) play an essential role in finding high-quality feasible solutions for real-world instances within acceptable computing times. Compared to procedures not based on optimization techniques, significant cost savings and a better utilization of the vehicle fleet can be achieved. In addition, by means of such planning software it is possible to improve the automation, standardization, and integration into the organizations' overall planning processes, leading to less time-consuming and more cost-efficient planning processes with respect to manual planning. Moreover, computerized planning allows planners to compare several different planning scenarios, and herewith to choose a best one through a careful and fast evaluation of cost and service-related performance indicators.

In the last years, software tools integrating telematics services (electronic data transmission between vehicles and planners) have been developed with the aim of enabling a faster reaction of planners to the dynamics of the transportation system and to possible disruptions caused by the failure of vehicles or by heavy traffic conditions on the roads. Of course, online planning tools are indispensable for real-time applications such as the control of *automatic guided vehicles*.

More than 50 years have elapsed since the 1959 paper by Dantzig and Ramser [40] which introduced the VRP (then called the *truck dispatching problem*) as a real-world application concerning the delivery of gasoline to gas stations. In this seminal paper the authors proposed the first mathematical programming formulation and algorithmic approach for the VRP. Again, approximately 50 years separate us from the famous paper by Clarke and Wright [33], who in 1964 proposed an effective greedy heuristic for the approximate solution of the VRP. Following such brilliant forerunners, a huge number of papers have been published during the past five decades in the international Operations Research and Transportation Science journals, presenting mathematical models and proposing exact and (meta)heuristic algorithms for the optimal and approximate solution of the different versions of the VRP. Among the journals that regularly publish papers on the VRP, we can mention *Operations Research, Transportation Science, Computers & Operations Research, European Journal of Operational Research, EURO Journal on Transportation and Logistics, Journal of the Operational Research Society, Transportation Research, Networks*, and *Journal of Heuristics*. Finally, in all the main international Operations Research conferences there is at least one stream (often containing several sessions) with presentations of new research and application results on the VRP.

The high interest of the international research community in the different variants of VRP is not only motivated by their notorious difficulty as combinatorial optimization problems but also, as previously mentioned, by their practical relevance. As a consequence, researchers from both the academic and the industrial world work on the subject. In North America, the majority of the academics and practitioners working on VRP are members of the *Transportation Science and Logistics society* (TSL) within INFORMS (the Institute for Operations Research and Management Science), while in Europe a new working group of EURO (the association of the European operational research societies), called VeRoLog (Vehicle Routing and Logistics optimization), has recently been founded. Its purpose consists in "favouring the development and application of Operations Research models, methods and tools in the field of vehicle routing and logistics" and in "encouraging the exchange of information among practitioners, end-users and researchers in the area of vehicle routing and logistics, stimulating the work on new and important problems with sound scientific methods"; see http://www.verolog.eu/.

It is finally worth noting that the mathematical models and the exact and metaheuristic algorithms proposed for the VRP and presented in the chapters of this book constitute a reference point not only for the research on this specific area but also for the general domain of combinatorial optimization. Indeed, the VRP, and its variants are often used as benchmarks for the development of new models and algorithmic techniques later successfully applied for the effective solution of other difficult combinatorial optimization problems.

1.2 ▪ The Capacitated Vehicle Routing Problem

The *Capacitated Vehicle Routing Problem* (CVRP) is the most studied version of the VRP, although it has primarily an academic relevance. For the sake of clarity, we start our illustration of the *VRP family* with this basic variant in order to introduce a concrete example. Notation and models discussed here will be helpful to introduce other variants and to clarify the differences between the diverse types of VRPs.

1.2.1 ▪ Problem Statement

In the CVRP, the transportation requests consist of the distribution of goods from a single *depot*, denoted as point 0, to a given set of n other points, typically referred to as *customers*, $N = \{1, 2, \ldots, n\}$. The amount that has to be delivered to customer $i \in N$ is the customer's *demand*, which is given by a scalar $q_i \geq 0$, e.g., the weight of the goods to deliver. The *fleet* $K = \{1, 2, \ldots, |K|\}$ is assumed to be *homogeneous*, meaning that $|K|$ vehicles are available at the depot, all have the same capacity $Q > 0$, and are operating at identical costs. A vehicle that services a customer subset $S \subseteq N$ starts at the depot, moves once to each of the customers in S, and finally returns to the depot. A vehicle moving from i to j incurs the *travel cost* c_{ij}.

The given information can be structured using an undirected or directed graph. Let $V = \{0\} \cup N = \{0, 1, \ldots, n\}$ be the set of *vertices* (or nodes). It is convenient to define $q_0 := 0$ for the depot. In the symmetric case, i.e., when the cost for moving between i and j does not depend on the direction, i.e., either from i to j or from j to i, the underlying graph $G = (V, E)$ is complete and undirected with edge set $E = \{e = \{i, j\} = \{j, i\} : i, j \in V, i \neq j\}$ and edge costs c_{ij} for $\{i, j\} \in E$. Otherwise, if at least one pair of vertices $i, j \in V$ has asymmetric costs $c_{ij} \neq c_{ji}$ then the underlying graph is a complete digraph $G = (V, A)$ with arc set $A = \{(i, j) \in V \times V : i \neq j\}$ and arc costs c_{ij} for $(i, j) \in A$. Note that $|E| = n(n+1)/2$ and $|A| = n(n+1)$ so that both graphs contain $\mathcal{O}(n^2)$ links. Overall, a CVRP instance is uniquely defined by a complete weighted graph $G = (V, E, c_{ij}, q_i)$ or digraph $G = (V, A, c_{ij}, q_i)$ together with the size $|K|$ of the vehicle fleet K and the vehicle capacity Q.

A *route* (or *tour*) is a sequence $r = (i_0, i_1, i_2, \ldots, i_s, i_{s+1})$ with $i_0 = i_{s+1} = 0$, in which the set $S = \{i_1, \ldots, i_s\} \subseteq N$ of customers is visited. The route r has cost $c(r) = \sum_{p=0}^{s} c_{i_p, i_{p+1}}$. It is *feasible* if the capacity constraint $q(S) := \sum_{i \in S} q_i \leq Q$ holds and no customer is visited more than once, i.e., $i_j \neq i_k$ for all $1 \leq j < k \leq s$. In this case, one says that $S \subseteq N$ is a *feasible cluster*. A solution to a CVRP consists of $|K|$ feasible routes, one for each vehicle $k \in K$. The routes $r_1, r_2, \ldots, r_{|K|}$ and corresponding clusters $S_1, S_2, \ldots, S_{|K|}$ provide a *feasible solution* to the CVRP if all routes are feasible and the clusters form a partition of N. Concluding, the CVRP consists of two interdependent tasks:

(i) the partitioning of the customer set N into feasible clusters $S_1, \ldots, S_{|K|}$;

(ii) the routing of each vehicle $k \in K$ through $\{0\} \cup S_k$.

The latter task requires the solution of a *Traveling Salesman Problem* (TSP) over $\{0\} \cup S_k$ (see, e.g., Lawler et al. [93] and Gutin and Punnen [77]). Both tasks are intertwined because the cost of a cluster depends on the routing, and the routing needs clusters as an input.

1.2.2 ▪ Models

In this section, we present four important mathematical programming formulations for the CVRP. We have selected these four models in order to provide different views on the CVRP and to modify and extend some parts of the models to better explain VRP variants in later sections. Our choice of models is not intended to provide a comprehensive survey of VRP models, but additional formulations such as commodity-flow formulations can be found in Laporte and Nobert [90] and Toth and Vigo [136].

Basic Notation.   Let $S \subseteq V$ be an arbitrary subset of vertices. For undirected graphs, the *cut set* $\delta(S) = \{\{i,j\} \in E : i \in S, j \notin S\}$ (set $E(S) = \{\{i,j\} \in E : i,j \in S\}$) is the set of edges with exactly one (both) endpoint(s) in S. For directed graphs $G = (V, A)$, the *in-arcs* and *out-arcs* of S are defined as $\delta^-(S) = \{(i,j) \in A : i \notin S, j \in S\}$ and $\delta^+(S) = \{(i,j) \in A : i \in S, j \notin S\}$, respectively. It has become a standard to define $\delta(i) := \delta(\{i\})$ for singleton sets $S = \{i\}$ (similarly, $\delta^+(i)$ and $\delta^-(i)$). Moreover, $A(S) = \{(i,j) \in A : i,j \in S\}$ is the set of all arcs connecting vertices in S.

We will use a condensed notation in models, where for any vector of variables or coefficients x indexed by $i \in J$ and any subset $I \subseteq J$, the term $x(I)$ means $x(I) = \sum_{i \in I} x_i$. However, for the sake of completeness, for the first model we will also describe it with the more traditional notation, which makes explicit use of summations. Both notations will be used throughout the book.

For a customer subset $S \subseteq N$, let $r(S)$ be the minimum number of vehicle routes needed to serve S. In the CVRP, the number $r(S)$ can be computed by solving a *bin packing problem* (see Martello and Toth [98]) with items N of weight $q_i, i \in N$, and bins of size Q. A lower bound, often used instead of $r(S)$, is given by $\lceil q(S)/Q \rceil$.

Compact Formulations.   We present here the classical *compact formulations* for the CVRP, i.e., (Mixed) Integer Programming (MIP and IP) models which have a polynomial number of variables with respect to $n = |N|$ and $|K|$. We start with two *vehicle-flow formulations* that have an exponential number of constraints, and we briefly discuss modeling techniques to reduce the cardinality of the constraint sets to a polynomial number. Compact models are particularly well suited to solve simple VRP variants with mathematical programming based techniques, i.e., the direct use of a MIP solver or Branch-and-Cut algorithms; see Chapter 2. By "simple variant" we refer to those variants of the VRP in which the objective function and the constraints are expressed as summations over the visited vertices and traversed links (as opposed to, e.g., load-dependent costs).

A first important class of models has integer decision variables x_{ij} for $\{i,j\} \in E$ (or $(i,j) \in A$) indicating how often a vehicle directly moves between i and j (from i to j). Since the variables have two indices, the formulations are known as *two-index (vehicle-flow) formulations*.

The model for directed CVRP, denoted here as VRP1, was introduced by Laporte, Mercure, and Nobert [89]. As previously discussed, we first give the model for directed graphs in the traditional notation, which reads as follows:

(1.1)   (VRP1) $\text{minimize} \displaystyle\sum_{(i,j)\in A} c_{ij} x_{ij}$

(1.2)   s.t. $\displaystyle\sum_{j\in\delta^+(i)} x_{ij} = 1$ $\hspace{3em} \forall i \in N,$

$\displaystyle\sum_{i\in\delta^-(j)} x_{ij} = 1$ $\hspace{3em} \forall j \in N,$

(1.3) $\displaystyle\sum_{j\in\delta^+(0)} x_{0j} = |K|,$

(1.4) $\displaystyle\sum_{(i,j)\in\delta^+(S)} x_{ij} \geq r(S)$ $\hspace{3em} \forall S \subseteq N, S \neq \emptyset,$

(1.5) $x_{ij} \in \{0,1\}$ $\hspace{3em} \forall (i,j) \in A.$

Next, we present the same model in more condensed form using vectors and the above condensed notation for summations. Moreover, we set formulation VRP1 side by side with the two-index formulation for the undirected CVRP (denoted here as VRP2) introduced by Laporte, Nobert, and Desrochers [92] in order to highlight their similarities and differences:

		(VRP1)	(VRP2)	
(1.1)		$\text{minimize } c^\top x$	$\text{minimize } c^\top x$	
(1.2)	s.t.	$x(\delta^+(i)) = 1$	$x(\delta(i)) = 2$	$\forall i \in N,$
		$x(\delta^-(j)) = 1$		$\forall j \in N,$
(1.3)		$x(\delta^+(0)) = |K|$	$x(\delta(0)) = 2|K|,$	
(1.4)		$x(\delta^+(S)) \geq r(S)$	$x(\delta(S)) \geq 2r(S)$	$\forall S \subseteq N, S \neq \emptyset,$
(1.5)		$x_a \in \{0,1\} \ \forall a \in A$	$x_e \in \{0,1,2\} \ \forall e \in \delta(0)$	
			$x_e \in \{0,1\} \ \forall e \in E \setminus \delta(0).$	

In both models, the objective (1.1) is the minimization of the overall routing costs. Constraints (1.2) state that in a route, each customer vertex is connected to two other vertices, which are its *predecessor* and *successor*. Similarly, constraints (1.3) ensure that exactly $|K|$ routes are constructed. Therefore, the depot has $|K|$ successor vertices and is connected to $2|K|$ customer vertices. If more vehicles than needed are available (i.e., $|K| > r(N)$), one can replace the equalities (1.3) with inequalities of type "$\leq$". Note that a solution with $|K| = r(N)$ may have larger routing costs than one where more routes are allowed. In fact, fleet size minimization and routing cost minimization are conflicting objectives. However, by adding fixed costs for routes (altering the cost coefficients c_{0i}), both objectives can be integrated.

Constraints (1.4) serve at the same time as *capacity constraints* and *Subtour Elimination Constraints* (SECs), which can be seen as follows: First, consider an infeasible route over the cluster $S \subseteq N$ with a demand $q(S) > Q$. Due to $r(S) > 1$, at least two routes must connect S with its complement $V \setminus S$, so that any capacity-infeasible route is excluded. Second, any subtour over a non-empty subset $S \subseteq N$ (S does not contain the depot) fulfills $x(\delta^{(+)}(S)) = 0$. Due to $r(S) \geq 1$ this subtour is also eliminated. In the directed case, the SEC $x(\delta^+(S)) \geq r(S)$ are equivalent to $x(\delta^-(S)) \geq r(S)$ and $x(A(S)) \leq |S| - r(S)$. In the undirected case, the SEC $x(\delta(S)) \geq 2r(S)$ are equivalent to $x(E(S)) \leq |S| - r(S)$. This can be proven easily by summing up the degree constraints (1.2) for $i \in S$ and $j \in S$. In all cases, the number of constraints grows exponentially with the number of vertices.

There are two basic techniques to handle this exponential number of constraints. On the one hand, leaving out all or some SEC gives a relaxation of the CVRP. The resulting LP-relaxation can be solved with cutting-plane algorithms in which at each iteration the violated SEC are identified (using a so-called *separation procedure*) and added to the relaxation until no more violated SEC are found. Branch-and-Cut algorithms include procedures for the separation of other classes of valid inequalities for the associated integer polyhedron (see Naddef and Rinaldi [104] and Chapter 2). On the other hand, when considering the directed model VRP1 the above SEC can be replaced by another set of constraints using additional variables. The model is known as the *MTZ-formulation* as introduced by Miller, Tucker, and Zemlin [99] for the TSP. More precisely, for the directed model VRP1, the additional variables $u = (u_1, \dots, u_n)^\top$ indicate the accumulated demand u_i already distributed by the vehicle when arriving at customer $i \in N$. Constraints (1.4) can be replaced by MTZ-specific SEC

$$(1.6) \qquad u_i - u_j + Q x_{ij} \leq Q - q_j \qquad \forall (i,j) \in A(N)$$

and capacity constraints

$$(1.7) \qquad q_i \leq u_i \leq Q \qquad \forall i \in N.$$

Note that $x_{ij} = 1$ implies $u_j \geq u_i + q_j > u_i$. Hence, the presence of a subtour $(i, j, \dots, i)$ not containing the depot leads to the contradiction $u_i > u_j > \cdots > u_i$. The advantage of the MTZ-formulation is that it has $\mathcal{O}(n^2)$ variables and constraints. On the downside, however, its linear relaxation generally produces a significantly weaker lower bound compared to those produced by model VRP1 (see, e.g., Desrochers and Laporte [45] and Padberg and Sung [108]). Thus, there is a clear trade-off between the strength of the linear relaxation and the size of the underlying MIP model.

The two-index formulations VRP1 and VRP2 model the underlying fleet only implicitly. Even if we know that a specific edge $\{i, j\}$ or arc (i, j) is used in a solution, it is not clear which vehicle k will travel between i and j. Hence, two-index formulations cannot model vehicle-specific characteristics such as different capacities, associated depots, and costs (see Section 1.3.4), which is a clear disadvantage. However, models VRP1 and VRP2 have the advantage of providing non-redundant representations, i.e., no symmetric solutions resulting from renumbering the vehicles, such that there is a one-to-one correspondence between feasible VRP solutions and feasible vectors x.

Next we present a *three-index (vehicle-flow) formulation*, which is based on a directed graph $G = (V, A)$ in which the depot 0 is replaced by two vertices o and d representing the startpoint and endpoint of a route. The new definition of the vertex and arc sets is

$$V := N \cup \{o, d\} \quad \text{and} \quad A := (V \setminus \{d\}) \times (V \setminus \{o\}).$$

As their name suggests, three-index formulations have binary variables of the form x_{ijk} modeling the movement of the vehicles over the arcs; in other words, $x_{ijk} = 1$ if and only if vehicle $k \in K$ moves over the arc $(i, j) \in A$. Moreover, binary variables y_{ik} indicate whether or not vehicle k visits the vertex $i \in V$. In this case, u_{ik} is (a lower bound on) the load in vehicle k directly before reaching i. For completeness, we define $q_o = q_d = 0$. The three-index formulation, originally proposed by Golden, Magnanti, and Nguyen [72] in a slightly different form, is

(1.8)   (VRP3) $\text{minimize} \sum_{k \in K} c^\top x_k$

(1.9)   s.t. $\sum_{k \in K} y_{ik} = 1$ $\forall i \in N,$

(1.10) $x_k(\delta^+(i)) - x_k(\delta^-(i)) = \left\{ \begin{array}{ll} 1, & i = o, \\ 0, & i \in N, \end{array} \right.$ $\forall i \in V \setminus \{d\}, k \in K,$

(1.11) $y_{ik} = x_k(\delta^+(i))$ $\forall i \in V \setminus \{d\}, k \in K,$

(1.12) $y_{dk} = x_k(\delta^-(d))$ $\forall k \in K,$

(1.13) $u_{ik} - u_{jk} + Q x_{ijk} \le Q - q_j$ $\forall (i,j) \in A, k \in K,$

(1.14) $q_i \le u_{ik} \le Q$ $\forall i \in V, k \in K,$

(1.15) $x = (x_k) \in \{0,1\}^{K \times A},$

(1.16) $y = (y_k) \in \{0,1\}^{K \times V}.$

The objective (1.8) minimizes the overall routing costs, and constraints (1.9) ensure that each customer is served exactly once. The path-flow constraints (1.10) imply that $x_k(\delta^+(d)) - x_k(\delta^-(d)) = -1$ holds. Therefore, for a fixed $k \in K$, the arc set $\{(i,j) \in A : x_{ijk} = 1\}$ induces an o-d-path, i.e., the route performed by vehicle k. Constraints (1.11)–(1.12) couple routing variables x_{ijk} with indicator variables y_{ik}. Vehicle-specific MTZ and capacity constraints are given by (1.13) and (1.14).

We will see in Section 1.3.4 that formulation VRP3 enables simple modifications to take vehicle-specific characteristics into account. However, in particular for large fleets K, the model suffers from its inherent symmetry with respect to the numbering of the vehicles, since any solution to VRP3 has $|K|!$ equivalent solutions permuting the vehicle indices. Even with the addition of symmetry breaking constraints, such as those proposed by Fischetti, Salazar-González, and Toth [58], the solution of VRP3 with enumerative or MIP-based approaches remains often intractable.

Extensive Formulation. An *extensive formulation* for the CVRP was first proposed by Balinski and Quandt [8] and is based on an extended set partitioning or set covering model. The idea is that feasible routes are the basic objects to work with, and a solution to the model directly decides about the routes to include in the solution. Let Ω be the set of feasible CVRP routes. Each route $r \in \Omega$ is of the form $r = (i_0, i_1, i_2, \ldots, i_s, i_{s+1})$ with $i_0 = o$ and $i_{s+1} = d$ so that we can assign the cost $c_r = \sum_{j=0}^s c_{i_j, i_{j+1}}$ to it. Furthermore, the coefficient $a_{ir} \in \{0,1\}$ is equal to 1 if and only if the route r visits customer $i \in N$, i.e., if $i \in \{i_1, i_2, \ldots, i_s\}$. Finally, binary variables λ_r indicate whether or not route $r \in \Omega$ is selected. The extensive formulation is then

(1.17)   (VRP4) $\text{minimize } c^\top \lambda$

(1.18)   s.t. $\sum_{r \in \Omega} a_{ir} \lambda_r = 1$ $\forall i \in N,$

(1.19) $\mathbb{1}^\top \lambda = |K|,$

(1.20) $\lambda \in \{0,1\}^\Omega.$

The cost of the selected routes is minimized by (1.17). The set partitioning constraints (1.18) stipulate that each customer is visited once by one route, and the constraint (1.19) requires that the complete fleet be utilized.

If the routing costs fulfill the triangle inequality, i.e., $c_{hi}+c_{ij} \geq c_{hj}$ for all $(h,i),(i,j) \in A$, then the set partitioning constraints (1.18) can be replaced by set covering constraints (i.e., $= 1$ becomes ≥ 1). The reason is that removing one or several customers from a feasible route results in a subroute that is feasible and less (or equally) costly. Moreover, if fewer than $|K|$ vehicles suffice to serve all customers, and leaving a vehicle unused is feasible, then one can replace $=$ by $\leq$ in (1.19).

The extensive formulation has two major advantages over compact formulations. First, it typically provides excellent lower bounds by solving its linear relaxation; see Bramel and Simchi-Levi [17]. Second, the costs and constraints that describe the feasibility of a route are implicitly hidden in the definition of the set Ω. Therefore, highly complex non-linear cost functions and any type of intra-route constraints (even beyond those discussed in Sections 1.3.3 and 1.3.6) can also be used in the definition of c_r and Ω. Moreover, model VRP4 does not suffer from symmetry problems. On the downside, the number $|\Omega|$ of feasible routes can easily grow to dimensions that prohibit the explicit instantiation and (even more) the direct solution of (1.17)–(1.20). Therefore, one has to rely on algorithms that implicitly work on the huge set Ω such as Lagrangian relaxation and column generation techniques (see Bramel and Simchi-Levi [18]). In fact, model VRP4 results from a Dantzig–Wolfe decomposition (Dantzig and Wolfe [41]) and a commodity aggregation of the three-index formulation VRP3 (see, e.g., Desaulniers et al. [44] and Lübbecke and Desrosiers [97]).

1.3 ▪ The Family of VRP

In this section, we give a broad overview of the most important variants of the VRP presented in the literature during the more than 50 years of its history. Far from being exhaustive, our presentation mainly aims at introducing a classification perspective which may help in identifying the specific characteristics of a VRP that one wants to model and solve. In particular, we classify problems according to

- the (road) network structure (Section 1.3.1),

- the type of transportation requests (Section 1.3.2),

- the constraints that affect each route individually (Section 1.3.3),

- the fleet composition and location (Section 1.3.4),

- the inter-route constraints (Section 1.3.5), and

- the optimization objectives (Section 1.3.6).

Finally, Section 1.3.7 reviews some further aspects arising when integrating additional logistics issues such as inventory management and synchronization.

1.3.1 ▪ Network Characteristics

In the CVRP, the transportation tasks are related to *points in space*, i.e., locations to which goods have to be delivered. Since we typically model these points as vertices of a graph, the corresponding VRP is a so-called *node routing problem*. In contrast, the tasks in *Arc Routing Problems* (ARP; see Dror [53], Wøhlk [141], and Corberán and Laporte [35]) are services to be performed on *street segments*, also called *connections* or *links*. Examples are the routing of vehicles for street sweeping and inspections as well as winter services with salt gritting and snow plowing or removal. In many urban areas, services such as mail delivery, garbage collection, and meter reading are often also considered as tasks on edges or arcs because households are typically located densely along a street segment. Even a mixture of tasks on vertices and arcs or edges is possible, leading to *General Routing Problems* (GRP; see Orloff [107]).

If the underlying data for travel costs are symmetric (more complex variants of VRP also consider travel times or more general resource consumptions), the problem is symmetric and it is possible to model it by means of an undirected graph. If the movement between two vertices is restricted to one direction or any relevant cost or resource consumption is asymmetric, the underlying graph is either a directed, or mixed, or *windy* (see Guan [75]) graph. Asymmetric models are suitable in these cases. But even in symmetric problems sometimes modeling or solution techniques themselves require the underlying graph being directed. An example is the MTZ-formulation for the CVRP discussed in Section 1.2.2.

A point that distinguishes VRP and ARP/GRP is the *granularity of the data and network*. Typically, edges and arcs in an ARP or GRP represent individual street segments. In VRP, by contrast, edges and arcs result from paths between two points, each consisting of a possibly larger number of street segments. Thus, the distance and travel time information for the paths are to be computed by the routing component of a *Geographical Information System* (GIS; see Chang [26]). Usually, these paths are computed as shortest paths with respect to a weighted combination of distance and travel time so that the resulting cost and travel time matrices do not need to satisfy the triangle inequality. Even worse, relying on a single shortest path means that many other paths with Pareto-optimal resource consumption may simply be neglected. For this reason, Garaix et al. [60] model and solve VRP on multi-graphs in which parallel arcs represent the different Pareto-optimal paths. Summarizing, the ARP and the GRP are closer to GIS raw data, and their underlying graphs are sparse. In contrast, VRPs often have dense or complete underlying graphs (as the CVRP and several other variants discussed in the following), where edges and arcs represent shortest paths. Furthermore, in a real-world setting, a shortest path between two points (with respect to travel time or cost) can heavily depend on the time of day when the vehicle drives between the points. In *dynamic VRP*, some data is not known in advance but becomes available during operation. Also, if some data (cost, travel time, demand, etc.) are not known in advance, but are described by a random variable with a given *probability distribution*, the VRP is *stochastic* (we come back to both aspects in the next section). Obviously, the CVRP is neither dynamic nor stochastic but *static* and *deterministic*.

In this book, we will restrict ourselves to node routing problems because the ARP and the GRP are covered by the companion volume edited by Corberán and Laporte [35].

1.3.2 ▪ Types of Transportation Requests

We have already described the CVRP as a variant of VRP in which all transportation requests consist of the *distribution of goods from a depot to customers*. We start by classifying other types of transportation requests in the node routing context.

Delivery and Collection. The counterpart of delivery to customers is *collection from customers*, where all tasks involve the movement of goods or waste from a point to the depot. Collections are often called *pickups*. Associated routing problems often occur either at the beginning of a supply chain, e.g., for raw-milk collection (see, e.g., Sankaran and Ubgade [120]), or at the very end of it, e.g., in reverse logistics where returned empties have to be collected, or waste has to be disposed (see Golden, Assad, and Wasil [71] and Chapter 15 in [35]). The equivalence of pure distribution and pure collection VRP turns out to be obvious when one reverses the routes so that collection becomes distribution and vice versa.

Different variants of VRP result when both *collection and distribution* (or *pickup and delivery*) of material occur together in a route. We still assume that distribution starts and collection ends at the depot. Therefore, problems with collection are also known as *many-to-one* VRP and problems with distribution as *one-to-many* VRP. The first and probably simplest variant is the *VRP with Backhauls* (VRPB; see Chapter 9). When, for example, transporting bulky material, all deliveries to the so-called *linehaul customers* must be performed first so that the vehicle is empty when it arrives at the first collection point, called a *backhaul customer*. Because any movement from a backhaul customer to a linehaul customer is forbidden, model VRP1 remains applicable if the corresponding arcs are removed from the arc set A (alternatively, one can set the costs of these arcs to a sufficiently large number M). Backhauling constraints result from the difficulty of rearranging the loaded items inside the vehicle. If the loading space allows rearrangements, e.g., because the vehicle can be loaded from rear and front or all sides, the resulting problem is a *VRP with mixed deliveries and collections*, or simply *Mixed VRPB* (MVRPB; see Wade and Salhi [139]). Here, the vehicle capacity must be checked for each edge (or arc) traversed; i.e., the load already collected from backhaul customers plus the load to be delivered to linehaul customers can never exceed the given capacity.

While each customer in the VRPB and MVRPB requires either a delivery or a collection, but not both, the *VRP with Simultaneous Pickup and Delivery* (VRPSPD; see Min [100] and Chapter 6) comprises two transportation requests for each customer, namely a delivery from depot to customer and a pickup from customer to depot. Both transportation requests must be performed by one vehicle in a single visit. This situation is common in several real-world applications, such as the delivery of beverages and simultaneous collection of empty bottles, and the bus transportation of newly arriving and leaving hotel guests between a local airport and several hotels (this is common practice of tour operators in many holiday regions). Again, the capacity constraint ensures that no vehicle is overloaded at any point. Interestingly, there exists a simple criterion to check whether or not a given customer set $S \subseteq N$ can be feasibly served by a single vehicle: A feasible route exists if neither the overall amount to be delivered nor that to be collected exceeds the vehicle capacity. One simply has to build the route so that customers with a higher amount to be delivered than to be collected are visited first in the route. A relaxation of the VRPSDP is the *VRP with Divisible Deliveries and Pickups* (VRPDDP). Here, delivery and pickup at a customer may be performed in a single visit or in two separate visits by the same vehicle. Since less capacity is needed when the delivery properly precedes the pickup, cost savings compared to the VRPSDP are possible in VRPDDP. The form of routes is more general and includes so-called *lasso routes* and others; see Gribkovskaia et al. [73]. The VRPSDP and VRPDDP should not be mixed up with the problems presented in a following paragraph on point-to-point transportation, which are also known as pickup-and-delivery problems.

Simple Visits and Vehicle Scheduling.   Sometimes the service is neither collection nor delivery, but merely consists of *visiting a customer* or location. For example, service technicians repair or install something, and nurses take care of elderly people at their homes. Another specific case is that route segments have to be followed in a priori given sequences (and schedules). This is the class of *Vehicle Scheduling Problems* (VSPs), which arise in public transport when scheduling buses, trams, trains, etc., in urban, local, and national transport systems. We exclude these problems from our consideration because the major routing decisions are determined beforehand. Surveys on VSP are compiled by Desrosiers et al. [46] and Bunte and Kliewer [22], while school bus routing and other VSP applications for passenger transportation are discussed in Chapter 7.

Alternative and Indirect Services.   There exist situations in which a service can be performed in alternative ways and modes. For example, the delivery of a parcel can either be made to a person's working place (during office hours), to his or her private home, or to an automated parcel terminal (see Drexl [49]). If the service provider can choose between different alternatives, the simultaneous service choice and routing of vehicles can help in consolidating tours (see Cardeneo [24]).

In the *Multi-Vehicle Covering Tour Problem* (MVCTP; see Hachicha et al. [78]), the service to customers consists of visiting a location that is close enough to the customer. Applications of the MVCTP are the simultaneous location of postboxes at given potential sites and the construction of optimal collection routes. Moreover, in some developing countries, mobile health care delivery teams operate at a selected number of sites, to which the supported people must travel. The arc routing pendant of the problem occurs in automatic meter reading (see Shuttleworth et al. [125] and Chapter 13 in [35]).

Point-to-Point Transportation.   Pickup-and-delivery problems are VRPs in which the transportation requests consist of *point-to-point transports*. More precisely, each transportation request consists of the movement of goods or people between two particular locations, one where someone or something is picked up, and a corresponding location for the delivery. Generally, neither of these locations is a depot so that these problems are also referred to as *many-to-many VRP*. The problem is called a *Pickup-and-Delivery Problem* (PDP; see Desaulniers et al. [43], Parragh, Doerner, and Hartl [109], and Chapter 6) in the context of goods transportation. Applications include freight transportation, like those described by Savelsbergh and Sol [122]. In the context of passenger transportation, the problem is known as the *Dial-a-Ride Problem* (DARP; see Chapter 7). There exist applications in bus routing for pupils, patients, handicapped persons, and elderly (between their individual homes and schools, care facilities, or hospitals). Almost all DARP variants include time-window constraints. Often service levels and user (in)convenience is taken into account either with constraints or in the objective function (see Sections 1.3.3 and 1.3.6).

Repeated Supply.   In the context of goods delivery, customers may require that they repeatedly receive deliveries of goods. Considering a longer planning horizon, a customer might be unconcerned about the specific day when receiving a shipment as long as he or she does not run out of stock. In the *Periodic VRP* (PVRP; see Cordeau, Gendreau, and Laporte [37], Mourgaya and Vanderbeck [103], and Chapter 9), there are two planning levels: At the first level, a feasible *visiting pattern* for each customer has to be selected from a given set of admissible ones. For example, a customer may have agreed to receive two or three deliveries per week, either on days Mon-Thu, Tue-Fri, Mon-Wed-Fri. Note that the customer demand may then depend on the particular visiting pattern and day. At the second level, a VRP must be solved for each day, where the subset of customers to be visited and their demands result from the first-level decisions.

Another form of repeated supply occurs in *Inventory Routing Problems* (IRPs; see Campbell, Clarke, and Savelsbergh [23], Bertazzi and Speranza [12], and Coelho, Cordeau, and Laporte [34]). The fundamental difference with respect to the VRP variants discussed so far is that there are no customer orders in an IRP. Instead, the delivery company decides when to visit a customer and how much to deliver so that no stock-out occurs. Based on these decisions and assuming that the planning horizon spans several days, an optimal routing of the vehicles has to be determined for every day. Thus, daily routing decisions depend on the selected subset of customers to visit and quantities to deliver on that particular day, and vice versa. Moreover, the amount to deliver to a customer may be restricted by a maximal storage level. The objective in IRP variants can differ, but it normally includes routing costs and inventory holding costs.

There is a growing interest in IRP in *Supply Chain Management* (SCM). A mean to reduce the bull-whip effect (see Lee, Padmanabhan, and Whang [94]) is *Vendor Managed Inventory* (VMI; see Disney and Towill [47]), where the supplier monitors the customer inventory levels and is responsible for the timely resupply. Advantages often cited in the SCM literature are a faster and more reliable information exchange between customer and supplier, shorter lead times, lower inventory levels, and higher service levels. Examples of real-world applications of IRP include deliveries to supermarkets or grocery stores, parts distribution in the automotive industries, refill of vending machines, e.g., with beverages, and fuel delivery to gas stations.

The *PVRP with service choice*, where the delivery frequency has an impact on the demand and the service level, as introduced by Francis, Smilowitz, and Tzur [59], can be seen as an intermediate between PVRP and IRP. Extensions of IRP cover aspects such as stochastic daily demand (for an overview on stochastic IRP see Bertazzi and Speranza [13]) and combined production planning and IRP as described by Adulyasak, Cordeau, and Jans [1].

Non-split and Split Services.   Until now, we have assumed that all service tasks are performed by a single vehicle in one service operation, i.e., services are *non-split*. However, there are two reasons for splitting some services: On the one hand, if demand exceeds the vehicle capacity, more than one visit is unavoidable. On the other hand, splitting services into several smaller service requests can produce significant cost savings. The *Split Delivery VRP* (SDVRP; see Dror and Trudeau [54], Dror and Trudeau [55], and Chapter 9) allows, in principle, that each demand be split into arbitrarily many smaller demands served by different vehicles.

Combined Shipment and Multi-modal Service.   Whereas in the SDVRP individual deliveries are split into smaller parts, *combined shipments* leave the individual shipments intact, but several vehicles transport the shipment from its supplier to the customer using intermediate transfer points or consolidation centers. This is a common practice in multi-modal transportation, where different types of vehicles and transportation modes are used at various steps: for example, large trucks for the long-distance full truckload transfer from factories to distribution centers and small trucks for the less-than-truckload transportation to final customers. Several variants arise depending on the specific distribution network structure, such as *hub-and-spoke* or *crossdocking*, and the presence of distribution routes or direct shipping for the *last mile* distribution. A recent survey on these problems can be found in Guastaroba, Speranza, and Vigo [76], whereas a thorough illustration of the case in which in-transit consolidation is carried out at intermediate consolidation centers is given by Song, Hsu, and Cheung [128]. Another example that has recently attracted some interest due to its applications in *city logistics* is the so-called *2-Echelon VRP* (2E-VRP; see Laporte and Nobert [91] and Perboli, Tadei, and Vigo [110]), where the delivery from a single depot to the customers is managed using *intermediate depots*, also called *satellites*.

Routing with Profits and Service Selection.   With a limited fleet (see Section 1.3.4) it may be impossible to service all transportation requests. Then, just a subset of requests has to be fulfilled. More generally, by optimizing routing and request selection simultaneously, a company may gain additional revenues compared to traditional two-staged decisions, where the request acceptance precedes the route planning step. In order to control this request selection process, one can either set constraints on service levels and costs, respectively, or penalize unmet requests or reward the ones met. Note that if penalties are chosen as negative profits, then both approaches are equivalent.

Most of the problems of this type were first introduced by considering the single-vehicle case, i.e., as variants of the TSP, and later extended to the VRP. In addition, these problems are known under a variety of names such as the *selective TSP/VRP* and the *Maximum Collection TSP/VRP*. Following the taxonomy introduced by Feillet, Dejax, and Gendreau [57], there are three categories of problems:

(i)   If routing costs and profits are combined into one objective, the single-vehicle routing problem is a *Profitable Tour Problem* (PTP; see Chapter 10). The VRP variant has no consistent naming, but can be found as *Capacitated PTP* (CPTP); see Archetti et al. [3].

(ii)   The variant where the route length is bounded from above and the objective is profit maximization is called the *Team Orienteering Problem* (TOP; see Archetti, Hertz, and Speranza [4] and Chapter 10). The single-vehicle case is known as the *Orienteering Problem* (OP) and is discussed in Chapter 10.

(iii)   Finally, if there exists a lower bound on the profit to be collected and the goal is to find a least cost routing, the problem is a *Prize-Collecting VRP* (PCVRP; see Tang and Wang [132]). The variant with only one vehicle, known as *Prize-Collecting TSP* (PCTSP), is analyzed in Chapter 10.

A related variant that attracted some interest in the recent literature is the so-called *VRP with Private fleet and Common carrier* (VRPPC; see Chu [32] and Potvin and Naud [112]), where customers may either be served by using owned vehicles with traditional routes or be assigned to a common carrier, which serves them directly at a prefixed cost. This problem can be easily seen as a PCVRP and takes into account the option of subcontracting unprofitable customers, as common in the parcel delivery service. The multiple depot case is studied by Stenger et al. [129].

The *Multiple Vehicle Traveling Purchaser Problem* (MV-TPP) (see Choi and Lee [30]) is yet another variant with service selection. We are given a set of marketplaces, where a set of goods is available at specified prices, and the demands are also known. The MV-TPP requires the determination of routes for capacitated vehicles that have to select and visit a subset of the marketplaces so that the required demand is collected and the overall sum of routing and purchase costs is minimized. Recently, the MV-TPP was used by Riera-Ledesma and Salazar-González [118] to model a school bus route design problem, whereas a distance-constrained variant of the problem is studied by Bianchessi, Mansini, and Speranza [15].

Dynamic and Stochastic Routing   Important VRP variants arise with the consideration of uncertainty and variability of system conditions. In general, a problem is

- *dynamic* if parts or all relevant information about the system conditions become available during operation;

- *stochastic* if system conditions are uncertain, but uncertainty is described by a given probability distribution.

In *dynamic VRP*, the information generally revealed over time consists of customers' locations and demands (see Tillman [135], Psaraftis [114], and Chapter 11): some of them are possibly known in advance, but for the remaining customers only probabilistic information is given. This type of problem is also called *online* when the emphasis is given to the development of heuristic methods with a performance guarantee (see Jaillet and Wagner [84] for an overview). Two other types of dynamic problems have been studied: The first one considers the time dependency of travel duration (see Haghani and Jung [79] and Chapter 11). In fact, vehicle speed along roads is not only affected by the variability of traffic volumes, which may change considerably during the day, but is also subject to irregular congestion phenomena caused, for example, by car accidents, street maintenance, and illegal parking of vehicles reducing the streets' capacity. In the second type, the availability of vehicles is a dynamic component of the problem, which requires rescheduling of the routes when vehicle breakdowns and delays occur (see, e.g., Li, Mirchandani, and Borenstein [96] and Chapter 11).

In *stochastic VRP*, some problem components, such as customer demand and travel times, are uncertain (see Bertsimas [14], Gendreau, Laporte, and Séguin [62], and Chapter 8) and described as random variables. As a result, planned routes may incur delayed service at customers and may be terminated prematurely when the vehicle capacity has been reached. The focus of stochastic VRP is, therefore, on analyzing the impact of uncertainty on the resulting costs and service levels.

1.3.3 ▪ Intra-route Constraints

A key aspect to consider when defining different VRP variants is the type of constraints that determine whether or not a route is *feasible*. In this section, we discuss issues related to loading, route length, reuse of vehicles, time schedules, and the various combinations of these types of constraints occurring in practice. Common to these so-called *intra-route constraints* (also called *local constraints*) is that they can be checked once an individual route (the vertex sequence) is known, independently from what the other routes are.

Loading. The *capacity constraints*, as presented for the CVRP, belong to the simplest type of constraints, since they can be written as an overall bound on a resource that is consumed at every vertex the vehicle reaches:

$$(1.21) \qquad\qquad \sum_{i \in V} q_i y_{ik} \leq Q \qquad \forall k \in K.$$

Note that we will refer to the three-index model VRP3 in the following, since many modifications and extensions can be described conveniently in that case. There may exist *several capacity constraints* relating to weight, space (e.g., pallets), and volume (e.g., m^3 for liquid goods) that can individually restrict the loading. Instead of (1.21), several constraints with their corresponding coefficients q_i and Q must then be added to the model.

More complex loading constraints occur when both the shipments and the cargo compartments are described either by 2-dimensional or 3-dimensional quantities. In these VRP variants, multi-dimensional packing problems and VRPs are combined. In the *CVRP with 2-dimensional Loading constraints* (2L-CVRP; see Iori, Salazar-Gonzalez, and Vigo [83]), shipments are rectangular items that have to be feasibly assigned to a rectangular compartment. Moreover, the items sent to the same customer must arrive with the same vehicle (*item clustering constraint*), and the items may or may not be rotated (*item orientation constraint*). When delivering an item, no items of customers served later along the route may lay, not even partially, in the rectangular area between that item and the door of the vehicle. This is a *sequential loading constraint*. As in the CVRP, a capacity constraint with respect to the weight (kg) has to be taken into account as well. The problem with *3-dimensional Loading constraints* is the 3L-CVRP as described by Gendreau et al. [61], where additional operational constraints are introduced to ensure the stability of stacked boxes, the secure transportation of fragile boxes, and the easy unloading of boxes at the customer locations. In the *Pallet-Packing VRP* (PPVRP; see Zachariadis, Tarantilis, and Kiranoudis [144]), the shipments are 3-dimensional boxes that have to be feasibly stacked onto pallets before they are loaded into the vehicles.

Interesting additional complications arise if vehicles have more than one compartment. In the *VRP with Compartments* (VRPC; see Derigs et al. [42]), the vehicle capacity, modeled as a one-dimensional quantity Q, is partitioned into a given number of smaller compartments. The compartments are either fixed units, such as tanks in a road tanker, or result from the use of dividers that split the available space. In the latter case, the divider's position may or may not be completely or partially flexible. The *compatibility constraints* refer to two aspects: Only compatible items can share a compartment (*item-item compatibility*), e.g., the smell of washing powder spoils some food, and items can only be assigned to compatible compartments (*item-compartment compatibility*), e.g., refrigerated cargo must go into cooled compartments. Compatibility between groups of items is also considered in some industrial variants (see, e.g., Xu et al. [143]).

Last-In-First-Out (LIFO) loading constraints mean that, by considering all shipments that are currently on board a vehicle, one must deliver the shipment that is most recently picked. LIFO loading constraints were mainly considered in the single-vehicle routing context (see Cordeau et al. [38]) but apply to the PDP as well (see Cherkesly, Desaulniers, and Laporte [29]). LIFO loading is intended to reduce loading and unloading times to a minimum because for vehicles that are loaded from one side very few rearrangements need to take place.

Route Length. Another simple type of constraints results from global bounds that limit the consumption of a resource consumed on edges or arcs. The addition of distance constraints to the CVRP yields the *Distance-constrained CVRP* (DCVRP; see Christofides, Mingozzi, and Toth [31] and Laporte, Desrochers, and Nobert [88]). In the following, we assume that $t_{ij} > 0$ is the distance between vertex i and vertex j for all $(i,j) \in A$. If $L > 0$ is a given upper bound on the length of a route, the distance constraints are

$$(1.22) \qquad \sum_{(i,j)\in A} t_{ij} x_{ijk} \leq L \qquad \forall k \in K.$$

Not only bounds on the *spatial distance* can be modeled, but so can constraints on the *route duration* (with travel times t_{ij}), on the routing costs, or on the number of connections with a certain property (with indicators $t_{ij} \in \{0,1\}$). For additional examples we refer the reader to Avella, Boccia, and Sforza [5].

Multiple Use of Vehicles. The standard assumption for many VRP variants is that each vehicle performs only one route over the planning horizon T. In the *VRP with Multiple use of vehicles* (VRPM; see Taillard, Laporte, and Gendreau [131]), vehicles may perform several routes. Given some routes, with durations $T_1, T_2, \ldots, T_p$, a single vehicle may perform them if $T_1 + T_2 + \cdots + T_p \leq T$ holds. In particular, if the vehicle capacity Q is relatively small or other constraints impose a small number of services per route, feasible solutions with a limited fleet of size $|K|$ can only be achieved when vehicles are reused. Note that routing with unlimited fleet first and subsequently packing the resulting routes into the planning horizon generally yields suboptimal solutions to VRPM. In some cases, overtime for the drivers is permitted with a penalty (see, e.g., Brandão and Mercer [19]).

The VRPM is also known as the *Multi-Trip VRP* (MTVRP) and recently attracted new research efforts in the context of real-world applications (see Battarra, Monaci, and Vigo [10]) and city logistics (see Cattaruzza et al. [25]). An exact algorithm has been proposed by Mingozzi, Roberti, and Toth [102], while metaheuristics have been recently proposed by Olivera and Viera [106], Petch and Salhi [111] and Alonso, Alvarez, and Beasley [2].

The large recent diffusion of alternative-fuel vehicles, e.g., electrical cars, has stimulated interest of the research community in the variants of VRP that incorporate the environmental issues. The main characteristic is the currently limited autonomy of vehicles due to restricted battery capacities, which forces the visit of refuel or recharge stations during the travel. VRP with the possibility or need of refueling during the trip have been studied by Erdogan and Miller-Hooks [56] and by Schneider, Stenger, and Goeke [123] for the case with time windows.

Time Windows and Scheduling Aspects.   Some very practically relevant constraints present in most VRP variants are those related to *scheduling*, i.e., requiring the consideration of *travel, service, and waiting times* together with *time-window* constraints. In the *VRP with Time Windows* (VRPTW; see Cordeau et al. [36] and Chapter 5), a traversal time t_{ij} for each arc $(i,j) \in A$ and a time window $[a_i, b_i]$ for each vertex $i \in V$ are given. A schedule, i.e., a combination of start times T_{ik} for the service at a vertex $i \in V$ when visited by vehicle $k \in K$, is considered feasible if

$$(1.23) \qquad\qquad a_i \leq T_{ik} \leq b_i \qquad \forall i \in V, k \in K$$

(if vehicle k does not visit vertex i, the time T_{ik} is irrelevant) and

$$(1.24) \qquad x_{ijk} = 1 \quad \Rightarrow \quad T_{ik} + t_{ij} \leq T_{jk} \qquad \forall (i,j) \in A, k \in K$$

holds. Model VRP3 along with (1.23)–(1.24) is a three-index formulation for the VRPTW. The latter constraints couple routing decisions with the time schedule. They can be linearized by means of MTZ-like constraints of the form

$$T_{ik} - T_{jk} + M x_{ijk} \leq M - t_{ij} \qquad \forall (i,j) \in A, k \in K.$$

Note that, with the above definition, time windows are asymmetric in the sense that arriving at vertex i before time a_i is allowed, in which case the vehicle has to wait until time a_i, while arriving later than time b_i is prohibited. Some authors also add *service times* s_i at vertices to their models. This is only a minor extension, since these can be included by properly redefining the travel times and time windows.

Time-window constraints can be altered and generalized in various ways: *Multiple time windows* mean that (1.23) has to be replaced by $T_{ik} \in \bigcup_{\ell=1}^{p}[a_i^\ell, b_i^\ell]$ when $[a_i^1, b_i^1], \ldots, [a_i^p, b_i^p]$ are p alternative possible intervals for the start of service at vertex i. Disjunctive constraints have to be used for linearizing the above conditions.

Travel times may depend on *the time of the day* so that t_{ij} have to be replaced by travel *time functions* $t_{ij}(T_i)$, also called *time-dependent travel times* (see, e.g., Haghani and Jung [79] and Chen, Hsueh, and Chang [28]). Note also that, in the dynamic context, time-dependent travel times may be present even without time windows.

The route performed by vehicle k *ends* at time T_{dk} and has a *duration* $T_{dk} - T_{ok}$ (both plus a possible service time at d). If $x_{ijk} = 1$, i.e., vehicle k visits vertex i directly before arriving at vertex j, then the *waiting time* at vertex j is $w_j = (a_j - T_{ik} - t_{ij})^+$, where $x^+ =$

$\max\{0, x\}$. Some interesting VRPTW variants result when these and other quantities related to the schedule are either bounded or if they contribute to the objective. In the *VRP with Soft Time Windows* (VRPSTW; see Taillard et al. [130]), a convex function p_i : $[a_i, b_i] \to \mathbb{R}$ is used to generate solutions in which service is provided at a time close to the minimum of p_i. The most studied variants correspond to linear penalty functions of the form $p_i(T_{ik}) = p(T_{ik} - b_i')^+$, where p is a penalty factor and $a_i \leq b_i' < b_i$ defines a linear *penalty on late services*, and $p_i(T_{ik}) = p(a_i' - T_{ik})^+ + p(T_{ik} - b_i')^+$ with $a_i < a_i' \leq b_i' < b_i$ for linear *penalties on both early and late services*. In the first case, the favored start time is in the time interval $[a_i, b_i']$, while in the second it is in $[a_i', b_i']$, and deviating from it is penalized linearly. Other VRP variants require waiting times to be bounded at each vertex, per vehicle, or over all vehicles, or waiting is penalized (see also Section 1.3.4). The route completion times T_{dk} can be bounded individually, and in other applications the *makespan* $\max_{k \in K} T_{dk}$ is a component of the objective.

Probably the most complex scheduling constraints are related to driving rules and schedule regulations. For example, the very advanced approach of Kok et al. [87] takes into account the driving rules appointed by European Regulation (EC) No. 561/2006, Directive 2002/15/EC, and several modified rules in these laws. These include rules that describe in detail restrictions on *driving periods* (max. 4.5 hours), *daily driving times* (max. 9 hours, but twice a week max. 10 hours), *weekly driving times* (max. 56 hours and max. 90 hours within two weeks), *breaks to end a driving period* (min. 45 minutes; also possible 30 minutes plus an earlier break of min. 15 minutes), *daily rest periods* (min. 11 hours; might be reduced to 9 hours under various preconditions), and *weekly rest periods* (min. 45 hours; might be reduced to 24 hours in every second week if compensated by an equally longer rest; max. 144 hours between two weekly rests). Obviously, the consideration of all these rules and, in particular, the possibilities to deviate from default requirement creates highly intricate VRPTW variants. Even checking the feasibility of a given route (as a vertex sequence), i.e., checking whether or not there exists a feasible time schedule for driving, breaks, and rests, is highly complex. Similar settings were handled in general and for Australia, Canada, Europe, the U.S., and in an international comparison by Goel [64], Goel, Archetti, and Savelsbergh [66], Goel and Rousseau [69], Goel [65], Prescott-Gagnon et al. [113], Goel and Kok [67], Rancourt, Cordeau, and Laporte [115], and Goel and Vidal [70].

The DARP (see Section 1.3.2 and Chapter 7) is the pickup-and-delivery problem for passenger transportation. A transportation request, given by a pair of pickup-and-delivery vertices $(i, i + n)$, comes with two time windows: one window $[a_i, b_i]$ for the pickup operation and one window $[a_{i+n}, b_{i+n}]$ for the delivery operation. The ride time R_i is therefore bounded by $b_{i+n} - a_i$ and may be further controlled by imposing a so-called *ride-time constraint* of the form $T_{i+n} - T_i \leq R_i$, or by penalizing user inconvenience through a non-decreasing function in $T_{i+n} - T_i$ in the objective.

Many relevant aspects related to timing problems, their modeling, and efficient algorithmic treatment are covered in a recent paper by Vidal et al. [138].

1.3.4 ▪ Fleet Characteristics

So far we have discussed VRPs with identical vehicles based at the same depot. In this section, we turn to fleets of vehicles that are stationed at different depots and vehicles having different characteristics concerning capacity, costs, speed, and the ability to load material and access locations. Moreover, VRP with trucks and trailers consider autonomous and non-autonomous vehicles, where the latter need to be pulled for moving from one place to another and can be decoupled to reach otherwise inaccessible locations.

Multiple Depot VRP.   If the fleet of vehicles is homogeneous, but vehicles start and end their routes at different depots, the resulting problem is known as the *Multi(ple) Depot VRP* (MDVRP; see Renaud, Laporte, and Boctor [116]). Vehicle specific locations for beginning and terminating a route can easily be incorporated in model VRP3: Replace the vertices o and d (or the corresponding entries in the cost matrix) of the single-depot case with vehicle specific vertices o_k and d_k (with new entries). In principle, every vehicle $k \in K$ may have its own starting and ending locations. However, in the MDVRP, groups of vehicles are assigned to a (smaller) number of depots. Depots may have a limited capacity and host a limited or an unlimited subfleet. Vidal et al. [137] present a review of the recent literature on the MDVRP as well as an effective heuristic for the problem and its periodic variants.

The variant of the MDVRP, where depots can act as intermediate replenishment facilities along the route of a vehicle, is considered by Crevier, Cordeau, and Laporte [39] and Tarantilis, Zachariadis, and Kiranoudis [134]. This problem is strongly related with the multiple use of vehicles discussed in Section 1.3.3.

Heterogeneous or mixed Fleet VRP.   The class of *Heterogeneous or mixed Fleet VRP* (HFVRP; see Baldacci, Battarra, and Vigo [6] and Chapter 9) considers groups or types of vehicles that can differ in capacity, variable and fixed costs, speeds, and the customers that they can access. The fleet K is partitioned into $|P|$ subsets of homogeneous vehicles $K = K^1 \cup K^2 \cup \cdots \cup K^{|P|}$ (also called *vehicle types*). All vehicles $k \in K^p$ from the pth type ($p = 1, \ldots, |P|$) are characterized by a capacity $Q_k = Q^p$, variable routing costs $c_{ijk} = c_{ij}^p$, *fixed costs* $FC_k = FC^p$, and the subset $N_k = N^p \subseteq N$ of *accessible* customers. Individual travel times $t_{ijk} = t_{ij}^p$ can replace standard travel times t_{ij}.

Fixed costs become relevant only if not all vehicles need to be utilized. In this case, fixed costs often reflect the effort required to provide a driver and to prepare and maintain the vehicle. The inclusion of overhead costs is critical because these cannot be assigned fairly to a single route according to the cost-by-cause principle. Generally, the magnitude of the fixed costs strongly depends on whether the fleet is owned by the decision maker or whether the transport service is performed by a third party.

It is easy to modify model VRP3 to take vehicle-specific attributes into account. One simply has to replace the general coefficients with vehicle-specific coefficients, e.g., the capacity Q with Q_k. *Vehicle-dependent routing costs* arise when c_{ij} is replaced with different c_{ijk} for all $(i,j) \in A$. Fixed costs can be incorporated into the routing costs by replacing c_{oj} with $c_{ojk} + FC_k$ for all $(o,j) \in \delta^+(o)$. Also, to model inaccessible customers $j \in N \setminus N_k$, one can set c_{ijk} to a sufficiently large number M for all $(i,j) \in \delta^-(j)$.

Another important aspect is whether the fleet and groups of vehicles are limited or not. In principle, model VRP3 is a limited-fleet model because the fleet consists of exactly $|K| = |K^1| + |K^2| + \cdots + |K^{|P|}|$ vehicles. However, using sufficiently large numbers for $|K^p|$, unlimited fleet models result.

A classification scheme for heterogeneous fleet VRP is offered by Baldacci, Battarra, and Vigo [6]. In *Heterogeneous VRP* (HVRP) and *site-dependent VRP*, the fleet is always limited, while for *Fleet Size and Mix VRP* (FSM) the fleet is unlimited. In FSM, the planning task is more strategic or tactical, since it is typically related to an optimal acquisition of vehicles. In the site-dependent VRP, site dependencies are the only characteristic in which vehicles differ. Therefore, no other vehicle-dependent costs or fixed costs are considered and vehicles have the same capacity. In contrast, HVRP and FSM exist in both variants where fixed costs and vehicle-dependent routing costs are either considered or not. More details on this class of problems, including the above-mentioned classification, can be found in Chapter 9.

Routing of Trucks and Trailers. The *Truck-and-Trailer Routing Problem* (TTRP; see Chao [27]) considers a fleet with at least two groups of vehicles: *Single Trucks* (ST), i.e., normal vehicles without trailer, and *Truck-and-Trailer Combinations* (TTC). While a TTC is attractive due to larger overall capacity, there exist site-dependency conditions stating that some customers are not accessible by a TTC. Insufficient maneuvering space at a customer location is a typical reason for this type of inaccessibility. Customers inaccessible by TTC are called *truck customers*. All others are *regular customers*. Hence, three types of routes are possible:

(i) an ST route is performed by a ST and visits any type of customer;

(ii) a pure TTC route is performed by a TTC and visits only regular customers;

(iii) a mixed TTC route is also performed by a TTC but visits both truck and regular customers. Here, the TTC travels on a main route between regular customers. In order to reach the truck customers, the truck performs one or several subtours, where the trailer is first decoupled at a regular customer, and the truck returns to the same place at the end of the subtour before continuing the main route with the coupled trailer. Note that the demand collected on a truck subtour must fit into the ST. Only at regular customers can the truck transfer all or some of its load into the trailer so that the joint capacity of the TTC can be used.

The TTRP is generalized by Drexl [48] with respect to three aspects: First, variable and fixed costs for trucks and trailers are considered as in the HFVRP. Second, in addition to locations of regular customers, there are optional locations for parking trailers and performing the load transfer. Third, time-window constraints have to be respected at all locations.

A much more intricate routing problem is the *VRP with Trailers and Transshipments* (VRPTT; see Drexl [51]). The main difference with respect to the TTRP is that in the VRPTT there is no fixed assignment of trailers to trucks. Instead, a trailer may be pulled by one or several trucks on parts of its itinerary. This offers the possibility that different trucks, even those that just meet but do not pull the trailer, transfer all or parts of their collected load into the trailer. Drexl [51] extends the VRPTT by integrating several other constraints and options arising, e.g., in applications such as raw-milk collection: There exist several classes of trucks and trailers either compatible or incompatible with respect to load transfer and/or the possibility that the truck moves the trailer. Moreover, *support trucks* cannot access any customers but serve as mobile depots to which other vehicle can transfer their load. The TTRP is a prime example of a VRP with synchronization constraints discussed later in Section 1.3.7.

1.3.5 ▪ Inter-route Constraints

Many VRP variants have the property that a given set of routes provides a feasible solution whenever the routes are feasible and the given transportation requests are partitioned accordingly. The only dependency between the routes consists of the partitioning of the transportation requests. Thus, feasibility exclusively depends on the intra-route constraints discussed in Section 1.3.3. In this section, we present examples of so-called *inter-route constraints* (also *global constraints*), where the feasibility of a solution depends on how the routes (and their schedules) are combined.

A first example are *balancing constraints* that often result from fairness considerations. They state, e.g., that the *difference between maximum and minimum route duration in a solution* should not exceed a given threshold (see, e.g., Bodin, Maniezzo, and Mingozzi [16]). Thus, the workload can be evenly distributed among the drivers. Balancing route duration also makes sense in connection with time windows. The constraints generalize simpler types of balancing, e.g., when the number of stops in a route, the route length, or the delivery amount of a route is taken as a balancing criterion.

A second example are *inter-route resource constraints* which occur in VRPs when different vehicles compete for globally limited resources (see Hempsch and Irnich [82]). The simplest type is a constraint limiting the number of available vehicles that can be allocated to a specific depot (the case of a limited fleet was discussed in Section 1.3.4). One may also wish to restrict the number of routes having a specific characteristic, such as the number of long routes (with respect to distance, number of stops, late arrival, etc.) or the number of routes crossing an area. Other interesting examples are a limited number of docks at a depot or hub (see also Rieck and Zimmermann [117]) and a limited processing capacity for incoming goods at a destination depot. The latter arises in mail collection, where mail is collected from customers and postboxes and is then processed and sorted at the depot before a given cut-off time (as for parcels). A limited processing capacity means that the vehicles need to arrive in a staggered sequence, where feasibility depends on both the arrival time and on the amount that remains to be processed before cut-off (this is the main application in [82]).

The third example is related to synchronization issues. Here also the routes and schedules of different vehicles are interdependent and need to be coordinated. A first systematic study on *VRP with Multiple Synchronization constraints* (VRPMS) was provided by Drexl [50]. He offers the following (not exclusive) classification of synchronization:

(i) *Task synchronization:* One must decide which vehicle or vehicles jointly fulfill each task. Basically, this is a problem of clustering tasks, but recall from Section 1.3.2 that the tasks may also be split by load or volume, as in the SDVRP, or by periods, as in the PVRP or the IRP, or may be transshipped between vehicles, as in the 2E-VRP.

(ii) *Operation synchronization:* Different vehicles at the same or different locations are needed to fulfill a task, where the operations performed by the individual vehicles need to take place either at the same time or with precedence, where so-called *dynamic time windows* may bound the time lag. For example, pairs of service technicians carry out the setup of some services for customers, where a first technician must establish the supply at an associated supply site of the provider before the second can establish the service by performing some operations at the customers site (see Goel and Meisel [68]).

(iii) *Movement synchronization:* Two or more vehicles must move along a part of their itinerary together. For example, a trailer has to be pulled by a truck (Section 1.3.4), and for winter road maintenance, when routing snow plows, the clearing of a two-lane or three-lane street requires that several snow plows follow one another to clear the snow away (see Salazar-Aguilar, Langevin, and Laporte [119]).

(iv) *Load synchronization:* The right amount of load is collected, delivered, and transshipped at all locations and for all vehicles and their combinations when interacting.

(v) *Resource synchronization:* At any point in time, the total utilization or consumption of resources by all vehicles must respect the available capacities. This is similar to the above-mentioned inter-route resource constraints.

Several forms of synchronization can occur together: In the VRPTT (Section 1.3.4), trailers are pulled by trucks (movement sync.), a truck transfers some or all of its load to a trailer (task and load sync.), at which time the trailer is exclusively available for this operation (operation and resource sync.), and the transfer time may depend on the amount that is transferred (also operation and resource sync.).

The most studied VRPMS are, according to Drexl [50], the N-echelon VRP and *location routing problems* (see Perboli, Tadei, and Vigo [110], Baldacci et al. [7], and Section 1.3.7), the PDPTW with transshipments (see Wen et al. [140]), and the simultaneous vehicle and crew routing and scheduling problems (see Drexl et al. [52]).

1.3.6 ▪ Objectives

We have introduced VRPs as pure routing cost minimization problems. However, the objectives may model various goals. In order to structure the discussion about the goals in vehicle routing, we will start with a discussion of single objectives before hierarchical objectives and multi-criteria problems are sketched.

Single Objective Optimization. The simplest modification to the objective (1.8) is setting some of the routing costs c_{ijk} to zero or to a sufficiently large number (such modifications using big-M exclude infeasible or undesirable arcs or edges). Recall that the VRPB and the site-dependent VRP can be transformed into a standard CVRP in this way. Moreover, the *Open VRP* (OVRP; see Li, Golden, and Wasil [95]) models that a vehicle does not return to the depot after servicing the last customer on the route, or likewise the return trip to the depot is not charged.

Whenever it is possible to select among the tasks being serviced, the objective can also include a profit component (to be maximized) for covered tasks, as discussed for the VRP with profits in Section 1.3.2.

For heterogeneous fleet VRP, we mentioned that not only variable routing costs c_{ijk} but also fixed costs FC_k can be relevant when deciding about how many vehicles to acquaint and to use for fulfilling the given transportation requests.

The costs for performing a route may have the following additional components: In the scheduling and time-windows context, i.e., when time schedules (T_{ik}) have to be determined, the costs for employing drivers may be relevant so that costs depend on the *route durations* $T_{dk} - T_{ok}$ (see Savelsbergh [121]) or the *time of finishing routes*, i.e., T_{dk} (see Jozefowiez, Semet, and Talbi [85]). In service industries, *customer satisfaction* is often crucial, and can be incorporated with a cost component depending on a (weighted) latency, i.e., a term of the form $p_i \max(T_i - a_i)^+$ (where the weight p_i is related to the importance of customer i and a_i is the earliest possible service time). Humanitarian applications provide another real-world example of a latency objective, where in case of a disaster, a limited fleet of vehicles has to reach cities or affected areas as early as possible (see Chapter 11). Also waiting times w_i at the customers might be undesirable and incur additional costs (see, e.g., Solomon [127] and Bräysy and Gendreau [20]). Another objective related to customer dissatisfaction is related to the use of soft time windows (for both waiting times and soft time windows, see Section 1.3.3). Finally, in the *delivery man problem with time windows* (see Heilporn, Cordeau, and Laporte [81]) the objective is minimizing the duration that every delivery request is on board a vehicle, i.e., $\sum_{k \in K} \sum_{i \in N} (T_{ik} - T_{ok})$. Applications include passengers or perishable goods distribution, school bus routing and scheduling, and the transportation of disabled persons.

As a substitute for balancing constraints (see Section 1.3.5), one may apply a *min-max objective*, e.g., in order to minimize the length (or duration, or workload) of the longest route (more common in the arc routing context; see Chapter 11 in [35]). Note that a balancing objective alone almost never makes sense because a perfectly balanced solution may contain highly inefficient routes.

From a more general point of view, costs can depend on any type of attribute related to a resource that is consumed or utilized when the vehicle travels. In the soft time-windows example, this resource is time and costs depend on the time of service. Obviously, considering resources such as load on board the vehicle and the distances traveled allow for complex objectives that model real-world *transport tariffs*. In these tariffs, costs often depend in a non-linear way on distances, load (weight, volume, number of pieces, etc.), time, and the particular itinerary. Non-trivial cost functions can also be found in hazardous materials transportation (see, e.g., Tarantilis and Kiranoudis [133]).

Many metaheuristics explicitly consider feasible and infeasible solutions in order to allow neighborhood operators to faster reach high-quality solutions (see Gendreau et al. [63]). *Penalties* are a means of guiding these metaheuristics towards feasible solutions. In the VRP context, these penalties generally depend on combinations of attributes changing along the traveled route. Thus, complex objectives may result from both complicated real-world cost functions and the VRP solution method.

A recently introduced and interesting area of research incorporates energy consumption and pollutant emission control in the routing context (see Bektaş and Laporte [11]). Generally, such aspects are incorporated in the problem as specific and complex objective functions. A comprehensive review of these problems, which are broadly called *green vehicle routing problems*, is given in Chapter 15.

Hierarchical Objectives.   Minimizing route length, duration, and completion time are generally conflicting objectives. Moreover, the minimization of the number of vehicles employed also conflicts with these goals. Since the use of vehicles and drivers typically incurs high route-related fixed costs, a common hierarchical way of optimizing is to minimize the number of vehicles first and then, with this number of vehicles fixed, a secondary objective is optimized. For example, the heuristics literature on VRPTW predominantly uses this hierarchical optimization model with route length minimization as the second objective (see Bräysy and Gendreau [20, 21] and Chapter 5).

Multi-criteria Optimization.   The survey by Jozefowiez, Semet, and Talbi [85] provides a comprehensive overview of multi-objective optimization in routing problems. The same authors apply in [86] a bi-objective optimization approach to find a compromise between the overall route length and balancing of routes.

1.3.7 ▪ Other Extensions

We have seen that interesting problems result from the simultaneous consideration of vehicle routing and other logistics activities: If the latter are replenishment activities, the class of IRP results (Section 1.3.2). The addition of production decisions leads to the class of production routing problems (see Adulyasak, Cordeau, and Jans [1]). Location Routing Problems (LRPs; see Min, Jayaraman, and Srivastava [101] and Nagy and Salhi [105]) simultaneously seek optimal location and routing decisions. We already discussed VRP with loading constraints in Section 1.3.3, where routing and packing are jointly taken into account.

An important class of VRP variants considers *route consistency* aspects when the service occurs repeatedly or regularly. For example, Gröer, Golden, and Wasil [74] define the *Consistent VRP* (ConVRP), where it is required that the same driver visit the same customers at roughly the same time on each day whenever this customer needs service. Smilowitz, Nowak, and Jiang [126] examine the interaction between routing and workforce management by taking into account drivers' familiarity with customers and delivery regions. Often, such consistency requirements are combined with those of *route compactness* (a.k.a. *visual beauty*), where compact *service territory* are determined and used as a base for daily route construction (see Wong and Beasley [142]). Within these settings, more recent articles examine the impact of demand and customer's location uncertainty on service territory construction (see Zhong, Hall, and Dessouky [145], Haugland, Ho, and Laporte [80], and Schneider et al. [124]). A different way to enforce route compactness is represented by the so-called *Clustered VRP* (CluVRP), where customers are partitioned into clusters: once a vehicle enters a cluster it must serve all its customers before leaving it (see, e.g., Battarra, Erdogan, and Vigo [9]).

It is clear that such a compilation of VRP variants and extensions can never be exhaustive: We had to select the most relevant material to include, and the research field is developing very fast. Even within the time between finishing our chapter and the publication of the book, new models and powerful solution approaches will certainly emerge. This is, to a large extent, driven by the need to better solve real-world VRPs: Important criteria are a *higher solution quality*, not only measured by the objective but also by realistically integrating all relevant aspects, and *faster solution algorithms* which are capable of solving *larger size instances*. For the sake of brevity, we almost completely omitted a discussion of algorithms here. The other chapters of this book will complement this introduction with additional descriptions and references.

We hope that the chapter at hand, with the presented topics and their classification, will prove a helpful introduction and will provide an informative overview of the many relevant aspects arising when modeling VRP. For the future, we expect to see new challenging combinations of problem features and further interesting extensions of the many facets describing the *Family of the Vehicle Routing Problems*.

Acknowledgment

We would like to thank Gilbert Laporte for carefully proofreading the manuscript of this chapter and pointing us to several important references.

Bibliography

[1] Y. ADULYASAK, J.-F. CORDEAU, AND R. JANS, *Formulations and branch-and-cut algorithms for multi-vehicle production and inventory routing problems*, Technical Report G-2012-14, GERAD, Montréal, Canada, 2012.

[2] F. ALONSO, M. J. ALVAREZ, AND J. E. BEASLEY, *A tabu search algorithm for the periodic vehicle routing problem with multiple vehicle trips and accessibility restrictions*, Journal of the Operational Research Society, 59 (2008), pp. 963–976.

[3] C. ARCHETTI, D. FEILLET, A. HERTZ, AND M. G. SPERANZA, *The capacitated team orienteering and profitable tour problems*, Journal of the Operational Research Society, 60 (2009), pp. 831–842.

[4] C. ARCHETTI, A. HERTZ, AND M. G. SPERANZA, *Metaheuristics for the team orienteering problem*, Journal of Heuristics, 13 (2007), pp. 49–76.

[5] P. AVELLA, M. BOCCIA, AND A. SFORZA, *Resource constrained shortest path problems in path planning for fleet management*, Journal of Mathematical Modelling and Algorithms, 3 (2004), pp. 1–17.

[6] R. BALDACCI, M. BATTARRA, AND D. VIGO, *Routing a heterogeneous fleet of vehicles*, in The Vehicle Routing Problem: Latest Advances and New Challenges, B. L. Golden, S. Raghavan, and E. A. Wasil, eds., vol. 43 of Operations Research/Computer Science Interfaces Series, Springer, New York, 2008, pp. 3–27.

[7] R. BALDACCI, A. MINGOZZI, R. ROBERTI, AND R. WOLFLER CALVO, *An exact algorithm for the two-echelon capacitated vehicle routing problem*, Operations Research, 61 (2013), pp. 298–314.

[8] M. L. BALINSKI AND R. E. QUANDT, *On an integer program for a delivery problem*, Operations Research, 12 (1964), pp. 300–304.

[9] M. BATTARRA, G. ERDOGAN, AND D. VIGO, *Exact algorithms for the clustered vehicle routing problem*, Operations Research, 62 (2014), pp. 58–71.

[10] M. BATTARRA, M. MONACI, AND D. VIGO, *An adaptive guidance approach for the heuristic solution of a multi-trip vehicle routing problem*, Computers & Operations Research, 36 (2009), pp. 3041–3050.

[11] T. BEKTAŞ AND G. LAPORTE, *The pollution-routing problem*, Transportation Research Part B: Methodological, 45 (2011), pp. 1232–1250.

[12] L. BERTAZZI AND M. G. SPERANZA, *Inventory routing problems: An introduction*, EURO Journal on Transportation and Logistics, 1 (2012), pp. 307–326.

[13] ——, *Inventory routing problems with multiple customers*, EURO Journal on Transportation and Logistics, 2 (2013), pp. 255–275.

[14] D. J. BERTSIMAS, *A vehicle routing problem with stochastic demand*, Operations Research, 40 (1992), pp. 574–585.

[15] N. BIANCHESSI, R. MANSINI, AND M. G. SPERANZA, *The distance constrained multiple vehicle traveling purchaser problem*, European Journal of Operational Research, 235 (2014), pp. 73–87.

[16] L. D. BODIN, V. MANIEZZO, AND A. MINGOZZI, *Street routing and scheduling problems*, in Handbook of Transportation Science, R. W. Hall, ed., vol. 23 of International Series in Operations Research & Management Science, Kluwer Academic Publishers, Norwell, MA, 1999, pp. 395–432.

[17] J. BRAMEL AND D. SIMCHI-LEVI, *On the effectiveness of set covering formulations for the vehicle routing problem with time windows*, Operations Research, 45 (1997), pp. 295–301.

[18] ——, *Set-covering-based algorithms for the capacitated VRP*, in The Vehicle Routing Problem, P. Toth and D. Vigo, eds., SIAM, Philadelphia, 2002, ch. 4, pp. 85–108.

[19] J. BRANDÃO AND A. MERCER, *A tabu search algorithm for the multi-trip vehicle routing and scheduling problem*, European Journal of Operational Research, 100 (1997), pp. 180–191.

[20] O. Bräysy and M. Gendreau, *Vehicle routing with time windows, Part I: Route construction and local search algorithms*, Transportation Science, 39 (2005), pp. 104–118.

[21] ——, *Vehicle routing with time windows, Part II: Metaheuristics*, Transportation Science, 39 (2005), pp. 119–139.

[22] S. Bunte and N. Kliewer, *An overview on vehicle scheduling models*, Public Transport, 1 (2009), pp. 299–317.

[23] A. M. Campbell, L. W. Clarke, and M. W. P. Savelsbergh, *Inventory routing in practice*, in The Vehicle Routing Problem, P. Toth and D. Vigo, eds., SIAM, Philadelphia, 2002, ch. 12, pp. 309–330.

[24] A. Cardeneo, *Modellierung und Optimierung des B2C-Tourenplanungsproblems mit alternativen Lieferorten und -zeiten*, Dissertation, Karlsruher Institut für Technologie KIT, 2005 (in German).

[25] D. Cattaruzza, N. Absi, D. Feillet, and T. Vidal, *A memetic algorithm for the multi trip vehicle routing problem*, European Journal of Operational Research, 236 (2014), pp. 833–848.

[26] K.-T. Chang, *Introduction to Geographic Information Systems*, McGraw–Hill, New York, NY, 4th ed., 2008.

[27] I.-M. Chao, *A tabu search method for the truck and trailer routing problem*, Computers & Operations Research, 29 (2002), pp. 33–51.

[28] H.-K. Chen, C.-F. Hsueh, and M.-S. Chang, *The real-time time-dependent vehicle routing problem*, Transportation Research Part E: Logistics and Transportation Review, 42 (2006), pp. 383–408.

[29] M. Cherkesly, G. Desaulniers, and G. Laporte, *Branch-price-and-cut algorithms for the pickup and delivery problem with time windows and LIFO loading*, Les Cahiers du GERAD G-2013-31, GERAD, Montréal, Canada, 2013.

[30] M.-J. Choi and S.-H. Lee, *The multiple traveling purchaser problem*, in 40th International Conference on Computers and Industrial Engineering (CIE), 2010, pp. 1–5.

[31] N. Christofides, A. Mingozzi, and P. Toth, *The vehicle routing problem*, in Combinatorial Optimization, N. Christofides, A. Mingozzi, P. Toth, and C. Sandi, eds., Wiley, Chichester, UK, 1979, ch. 11, pp. 315–338.

[32] C.-W. Chu, *A heuristic algorithm for the truckload and less-than-truckload problem*, European Journal of Operational Research, 165 (2005), pp. 657–667.

[33] G. Clarke and J. W. Wright, *Scheduling of vehicles from a central depot to a number of delivery points*, Operations Research, 12 (1964), pp. 568–581.

[34] L. C. Coelho, J.-F. Cordeau, and G. Laporte, *Thirty years of inventory routing*, Transportation Science, 48 (2014), pp. 1–19.

[35] Á. Corberán and G. Laporte, eds., *Arc Routing: Problems, Methods, and Applications*, MOS-SIAM Series on Optimization, SIAM, Philadelphia, 2014.

[36] J.-F. CORDEAU, G. DESAULNIERS, J. DESROSIERS, M. M. SOLOMON, AND F. SOUMIS, *VRP with time windows*, in The Vehicle Routing Problem, P. Toth and D. Vigo, eds., SIAM, Philadelphia, 2002, ch. 7, pp. 155–194.

[37] J.-F. CORDEAU, M. GENDREAU, AND G. LAPORTE, *A tabu search heuristic for periodic and multi-depot vehicle routing problems*, Networks, 30 (1997), pp. 105–119.

[38] J.-F. CORDEAU, M. IORI, G. LAPORTE, AND J.-J. SALAZAR-GONZÁLEZ, *A branch-and-cut algorithm for the pickup and delivery traveling salesman problem with LIFO loading*, Networks, 55 (2010), pp. 46–59.

[39] B. CREVIER, J.-F. CORDEAU, AND G. LAPORTE, *The multi-depot vehicle routing problem with inter-depot routes*, European Journal of Operational Research, 176 (2007), pp. 756–773.

[40] G. B. DANTZIG AND J. H. RAMSER, *The truck dispatching problem*, Management Science, 6 (1959), pp. 80–91.

[41] G. B. DANTZIG AND P. WOLFE, *Decomposition principle for linear programs*, Operations Research, 8 (1960), pp. 101–111.

[42] U. DERIGS, J. GOTTLIEB, J. KALKOFF, M. PIESCHE, F. ROTHLAUF, AND U. VOGEL, *Vehicle routing with compartments: Applications, modelling and heuristics*, OR Spectrum, 33 (2011), pp. 885–914.

[43] G. DESAULNIERS, J. DESROSIERS, A. ERDMANN, M. M. SOLOMON, AND F. SOUMIS, *VRP with pickup and delivery*, in The Vehicle Routing Problem, P. Toth and D. Vigo, eds., SIAM, Philadelphia, 2002, ch. 9, pp. 225–242.

[44] G. DESAULNIERS, J. DESROSIERS, I. IOACHIM, M. M. SOLOMON, F. SOUMIS, AND D. VILLENEUVE, *A unified framework for deterministic time constrained vehicle routing and crew scheduling problems*, in Fleet Management and Logistics, T. Crainic and G. Laporte, eds., Kluwer Academic Publishers, Boston, Dordrecht, London, 1998, ch. 3, pp. 57–93.

[45] M. DESROCHERS AND G. LAPORTE, *Improvements and extensions to the Miller-Tucker-Zemlin subtour elimination constraints*, Operations Research Letters, 10 (1991), pp. 27–36.

[46] J. DESROSIERS, Y. DUMAS, M. M. SOLOMON, AND F. SOUMIS, *Time constrained routing and scheduling*, in Network Routing, M. O. Ball, T. L. Magnanti, C. L. Monma, and G. L. Nemhauser, eds., vol. 8 of Handbooks in Operations Research and Management Science, Elsevier, Amsterdam, 1995, ch. 2, pp. 35–139.

[47] S. M. DISNEY AND D. R. TOWILL, *Vendor-managed inventory and bullwhip reduction in a two-level supply chain*, International Journal of Operations and Production Management, 23 (2003), pp. 625–651.

[48] M. DREXL, *Branch-and-price and heuristic column generation for the generalized truck-and-trailer routing problem*, Journal of Quantitative Methods for Economics and Business Administration, 12 (2011), pp. 5–38.

[49] ——, *Rich vehicle routing in theory and practice*, Logistics Research, 5 (2012), pp. 47–63.

[50] ———, *Synchronization in vehicle routing—A survey of VRPs with multiple synchronization constraints*, Transportation Science, 46 (2012), pp. 297–316.

[51] ———, *Applications of the vehicle routing problem with trailers and transshipments*, European Journal of Operational Research, 227 (2013), pp. 275–283.

[52] M. DREXL, J. RIECK, T. SIGL, AND B. PRESS, *Simultaneous vehicle and crew routing and scheduling for partial- and full-load long-distance road transport*, BuR – Business Research, 6 (2013), pp. 242–264.

[53] M. DROR, *Arc Routing: Theory, Solutions and Applications*, Kluwer, Boston, 2000.

[54] M. DROR AND P. TRUDEAU, *Savings by split delivery routing*, Transportation Science, 23 (1989), pp. 141–145.

[55] ———, *Split delivery routing*, Naval Research Logistic Quarterly, 37 (1990), pp. 382–402.

[56] S. ERDOGAN AND E. MILLER-HOOKS, *A green vehicle routing problem*, Transportation Research Part E: Logistics and Transportation Review, 48 (2012), pp. 100–114.

[57] D. FEILLET, P. DEJAX, AND M. GENDREAU, *Traveling salesman problems with profits*, Transportation Science, 39 (2005), pp. 188–205.

[58] M. FISCHETTI, J.-J. SALAZAR-GONZÁLEZ, AND P. TOTH, *Experiments with a multi-commodity formulation for the symmetric capacitated vehicle routing problem*, in Proceedings of the 3rd Meeting of the EURO Working Group on Transportation, 1995.

[59] P. FRANCIS, K. SMILOWITZ, AND M. TZUR, *The period vehicle routing problem with service choice*, Transportation Science, 40 (2006), pp. 439–454.

[60] T. GARAIX, C. ARTIGUES, D. FEILLET, AND D. JOSSELIN, *Vehicle routing problems with alternative paths: An application to on-demand transportation*, European Journal of Operational Research, 204 (2010), pp. 62–75.

[61] M. GENDREAU, M. IORI, G. LAPORTE, AND S. MARTELLO, *A tabu search algorithm for a routing and container loading problem*, Transportation Science, 40 (2006), pp. 342–350.

[62] M. GENDREAU, G. LAPORTE, AND R. SÉGUIN, *Stochastic vehicle routing*, European Journal of Operational Research, 88 (1996), pp. 3–12.

[63] M. GENDREAU, J.-Y. POTVIN, O. BRÄYSY, G. HASLE, AND A. LØKKETANGEN, *Metaheuristics for the vehicle routing problem and its extensions: A categorized bibliography*, in The Vehicle Routing Problem: Latest Advances and New Challenges, B. L. Golden, S. Raghavan, and E. A. Wasil, eds., vol. 43 of Operations Research Computer Science Interfaces Series, Springer, New York, 2008, pp. 143–169.

[64] A. GOEL, *Vehicle scheduling and routing with drivers' working hours*, Transportation Science, 43 (2009), pp. 17–26.

[65] ———, *Truck driver scheduling in the European Union*, Transportation Science, 44 (2010), pp. 429–441.

[66] A. GOEL, C. ARCHETTI, AND M. SAVELSBERGH, *Truck driver scheduling in Australia*, Computers & Operations Research, 39 (2012), pp. 1122–1132.

[67] A. GOEL AND L. KOK, *Truck driver scheduling in the United States*, Transportation Science, 46 (2012), pp. 317–326.

[68] A. GOEL AND F. MEISEL, *Workforce routing and scheduling for electricity network maintenance with downtime minimization*, European Journal of Operational Research, 231 (2013), pp. 210–228.

[69] A. GOEL AND L.-M. ROUSSEAU, *Truck driver scheduling in Canada*, Journal of Scheduling, 15 (2012), pp. 783–799.

[70] A. GOEL AND T. VIDAL, *Hours of service regulations in road freight transport: An optimization-based international assessment*, Transportation Science, 48 (2014), pp. 391–412.

[71] B. L. GOLDEN, A. A. ASSAD, AND E. A. WASIL, *Routing vehicles in the real world: Applications in the solid waste, beverage, food, dairy, and newspaper industries*, in The Vehicle Routing Problem, P. Toth and D. Vigo, eds., SIAM, Philadelphia, 2002, ch. 10, pp. 245–286.

[72] B. L. GOLDEN, T. L. MAGNANTI, AND H. Q. NGUYEN, *Implementing vehicle routing algorithms*, Networks, 7 (1977), pp. 113–148.

[73] I. GRIBKOVSKAIA, Ø. HALSKAU SR., G. LAPORTE, AND M. VLČEK, *General solutions to the single vehicle routing problem with pickups and deliveries*, European Journal of Operational Research, 180 (2007), pp. 568–584.

[74] C. GRÖER, B. L. GOLDEN, AND E. A. WASIL, *The consistent vehicle routing problem*, Manufacturing and Service Operations Management, 11 (2009), pp. 630–643.

[75] M. GUAN, *On the windy postman problem*, Discrete Applied Mathematics, 9 (1984), pp. 41–46.

[76] G. GUASTAROBA, M. G. SPERANZA, AND D. VIGO, *Designing service networks with intermediate facilities: An overview*, Technical Report, Department of Quantitative Methods, University of Brescia, Italy, 2013.

[77] G. GUTIN AND A. P. PUNNEN, eds., *The Traveling Salesman Problem and Its Variations*, vol. 12 of Combinatorial Optimization, Kluwer Academic Publishers, Dordrecht, 2002.

[78] M. HACHICHA, M. J. HODGSON, G. LAPORTE, AND F. SEMET, *Heuristics for the multi-vehicle covering tour problem*, Computers & Operations Research, 27 (2000), pp. 29–42.

[79] A. HAGHANI AND S. JUNG, *A dynamic vehicle routing problem with time-dependent travel times*, Computers & Operations Research, 32 (2005), pp. 2959–2986.

[80] D. HAUGLAND, S. C. HO, AND G. LAPORTE, *Designing delivery districts for the vehicle routing problem with stochastic demands*, European Journal of Operational Research, 180 (2007), pp. 997–1010.

[81] G. HEILPORN, J.-F. CORDEAU, AND G. LAPORTE, *The delivery man problem with time windows*, Discrete Optimization, 7 (2010), pp. 269–282.

[82] C. HEMPSCH AND S. IRNICH, *Vehicle-routing problems with inter-tour resource constraints*, in The Vehicle Routing Problem: Latest Advances and New Challenges, B. L. Golden, S. Raghavan, and E. A. Wasil, eds., vol. 43 of Operations Research/Computer Science Interfaces Series, Springer, New York, 2008, pp. 421–444.

[83] M. IORI, J.-J. SALAZAR-GONZALEZ, AND D. VIGO, *An exact approach for the vehicle routing problem with two dimensional loading constraints*, Transportation Science, 41 (2007), pp. 253–264.

[84] P. JAILLET AND M. R. WAGNER, *Online vehicle routing problems: A survey*, in The Vehicle Routing Problem: Latest Advances and New Challenges, B. L. Golden, S. Raghavan, and E. A. Wasil, eds., vol. 43 of Operations Research/Computer Science Interfaces Series, Springer, New York, 2008, pp. 221–237.

[85] N. JOZEFOWIEZ, F. SEMET, AND E.-G. TALBI, *Multi-objective vehicle routing problems*, European Journal of Operational Research, 189 (2008), pp. 293–309.

[86] ———, *An evolutionary algorithm for the vehicle routing problem with route balancing*, European Journal of Operational Research, 195 (2009), pp. 761–769.

[87] A. L. KOK, C. M. MEYER, H. KOPFER, AND J. M. J. SCHUTTEN, *A dynamic programming heuristic for the vehicle routing problem with time windows and European Community social legislation*, Transportation Science, 44 (2010), pp. 442–454.

[88] G. LAPORTE, M. DESROCHERS, AND Y. NOBERT, *Two exact algorithms for the distance-constrained vehicle routing problem*, Networks, 14 (1984), pp. 161–172.

[89] G. LAPORTE, H. MERCURE, AND Y. NOBERT, *An exact algorithm for the asymmetrical capacitated vehicle routing problem*, Networks, 16 (1986), pp. 33–46.

[90] G. LAPORTE AND Y. NOBERT, *Exact algorithms for the vehicle routing problem*, Annals of Discrete Mathematics, 31 (1987), pp. 147–184.

[91] ———, *A vehicle flow model for the optimal design of a two-echelon distribution system*, in Advances in Optimization and Control, H. A. Eiselt and G. Pederzoli, eds., vol. 302 of Lecture Notes in Economics and Mathematical Systems, Springer, Berlin, Heidelberg, 1988, pp. 158–173.

[92] G. LAPORTE, Y. NOBERT, AND M. DESROCHERS, *Optimal routing under capacity and distance restrictions*, Operations Research, 33 (1985), pp. 1050–1073.

[93] E. L. LAWLER, J. K. LENSTRA, A. H. G. RINNOOY KAN, AND D. B. SHMOYS, eds., *The Traveling Salesman Problem. A Guided Tour of Combinatorial Optimization*, Wiley-Interscience Series in Discrete Mathematics, Wiley, Chichester, UK, 1985.

[94] H. L. LEE, V. PADMANABHAN, AND S. WHANG, *The bullwhip effect in supply chains*, Sloan Management Review, 38 (1997), pp. 93–102.

[95] F. LI, B. L. GOLDEN, AND E. A. WASIL, *The open vehicle routing problem: Algorithms, large-scale test problems, and computational results*, Computers & Operations Research, 34 (2007), pp. 2918–2930.

[96] J.-Q. LI, P. B. MIRCHANDANI, AND D. BORENSTEIN, *Real-time vehicle rerouting problems with time windows*, European Journal of Operational Research, 194 (2009), pp. 711–727.

[97] M. E. LÜBBECKE AND J. DESROSIERS, *Selected topics in column generation*, Operations Research, 53 (2005), pp. 1007–1023.

[98] S. MARTELLO AND P. TOTH, *Knapsack Problems: Algorithms and Computer Implementations*, Wiley Interscience Series in Discrete Mathematics and Optimization, Wiley, New York, 1990.

[99] D. L. MILLER, A. W. TUCKER, AND R. A. ZEMLIN, *Integer programming formulations of traveling salesman problems*, Journal of the ACM, 7 (1960), pp. 326–329.

[100] H. MIN, *The multiple vehicle routing problem with simultaneous delivery and pick-up points*, Transportation Research Part A: Policy and Practice, 23 (1989), pp. 377–386.

[101] H. MIN, V. JAYARAMAN, AND R. SRIVASTAVA, *Combined location-routing problems: A synthesis and future research directions*, European Journal of Operational Research, 108 (1998), pp. 1–15.

[102] A. MINGOZZI, R. ROBERTI, AND P. TOTH, *An exact algorithm for the multitrip vehicle routing problem*, INFORMS Journal on Computing, 25 (2013), pp. 193–207.

[103] M. MOURGAYA AND F. VANDERBECK, *The periodic vehicle routing problem: Classification and heuristic*, RAIRO – Operations Research, 40 (2006), pp. 169–194.

[104] D. NADDEF AND G. RINALDI, *Branch-and-cut algorithms for the capacitated VRP*, in The Vehicle Routing Problem, P. Toth and D. Vigo, eds., SIAM, Philadelphia, 2002, ch. 3, pp. 53–84.

[105] G. NAGY AND S. SALHI, *Location-routing: Issues, models and methods*, European Journal of Operational Research, 177 (2007), pp. 649–672.

[106] A. OLIVERA AND O. VIERA, *Adaptive memory programming for the vehicle routing problem with multiple trips*, Computers & Operations Research, 34 (2007), pp. 28–47.

[107] C. S. ORLOFF, *Routing a fleet of m vehicles to/from a central facility*, Networks, 4 (1974), pp. 147–162.

[108] M. PADBERG AND T.-Y. SUNG, *An analytical comparison of different formulations of the travelling salesman problem*, Mathematical Programming, 52 (1991), pp. 315–357.

[109] S. PARRAGH, K. F. DOERNER, AND R. F. HARTL, *A survey on pickup and delivery problems*, Journal für Betriebswirtschaft, 58 (2008), pp. 81–117.

[110] G. PERBOLI, R. TADEI, AND D. VIGO, *The two-echelon capacitated vehicle routing problem: Models and math-based heuristics*, Transportation Science, 45 (2011), pp. 364–380.

[111] R. J. PETCH AND S. SALHI, *A multi-phase constructive heuristic for the vehicle routing problem with multiple trips*, Discrete Applied Mathematics, 133 (2004), pp. 69–92.

[112] J.-Y. POTVIN AND M.-A. NAUD, *Tabu search with ejection chains for the vehicle routing problem with private fleet and common carrier*, Journal of the Operational Research Society, 62 (2011), pp. 326–336.

[113] E. PRESCOTT-GAGNON, G. DESAULNIERS, M. DREXL, AND L.-M. ROUSSEAU, *European driver rules in vehicle routing with time windows*, Transportation Science, 44 (2010), pp. 455–473.

[114] H. N. PSARAFTIS, *Dynamic vehicle routing problems*, in Vehicle Routing: Methods and Studies, B. L. Golden and A. A. Assad, eds., North–Holland, Amsterdam, 1988, pp. 223–248.

[115] M.-E. RANCOURT, J.-F. CORDEAU, AND G. LAPORTE, *Long-haul vehicle routing and scheduling with working hour rules*, Transportation Science, 47 (2013), pp. 81–107.

[116] J. RENAUD, G. LAPORTE, AND F. F. BOCTOR, *A tabu search heuristic for the multi-depot vehicle routing problem*, Computers & Operations Research, 23 (1996), pp. 229–235.

[117] J. RIECK AND J. ZIMMERMANN, *A new mixed integer linear model for a rich vehicle routing problem with docking constraints*, Annals of Operations Research, 181 (2010), pp. 337–358.

[118] J. RIERA-LEDESMA AND J.-J. SALAZAR-GONZÁLEZ, *Solving school bus routing using the multiple vehicle traveling purchaser problem: A branch-and-cut approach*, Computers & Operations Research, 39 (2012), pp. 391–404.

[119] M. A. SALAZAR-AGUILAR, A. LANGEVIN, AND G. LAPORTE, *Synchronized arc routing for snow plowing operations*, Computers & Operations Research, 39 (2012), pp. 1432–1440.

[120] J. K. SANKARAN AND R. R. UBGADE, *Routing tankers for dairy milk pickup*, Interfaces, 24 (1994), pp. 59–66.

[121] M. W. P. SAVELSBERGH, *The vehicle routing problem with time windows: Minimizing route duration*, ORSA Journal on Computing, 4 (1992), pp. 146–154.

[122] M. W. P. SAVELSBERGH AND M. SOL, *The general pickup and delivery problem*, Transportation Science, 29 (1995), pp. 17–29.

[123] M. SCHNEIDER, A. STENGER, AND D. GOEKE, *The electric vehicle routing problem with time windows and recharging stations*, Transportation Science, forthcoming (2014).

[124] M. SCHNEIDER, A. STENGER, F. SCHWAHN, AND D. VIGO, *Territory-based vehicle routing in the presence of time window constraints*, Transportation Science, forthcoming (2014).

[125] R. SHUTTLEWORTH, B. L. GOLDEN, S. SMITH, AND E. A. WASIL, *Advances in meter reading: Heuristic solution of the close enough traveling salesman problem over a street network*, in The Vehicle Routing Problem: Latest Advances and New Challenges, B. L. Golden, S. Raghavan, and E. A. Wasil, eds., vol. 43 of Operations Research/Computer Science Interfaces, Springer, New York, 2008, pp. 487–501.

[126] K. SMILOWITZ, M. NOWAK, AND T. JIANG, *Workforce management in periodic delivery operations*, Transportation Science., (2013), pp. 214–230.

[127] M. M. SOLOMON, *Algorithms for the vehicle routing and scheduling problem with time window constraints*, Operations Research, 35 (1987), pp. 254–265.

[128] H. SONG, V. N. HSU, AND R. K. CHEUNG, *Distribution coordination between suppliers and customers with a consolidation center*, Operations Research, 56 (2008), pp. 1264–1277.

[129] A. STENGER, D. VIGO, S. ENZ, AND M. SCHWIND, *A variable neighborhood search algorithm for a vehicle routing problem arising in small package shipping*, Transportation Science, 47 (2013), pp. 64–80.

[130] É. D. TAILLARD, P. BADEAU, M. GENDREAU, F. GUERTIN, AND J.-Y. POTVIN, *A tabu search heuristic for the vehicle routing problem with soft time windows*, Transportation Science, 31 (1997), pp. 170–186.

[131] É. D. TAILLARD, G. LAPORTE, AND M. GENDREAU, *Vehicle routeing with multiple use of vehicles*, Journal of the Operational Research Society, 47 (1996), pp. 1065–1070.

[132] L. TANG AND X. WANG, *Iterated local search based on very large-scale neighborhood for prize-collecting vehicle routing problem*, The International Journal of Advanced Manufacturing Technology, 29 (2006), pp. 1246–1258.

[133] C. D. TARANTILIS AND C. T. KIRANOUDIS, *Using the vehicle routing problem for the transportation of hazardous materials*, Operational Research. An International Journal, 1 (2001), pp. 67–78.

[134] C. D. TARANTILIS, E. E. ZACHARIADIS, AND C. T. KIRANOUDIS, *A hybrid guided local search for the vehicle routing problem with intermediate replenishment facilities*, INFORMS Journal on Computing, 20 (2008), pp. 154–168.

[135] F. A. TILLMAN, *The multiple terminal delivery problem with probabilistic demands*, Transportation Science, 3 (1969), pp. 192–204.

[136] P. TOTH AND D. VIGO, *An overview of vehicle routing problems*, in The Vehicle Routing Problem, P. Toth and D. Vigo, eds., SIAM, Philadelphia, 2002, ch. 1, pp. 1–26.

[137] T. VIDAL, T. G. CRAINIC, M. GENDREAU, N. LAHRICHI, AND W. REI, *A hybrid genetic algorithm for multidepot and periodic vehicle routing problems*, Operations Research, 60 (2012), pp. 611–624.

[138] T. VIDAL, T. G. CRAINIC, M. GENDREAU, AND C. PRINS, *A unifying view on timing problems and algorithms*, Technical Report 2011-43, CIRRELT, Montréal, Canada, 2011.

[139] A. C. WADE AND S. SALHI, *An investigation into a new class of vehicle routing problem with backhauls*, Omega, 30 (2002), pp. 479–487.

[140] M. WEN, J. LARSEN, J. CLAUSEN, J.-F. CORDEAU, AND G. LAPORTE, *Vehicle routing with cross-docking*, Journal of the Operational Research Society, 60 (2009), pp. 1708–1718.

[141] S. WØHLK, *A decade of capacitated arc routing*, in The Vehicle Routing Problem: Latest Advances and New Challenges, B. L. Golden, S. Raghavan, and E. A. Wasil, eds., vol. 43 of Operations Research/Computer Science Interfaces Series, Springer, New York, 2008, pp. 29–48.

[142] K. F. WONG AND J. E. BEASLEY, *Vehicle routing using fixed delivery areas*, Omega, 12 (1984), pp. 591–600.

[143] H. XU, Z.-L. CHEN, S. RAJAGOPAL, AND S. ARUNAPURAM, *Solving a practical pickup and delivery problem*, Transportation Science, 37 (2003), pp. 347–364.

[144] E. E. ZACHARIADIS, C. D. TARANTILIS, AND C. T. KIRANOUDIS, *The pallet-packing vehicle routing problem*, Transportation Science, 46 (2012), pp. 341–358.

[145] H. ZHONG, R. W. HALL, AND M. DESSOUKY, *Territory planning and vehicle dispatching with driver learning*, Transportation Science, 41 (2007), pp. 74–89.

Part I

The Capacitated Vehicle Routing Problem

Chapter 2

Classical Exact Algorithms for the Capacitated Vehicle Routing Problem

Frédéric Semet
Paolo Toth
Daniele Vigo

2.1 ▪ Introduction

In this chapter we present an overview of the early exact methods used for the solution of the *Capacitated Vehicle Routing Problem* (CVRP). The CVRP is an extension of the well-known *Traveling Salesman Problem* (TSP), calling for the determination of a Hamiltonian circuit with minimum cost visiting exactly once a given set of points. Therefore, the foundation of many exact approaches for the CVRP were derived from the extensive and successful work done for the exact solution of the TSP. However, even if tremendous progress has been made with respect to the first algorithms, such as the tree search method by Christofides and Eilon [17], the CVRP is still far from being satisfactorily solved. Our analysis encompasses more than three decades of research and examines the main families of approaches, from direct tree search methods based on Branch-and-Bound to column generation and Branch-and-Cut algorithms presented around the year 2000. The wide variety and richness of methods proposed in these early decades of CVRP history is witnessed by the good number of survey works that analyzed the relevant literature. Following the first comprehensive work of Laporte and Nobert [39], several review papers were devoted to the analysis of exact algorithms for the VRP as those of Laporte [36], Toth and Vigo [52, 53], Bramel and Simchi-Levi [15], Naddef and Rinaldi [47], Cordeau et al. [20], and Baldacci, Toth, and Vigo [10, 11]. More recent Branch-and-Cut-and-Price algorithms, which have successfully combined and enhanced those described in the following, will be covered in detail in Chapter 3.

We will denote with SCVRP and ACVRP the symmetric and asymmetric CVRP, respectively. When the explicit distinction between the two versions is not needed, we simply use CVRP. Moreover, throughout this chapter the graphs, both directed or undirected, are assumed to be complete. In this chapter we extensively refer to the basic notation and to the models presented in Chapter 1, but we recall the most used elements to facilitate the reading. The computational testing of the algorithms is generally performed by using a rather limited set of benchmark instances from the literature. To identify such test instances we adopt the naming convention described in Chapter 1 of Toth and Vigo [54]. Moreover, when available we also report the hardware used in the tests and its relative speed in Mflops given by Dongarra [24].

The chapter is subdivided in three sections, each devoted to the main approaches used in early algorithms. In particular, in Section 2.2 we examine the Branch-and-Bound algorithms and in Section 2.3 the column generation methods based on the set partitioning formulation. Finally, in Section 2.4 we present the algorithms based on Branch-and-Cut paradigms.

2.2 ▪ Branch-and-Bound Algorithms

Up to the late eighties, the most effective exact approaches for the CVRP were mainly direct tree search algorithms based on Branch-and-Bound. Similarly to what was proposed earlier for the TSP (see Little et al. [42]) they incorporated basic combinatorial relaxations based either on the *Assignment Problem* (AP) or the *Shortest Spanning Tree* (SST). Such initial approaches were able to solve to optimality instances with some tens of customers, even with the relatively limited computing hardware available at that time. Given such encouraging results, at the end of the nineties more sophisticated bounds were proposed, such as those based on Lagrangian relaxations and the additive approach, which brought direct tree search methods to their best possible performance prior to the systematic introduction of cutting planes.

To illustrate the relaxations that provide the fundamental component of Branch-and-Bound methods we refer to the two-index vehicle flow formulations of the VRP, which are described in detail in Chapter 1 and briefly summarized hereafter to facilitate the reading. The ACVRP is defined on a complete directed graph $G = (V, A)$, where $V = \{0\} \cup N = \{0, 1, \ldots, n\}$ is the set of vertices representing the depot (vertex 0) and the customers, which are the vertices in $N = \{1, \ldots, n\}$. Set A includes the directed arcs, whereas, in the symmetric case, the undirected graph $G = (V, E)$ contains the set E of undirected edges. The cost matrix c, associated with the arcs or edges, is either asymmetric (for ACVRP) or symmetric (for SCVRP). Finally, with each customer $i \in N$ is associated a demand $q_i \geq 0$ and a fleet K of identical vehicles, each having capacity Q available at the depot. Further details on the notation are given in Section 1.2. The models, denoted as VRP1 for the ACVRP and as VRP2 for the SCVRP, are reported side by side in the following:

		(VRP1)	(VRP2)	
(2.1)		minimize $c^\top x$	minimize $c^\top x$	
(2.2)	s.t.	$x(\delta^+(i)) = 1$	$x(\delta(i)) = 2$	$\forall i \in N,$
		$x(\delta^-(j)) = 1$		$\forall j \in N,$
(2.3)		$x(\delta^+(0)) = |K|$	$x(\delta(0)) = 2|K|,$	
(2.4)		$x(\delta^+(S)) \geq r(S)$	$x(\delta(S)) \geq 2r(S)$	$\forall S \subseteq N, S \neq \emptyset,$
(2.5)		$x_a \in \{0,1\} \; \forall a \in A$	$x_e \in \{0,1,2\} \; \forall e \in \delta(0),$	
			$x_e \in \{0,1\} \; \forall e \in E \setminus \delta(0).$	

In this formulation, constraints (2.2) and (2.3) are the degree restrictions at the customers and at the depot. Constraints (2.4) impose the capacity constraints. In their expression, $r(S)$ represents the number of vehicles required to serve the vertices in S. Such constraints, called *Generalized Subtour Elimination Constraints* (GSEC), also impose solution connectivity since $r(S) \geq 1$. Finally, constraints (2.5) define the decision variables.

Generally these are binary variables, indicating whether the corresponding arcs or edges are used or not in the optimal CVRP solution; however, for the SCVRP the variables associated with the edges incident into the depot may also take value two, thus representing a route which includes only one customer.

In the following we review the main ingredients of Branch-and-Bound algorithms proposed for the CVRP. We first examine the basic combinatorial relaxations obtained by algebraic manipulation of the above formulation. The quality of the resulting lower bounds is generally poor, and substantial efforts are needed to improve them, as in the bounding procedures based on Lagrangian and additive approaches presented next. The section is concluded examining branching a reduction strategy commonly implemented in the literature.

2.2.1 ▪ Bounds Based on Assignment and Matching

The first type of combinatorial relaxation for the CVRP is an extension of the method proposed by Little et al. [42] for the TSP: it is obtained by considering the directed model VRP1 and dropping the GSECs (2.4). The resulting problem is a *Transportation Problem* (TP), calling for a min-cost collection of circuits of G visiting once all the vertices in N, and $|K|$ times vertex 0. The solution of the relaxed problem can be infeasible for the CVRP since

(i) the total customer demand on a route may exceed the vehicle capacity;

(ii) there may exist "isolated" routes, i.e., subtours not visiting the depot.

To efficiently solve the TP, early algorithms transformed it into an equivalent AP defined on an extended complete directed graph $G' = (V', A')$, obtained by adding $|K| - 1$ copies of the depot vertex to V. More precisely, $V' := N \cup W$, where $W := \{0\} \cup \{n+1, \ldots, n+|K|-1\}$ is the set of the $|K|$ vertices of G' associated with the depot and each customer in N is connected to each such copy. Moreover, the cost c'_{ij} of each arc in A' is defined as follows:

$$(2.6) \qquad c'_{ij} := \begin{cases} c_{ij} & \text{for } i, j \in N; \\ c_{i0} & \text{for } i \in N, j \in W; \\ c_{0j} & \text{for } i \in W, j \in N; \\ \lambda & \text{for } i, j \in W, \end{cases}$$

where λ is a parameter whose value influences the number of vehicles used by the solution. In particular, when $\lambda = +\infty$, the model imposes using all the $|K|$ available vehicles, as is typically required in CVRP. Note, however, that other values of the parameter permit using fewer vehicles if convenient. For example, $\lambda = 0$ leads to the min-cost solution using *at most* $|K|$ routes, whereas defining $\lambda = -\infty$ leads to the min-cost solution using exactly $r(N)$ routes.

This type of relaxation was used in the first Branch-and-Bound for SCVRP by Christofides and Eilon [17], who also used a simple bound based on SST and were able to solve two small problems with 6 and 13 customers on an IBM 7090. The same relaxation was used by Laporte, Mercure, and Nobert [37] within a Branch-and-Bound algorithm for the ACVRP which, thanks to the much better quality of the AP relaxation for asymmetric problems, was able to solve randomly generated instances with some tens of customers and up to four vehicles VAX 11/780 (0.14 Mflops).

The same type of relaxation when performed on the symmetric model VRP2 amounts to a so-called b-Matching problem and requires the determination of a min-cost collection of tours covering all the vertices and such that the degree of each vertex i is equal to b_i, where $b_i = 2$ for all the customer vertices, and $b_0 = 2|K|$ for the depot vertex. This relaxation was used by Miller [45] as a base for an effective Lagrangian bound presented in Section 2.2.3. Also in this case it is possible to transform the problem by adding $|K|-1$ copies of the depot, thus obtaining an equivalent 2-Matching relaxation.

According to the experimental evaluation performed in Toth and Vigo [53] on nine test instances including between 44 and 199 customers, the b-Matching is the best relaxation of this type with an average ratio of 76.7% of the corresponding lower bound with respect to the best known feasible solution value. The simpler AP bound has a worse performance on symmetric instances, with a ratio of 67.4%. However, a similar evaluation on asymmetric instances with up to 70 customers shows, not surprisingly, a much better ratio of 91.3%.

We finally mention that Fischetti, Toth, and Vigo [27] used an AP relaxation as the base for a bounding procedure based on a disjunction on infeasible arc subsets. Given a set B of arcs that may not be used all together in any feasible solution of CVRP, a valid lower bound can be computed as the minimum value of the bounds computed by excluding in turn each such arc from the solution. This is easily done by setting to ∞ the cost of an arc in B and determining the corresponding bound. In [27] the AP relaxation was used both to detect possible sets B and to compute the overall bound which resulted slightly better than that obtained with the AP relaxation for ACVRP instances.

2.2.2 ▪ Bounds Based on Spanning Trees and Shortest Paths

Several relaxations based on the solution of the SST, i.e., the problem of finding a minimum-cost subset of edges connecting all vertices of the graph, were proposed for SCVRP. These relaxations are obtained by weakening the GSECs so as to impose only the connectivity of the solution and by ignoring part of the degree requirements of the vertices.

The first attempt in this direction was performed by Christofides and Eilon [17], who directly used the well-known 1-tree relaxation introduced by Held and Karp [34] for the TSP. The 1-tree is a subset of $|V| = n+1$ edges obtained by adding to the n edges, forming an SST over G, the minimum cost edge not belonging to the SST. The direct extension of the 1-tree relaxation to SCVRP was performed by Fisher [28], who used as a basic relaxation a K-tree, defined as a set of $n + |K|$ edges spanning the graph. Fisher modeled the SCVRP as the problem of determining a K-tree with degree equal to $2|K|$ at the depot vertex, and with additional constraints imposing (i) the vehicle capacity requirements, and (ii) the degree of each customer vertex, which must be equal to 2. The determination of a K-tree with degree $2K$ at the depot requires $O(n^3)$ time. Although the quality of the K-tree bound is experimentally quite poor (see Toth and Vigo [53], who report a ratio of 74.7%), it was incorporated into an effective Lagrangian bound described in Section 2.2.3. Similar relaxations may be derived for the ACVRP as described in Toth and Vigo [53].

A different tree-based relaxation for SCVRP was introduced by Christofides, Mingozzi, and Toth [18], who defined the so-called k-Degree Center Tree (k-DCT) as a tree with degree k at vertex 0, where $|K| \leq k \leq 2|K|$. The overall relaxation is obtained by adding to the k-DCT a set of $2|K| - k$ edges with minimum total cost. An overall Lagrangian bound was then derived by dualizing the degree constraints at the customers (2.2) and at the depot (2.3). The resulting bound incorporated into a Branch-and-Bound algorithm was able to solve instances with up to 20 customers on a CDC 7600 (3.3 Mflops).

Another very important relaxation of SCVRP, first described in Christofides, Mingozzi, and Toth [18], is based on the so-called q-route, which is not necessarily a simple route with total load equal to q and not including loops with two vertices. In [18] an efficient procedure is given to compute a q-route as a q-path from the depot to customer i plus the arc from i to the depot. The q-routes are used to derive an overall bound that proved to be tighter than the k-DCT one and allowed solving instances with up to 25 customers. In Hadjiconstantinou, Christofides, and Mingozzi [33] an enhanced version of this method, called *through q-route*, is obtained by combining the two shortest paths from the depot to customer i. The q-route concept is also extensively used in recent Branch-and-Cut-and-Price algorithms that are the current best exact approaches for the CVRP and are described in detail in Chapter 3.

2.2.3 ▪ Improved Bounds: Lagrangian and Additive Approaches

As discussed in the previous sections, the basic combinatorial relaxations available for both ACVRP and SCVRP have a poor quality and, when used within Branch-and-Bound approaches, they allow for the optimal solution of small instances only. To considerably increase the solution effectiveness of the Branch-and-Bound algorithms more sophisticated bounding techniques are needed.

Fisher [28] and Miller [45] proposed strengthening the basic SCVRP relaxations by dualizing, in a Lagrangian fashion, some of the relaxed constraints. In particular, Fisher started from his K-tree relaxation and included in the objective function the degree constraints (2.2) and some of the GSECs (2.4) in the cut form called *Capacity Cut Constraints* (CCCs), whereas Miller included a subset of the GSECs that were removed to obtain the b-matching relaxation. As in related problems, good values for the Lagrangian multipliers associated with the relaxed constraints are determined by using a standard subgradient optimization procedure (see, e.g., Held, Wolfe, and Crowder [35]).

The main difficulty associated with these relaxations is represented by the exponential cardinality of the set of relaxed constraints (i.e., the CCCs and GSECs), which does not allow for the explicit inclusion of all of them into the objective function. To this end, both authors proposed including only a limited family $\mathscr{F}$ of CCCs or GSECs and iteratively adding to the Lagrangian relaxation the constraints which are violated by the current solution of the Lagrangian problem. In particular, at each iteration of the subgradient optimization procedure, the edges incident to the depot in the current Lagrangian solution are removed. Violated constraints (i.e., CCCs or GSECs, depending on the approach), if any, are *separated* (i.e., detected) by examining the connected components obtained in this way. This separation routine is exact; i.e., if a constraint associated with, say, vertex set S is violated by the current Lagrangian solution, then there is a connected component of that solution spanning all the vertices in S and violating the constraint. The new constraints are added to the Lagrangian problem, i.e., to $\mathscr{F}$, with an associated multiplier, and the process is iterated until no violated constraint is detected (hence the Lagrangian solution is feasible) or a prefixed number of subgradient iterations has been executed. Slack constraints are periodically *purged* (i.e., removed) from $\mathscr{F}$.

Fisher [28] initialized $\mathscr{F}$ with an explicit set of constraints containing the customer subsets nested around $|K|+3$ *seed* customers. The seeds were chosen as the $|K|$ customers farthest from the depot in the routes corresponding to an initial feasible solution, whereas the last three customers were the ones maximally distant from the depot and the other seeds. For each seed, 60 sets were generated by including customers according to increasing distances from the seed. New violated CCCs, if any, are identified and added every 50 subgradient iterations. The number of iterations of the subgradient optimization procedure performed at the root node of the Branch-and-Bound algorithm ranged between 2000 and 3000. The overall Lagrangian bound considerably improved the basic K-tree relaxation and was, on average, larger than 99% of the optimal solution value for the three

Euclidean instances with $n \leq 100$ solved to optimality in Fisher [28] within 60,000 seconds on an Apollo Domain 3000 (0.071 Mflops).

Miller [45] initialized $\mathscr{F}$ as the empty set and at each iteration of the subgradient procedure detected violated GSECs, if any. The iteration was stopped when no improvement was obtained over 50 subgradient iterations. Also in this case the final Lagrangian bound is considerably tight, being on average 98% of the optimal solution value for the eight problems with $n \leq 50$ solved in [45] within 15,000 seconds on a Sun Sparc 2 (4 Mflops).

The relax-and-cut algorithm by Martinhon, Lucena, and Maculan [44] generalizes these Lagrangian-based approaches by considering also comb and multistar inequalities, and they were able to moderately improve the quality of the overall Lagrangian bound.

As to the ACVRP, Fischetti, Toth, and Vigo [27] obtained an improvement with respect to the AP bound by combining several different relaxations into an overall *additive* bounding procedure. The additive approach was proposed by Fischetti and Toth [26] and allows for the combination of different lower bounding procedures, each exploiting different substructures of the considered problem. The bounding procedures are applied in sequence, and the overall additive lower bound is given by the sum of the lower bounds obtained in this way. The bounding procedures that are combined in [27] are the disjunctive relaxation described in Section 2.2.1 and a new one based on min-cost flow computation that permits imposing a subset of GSECs. The resulting additive bound is considerably better than the AP bound and when used in a Branch-and-Bound algorithm permitted to solve random ACVRP instances with up to 300 customers and four vehicles within 1000 CPU seconds on a DECstation 5000/240 (5.3 Mflops). In [27] some real-world ACVRP instances are also solved by using the additive bound, whereas in Toth and Vigo [53] its successful application to SCVRP instances with up to 47 vertices is reported.

An interesting additive approach was also adopted by Hadjiconstantinou, Christofides, and Mingozzi [33] to compute a lower bound to CVRP. In particular, they considered the set partitioning formulation VRP4 of CVRP (see Section 2.3) and the dual of the corresponding linear programming relaxation. It is clear that, by linear programming duality, any feasible solution to such a dual problem provides a valid lower bound for CVRP. Therefore, they combined different relaxations based on q-routes and shortest paths to compute feasible solutions of the dual problem. The resulting lower bound was very tight and permitted to solve problem instances with up to 50 customers within a time limit of 12 hours on a Silicon Graphics Indigo R4000 (12 Mflops).

2.2.4 ▪ Structure of the Branch-and-Bound Algorithms

In addition to the bounding procedures, several other ingredients are crucial for a successful implementation of a Branch-and-Bound algorithm, and CVRP is not an exception. Most of the issues we analyze below are relevant also for other approaches which use the implicit enumeration scheme of Branch-and-Bound, such as Branch-and-Cut and Branch-and-Cut-and-Price.

Branching Scheme. Many branching schemes were used for SCVRP, and almost all are extensions of those used for the TSP. The first scheme we consider, proposed in Christofides [17], is known as *branching on arcs*, and proceeds by extending partial paths, starting from the depot and finishing at a given vertex. At each node of the branch-decision tree, an edge (i, j) is selected to extend the current partial path, and two descendant nodes are generated: the first node is associated with the inclusion of the selected arc in the solution (i.e., $x_{ij} = 1$), while in the second node the arc is excluded (i.e., $x_{ij} = 0$). In Miller [45] the arc selected for branching is determined as that expanding the current partial path in the best Lagrangian solution. When no such partial path exists (e.g., at the root node of the branch-decision tree) the selected arc is that connecting the depot with the unserved customer with the largest demand.

Fisher [28] used a mixed scheme where branching on arcs is used whenever no partial path is present in the current subproblem. In this case the currently unserved customer i with the largest demand is chosen and the arc (i, j) is used for branching, where j is the unserved customer closest to i. At the node where arc (i, j) is excluded from the solution, branching on arcs is again used, whereas at the second node the scheme known as *branching on customers* is used. One of the two ending customers, say v, of the currently imposed sequence of customers is chosen, and branching is performed by enumerating the customers which may be appended to that end of the sequence. A subset T of currently unserved customers is selected, e.g., that including the unserved customers closest to v, and $|T| + 1$ nodes are generated. Each of the first $|T|$ nodes corresponds to the inclusion in the solution of a different arc (v, j), $j \in T$, while in the last node all the arcs (v, j), $j \in T$, are excluded. The mixed branching scheme was used by Fisher to attempt the solution of Euclidean CVRP instances with real distances and about 100 customers, but it was unsuccessful. In fact, Fisher observed that in instances where many small clusters of close customers exist (as is the case of several instances from the literature) any solution in which these customers are served contiguously in the same route has almost the same cost. Thus, when the sequence of these customers has to be determined through branching, unless an extremely tight bound is used, it would be very difficult to fathom many of the resulting nodes. Therefore, in [28] an alternative branching scheme based on specific GSECs is proposed, aiming at exploiting macro properties of the optimal solution whose violation would have a large impact on the cost, thus allowing the fathoming of the corresponding nodes. To this end a subset T of currently unserved customers is selected and two descendant nodes are created: at the first node the additional constraint $\sum_{e \in \delta(T)} x_e = 2\lceil q(T)/C \rceil$ is added to the current problem, while at the second node the constraint $\sum_{e \in \delta(T)} x_e \geq 2\lceil q(T)/C \rceil + 2$ is imposed. Some ways of identifying suitable subsets, as well as additional dominance rules, are described in [28].

The two algorithms proposed for ACVRP by Laporte, Mercure, and Nobert [37] and by Fischetti, Toth, and Vigo [27] adopted the same branching rule related to the *subtour elimination* scheme used for the asymmetric TSP. At a node v of the branch-decision tree, let I_v and F_v contain the arcs imposed and forbidden in the current solution, respectively (with $I_v = \emptyset$ and $F_v = \emptyset$ if v is the root node). Given the set A^* of arcs corresponding to the optimal solution of the current relaxation, a non-imposed arc subset $B := \{(a_1, b_1), (a_2, b_2), \ldots, (a_h, b_h)\} \subset A^*$ on which to branch is chosen. In [27] set B is defined by considering the subset of A^* with the minimum number of non-imposed arcs among those defining a path or a circuit which is infeasible because it is overloaded or disconnected from the depot. Then $h = |B|$ descendant nodes are generated. The subproblem associated with node v_i, $i = 1, \ldots, h$, is defined by excluding the ith arc of B and by imposing the arcs up to $i - 1$:

$$(2.7) \qquad I_{v_i} := I_v \cup \{(a_1, b_1), \ldots, (a_{i-1}, b_{i-1})\},$$

$$(2.8) \qquad F_{v_i} := F_v \cup \{(a_i, b_i)\},$$

where $I_{v_1} := I_v$.

Laporte, Mercure, and Nobert [37] defined B as an infeasible subtour according to conditions (i) and (ii) of Section 2.2.1, and used a more complex branching rule in which, at each descendant node, at most r arcs of B are simultaneously excluded, where $r := \lceil q(S)/C \rceil$, S is the set of vertices spanned by B, and $q(S)$ represents the sum of the demands of the vertices in S. In this case, since at most $\binom{|B|}{r}$ descendant nodes may be generated, the set B is chosen as the one minimizing $\binom{|B|}{r}$.

The algorithms for CVRP generally adopt a *best-bound-first* search strategy; i.e., branching is always executed on the pending node of the branch-decision tree with the smallest lower bound value. This rule allows for the minimization of the number of subproblems solved at the expense of larger memory usage, and computationally proved to be more effective than the *depth-first* strategy, where the branching node is selected according to a *Last-In-First-Out* (LIFO) rule.

Reduction and Dominance.   Several rules may be used to possibly remove some arcs which cannot belong to an optimal solution, hence forbidding their use in the computation of bounds and allowing for the early detection of infeasibilities and dominance relations, thus speeding up the solution of CVRP. Many of these rules are inspired by the work done on the TSP. In the following we refer, for short, to the more general case of the ACVRP and we explicitly remove arcs from A even if often, to preserve graph completeness, such a removal is implemented by setting the cost of the arcs to be removed equal to a very large positive value equivalent to $+\infty$.

The reduction rules may be applied either to the original problem or to a subproblem associated with a node of the branch-decision tree, where arcs of a given subset I are imposed in the solution, as it happens in Branch-and-Bound and Branch-and-Cut algorithms. In this latter case the arcs of I define complete routes and paths, some of which may enter or leave the depot. For reduction purposes it is convenient to create a reduced graph $\tilde{G} = (\tilde{V}, \tilde{A})$ in which all the customers belonging to the complete routes induced by I are removed from $\tilde{V}$ and the set $\tilde{K}$ of available vehicles is updated accordingly. In addition, all paths induced by I are replaced in $\tilde{G}$ by single vertices with demand equal to the total demand of the vertices in the path. The costs of the arcs entering and leaving each such representative vertex are defined as those of the arcs entering the first and leaving the last vertex in the path, respectively.

The first type of reduction rules tries to remove from $\tilde{A}$ all the arcs that, if used, would produce infeasible solutions, namely those of each pair $i, j \in \tilde{V}$ such that $q_i + q_j > Q$. The second type of reduction rules tries to remove for $\tilde{A}$ the arcs that, if used, would not improve the currently best known solution. For example, let L and U be a lower and an upper bound on the optimal ACVRP solution value, respectively. For each $(i, j) \in \tilde{A}$ let $\bar{c}_{ij}$ be the reduced cost of arc (i, j) associated with the lower bound L. It is well known that the reduced cost of an arc represents a lower bound on the increase of the optimal solution value if this arc is used. Therefore, for each $(i, j) \in \tilde{A}$, if $L + \bar{c}_{ij} \geq U$, we may remove (i, j) from $\tilde{A}$. Whenever a customer has only one entering or leaving arc belonging to $\tilde{A}$, we may impose this arc (by adding it to I), redefine the graph $\tilde{G}$, and execute again the reductions above.

The performance of the branching schemes may be enhanced by means of a dominance test proposed by Fischetti and Toth [25]. A node of the branch-decision tree where a partial sequence of customers $v, \ldots, w$ is fixed can be fathomed if there exists a lower cost ordering of the customers in the sequence starting with v and ending with w. The improved ordering may be heuristically determined, e.g., by means of insertion and exchange procedures.

Finally, several Branch-and-Bound algorithms include the use of heuristic algorithms which exploit the information associated with the current relaxed problems to obtain feasible solutions which may possibly improve the current incumbent solution.

2.3 ▪ Early Set Partitioning Algorithms

An alternative formulation that has been widely used to model CVRP and its variants is that based on *Set Partitioning* (SP) or *Set Covering* (SC). The formulation was originally proposed by Balinski and Quandt [12] and uses a possibly exponential number of binary variables. The formulation VRP4, presented in Chapter 1, is recalled here to facilitate the reading. Let Ω denote the collection of all the circuits of G, corresponding to feasible CVRP routes. Each route $r \in \Omega$ has an associated cost c_r, and let a_{ir} be a binary coefficient which takes value 1 if vertex $i \in N$ is visited (i.e., *covered*) by route r, and 0 otherwise. The binary variable λ_r, $r \in \Omega$, is equal to 1 if and only if route r is selected in the optimal CVRP solution. The resulting extensive model for CVRP is then

$$(2.9) \qquad \text{(VRP4)} \qquad \text{minimize } c^\top \lambda$$

$$(2.10) \qquad \text{s.t.} \quad \sum_{r \in \Omega} a_{ir} \lambda_r = 1 \qquad\qquad \forall i \in N,$$

$$(2.11) \qquad\qquad \mathbb{1}^\top r = |K|,$$

$$(2.12) \qquad\qquad \lambda \in \{0,1\}^r.$$

Constraints (2.10) impose that each customer i is covered by exactly one of the selected routes, and (2.11) requires that $|K|$ routes be selected. As route feasibility is implicitly considered in the definition of set Ω, this is a very general model which may easily take into account additional constraints. Moreover, when the cost matrix satisfies the triangle inequality (i.e., $c_{ij} \le c_{ik} + c_{kj}$ for all $i, j, k \in V$), the SP model VRP4 may be transformed into an equivalent covering model, VRP4$_\ge$, by replacing equality with the inequality "$\ge$" in (2.11). Any feasible solution to model VRP4 is clearly feasible for VRP4$_\ge$, and any feasible solution to VRP4$_\ge$ may be transformed into a feasible solution of VRP4 of not greater cost. Indeed, if one customer is visited more than once in a VRP4$_\ge$ solution, it may be removed from all but one of the routes where it is served by applying shortcuts that will not increase the solution cost because of the triangle inequality. The main advantage of using the VRP4$_\ge$ formulation with respect to the VRP4 one is that in the former only inclusion-maximal feasible circuits, among those with the same cost, need to be considered in the definition of Ω. This considerably reduces the number $|\Omega|$ of variables. In addition, when using the VRP4$_\ge$ formulation the dual solution space is considerably reduced since dual variables are restricted to non-negative values only.

One of the main drawbacks of models VRP4 and VRP4$_\ge$ is represented by the huge number of variables, which, in non–tightly constrained instances with dozens of customers, may easily run into the billions. Thus, one has to resort to a *Column Generation* (CG) approach to solve the linear programming relaxation of these models, as described in detail in Bramel and Simchi-Levi [15]. The CG method starts from a small subset of routes Ω' and solves the linear relaxation of the corresponding reduced model VRP4' (or VRP4'$_\ge$) deriving the optimal dual variables associated with the constraints. Given the dual information, the CG problem (also called the *pricing* problem) amounts to finding the route not in Ω' with the most negative reduced cost or proving that no such route exists. In this latter case the current solution to the linear model VRP4' is the optimal linear relaxation of VRP4 as well and the process terminates. Otherwise the route returned by CG is added to Ω' and a new iteration is performed. The resulting bound is typically very tight, and this motivated recent extensive research on this approach, leading to the current Branch-and-Cut-and-Price algorithms, which greatly outperformed the early approaches described in this chapter. Since Branch-and-Cut-and-Price algorithms for the CVRP are described in Chapter 3 we limit our exposition here to the first seminal papers that opened this fruitful research direction.

The first of these approaches is due to Agarwal, Mathur, and Salkin [3], who considered a relaxation of model VRP4 with an unlimited number of vehicles, i.e., not including constraint (2.11). To solve the resulting model, they implemented a CG approach in which the pricing problem is faced through a dedicated Branch-and-Bound algorithm. Within this algorithm, a lower bound on the reduced cost of the route is obtained by solving an appropriate knapsack problem. Agarwal, Mathur, and Salkin used their algorithm to solve seven Euclidean CVRP instances with up to 25 customers on a IBM 370 (0.2-0-4 Mflops). Interesting alternative ways of computing the lower bounds used within the CG approach were proposed by Bixby, Coullard, and Simchi-Levi [13], who used a cutting plane algorithm for a suitably defined Prize-Collecting TSP, as well as the famous approach based on dynamic programming by Desrochers, Desrosiers, and Solomon [23], which inspired most pricing schemes used in current Branch-and-Cut-and-Price algorithms. Finally, we recall here the additive approach by Hadjiconstantinou, Christofides, and Mingozzi [33], presented in Section 2.2.3, used to compute approximate solutions of the dual problem associated with model VRP4 that yields tight lower bounds for SCVRP.

2.4 ▪ Branch-and-Cut Algorithms

In this section we review the main research works on the Branch-and-Cut algorithms for the SCVRP realized from 1980 to 2005. They are based on the seminal work by Laporte, Nobert, and Desrochers [40], who introduced the two-index formulation VRP2 of the SCVRP reported in Section 2.2 and described a first Branch-and-Cut algorithm for its solution. In this section, we build upon a brief review of their work to describe the developments proposed during the following 20 years. We mainly describe additional cuts and the associated separation procedures.

In their article, Laporte, Nobert, and Desrochers [40] consider a relaxation of model VRP2 in which the GSECs (2.4), which impose the capacity requirements, are removed together with the restrictions on the integrality of the variables. Given the optimal solution of the relaxation, either the solution is feasible for the SCVRP and the algorithm terminates, or it is a non-feasible solution. Thanks to heuristic separation procedures (see below), they identify violated capacity inequalities (2.4). They compute $r(S)$ as the smallest number of vehicles required to serve vertices in S : $r(S) = \lceil \sum_{i \in S} q_i / Q \rceil$. At the root node, Gomory cuts are also introduced. Next, they add all generated constraints to the relaxed model and reiterate. When they are not able to detect such constraints, they create subproblems by branching on a fractional variable.

In 1995, Augerat described a Branch-and-Cut algorithm in his PhD thesis [7], which included for the first time valid inequalities not present in the model. Augerat separated four families of inequalities (see below): (i) the rounded capacity inequalities; (ii) the generalized capacity constraints; (iii) the comb inequalities; (iv) the hypotour inequalities. Moreover, a tabu search–based heuristic was used to generate an initial upper bound and to update it on the basis of the fractional solutions visited within the course of the algorithm. Last, Augerat explored various branching schemes based on constraints. Given a fractional solution, he identified a subset of vertices S such that $x(\delta(S)) \approx 2k + 1 + \epsilon$, where k is integer and ϵ takes a real value. Then, two subproblems were created by imposing $x(\delta(S)) \leq 2k$ and $x(\delta(S)) \geq 2(k + 1)$. Augerat performed extensive computational experiments to determine the best value for ϵ setting $k = 1$. He also considered strategies where ϵ is within some interval and some additional criterion is considered, such as the cardinality of S, the total demand of the vertices in S, and the distance from S to the depot. He concluded that the best strategy consists of selecting the best set S among those identified by applying independently each simple strategy considered. His computational results illustrated that this strategy clearly outperformed the simple strategy based on a single variable. However, a strategy mixing branching on variables and branching on constraints (as described before) led to similar results.

In 2003, Ralphs et al. [51] proposed a Branch-and-Cut algorithm following a different approach. First, they separated the capacity constraints thanks to three heuristics. When the heuristics failed to identify a violated inequality, they proposed using a decomposition algorithm to find additional constraints. First, the original network is expanded through the addition of $|K|-1$ copies of the depot with the corresponding edges as described in Section 2.2. Given a fractional solution on the extended graph, the decomposition algorithm aims to determine whether or not this solution can be written as a convex combination of Hamiltonian cycles. When it succeeds, the Hamiltonian cycles are inspected to find violated capacity inequalities. When it fails, the branching step is invoked. Finally, when the current fractional solution cannot be decomposed, a Farkas inequality is generated. The decomposition algorithm requires knowing a priori the set of the Hamiltonian cycles defined on the extended network. First, an enumerative search is used to generate a preset number of cycles. Then a column generation algorithm is invoked to generate dynamically additional cycles. Ralphs et al. [51] found that the most efficient branching scheme was to branch on variables using a strong branching strategy.

A new Branch-and-Cut algorithm for the CVRP was introduced by Lysgaard, Letchford, and Eglese [43] in 2004. Their algorithm relies on new separation procedures for valid inequalities already known: (i) the rounded capacity inequality; (ii) the framed capacity inequalities; (iii) the strengthened comb inequalities; (iv) the hypotour inequalities. Homogeneous multistar and partial multistar inequalities are also separated according to heuristics described in a previous paper (Letchford, Eglese, and Lysgaard [41]). To perturb the current fractional solution, mixed-integer Gomory cuts are also introduced once at the root node. This may lead to separate additional inequalities thanks to the separation heuristics used. The branching scheme is analogous to the one considered in Augerat [7]. Several sets S_i such that $x(\delta(S_i)) \approx 3$ for $i = 1, \ldots, t$ are identified and ordered according to $|x(\delta(S)) - 3| / \sum_{j \in S} q_j$. Following this order, the lower bounds LB_1 and LB_2 associated with the branches $x(\delta(S)) \leq 2$ and $x(\delta(S)) \geq 4$ are computed. The set leading to the best lower bound $(min(LB_1, LB_2))$ is selected. $max(LB_1, LB_2)$ is used to break a tie.

Other Branch-and-Cut algorithms were proposed during this period. Achuthan, Caccetta, and Hill [1] developed a method which relies on the separation of rounded capacity constraints thanks to heuristics. An improved version based on the separation of additional inequalities related to multistar constraints was described in a subsequent paper by Achuthan, Caccetta, and Hill [2]. Blasum and Hochstättler [14] studied three families of valid inequalities for which separation procedures were presented: (i) the multistar inequalities; (ii) the pathbin inequalities; (iii) the hypotour inequalities. However, most inequalities considered in these classes are less general than those considered in the papers described above and the implemented Branch-and-Cut algorithm is very similar to the approach by Augerat [7]. Baldacci, Hadjiconstantinou, and Mingozzi [9] presented a new integer programming formulation based on a two-commodity network flow approach. Since flow variables can be expressed in terms of arc variables, all valid inequalities for model VRP2 are also valid for the new model. Thus, the authors derived capacity constraints, separated as in [7]. Finally, when all customers have unit demand, two Branch-and-Cut algorithms were described by Araque et al. [6] and by Ghiani, Laporte, and Semet [30].

2.4.1 ▪ Families of Cuts

In this section, we present valid inequalities for the polytope of the CVRP defined as the convex hull of all solutions of the CVRP described by the two-index flow formulation. Since the dimension of the polytope is a complex function of the parameters of the problem, it is difficult to prove whether these valid constraints are facet-defining (see, e.g., Campos, Corberán, and Mota [16] when all the demands are equal). To derive facial properties, relaxations of the polytope of the CVRP are frequently considered. The most common relaxation is the graphical relaxation introduced by Cornuéjols and Harche [21]. The *Graphical Vehicle Routing Problem* (GrVRP) is as the CVRP except that each customer is visited on at least one route but served on exactly one route. The polytope of the CVRP is a face of the polyhedron of the GrVRP. Several valid inequalities described below have been proved to be facet-defining for this polyhedron (see, e.g., Cornuéjols and Harche [21], De Vitis, Harche, and Rinaldi [22], and Blasum and Hochstättler [14]).

TSP-Related Valid Inequalities. A first attempt to propose valid inequalities was done by generalizing constraints which were first developed for the TSP. A general result was obtained by Naddef and Rinaldi [46]. They showed that every constraint valid for the polytope of the symmetric TSP, put into a tight triangular form, is a valid inequality for the CVRP. For $a, x \in \Re^{|E|}$ and $b \in \Re$, an inequality $ax \geq b$ is said to be in tight triangular form if for every triplet of edges $((i,j),(i,k),(j,k))$, the triangle inequality is satisfied, $a_{(i,j)} \leq a_{(i,k)} + a_{(k,j)}$, and for all $i \in V$, two vertices $i_1, i_2 \in V$ exist such that $a_{(i_1,i_2)} = a_{(i,i_1)} + a_{(i,i_2)}$. Naddef and Rinaldi [46] showed that every constraint valid for the symmetric TSP can be put into the tight triangular form.

Capacity Constraints. The capacity constraints are constraints which can be expressed as in (2.4). Depending on how $r(S)$ is computed, this set of constraints has different names. We will see in the next section that these different classes can be more or less easy to separate. If we consider the lowest value for the right-hand side, $r(S) = \sum_{i \in S} q_i / Q$, they are named *fractional capacity inequalities*. When the previous value is rounded up, $r(S) = \lceil \sum_{i \in S} q_i / Q \rceil$, we obtain the *rounded capacity inequalities*. Given S, a valid value for $r(S)$ is the optimal value of the bin-packing problem for which the weights of the objects are equal to q_i for $i \in S$ and the bin capacity is set to Q. These constraints are called *weak capacity inequality*. In general, they are not supporting hyperplanes since the size of the fleet and the demand outside S are not considered. Last, if $r(S)$ is equal to the minimum number of vehicles required to serve the vertices of S when $|K|$ vehicles are available, the resulting inequalities are *global capacity constraints*.

More formally, $\mathbf{P} = \{P_1, \ldots, P_{|K|}\}$ is a $|K|$-partition of N if $\mathbf{P}$ is a partition of N satisfying $\sum_{j \in P_i} q_j \leq Q$ for $1 \leq i \leq |K|$. Let $\mathscr{P}$ be the set of the $|K|$-partitions. For every non-empty subset S of N and for every $\mathbf{P} \in \mathscr{P}$, we define $\beta_P(S)$ as the number of vehicles required to serve the vertices in S according to $\mathbf{P}$, i.e.,

$$\beta(\mathbf{P}, S) = |\{i : 1 \leq i \leq |K|, P_i \cap S \neq \emptyset\}|.$$

Then, we obtain $r(S) = \min_{P \in \mathscr{P}} (\beta(\mathbf{P}, S))$. Clearly, these four expressions provide increasing values for $r(S)$. Cornuéjols and Harche [21] provide an example where the inequalities are strict for the last three expressions. By construction, the global capacity constraints are supporting hyperplanes.

Augerat [7] considers an extended version of the global capacity constraints: the *generalized capacity constraints* (see also De Vitis, Harche, and Rinaldi [22]). Let $\mathbf{S} = \{S_1, \ldots, S_t\}$

be a set of t disjoint subsets of N; then the following inequality is valid for the CVRP:

$$(2.13) \qquad \sum_{i=1}^{t} x(\delta(S_i)) \geq 2 \min_{P \in \mathscr{P}} \left(\sum_{i=1}^{t} \beta(\mathbf{P}, S_i) \right).$$

The generalized capacity constraint (2.13) dominates the aggregation of the global capacity constraints defined on S_i for $i = 1, \ldots, t$. Since the right-hand side is difficult to evaluate, a weaker form of the generalized capacity constraint (2.13) has been considered in Augerat [7]. It consists of replacing the right-hand side with a lower bound based on the solution of a bin-packing problem. Let $\mathbf{S}$ be such that $\sum_{j \in S_i} q_j \leq Q$ for $1 \leq i \leq t$. $r(N|S_1, \ldots, S_t)$ is the optimal value of the bin-packing problem where the bin capacity is Q and where an object is associated with each S_i with a weight equal to the total demand of the vertices of S_i, and an object is associated with each vertex j in $N \backslash (\bigcup_{i=1}^{t} S_i)$ with a weight equal to q_j. Then the following value is a lower bound on the right-hand side of (2.13):

$$2(t + r(N|S_1, \ldots, S_t) - |K|).$$

Framed Capacity Inequalities. These constraints, introduced by Augerat [7], are an extension of the weak generalized capacity constraints. This class of constraints is defined for all $H, S_1, \ldots, S_t \subseteq N$ such that

$$(2.14) \qquad S_i \subset H, \qquad\qquad i = 1, \ldots, t,$$
$$(2.15) \qquad S_i \cap S_j = \emptyset, \qquad i \neq j \; i, j = 1, \ldots, t,$$
$$(2.16) \qquad \sum_{k \in S_i} q_k \leq Q, \qquad\quad i = 1, \ldots, t.$$

The framed capacity inequality is expressed as follows:

$$(2.17) \qquad x(\delta(H)) + \sum_{i=1}^{t} x(\delta(S_i)) \geq 2(t + r(H|S_1, \ldots, S_t)),$$

where $r(H|S_1, \ldots, S_t)$ is defined as above for set H. Intuitively, this constraint means that if the capacity constraints are tight for each $S_i \subset H$, then $r(H|S_1, \ldots, S_t)$ is a lower bound on the number of vehicles required to serve the vertices in H. Lysgaard, Letchford, and Eglese [43] considered an extended variant of these constraints in which the hypothesis on the total demand for set S_i (2.16) is relaxed. Then, the right-hand side of the framed capacity inequality becomes

$$2(r(H|S_1, \ldots, S_t) + \textstyle\sum_{i=1}^{t} r(S_i)).$$

Last, Augerat [7] proposed additional but more complex constraints of this family.

Comb Inequalities. This class of constraints was proposed by Chvàtal [19] and Grötschel and Padberg [32] for the symmetric TSP. Since comb inequalities can be put into the tight triangular form, they are valid inequalities for the CVRP (see above). Considering the CVRP, they can be expressed as follows. Let $H, T_1, \ldots, T_t \subset N$ be a handle and the associated teeth such that $t \geq 3$ and odd, $H \cap T_i \neq \emptyset$ and $T_i \backslash H \neq \emptyset$ for $i = 1, \ldots, t$, and $T_i \cap T_j = \emptyset$ for $i, j = 1, \ldots, t, i \neq j$. Then the comb inequality is (see Augerat [7])

$$(2.18) \qquad x(\delta(H)) \geq (t + 1) - \sum_{i=1}^{t} (x(\delta(T_i)) - 2).$$

Assuming that the subtour elimination constraints: $x(\delta(T_i)) \geq 2$ are tight for $i = 1,\ldots,t$, the comb inequality states that $x(\delta(H))$ is at least $(t + 1)$. To derive such an inequality, subtour elimination constraints on $T_i, T_i \setminus H, T_i \cap H$ for $i = 1,\ldots,t$ are summed up as well as degree and non-negativity inequalities. Comb inequalities can be strengthened in the CVRP case by taking into account the packing structure. Thus, Laporte and Nobert [38] consider the capacity constraints (2.4) instead of the subtour elimination constraints. Assuming that for every tooth T_i the following inequality holds, $r(T_i \setminus H) + r(T_i \cap H) > r(T_i)$, they obtain the strengthened comb inequality

$$(2.19) \qquad x(\delta(H)) \geq (t+1) - \sum_{i=1}^{t}(x(\delta(T_i)) - 2r(T_i)).$$

The explanation of the constraint is as in the previous case. Constraint (2.19) is more or less strong depending on how $r(S)$ is computed (see above) for $S = H, T_1, \ldots, T_t$. In their article, Laporte and Nobert [38] consider a lower bound or the optimal value of the bin-packing problem associated with S.

Other adaptations of the comb inequalities have been proposed. Araque [4] proposed several extended comb inequalities for the CVRP with unit demands. Some of them were adapted to the general case by Lysgaard, Letchford, and Eglese [43].

Hypotour Inequalities. Also these inequalities were introduced by Augerat [7]. This family of constraints aims to identify subnetworks of G which cannot include feasible solutions for the CVRP. More formally, let $G' = (V', E')$ be a subgraph of G such that G' does not contain any feasible solution for the CVRP, but a solution can be identified on $G' \cup \{e\}$ for all $e \in E \setminus E'$. Since the number of edges in any solution of the CVRP is $|N| + |K|$, the *hypotour inequality* $x(E') \leq |N| + |K| - 1$ is valid for the CVRP. Let $F = E \setminus E'$ define a subset of edges such that any feasible solution contains at least one edge of F. The hypotour inequality can be rewritten as $x(F) \geq 1$. Augerat [7] proposed various families of subsets F. The simplest one is as follows. Consider a vertex v and a tree T such that v is not a leaf of T and the total demand on every path including v between two leaves strictly exceeds Q. If $E(\bar{T})$ is the set of edges not present in T and incident with non-pendant vertices of T, then $x(E(\bar{T})) \geq 1$ is a valid inequality.

Augerat [7] introduced also the *extended hypotour inequalities*. Lysgaard, Letchford, and Eglese [43] considered one of them, called a *2-edges extended hypotour inequality*. For any subset $W \subset N$, $e_1, e_2 \in \delta(W)$, and $F \subset E$ as above, the following inequality is valid:

$$(2.20) \qquad x(\delta(W)) + 2x(F) \geq 2(x_{e_1} + x_{e_2}).$$

This constraint states that at least one edge from F must be present in any feasible solution as soon as $x_{e_1} = x_{e_2} = 1$.

Multistar Inequalities. These constraints were proposed by Araque, Hall, and Magnanti [5] in the case of the CVRP with unit demands. Let S and T be two set of vertices with $S \subset N$ and $T \subset N \setminus S$. For α, β, γ given, any valid inequality of the form

$$(2.21) \qquad \alpha x(E(S)) + \beta x(\delta(S) \cap \delta(T)) \leq \gamma$$

is called a *multistar inequality*. Araque, Hall, and Magnanti [5] identify three types of such constraints. In their paper, Letchford, Eglese, and Lysgaard [41] propose extended versions of this class of inequalities for the case of general demands. The main idea consists of first computing an upper bound on the value of $x(E(S))$. Then, for each value of

$x(E(S))$, an upper bound on the $x(\delta(S) \cap \delta(T))$ is evaluated. The multistar inequalities are obtained by determining the convex hull of the resulting set of points in the $(x(E(S)), x(\delta(S) \cap \delta(T)))$ plane. Note that related inequalities called *generalized large multistar inequalities* were proposed by Gouveia [31].

2.4.2 ▪ Separation Procedures

TSP-Related Valid Inequalities. Separation procedures for constraints that generalized TSP valid inequalities are analogous to those proposed in this context. Naddef and Thienel [48, 49] present a detailed description of them.

Capacity Constraints. Fractional capacity constraints are simple to separate. Indeed, as subtour elimination constraints for the TSP, they can be identified by solving flow problems. Separating all other types of capacity constraints is much more challenging since the separation problem turns to be NP-complete (see Augerat [7]). This is the reason why various heuristics have been suggested.

A first approach introduced by Laporte, Nobert, and Desrochers [40] is based on the decomposition of a support graph in connected components. More precisely, consider the support graph $G' = (N, E')$, where E' is the set of all edges e in $E \setminus \delta(0)$ for which $x_e > 0$. For each connected component S of G', they checked whether the associated rounded capacity inequality is violated or not. If it is not the case, the vertex, which can lead to a violation, is excluded from S and a new check is performed while there are vertices remaining in S.

Several heuristics are based on *contracted* graphs. Given the support graph, a contracted graph is obtained by identifying a subset $S \subset N$ of adjacent vertices and by replacing S by a *supervertex* s. The demand of s is equal to the total demand of the vertices of S, and for $i \in S$, $j \notin S$ edges (i, j) are replaced by an edge (s, j) with a weight equal to $x(\delta(S) \cap \delta(j))$. A simple greedy heuristic proposed by Augerat [7] consists of building a series of contracted graphs obtained by successive edge contractions ($|S| = 2$). The selected edge is chosen among the edges non-incident to the depot and with a weight greater than or equal to one. After each contraction, the rounded capacity constraint associated with the supervertex formed is checked to be violated or not. Variants of this method were proposed by Ralphs et al. [51] and by Lysgaard, Letchford, and Eglese [43]. Based on this heuristic, Augerat et al. [8] proposed an adaptation of the tabu metaheuristic for identifying violated rounded capacity constraints.

The identification of *generalized capacity* and *framed capacity constraints* (see below) is NP-hard, as the computation of the right-hand sides requires the solution of bin-packing problems. Augerat [7] developed greedy procedures to separate generalized capacity constraints. These routines are based on contracted graphs. At each step, a partition of N is obtained by considering the nodes of the current contracted graph, and the identification of a violated inequality is performed by solving the associated bin-packing problem.

Framed Capacity Inequalities. Augerat [7] described a greedy separation procedure for the class of framed capacity constraints based on the routines developed for the generalized capacity constraints. The main difference lies in the selection of the initial set of vertices used to generate contracted graphs. While N is the set considered in the previous case, here the vertices are randomly selected to constitute the set H of vertices of a framed capacity inequality.

Lysgaard, Letchford, and Eglese [43] implemented an enumeration procedure to separate these constraints. First, they consider as H a connected component, or a subset of such a component, of the support graph of the current solution. Next, they enumerate partitions of this set through a search tree by imposing and forbidding edge contractions. The bin-packing problem associated with each partition is solved to identify violated inequalities.

Comb Inequalities. The complexity of comb inequality separation remains unknown for the CVRP. Comb inequalities were identified by Augerat [7] and by Lysgaard, Letchford, and Eglese [43] using similar heuristics to that proposed by Padberg and Rinaldi [50] for the TSP. In both cases, the value of $r(S)$ (see (2.19)) was given by $r(S) = \lceil \sum_{i \in S} q_i / Q \rceil$. For the TSP, more effective approximate separation routines were proposed by Naddef and Thienel [48] as well as a polynomial separation algorithm for a subclass of comb inequalities by Fleischer, Letchford, and Lodi [29].

Hypotour Inequalities. For the hypotour inequalities based on trees described above, Augerat [7] proposed an enumerative procedure. Given the support graph of the current solution and a vertex v, all paths including v with a total demand less than Q are identified. If none of them is connected with the depot, a violated hypotour constraint is identified since at least one edge not present in the current solution must be included in a feasible solution. This set of edges can be refined by considering only edges, not present in the current solution, which provide a feasible extension to one of the enumerated paths. Since this procedure relies on a complete enumerative scheme, it may require significant computation times. However, the number of edges in the support graph is usually smaller than in the original graph, and Augerat [7] indicated a number of enhancements to increase the efficiency of the algorithm. To separate the 2-edges extended hypotour inequality (2.20), Lysgaard, Letchford, and Eglese [43] developed a multi-step heuristic. The main steps are the choice of the set W as well as edges $e_1, e_2 \in \delta(W)$ thanks to a greedy heuristic and the solution of an assignment problem to build a set F.

Multistar Inequalities. The complexity of multistar inequality separation remains unknown. To identify violated multistar inequalities, Letchford, Eglese, and Lysgaard [41] proposed greedy heuristics. Following the notation given above, they first build sets S by including additional vertices in S, as can be done when contracted graphs are built (see Augerat [7]). For S given, they build T by removing from $N \setminus T$ the vertex i with the minimum value of $x(\delta(S) \cap \delta(i))$ repeatedly. Next, for each pair (S, T), they proceed as explained in the constraint description to obtain a set of multistar inequalities. Note that the *generalized large multistar inequalities* can be separated in polynomial time thanks to a network flow algorithm Blasum and Hochstättler [14].

2.4.3 ▪ Comparison of Branch-and-Cut Methods: Do We Have the Appropriate Model?

It is difficult to draw a comparison between the approaches proposed by Augerat [7], Lysgaard, Letchford, and Eglese [43], and Ralphs et al. [51]. Hardware and implementation vary from one algorithm to the other, and results on the same testbed are not available. However, it is possible to give some general trends. It seems that the Branch-and-Cut algorithm developed by Augerat [7] is the least efficient one. Instances proposed by Christofides and Eilon or by Fisher are not frequently solved to optimality. When optimal solutions are obtained, the Branch-and-Cut method due to Lysgaard, Letchford, and Eglese [43] produce them in shorter computation times. The algorithm due to Lysgaard, Letchford, and Eglese [43] compares favorably with the method developed by Ralphs et al. [51]. Indeed, by considering the 50 instances of type A and B originally proposed by Augerat [7], 37 of them are solved optimality by Lysgaard, Letchford, and Eglese [43], while the optimal solutions of 24 instances are obtained by Ralphs et al. [51]. The largest instances solved by both algorithms involved 100 customers and 7 to 8 vehicles, although this was obtained sometimes with prohibitive computation times. For example, the instance E076-08s (also known as E-n76-k8) was solved to optimality by the parallel code developed by Ralphs et al. [51] on a 700MHz Intel Pentium III Xeon platform in almost two million seconds.

Such results lead to the following comments. Over 25 years, researchers mainly focused their works on the two-index flow formulation. This model is an extended version of the classical formulation for the TSP, and many cuts have been proposed which are TSP-like cuts. While the Branch-and-Cut algorithms based on strong polyhedral results turn out to be very effective for the TSP, the results were somehow deceiving with respect to the CVRP. Clearly, the research efforts were not comparable. But, are there more fundamental questions? When the two-index flow model is used for the CVRP, the problem is viewed mainly as a sequencing problem, whereas other points of view could be consider. For instance, the CVRP could be studied from the viewpoint of a packing problem. More generally speaking, did we work with the appropriate model?

2.5 ▪ Conclusions and Future Research Directions

In this chapter we reviewed the most important exact methods that were proposed for the solution of CVRP during more than 30 years from the seventies out to about 2005. This was a very active research period in which all new techniques developed for the exact solution of hard combinatorial problems, from Branch-and-Bound to column generation and to Branch-and-Cut, were fruitfully applied to CVRP. Although the current best algorithms, all belonging to the Branch-and-Cut-and-Price family, greatly improved the performance of the methods presented in this chapter, such a huge research body still has a great value. This is not only because Branch-and-Cut-and-Price combines and improves what was done by previous methods, but also because early algorithms, particularly those using combinatorial relaxations, remain somehow much simpler to implement, thus still being very useful to define viable approaches for novel variants of VRP or large-scale instances that are intractable for linear programming–based approaches. In addition, some space for further improvement remains with respect to Branch-and-Bound and Branch-and-Cut methods. Two main research avenues are, in fact, still open: (i) the development of new models, and (ii) the identification of new cut families, as well as more effective separation procedures and branching strategies that will allow one to compute better bounds more quickly. We also note that the complexity of the separation problem for most of the cut families we presented has not been studied so far. The potential of the classical approaches we examined in this chapter is therefore far from being fully exhausted.

Bibliography

[1] N. R. ACHUTHAN, L. CACCETTA, AND S. P. HILL, *A new subtour elimination constraint for the vehicle routing problem*, European Journal of Operational Research, 91 (1996), pp. 573–586.

[2] ———, *An improved branch-and-cut algorithm for the capacitated vehicle routing problem*, Transportation Science, 37 (2003), pp. 153–169.

[3] Y. AGARWAL, K. MATHUR, AND H. SALKIN, *A set-partitioning-based exact algorithm for the vehicle routing problem*, Networks, 19 (1989), pp. 731–749.

[4] J. R. ARAQUE, *Contributions to the polyhedral approach to vehicle routing*, Discussion Paper 90-74, CORE, Catholic University of Louvain La Neuve, Belgium, 1990.

[5] J. R. ARAQUE, L. HALL, AND T. L. MAGNANTI, *Capacitated trees, capacitated routing and associated polyhedra*, Discussion Paper 90-61, CORE, Catholic University of Louvain La Neuve, Belgium, 1990.

[6] J. R. ARAQUE, G. KUDVA, T. L. MORIN, AND J. F. PEKNY, *A branch-and-cut algorithm for vehicle routing problems*, Annals of Operations Research, 50 (1994), pp. 37–59.

[7] P. AUGERAT, *Approche Polyèdrale du Problème de Tournées de Véhicules*, PhD thesis, Institut National Polytechnique de Grenoble, Grenoble, France, 1995.

[8] P. AUGERAT, J. BELENGUER, E. BENAVENT, A. CORBERÁN, AND D. NADDEF, *Separating capacity inequalities in the CVRP using tabu search*, European Journal of Operational Research, 106 (1998), pp. 546–557.

[9] R. BALDACCI, E. HADJICONSTANTINOU, AND A. MINGOZZI, *An exact algorithm for the capacitated vehicle routing problem based on a two-commodity network flow formulation*, Operations Research, 52 (2004), pp. 723–738.

[10] R. BALDACCI, P. TOTH, AND D. VIGO, *Exact algorithms for routing problems under vehicle capacity constraints*, Annals of Operations Research, 175 (2010), pp. 213–245.

[11] ——, *Exact solution of the capacitated vehicle routing problem*, in Wiley Encyclopedia in Operations Research and Management Science, vol. 3, Wiley, New York, 2011, pp. 1795–1807.

[12] M. BALINSKI AND R. QUANDT, *On an integer program for a delivery problem*, Operations Research, 12 (1964), pp. 300–304.

[13] A. BIXBY, C. COULLARD, AND D. SIMCHI-LEVI, *The capacitated prize-collecting traveling salesman problem*, Working Paper, Department of Industrial Engineering and Engineering Management, Northwestern University, Evanston, IL, 1997.

[14] U. BLASUM AND W. HOCHSTÄTTLER, *Application of the branch and cut method to the vehicle routing problem*, Technical Report ZPR2000-386, ZPR, Universität zu Köln, Germany, 2000.

[15] J. BRAMEL AND D. SIMCHI-LEVI, *Set-covering-based algorithms for the capacitated VRP*, in The Vehicle Routing Problem, P. Toth and D. Vigo, eds., SIAM, Philadelphia, 2002, ch. 4, pp. 85–108.

[16] V. CAMPOS, A. CORBERÁN, AND E. MOTA, *Polyhedral results for a vehicle routing problem*, European Journal of Operational Research, 52 (1991), pp. 75–85.

[17] N. CHRISTOFIDES AND S. EILON, *An algorithm for the vehicle dispatching problem*, Operational Research Quarterly, 20 (1969), pp. 309–318.

[18] N. CHRISTOFIDES, A. MINGOZZI, AND P. TOTH, *Exact algorithms for the vehicle routing problem based on the spanning tree and shortest path relaxations*, Mathematical Programming, 20 (1981), pp. 255–282.

[19] V. CHVÀTAL, *Edmonds polytopes and weakly Hamiltonian graphs*, Mathematical Programming, 5 (1973), pp. 29–40.

[20] J.-F. CORDEAU, G. LAPORTE, M. W. P. SAVELSBERGH, AND D. VIGO, *Vehicle routing*, in Transportation, C. Barnhart and G. Laporte, eds., vol. 14 of Handbooks in Operations Research and Management Science, Elsevier, 2007, ch. 6, pp. 367–428.

[21] G. CORNUÉJOLS AND F. HARCHE, *Polyhedral study of the capacitated vehicle routing problem*, Mathematical Programming, 60 (1993), pp. 21–52.

[22] A. DE VITIS, F. HARCHE, AND G. RINALDI. *Generalized capacity inequalities for vehicle routing problems.* unpublished manuscript, 2000.

[23] M. DESROCHERS, J. DESROSIERS, AND M. SOLOMON, *A new optimization algorithm for the vehicle routing problem with time windows*, Operations Research, 40 (1992), pp. 342–354.

[24] J. DONGARRA, *Performance of various computers using standard linear equations software*, Technical Report CS-89-85, University of Tennessee, Knoxville, TN, 2013.

[25] M. FISCHETTI AND P. TOTH, *A new dominance procedure for combinatorial optimization*, Operations Research Letters, 7 (1988), pp. 181–187.

[26] ——, *An additive bounding procedure for combinatorial optimization problems*, Operations Research, 37 (1989), pp. 319–328.

[27] M. FISCHETTI, P. TOTH, AND D. VIGO, *A branch-and-bound algorithm for the capacitated vehicle routing problem on directed graphs*, Operations Research, 42 (1994), pp. 846–859.

[28] M. FISHER, *Optimal solution of vehicle routing problems using minimum k-trees*, Operations Research, 42 (1994), pp. 626–642.

[29] L. FLEISCHER, A. LETCHFORD, AND A. LODI, *Polynomial-time separation of a superclass of simple comb inequalities*, Mathematics of Operations Research, 31 (2006), pp. 696–713.

[30] G. GHIANI, G. LAPORTE, AND F. SEMET, *The black and white traveling salesman problem*, Operations Research, 54 (2006), pp. 366–378.

[31] L. GOUVEIA, *A result on projection for the vehicle routing problem*, European Journal of Operational Research, 85 (1995), pp. 610–624.

[32] M. GRÖTSCHEL AND M. W. PADBERG, *On the symmetric travelling salesman problem I: Inequalities*, Mathematical Programming, 16 (1979), pp. 265–280.

[33] E. HADJICONSTANTINOU, N. CHRISTOFIDES, AND A. MINGOZZI, *A new exact algorithm for the vehicle routing problem based on q-paths and k-shortest paths relaxations*, Annals of Operations Research, 61 (1995), pp. 21–43.

[34] M. HELD AND R. KARP, *The traveling salesman problem and minimum spanning trees: Part II*, Mathematical Programming, 1 (1971), pp. 6–25.

[35] M. HELD, P. WOLFE, AND M. CROWDER, *Validation of the subgradient optimization*, Mathematical Programming, 6 (1974), pp. 62–88.

[36] G. LAPORTE, *The vehicle routing problem: An overview of exact and approximate algorithms*, European Journal of Operational Research, 59 (1992), pp. 345–358.

[37] G. LAPORTE, H. MERCURE, AND Y. NOBERT, *An exact algorithm for the asymmetrical capacitated vehicle routing problem*, Networks, 16 (1986), pp. 33–46.

[38] G. LAPORTE AND Y. NOBERT, *Comb inequalities for the vehicle routing problem*, Methods of Operations Research, 51 (1984), pp. 271–276.

[39] ———, *Exact algorithms for the vehicle routing problem*, Annals of Discrete Mathematics, 31 (1987), pp. 147–184.

[40] G. LAPORTE, Y. NOBERT, AND M. DESROCHERS, *Optimal routing under capacity and distance restrictions*, Operations Research, 33 (1985), pp. 1050–1073.

[41] A. N. LETCHFORD, R. W. EGLESE, AND J. LYSGAARD, *Multistars, partial multistars and the capacitated vehicle routing problem*, Mathematical Programming, 94 (2002), pp. 21–40.

[42] J. D. C. LITTLE, K. G. MURTY, D. W. SWEENEY, AND C. KAREL, *An algorithm for the traveling salesman problem*, Operations Research, 11 (1963), pp. 972–989.

[43] J. LYSGAARD, A. N. LETCHFORD, AND R. W. EGLESE, *A new branch-and-cut algorithm for the capacitated vehicle routing problem*, Mathematical Programming, 100 (2004), pp. 423–445.

[44] C. MARTINHON, A. LUCENA, AND N. MACULAN, *A relax and cut algorithm for the vehicle routing problem*, Technical Report RT-05/00, Universidade Federal Fluminense, Niterói, Brasil, 2000.

[45] D. MILLER, *A matching based exact algorithm for capacitated vehicle routing problems*, ORSA Journal on Computing, 7 (1995), pp. 1–9.

[46] D. NADDEF AND G. RINALDI, *The graphical relaxation: A new framework for the symmetric traveling salesman polytope*, Mathematical Programming, 58 (1993), pp. 53–88.

[47] ———, *Branch-and-cut algorithms for the capacitated VRP*, in The Vehicle Routing Problem, P. Toth and D. Vigo, eds., SIAM, Philadelphia, 2002, ch. 3, pp. 53–84.

[48] D. NADDEF AND S. THIENEL, *Efficient separation routines for the symmetric traveling salesman problem I: General tools and comb separation*, Mathematical Programming, 92 (2002), pp. 237–255.

[49] ———, *Efficient separation routines for the symmetric traveling salesman problem II: Separating multi-handle inequalities*, Mathematical Programming, 92 (2002), pp. 257–285.

[50] M. PADBERG AND G. RINALDI, *Facet identification for the symmetric traveling salesman polytope*, Mathematical Programming, 47 (1990), pp. 219–257.

[51] T. K. RALPHS, L. KOPMAN, W. R. PULLEYBLANK, AND L. E. TROTTER, *On the capacitated vehicle routing problem*, Mathematical Programming, 94 (2003), pp. 343–359.

[52] P. TOTH AND D. VIGO, *Models, relaxations and exact approaches for the capacitated vehicle routing problem*, Discrete Applied Mathematics, 123 (2002), pp. 487–512.

[53] ———, *Branch-and-Bound algorithms for the capacitated VRP*, in The Vehicle Routing Problem, P. Toth and D. Vigo, eds., SIAM, Philadelphia, 2002, ch. 2, pp. 29–51.

[54] ———, eds., *The Vehicle Routing Problem*, SIAM, Philadelphia, 2002.

Chapter 3

New Exact Algorithms for the Capacitated Vehicle Routing Problem

Marcus Poggi
Eduardo Uchoa

3.1 ▪ Introduction

Since the seminal work by Desrosiers, Soumis, and Desrochers [15], column generation has been the dominant approach for building exact algorithms for the *Vehicle Routing Problem with Time Windows* (VRPTW). This technique performed very well on tightly constrained instances (those with narrow time windows). As the *Capacitated Vehicle Routing Problem* (CVRP) can be regarded as the particular case of VRPTW where time windows are arbitrarily large, column generation was viewed as a non-promising approach for the problem. In fact, in the early 2000's, the best performing algorithms for the CVRP were Branch-and-Cut algorithms that separated quite complex families of cuts identified by polyhedral investigation (see Naddef and Rinaldi [31] and Chapter 2). In spite of their sophistication, some instances from the literature with only 50 customers could not be solved to optimality. At that moment, the *Branch-and-Cut-and-Price algorithm* (BCP) by Fukasawa et al. [19] showed that the combination of cut and column generation could be much more effective than each of those techniques taken alone. Since then, the most performing exact algorithms proposed for the CVRP are based on that combination.

According to the classification proposed in Poggi de Aragão and Uchoa [36], the BCP algorithm in Fukasawa et al. [19] only uses *robust cuts*. A cut is said to be robust when the value of the dual variable associated with it can be translated into costs in the pricing subproblem. Therefore, the structure and the size of that subproblem remain unaltered, regardless of the number of robust cuts added. On the other hand, *non-robust cuts* are those that change the structure and/or the size of the pricing subproblem; each additional cut makes it harder.

Robustness is a desirable property of a BCP. There is an asymmetry between the cutting and pricing operations in that kind of algorithm. If the separation subproblem for some family of cuts happens to be intractable, heuristics may be used. When the heuristics fail, some violated cuts may have been missed, but one certainly has a valid dual bound. There is no need to ever call an exact separation. On the other hand, even with good pricing heuristics, at least one call to the exact pricing is necessary to establish a valid dual bound. With the addition of non-robust cuts, there is a risk of having to solve to optimality an intractable subproblem.

Nevertheless, since Jepsen et al. [25] and Baldacci, Christofides, and Mingozzi [3] it has been known that some non-robust cuts can be effectively used, at least if they are separated in a controlled way, avoiding an excessive impact on the pricing. While the adjective *non-robust* focuses on their negative aspect, some authors mentioned in this chapter call them *strong cuts*, focusing on their positive aspect, a greater potential for significantly reducing the integrality gaps. In fact, some of the best algorithms available today rely heavily on them.

3.2 • Main Exact Approaches

We briefly review the most recent works proposing exact algorithms for the CVRP; all of them are based on the combination of column and cut generation. The main elements of those algorithms will be described in more depth (and even some terms will only be properly defined) in the remaining sections of this chapter.

1. **Fukasawa et al. [19]** presented a BCP algorithm having the following features:

   - The columns are associated with the q-routes without k-cycles, a relaxation of the elementary routes that allow multiple visits to a customer, on the condition that at least k other costumers are visited between successive visits. The separated cuts are the same used in previous Branch-and-Cut algorithms over the two-index CVRP formulation. Those cuts are robust with respect to q-route pricing.

   - If the column generation at the root node is found to be too slow, the algorithm automatically switches to a Branch-and-Cut.

   All benchmark instances from the literature with up to 135 vertices could be solved to optimality, a significant improvement over previous methods.

2. **Baldacci, Christofides, and Mingozzi [3]** presented an algorithm based on column and cut generation with the following features:

   - The columns are associated with elementary routes. Besides cuts for the two-index formulation, Strengthened Capacity Cuts and Clique Cuts are separated. Those latter cuts are effective but non-robust; they make the pricing harder.

   - A sequence of cheaper lower bounding procedures produces good estimates of the optimal dual variable values. The potentially expensive pricing is only called in the last stage, with the dual variables bounded to be above the estimates, obtaining convergence (to a bound slightly below the theoretical optimal) in fewer iterations.

   - Instead of branching, the algorithm finishes in the root node (therefore, it is not a BCP) by enumerating all elementary routes with reduced cost smaller than the duality gap. A set-partitioning problem containing all those routes is then given to a *Mixed-Integer Programming* (MIP) solver.

The algorithm could solve almost all instances solved in [19], usually taking much less time. However, the exponential nature of some algorithmic elements, in particular the route enumeration, made it fail on some instances with many customers per vehicle.

3. **Pessoa, Poggi de Aragão, and Uchoa [33]** presented some improvements over Fukasawa et al. [19]:

   - Cuts from an extended formulation with capacity indices were also separated. Those cuts do not change the complexity of the pricing of q-routes by dynamic programming.
   - The idea of performing elementary route enumeration and MIP solving to finish a node was borrowed from [3]. However, in order to avoid a premature failure when the root gap is too large, it was hybridized with traditional branching.

4. **Baldacci, Mingozzi, and Roberti [4]** improves upon Baldacci, Christofides, and Mingozzi [3] by the introduction of the following elements:

   - The ng-routes, a newly proposed relaxation that is more effective than the q-routes without k-cycles, are used in the earlier bounding procedures and also to accelerate the pricing/enumeration of elementary routes. This latter part of the algorithm is also enhanced by considering multiple dual solutions. A route having reduced cost larger, with respect to any dual solution, than the current duality gap can be discarded.
   - Subset Row Cuts and Weak Subset Row Cuts, which have less impact on the pricing with respect to Clique Cuts, are separated.

   The resulting algorithm is not only faster on average, but it is much more stable than the algorithm in [3], being able to solve even some instances with many customers per vehicle.

5. **Contardo [12]** introduced new twists on the use of non-robust cuts and on route enumeration:

   - The columns are associated with q-routes without 2-cycles, a relatively poor relaxation. The partial elementarity of the routes is enforced by non-robust Strong Degree Cuts. Robust cuts from two-index formulation and non-robust Strengthened Capacity Cuts and Subset Row Cuts are also separated.
   - The enumeration of elementary routes is directed to a pool of columns. As soon as the duality gap is sufficiently small to produce a pool with reasonable size (a few million routes), the pricing starts to be performed by inspection. From this point, an aggressive separation of non-robust cuts can be performed, leading to very small gaps.

   The reported computational results are very consistent. In particular, a hard instance from the literature, M-n151-k12 (151 vertices, 12 vehicles), was solved to optimality, setting a new record.

6. **Røpke [39]** went back to robust BCP. The main differences from [19] are the following:

   - Instead of q-routes without k-cycles, the more effective ng-routes are used.

- A sophisticated and aggressive strong branching is performed, drastically reducing the average size of the enumeration trees.

The overall results are comparable with the results in [12] and [4]. A long run of that algorithm could also solve M-n151-k12.

7. **Contardo and Martinelli [14]** improved upon Contardo [12]:

- Instead of q-routes without 2-cycles, ng-routes are used. Moreover, the performance of the dynamic programming pricing was enhanced by the use of the DSSR technique [38].
- Edge variables are fixed by reduced costs, using the procedure proposed in [23].

8. Finally, **Pecin et al. [32]** proposed a BCP that incorporates elements from all the previous algorithms, usually enhanced and combined with new elements:

- The most important original contribution is the introduction of the limited memory Subset Row Cuts. They are a weakening of the traditional Subset Row Cuts that can be dynamically adjusted, making them much less costly in the pricing, and yet without compromising their effectiveness.
- It uses capacity indices [33] in its underlying formulation. This allows a fixing of variables by reduced costs that is superior to fixing in [23].
- The columns in the BCP are associated with ng-routes. The dynamic programming pricing uses bidirectional search that differs a little from that proposed in [37] because the concatenation phase is not necessarily performed at half of the capacity. A column generation stabilization by dual smoothing [34] may also be employed.
- The BCP hybridizes branching with route enumeration. Actually, it performs an aggressive hierarchical strong branching.
- When the addition of a round of non-robust cuts makes the pricing too slow, the BCP performs a rollback. The offending cuts are removed even if the lower bound of the node decreases.

The algorithm could solve all the classical instances from the literature with up to 200 vertices in reasonable times, including the hard instances M-n200-k17 and M-n200-k16.

3.3 ▪ Formulations

This section revisits the CVRP formulations that may be considered as starting points in the design of the algorithms mentioned in Section 3.2. Let $G = (V, E)$ be a complete graph where $V = \{0, \ldots, n\}$ is the vertex set and E is the arc set. Vertices in $N = \{1, \ldots, n\}$ correspond to the *customers*, whereas vertex 0 corresponds to the *depot*. A non-negative cost, c_{ij}, is associated with each edge $(i, j) \in E$ and represents the *travel cost* spent to go from vertex i to vertex j. Edges are also indicated through a single index $e = (i, j)$. Let q_i denote the *demand* of customer i for $i \in N$, q_0 is defined as 0, and Q denote the *capacity* of each vehicle k in the set K of available vehicles. Given a vertex set $S \subseteq V$, let $\delta(S)$ denote the set of edges $e \in E$ which have only one endpoint in S. As usual, when a single vertex $i \in V$ is considered, we write $\delta(i)$ rather than $\delta(\{i\})$. For any set $S \subseteq V$, let $q(S) = \sum_{i \in S} q_i$, and $r(S) = \lceil q(S)/Q \rceil$. Further details on formulations for the CVRP can be found in Chapters 1 and 2.

3.3.1 ▪ The Two-Index Formulation

The classical *two-index formulation* is also known as *edge formulation* and is denoted as VRP2 in Chapter 1. The formulation, proposed by Laporte and Nobert [26], uses variables x_e that indicate how many times an edge $e \in E$ is traversed. In its summation form it is defined as

$$(3.1) \qquad (\text{VRP2}) \quad \text{minimize} \sum_{e \in E} c_e x_e$$

$$(3.2) \qquad \text{s.t.} \quad \sum_{e \in \delta(i)} x_e = 2 \qquad\qquad \forall\, i \in N,$$

$$(3.3) \qquad\qquad\quad \sum_{e \in \delta(0)} x_e = 2|K|,$$

$$(3.4) \qquad\qquad\quad \sum_{e \in \delta(S)} x_e \geq 2r(S) \qquad\qquad \forall\, S \subseteq N, S \neq \emptyset,$$

$$(3.5) \qquad\qquad\quad x_e \in \{0,1\} \qquad\qquad \forall\, e \in E \setminus \delta(0),$$

$$(3.6) \qquad\qquad\quad x_e \in \{0,1,2\} \qquad\qquad \forall\, e \in \delta(0).$$

Constraints (3.2) and (3.3) are the degree constraints for the customers and the depot, respectively. Constraints (3.4) are the so-called *Rounded Capacity Cuts* (RCCs). In spite of the fact that $r(S)$ is only a lower bound for the minimum number of vehicles that must visit S, VRP2 is not a relaxation but a complete formulation for the CVRP.

3.3.2 ▪ The Set Partitioning Formulation

Balinski and Quandt [5] proposed a formulation where a binary variable is associated with each possible route respecting the capacity constraint. Let Ω be the set of routes, given by a sequence of edges that describe a path from the depot to the customers and back. The cost c_r of a route $r \in \Omega$ is given by the sum of the cost of the edges in its path. Let a_{ir} represent the number of the times customer i is visited by route r. Finally, variable λ_r indicates whether a route $r \in \Omega$ is used or not. The *set partitioning formulation* is denoted as VRP4 in Chapter 1 and, in its summation form, is as follows:

$$(3.7) \qquad (\text{VRP4}) \quad \text{minimize} \sum_{r \in \Omega} c_r \lambda_r$$

$$(3.8) \qquad \text{s.t.} \quad \sum_{r \in \Omega} a_{ir} \lambda_r = 1 \qquad\qquad \forall\, i \in N,$$

$$(3.9) \qquad\qquad\quad \sum_{r \in \Omega} \lambda_r = |K|,$$

$$(3.10) \qquad\qquad\quad \lambda_r \in \{0,1\} \qquad\qquad \forall\, r \in \Omega.$$

The objective function (3.7) minimizes the overall cost of the selected routes. Constraints (3.8) guarantee that each customer is serviced by exactly one route. Constraint (3.9) imposes the use of $|K|$ vehicles. Finally, constraints (3.10) force the variables to be binary.

Balinski and Quandt defined Ω as the set of elementary routes, those visiting a customer at most once. Therefore, in this definition, coefficients a_{ir} are always binary. Formulation VRP4 has two important characteristics. The positive one is its gap of integrality, which is consistently observed to be reasonably small. The negative one is the exponential growth of the number of columns with the number of customers. Researchers seeking to take advantage of the former characteristic had to deal with the huge number of columns that instances with a few dozen customers already have. This led to the, today central in vehicle routing exact algorithms, use of column generation techniques. The pricing subproblem requires finding minimum cost capacitated elementary routes over a graph with both positive and negative edge costs, a strongly NP-hard problem. The cost of an edge $e = (u, v) \in E$ in those problems is given by its reduced cost $\bar{c}_e = c_e - (\pi_u + \pi_v)/2$, where π_i, $i \in N$, are the dual variables of constraints (3.8) and π_0 is the dual variable of constraint (3.9).

In the early 1980's, it was already observed that the formulation VRP4 could be turned into a more practical tool by changing the definition of the Ω set in order to obtain a more tractable pricing problem. Suppose that Ω is enlarged to contain all walks leaving and returning to the depot such that the sum of the demands of the visits of the walk does not exceed the vehicle capacity. The same customer can be visited more than once, but its demand will be counted again in each additional visit. The coefficient a_{ir} would assume a value equal to the number of visits to customer i in the walk. These walks were coined q-routes and used in the Lagrangian method proposed in Christofides, Mingozzi, and Toth [10]. The associated pricing problem is now weakly NP-hard, and a pseudo-polynomial dynamic programming algorithm for its resolution is available. This enlargement of Ω still lets VRP4 be a formulation for the CVRP, since constraints (3.8) forbid $\lambda_r = 1$ for any route r with a coefficient $a_{ir} > 1$. However, the lower bound given by the linear relaxation of the new model can be significantly worse.

The choice of set Ω plays an important role in the design of algorithms based on formulation VRP4. One idea is to impose some controlled amount of partial elementarity on the routes, in order to obtain bounds as close as possible to the elementary route bound, while still keeping the associated pricing problem tractable. There is an alternative that only allows q-routes without k-cycles, subcycles of length k or less, for some chosen k, as will be described in Section 3.5.2. Another more recent alternative is working with the so-called ng-routes, discussed in Section 3.5.3.

Anyway, even if Ω only contains elementary routes, the bounds given by formulation VRP4 (gaps between 1% and 4% are typical in the instances from the literature) are *not* good enough to be the basis of efficient exact algorithms. For this purpose, VRP4 must be reinforced with additional cuts. In fact, all the recent methods mentioned in Section 3.2 perform some kind of combined generation of both columns and cuts. Section 3.4 will present the main families of cuts used in those methods. Some of those cuts are robust; they do not change the structure of the pricing. However, the stronger cuts are non-robust. In this last case, the algorithm designer has to balance the strength of the cut (its capacity of improving the bounds) against its impact in the pricing complexity.

3.3.3 ▪ Capacity-Indexed Formulation

This extended formulation for the *Asymmetric CVRP* (ACVRP) (therefore valid for the CVRP) was presented in Pessoa, Poggi de Aragão, and Uchoa [33] (also in Godinho et al. [20] for the unitary demand case). Now $G_D = (V, A)$ is a complete directed graph with arc costs c_a, $a \in A$. For any set $S \subseteq V$, $\delta^-(S) = \{(i, j) \in A : i \in V \setminus S, j \in S\}$, and $\delta^+(S) = \{(i, j) \in A : i \in S, j \in V \setminus S\}$. Define binary variables x_a^q indicating that arc $a = (i, j)$ belongs to a route and that the total demand of j and of the vertices following j in the route is exactly q. The arcs returning to the depot must have $q = 0$. The

Capacity-Indexed Formulation (CIF) is

$$(3.11) \qquad \text{(CIF)} \qquad \text{minimize} \sum_{a \in A} c_a \sum_{q=0}^{Q} x_a^q$$

$$(3.12) \qquad \text{s.t.} \qquad \sum_{a \in \delta^-(\{i\})} \sum_{q=1}^{Q} x_a^q = 1 \qquad \forall\, i \in N,$$

$$(3.13) \qquad \sum_{q=1}^{Q} \sum_{i \in N} x_{0i}^q = |K|,$$

$$(3.14) \qquad \sum_{a \in \delta^-(\{i\})} x_a^q - \sum_{a \in \delta^+(\{i\})} x_a^{q-q_i} = 0 \qquad \forall\, i \in N,\ q = q_i, \dots, Q,$$

$$(3.15) \qquad x_a^q \in \{0,1\} \qquad \forall\, a \in A,\ q = 1, \dots, Q,$$

$$(3.16) \qquad x_{(i,0)}^q = 0 \qquad \forall\, i \in N,\ q = 1, \dots, Q.$$

Equations (3.12) and (3.13) are customer and depot degree constraints. Balance equations (3.14) state that if an arc with index q enters vertex i, then an arc with index $q - q_i$ must leave i. This both prevents cycles and routes with total demand greater than Q. The interest in the CIF lies in the fact that cuts expressed over its variables can be added into formulation VRP4 in a robust way, at least if the pricing is being solved by dynamic programming, as will be shown in Section 3.4.6.

3.4 ▪ Valid Cuts

This section discusses the main families of inequalities that have been used in recent works in order to reinforce formulation VRP2.

3.4.1 ▪ Cuts over the Edge Variables

A general inequality $\sum_{e \in E} \alpha_e x_e \geq b$ for model VRP2 can be included in the formulation VRP4 as $\sum_{r \in \Omega} (\sum_{e \in E} \alpha_e a_{er}) \lambda_j \geq b$, where a_{er} is the number of times that edge e appears in route r. In a column generation approach for solving the linear relaxation of VRP4, this additional cut contributes to the computation of reduced cost $\bar{c}_e$ of an edge $e \in E$ with the value $-\alpha_e \beta$, where β is the corresponding dual variable. Therefore, the structure of the pricing problem is not changed.

Besides RCCs (3.4), there are several known families of valid CVRP cuts over the edge variables. The package CVRPSEP (see Lysgaard [28]) contains effective heuristic separation procedures for the following families of cuts: rounded capacity, framed capacities, strengthened combs, multistars, and extended hypotours. While all those families may play a significant role in Branch-and-Cut algorithms over the formulation VRP2 (like Lysgaard, Letchford, and Eglese [29]), only RCCs and (by a much smaller degree) strengthened comb cuts can strengthen VRP4 in a significant way (see [19]). It seems that most cuts in the other families are already implicitly given by constraints (3.8)–(3.9). Indeed, Letchford and Salazar González [27] proved that all generalized large multistar cuts are implied by those constraints even if the definition of Ω includes all q-routes.

3.4.2 ▪ Strengthened Capacity Cuts

Baldacci, Christofides, and Mingozzi [3] introduced a family of cuts defined over the variables of the formulation VRP4. For each $S \subseteq N$ and for each $r \in \Omega$, define binary coefficient ζ_{Sr} as 1 if and only if the route r visits at least one vertex in S. The *Strengthened Capacity Cuts* (SCCs) are

$$(3.17) \qquad \sum_{r \in \Omega} \zeta_{Sr} \lambda_r \geq r(S) \qquad \forall\, S \subseteq N.$$

Remark that an RCC over S corresponds to $\sum_{r \in \Omega} a_{Sr} \lambda_r \geq r(S)$, where a_{Sr} counts how many times the route r enters (or leaves) S. Inequalities (3.17) are stronger because they are not "fooled" by routes that enter and leave S more than once. On the other hand, SCCs are non-robust, since they change the pricing subproblem for the linear relaxation of VRP4. In order to continue solving it by dynamic programming, it is necessary to include an additional binary dimension in each label indicating whether a partial route has already visited S or not. Notwithstanding this fact, the algorithms in [3, 4] and [12, 14] have successfully used (in a controlled way, limiting the number of separated cuts) SCCs. It can also be observed that if there exists a set $S' \subset S$ where $r(S') = r(S)$, then the SCC corresponding to S' dominates the SCC corresponding to S. This means that only the cuts corresponding to minimal sets (with respect to the function $r(\cdot)$) need to be separated.

3.4.3 ▪ Subset Row Cuts

Jepsen et al. [25] introduced the following family of cuts defined over the variables of the formulation VRP4. Given a set $C \subseteq N$ and a multiplier p, $0 < p < 1$, the (C, p)-Subset Row Cut (SRC)

$$(3.18) \qquad \sum_{r \in \Omega} \left\lfloor p \sum_{i \in C} a_i^r \right\rfloor \lambda_r \leq \lfloor p|C| \rfloor$$

is valid, since it can be obtained by a Chvátal–Gomory rounding of the corresponding constraints in (3.8).

The cuts where $|C| = 3$ and $p = 1/2$ are called *3-Subset Row Cuts* (3SRCs). The earlier algorithm by Baldacci, Christofides, and Mingozzi [3] used clique cuts, which are more general than 3SRCs. However, 3SRCs are favored in more recent works [4] and [12, 14], since they seem to be more suited to a column generation context, having less impact in the pricing subproblem. Each 3SRC requires an additional binary dimension in each dynamic programming label to indicate the parity of the number of visits made by a route to a vertex in a triplet C. The successful use of those cuts depends on a carefully controlled separation, in order to avoid pricing intractability.

Baldacci, Mingozzi, and Roberti [4] have also used a weakened variant of the 3SRCs for the case where Ω only contains elementary routes. Once a binary coefficient ξ_{Cr} is defined as 1 if and only if the route r visits at least one edge with both endpoints in C, the corresponding weakened 3SRC is

$$(3.19) \qquad \sum_{r \in \Omega} \xi_{Cr} \lambda_r \leq 1.$$

Since an elementary route can only visit two such edges if they are consecutive, it is easy to include the effect of the dual variables of weakened 3SRCs in a dynamic programming pricing. Therefore, there are no restrictions to its separation. A variant of this cut, suitable to the case where Ω contains non-elementary routes, needs to define ξ_{Cr} as the number of times that a route r visits non-consecutive edges with both endpoints in C.

3.4.4 ▪ Strong Degree Cuts

Contardo, Cordeau, and Gendron [13] introduced a family of cuts that correspond to SCCs over sets S of cardinality 1. Given a vertex $i \in N$ and for $r \in \Omega$, define binary coefficient ζ_{ir} as 1 if and only if route r visits i. The *Strong Degree Cut* (SDC) is

$$(3.20) \qquad \sum_{r \in \Omega} \zeta_{ir} \lambda_r \geq 1.$$

An SDC can only be non-redundant if the definition of set Ω allows non-elementary routes. In this case, the effect of SDCs is forbidding routes with cycles at certain vertices. Therefore, they are alternative ways of enforcing partial route elementarity. As happens with the more general SCCs, each SDC requires one additional binary dimension in the dynamic programming labels.

Contardo [12] and Contardo and Martinelli [14] also used weakened versions of the strong degree cuts. For an integer k, $i \in N$ and for $r \in \Omega$, let ν_{irk} be the number of times that route r visits i with at least k vertices between two consecutive visits. The corresponding *k-Cycle Elimination Cut* (k-CEC) is

$$(3.21) \qquad \sum_{r \in \Omega} \nu_{irk} \lambda_r \geq 1.$$

The effect of this cut is forbidding routes that have cycles over i of length k or less. The k-CECs have less impact in the dynamic programming pricing than the SDCs. The information that a vertex i was already visited is "forgotten" after k visits to other vertices.

3.4.5 ▪ Limited Memory Subset Row Cuts

Introduced in Pecin et al. [32], this generalization of a (C, p)-SRC requires an additional set M, $C \subseteq M \subseteq N$. The limited memory (C, M, p)-Subset Row Cut (lm-SRC) can be written as

$$(3.22) \qquad \sum_{r \in \Omega} \alpha(C, M, p, r) \lambda_r \leq \lfloor p|C| \rfloor,$$

where the coefficients α are computed by the following procedural function:

Function $\alpha(C, M, p, r)$
$coeff \leftarrow 0$, $state \leftarrow 0$
for every vertex $i \in r$ (in order) **do**
  if $i \notin M$ **then**
    $state \leftarrow 0$
  else if $i \in C$ **then**
    $state \leftarrow state + p$
    if $state \geq 1$ **then**
      $coeff \leftarrow coeff + 1$, $state \leftarrow state - 1$
return $coeff$

When $M = N$, the Function α will return $\lfloor p \sum_{i \in C} a_i^r \rfloor$ and the lm-SRC will be identical to an SRC. On the other hand, when M is not equal to N, the lm-SRC may be a weakening of its corresponding SRC. This happens because every time the route r leaves M, the variable $state$ is reset to zero, potentially decreasing the returned coefficient.

The potential advantage of the lm-SRCs over classical SRCs is their much reduced impact on the labeling algorithms used in the pricing, when $|M| \ll |N|$. The reasons for that reduction will be explained in Section 3.5.4. In order to obtain small memory sets, the separation of lm-SRCs presented in [32] used the following strategy. First, it identifies a violated (C, p)-SRC. Then, it determines a *minimal set M such that the lm-(C, M, p)-SRC has the same violation*. In practice, even on instances with hundreds of customers, those minimal sets seldom have cardinality larger than 15.

Given a base set $C \subseteq N$, for each integer d, $1 \leq d \leq n$, define a non-negative integer variable y_C^d as the sum of all variables λ_r such that $\sum_{i \in C} a_i^r = d$. Variables with $d > |C|$ can only be non-zero if Ω contains non-elementary routes. The interesting SRCs (and the corresponding lm-SRCs), for sets C with cardinality up to 5, are the following:

- The cuts where $|C| = 3$ and $p = 1/2$, the 3SRCs, can be expressed as $y_C^2 + y_C^3 + 2y_C^4 + 2y_C^5 + \cdots \leq 1$. The weak 3SRCs in [4] are equivalent (when all routes are elementary) to lm-3SRCs with $M = C$.

- Taking $|C| = 1$ and $p = 1/2$, the 1-Subset Row Cuts (1SRCs) $y_C^2 + y_C^3 + 2y_C^4 + \cdots \leq 0$ are obtained. They are equivalent to the SDCs $y_C^1 \geq 1$; both families forbid cycles over a vertex i ($C = \{i\}$). The weaker k-CECs [12, 14] only forbid cycles over i of size k or less. An lm-1SRC is a different kind of weakening; it forbids cycles over i contained in the set M. Of course, all these cuts can only be useful when the Ω set contains non-elementary routes.

- The cuts where $|C| = 4$ and $p = 2/3$ are 4-Subset Row Cuts (4SRCs), expressed as $y_C^2 + 2y_C^3 + 2y_C^4 + 3y_C^5 + 4y_C^6 + \cdots \leq 2$.

- There are two interesting families of cuts with $|C| = 5$. Those with $p = 1/3$ will be called 5,1SRCs, $y_C^3 + y_C^4 + y_C^5 + 2y_C^6 + \cdots \leq 1$; whereas those with $p = 1/2$ are 5,2SRCs, having the format $y_C^2 + y_C^3 + 2y_C^4 + 2y_C^5 + 3y_C^6 + \cdots \leq 2$. The latter family was already used in [12, 14].

3.4.6 ▪ Cuts over the CIF Variables

A general inequality $\sum_{a \in A} \sum_{q=0}^{Q} \alpha_a^q x_a^q \geq b$ for the CIF can be included in VRP4 as $\sum_{r \in \Omega} (\sum_{a \in A} \sum_{q=0}^{Q} \alpha_a^q a_{ar}^q) \lambda_j \geq b$, where a_{ar}^q is the number of times (always 0 or 1) that arc a appears with capacity q in route r. This additional cut contributes to the computation of an arc-capacity reduced cost $\bar{c}_a^q$ with the value $-\alpha_a^q \beta$, where β is the corresponding dual variable. This more general reduced cost structure does not affect the complexity of the dynamic programming algorithms usually employed for pricing routes. This happens because those algorithms already have a dimension for the capacity consumption in their labels.

There are potential gains of using cuts defined over the capacity-indexed extended variable space. For example, the reasoning behind the RCCs and SCCs assumes that every route visiting a set S can contribute with up to Q units for satisfying the demand $q(S)$. However, if the route also visits other vertices outside S, the actual contribution can be much smaller than Q. The *Extended Capacity Cuts* (ECCs) (see Pessoa, Poggi de Aragão, and Uchoa [33] and Pessoa, Uchoa, and Poggi de Aragão [35]) are a generalization of the RCCs devised to capture that fact, taking advantage of the knowledge of the actual route capacity when entering and leaving S.

As surveyed in Uchoa [41], ECCs (and some other families of cuts related to the CIF) have been successfully used in a number of vehicle routing variants, on parallel machine scheduling, and even on network design problems. However, the results for the CVRP reported in Pessoa, Poggi de Aragão, and Uchoa [33] and Uchoa [41] are more modest. In this case, the inclusion of ECCs in a robust BCP algorithm provides root bounds that are inferior to those that can be obtained by adding non-robust SCCs and SRCs, not good enough for solving instances larger than those already solved by previous algorithms. Nevertheless, we still believe in the possibility that cuts over the CIF can be a useful addition to future CVRP algorithms.

3.5 ▪ Pricing

This section presents the main ideas proposed in the literature for pricing routes, relaxed or not, for the formulation VRP4, with or without additional non-robust cuts.

3.5.1 ▪ q-Routes and Elementary Routes

The pricing subproblem for solving the linear relaxation of VRP4 by column generation, defined in Section 3.3, has been modeled by many authors as a *Shortest Path Problem with Resource Constraints* (SPPRC). The problem is defined over directed graphs ($V \cup \{n + 1\}, A$), where vertex $n + 1$ is a copy of the depot vertex 0; and A is the complete arc set minus $\{(0, n + 1), (n + 1, 0)\}$. For each $(i, j) \in A$, there are costs $\overline{c}_{ij}$, unrestricted in sign. Define a set H of resources, numbered from 0 to $|H| - 1$. Each $h \in H$ has a positive integer availability C^h. For each arc $(i, j) \in A$ and each $h \in H$, d_{ij}^h are non-negative integer resource consumption quantities. The SPPRC has the objective of finding the shortest, not necessarily elementary, path $P = (0, i_1, i_2, \ldots, i_p, n + 1)$ such that, for each resource $h \in H$, the total consumption T^h does not exceed C^h. The pricing of q-routes corresponds to an SPPRC with a single resource 0, such that $C^0 = Q$. For each $(i, j) \in A$, $d_{ij}^0 = q_j$ and $\overline{c}_{ij}$ is equal to the reduced cost of the corresponding edge. However, the pricing of elementary routes requires the definition of n additional resources, such that $C^i = 1$ for each $i \in N$. For each arc $(u, v) \in A$ and each $i \in N$, d_{uv}^i is defined as 1 if $i = v$ and 0 otherwise.

A forward dynamic programming label setting algorithm for the SPPRC represents a path P starting at 0 and ending at vertex $i \in V \cup \{n + 1\}$ as a label $L(P) = (i, \overline{c}(P), T^0(P), \ldots, T^{|H|-1}(P))$, where $\overline{c}(P)$ is the cost and $T^h(P)$, $h \in H$, are the resource consumptions in the path. Given a label $L = L(P)$, we define $v(L)$ as the ending vertex of P, $\overline{c}(L) = \overline{c}(P)$ and $T^h(L) = T^h(P)$, for each $h \in H$, and $P(L) = P$. The list of labels $\mathscr{L}$ is initialized with a single label representing a null path that ends at the depot with zero cost and no consumption of resources. As it iterates, the algorithm tries to extend a label $L \in \mathscr{L}$, creating new labels corresponding to the paths obtained by adding a vertex in $(N \cup \{n+1\}) \setminus \{v(L)\}$ to $P(L)$, being successful only when all resource constraints are still satisfied. Each newly created label L', if any, is added to $\mathscr{L}$ and linked to L by backward pointers; we say that $pred(L') = L$. When it is not possible to extend any label in $\mathscr{L}$, the minimum cost labels ending at $n + 1$ correspond to optimal solutions of the SPPRC.

When designing a dynamic programming label setting algorithm for the SPPRC, the main concern is how to mitigate the efficiency deterioration as the number of resources grows (the "curse of dimensionality"), keeping the list of labels $\mathscr{L}$ as small as possible. It is clear that if two labels ending at the same vertex also have the same resource consumption, only the least costly one (ties broken arbitrarily) should be kept. For pricing q-routes, when the algorithm considers only the capacity resource, this basic dominance rule is enough to show that at most $n(Q + 1)$ labels need to kept, giving a total complexity of

$O(n^2 Q)$ operations. This pseudo-polynomial complexity is not satisfactory when Q is very large. In those cases, it is recommended to scale the capacity and the demands, as done in Subramanian et al. [40]. However, for pricing elementary routes, this reasoning would only prove an exponential complexity of $O(2^n Q)$ for the label setting algorithm. We now present some of the ideas that have been introduced (Boland, Dethridge, and Dumitrescu [6], Chabrier [7], Feillet et al. [17], Righini and Salani [37], and Righini and Salani [38]) to tame the worst-case complexity of that algorithm, enough to turn it into a practical tool for pricing elementary routes on many instances.

1. **Stronger Dominance Rules.** The basic dominance rule can be improved by considering the monotonic effect of the resource consumption in the possible extensions of a label. If two labels L and L', both ending at the same vertex i, have $\bar{c}(L) \leq \bar{c}(L')$ and $T^h(L) \leq T^h(L')$, for $h \in H$, then L dominates L'. Therefore, L' can be removed from the set of labels. Whenever a new label L is created, it is interesting to check whether L is not dominated by some other label already in $\mathscr{L}$ or whether L dominates some labels in $\mathscr{L}$. While this is conceptually simple, efficient implementation may require proper data structures.

2. **Bidirectional Search.** The 2^n factor in the complexity is far too pessimistic for typical CVRP instances, when the capacity Q limits the number of customers in most routes to a value that is not much larger than $n/|K|$. In fact, the ratio $n/|K|$ is indeed a strong predictor of the algorithm performance. In order to better profit from this effect, Righini and Salani [37] proposed not expanding the labels with a capacity consumption larger than $\lceil Q/2 \rceil$. This usually reduces a lot the final size of $\mathscr{L}$. The remaining routes can be obtained by a concatenation phase, taking advantage of the symmetry of the CVRP. Each pair of labels L_1 and L_2 in $\mathscr{L}$ such that $T^h(L_1) + T^h(L_2) \leq C^h$, for each $h \in H$, corresponds to a route with cost $\bar{c}(L_1) + \bar{c}(L_2) + \bar{c}_{v(L_1)v(L_2)}$. Actually, since the concatenation phase can be costly, it is not clear that bidirectional search is always superior to the traditional forward search. For example, in the recent work by Martinelli, Pecin, and Poggi [30] the latter alternative was preferred.

3. **Completion Bounds.** In the pricing context, one is only interested in SPPRC solutions corresponding to routes with negative reduced cost. Given a label $L \in \mathscr{L}$, let $CB(L)$ be a completion bound on L, i.e., a lower bound on the cost of all routes that can be obtained as extensions of L. If $CB(L) \geq 0$, then L can be removed from $\mathscr{L}$. For example, labels from a previous run of the dynamic programming that considers only a subset of the resources in H provide valid completion bounds. A good completion bound can reduce considerably the final size of $\mathscr{L}$.

4. *Decremental State Space Relaxation* (DSSR). This technique was proposed independently by Boland, Dethridge, and Dumitrescu [6] (where it is called State Space Augmenting Algorithm) and by Righini and Salani [38] (where the name DSSR was proposed). It starts by running the label setting algorithm by only considering the capacity resource (i.e., pricing q-routes). The obtained solution is likely to contain cycles. The resources for the vertices that are visited more than once in that solution are then included in the next run of the label setting algorithm. As the algorithm progresses, the set of customers that cannot be revisited is enlarged, until elementary routes are obtained.

3.5.2 ▪ *q*-Routes with *k*-Cycle Elimination

The q-routes with k-cycle elimination are those where multiple visits to a vertex $i \in N$ are only allowed if at least k other costumers are visited between successive visits. Christofides, Mingozzi, and Toth [11] already realized that 2-cycle elimination in the definition of Ω improves significantly the bounds obtainable from the VRP4 relaxation, with a modest impact in the complexity of the pricing subproblem, as explained next. Each pair (i,q) where $i \in N$ and $q \in \{0, 1, \ldots, Q\}$ defines a bucket $B(i,q)$. A label $L \in \mathcal{L}$ is associated with $B(i,q)$ if $v(L) = i$ and $T^0(L) = q$. When pricing ordinary q-routes it is clear that the basic dominance rule assures that each bucket contains at most one label in $\mathcal{L}$. However, when performing 2-cycle elimination, a second label may need to be kept in each bucket. In this case, the least cost label L in a bucket does not dominate the second least costly label L' in the same bucket if $pred(L) = j$ and $pred(L') \neq j$ because, as the extension of L to j is forbidden, it is not possible to prove that the extension of L' to j is not part of the optimal solution.

However, when performing k-cycle elimination for $k \geq 3$, deciding which labels need to be kept in each bucket becomes quite intricate. Irnich and Villeneuve [24] propose an Intersection Algorithm for that purpose. Their conjecture (proved only for $k = 3$) is that this algorithm will keep at most $k!$ labels per bucket. Anyway, the average number of labels per bucket actually kept during a run may vary a lot, depending on the instance. Hoshino and de Souza [22] proposed a more efficient implementation where non-dominated labels are identified by deterministic finite automata; there is an automaton for each value of k.

Experiments with q-routes with k-cycle elimination in Fukasawa et al. [19] suggest that it is not worthy to use values of k above four. From this point, the pricing indeed becomes much slower. Unhappily, 4-cycle elimination is not enough to obtain bounds really close to the elementary bounds in some instances.

3.5.3 ▪ *ng*-Routes

A simple alternative to cycle elimination for obtaining partial elementarity in the routes would be to select a subset S of customers to forbid being revisited. The label setting algorithm for the SPPRC would be run defining unitary resources only for those customers. However, this is not likely to produce near-elementary routes if $|S|$ is significantly smaller than $|N|$. Baldacci, Mingozzi, and Roberti [4] devised a better way to impose partial elementarity. Instead of carrying the information that a certain vertex was already visited by a path in all its extensions, a more limited memory mechanism is proposed. It takes advantage of the fact that the cycles that are likely to appear when pricing non-elementary routes, even when they are too long to be efficiently avoided by k-cycle elimination, only use edges with small cost and are confined to relatively small "neighborhoods" in the graph.

For each customer $i \in N$, let $N_i \subseteq N$ be the ng-set of i, defining its neighborhood. This may stand for the $|N_i|$ (this cardinality is decided a priori) closest customers and includes i itself. When extending a label L_1 with $v(L_1) = i$ to a customer j, the visit to i is only remembered in the new label L_2 if i belongs to N_j. In this case, L_2 cannot be extended back to i. However, the extension of L_2 to another vertex k will only carry the information about i in the new label L_3 if i also belongs to N_k. At some point, the visit to i is "forgotten" and further extensions may visit i again, forming a cycle. In other words, an ng-route allows a cycle over a vertex i only if this cycle passes by a vertex v such that $i \notin N_v$.

The modified label setting algorithm for pricing ng-routes works as follows. Let $P = (0, i_1, i_2, \ldots, i_p)$ be a path associated with a label $L(P)$ and having $V(P)$ as the set of visited customers. Define the set of forbidden extensions of P as

$$\Pi(P) = \left\{ i_k \in V(P) \backslash \{i_p\} : i_k \in \bigcap_{s=k+1}^{p} N_{i_s} \right\} \cup \{i_p\}.$$

The label $L(P)$ can be extended to a customer i_{p+1} only when $i_{p+1} \notin \Pi(P)$ and all the resource constraints are satisfied. The extension to customer i_{p+1} creates a new label $L(P')$ corresponding to path $P' = (0, \ldots, i_p, i_{p+1})$ and forbidden extensions $\Pi(P') = (\Pi(P) \cap N_{i_{p+1}}) \cup \{i_{p+1}\}$. The routes that can be obtained by this algorithm are ng-routes. A label $L(P_1)$ can only dominate label $L(P_2)$ if the additional condition $\Pi(P_1) \subseteq \Pi(P_2)$ is satisfied. Assuming that the only resource considered is the capacity, the maximum number of labels in a bucket $B(i, q)$ is limited to $2^{|N_i|-1}$. Therefore, the algorithm remains pseudo-polynomial when the ng-sets have a fixed size.

In order to handle ng-sets with large cardinality, Martinelli, Pecin, and Poggi [30] proposed an adaptation of the DSSR concept. The labeling algorithm starts with the ng-sets relaxed into singletons. If a forbidden cycle over a vertex i is detected in the solution, i is restored in the ng-sets of the vertices in the cycle and the labeling is run again. The labels of the previous run are used for computing completion bounds. This is done until the desired ng-route is found. In that way, ng-sets with size 64, large enough to enforce elementarity, can be efficiently handled on most instances with up to 200 customers.

3.5.4 ▪ Pricing with Non-robust Cuts

The adaptation of the label setting algorithm for handling non-robust cuts requires additional dimensions in the labels, typically one dimension per cut. In the literature (for instance, Jepsen et al. [25]), those dimensions are still called "resources". However, they are not resources in the strict sense defined in Section 3.5.1. First, in the SPPRC definition, the consumption of a resource is monotonically non-decreasing along a path. This is not the case, for example, of the dimensions added for handing SRCs. More importantly, while the exhaustion of a resource forbids certain path extensions, dimensions corresponding to cuts never forbid path extensions. Actually, such a label dimension is only used to reward/penalize certain extensions, determining whether the value of the dual variable associated with a cut should be subtracted from the cost of the new labels.

Assume that the formulation VRP4 was enhanced with (i) the SCCs associated with the customer sets S in a collection $\mathscr{S}$, having dual variables π_S; (ii) the 3SRCs associated with the triplets C in a collection $\mathscr{C}$, having dual variables σ_C. Remark that some of the SCCs can be SDCs. A label L has additional binary dimensions D_S, $S \in \mathscr{S}$, with value 1 if $P(L)$ already visited some vertex in S and 0 otherwise; and binary dimensions D_C, $C \in \mathscr{C}$, indicating the parity of the number of visits made by $P(L)$ to a vertex in C. Whenever a label L is extended to a vertex j belonging to a set S where $D_S(L) = 0$, the value π_S is subtracted from the cost of the new label. As $\pi_S \geq 0$, this means that this particular extension is being rewarded. In a similar way, when L is extended to a vertex j that belongs to a triplet C where $D_C(L) = 1$, σ_C is subtracted from the cost of the new label. As $\sigma_C \leq 0$, this corresponds to a penalization. The dominance rule in the label setting algorithm also needs to be modified. Given labels L and L' with $v(L) = v(L')$ and $T^h(L) \leq T^h(L')$, for $h \in H$, it can be shown that L dominates L' if the following additional condition is satisfied:

$$(3.23) \qquad \bar{c}(L) \leq \bar{c}(L') - \sum_{S \in \mathscr{S}: D_S(L) > D_S(L')} \pi_S + \sum_{C \in \mathscr{C}: D_C(L) > D_C(L')} \sigma_C.$$

The second and third terms in the right-hand side are lower bounds on what the extensions of L' can gain over the extensions of L, by still being able to collect the rewards of visiting some sets in $\mathscr{S}$ already visited by $P(L)$ or by avoiding some penalizations that the extensions of L may incur by revisiting some triplets in $\mathscr{C}$ already visited by $P(L)$ an odd number of times. In general, the presence of too many such non-robust cuts ruins the performance of the labeling algorithm. This happens because the dual variable of each additional cut may interfere with condition (3.23), making it less likely to be satisfied. At some point, an exponential explosion in the number of non-dominated labels is observed.

Anyway, Contardo, Cordeau, and Gendron [13] showed that the addition of non-robust SDCs may be a more efficient way of imposing route elementarity than the traditional definition of resources for avoiding vertex revisitation. In the latter case, if a path P visits any vertex not visited by path P', $L(P)$ can never dominate $L(P')$. On the other hand, when SDCs are used, $L(P)$ can dominate $L(P')$ if $\bar{c}(L(P))$ is sufficiently smaller than $\bar{c}(L(P'))$ to compensate the dual variables corresponding to the vertices not visited in P'. The algorithm in Contardo [12] takes this idea to its logical extreme, only pricing q-routes without 2-cycles. The partial elementarity is obtained by the gradual introduction of non-robust k-CECs and SDCs. However, the algorithm in Contardo and Martinelli [14] obtained better results by pricing ng-routes with ng-sets of size 8 and only using k-CECs and SDCs for removing the larger cycles.

The labels from a previous run of the dynamic programming that does not consider the dimensions corresponding to non-robust cuts may not provide valid completion bounds, unless some adaptations are made. While the dual variables with negative values can be safely ignored, dual variables with positive values must be taken into account. Contardo, Cordeau, and Gendron [13] proposed underestimating the impact of dual variable π_S, $S \in \mathscr{S}$, by subtracting $\pi_S/2$ from the cost of all edges in $\delta(S)$. In order to obtain stronger completion bounds, [12] and [14] actually consider the dimensions corresponding to some of the cuts (those with larger absolute dual variable values) in the dynamic programming run.

Now we explain why the lm-SRCs introduced in [32], when their memory sets are small, can be much better handled by the labeling algorithms than the traditional SRCs. For simplicity, assume that the only non-robust cuts are the 3SRCs associated with pairs (C, M) in a collection $\mathscr{C}$, where C is a triplet and $M \supseteq C$ is a memory set, and having dual variables $\sigma_{C,M}$. As before, the labels have binary dimensions $D_{C,M}$ for each $(C, M) \in \mathscr{C}$. However, $L(D_{C,M})$ now indicates the parity of the number of visits made by $P(L)$ to a vertex in C, but only since $P(L)$ entered M for the last time. By definition, $L(D_{C,M})$ is zero on all labels L where $v(L) \notin M$. This means that only the dual variables $\sigma_{C,M}$ such that $v(L) \in M$ may interfere in the dominance condition:

$$(3.24) \qquad \bar{c}(L) \leq \bar{c}(L') + \sum_{(C,M) \in \mathscr{C}: D_{C,M}(L) > D_{C,M}(L')} \sigma_{C,M}.$$

In practice, this means that many more cuts can be added before the explosion in the number of non-dominated labels is observed.

3.5.5 ▪ Column Generation Issues

Column generation is a technique for solving linear programs with a huge number of variables. Of course, there is no magic; its efficiency still depends on "how huge" this number is. Since the number of variables in VRP4 is roughly proportional to $\binom{n}{n/|K|}$, it is quite expected that the number of pricing iterations until convergence depends significantly on the average number of customers per route. This is clearly observed in practice. On instances from the literature where $n/|K|$ is small, convergence of the standard column generation procedure typically occurs in a few dozen iterations. On the other hand, on instances where is $n/|K|$ is large, convergence may require hundreds of iterations. Since those are precisely the instances where each pricing iteration is slower (especially with elementary or near-elementary routes), some kind of dual stabilization (see Du Merle et al. [16] or Pessoa et al. [34]) may help to mitigate the convergence problems.

Anyway, it is always advisable to accelerate the column generation by devising faster pricing heuristics. The relatively expensive exact pricing is only called when the pricing heuristics cannot find routes with negative reduced cost. Such effective heuristics (for example, see Fukasawa et al. [19]) can be obtained by modifying the label setting algorithm. The *scaling* technique consists of running the algorithm with modified demands $q'_i = \lceil q_i/g \rceil$ for $i \in N$ and modified capacity $Q' = \lfloor Q/g \rfloor$ for a chosen factor g. The *sparsification* technique consists of only extending labels by considering a chosen subset of the edges, those that are more likely to appear in routes with negative reduced cost. Finally, the *bucket pruning* technique consists of storing only a chosen (small) number of labels per bucket. This means that many non-dominated labels, those that are less likely to lead to optimal solutions, may be dropped.

3.6 ▪ Branching vs. Route Enumeration

The column and cut generation over the formulation VRP4 may finish with a duality gap, either because it is not possible to find additional violated cuts or because tailing-off, minimal improvements in the lower bound after a number of cut rounds, is detected. In the case of an algorithm that uses non-robust cuts there is another possibility: additional separation would make the pricing excessively expensive. The algorithms in Fukasawa et al. [19] and in Røpke [39] try to close that gap by following the classical BCP paradigm; a branching would be performed, and each child node would be solved by column and cut generation. On the other hand, the algorithms in Baldacci, Christofides, and Mingozzi [3], Baldacci, Mingozzi, and Roberti [4], Contardo [12], and Contardo and Martinelli [14] do not perform branching, at least not directly. Instead, they proceed to enumerate all elementary routes that may belong to an optimal solution, using reduced cost information. A set-partitioning problem with those routes is then solved by the generic Branch-and-Cut algorithm of a MIP solver.

3.6.1 ▪ Branching

Branching directly over the VRP4 variables is non-robust and would complicate the pricing subproblem in the children nodes. On the other hand, it is possible to perform the branching on the edge variables of the formulation VRP2. This may be done either over individual edges $e \in E$, imposing disjunctions like $(x_e \leq 0) \vee (x_e \geq 1)$, or, more generally, over sets $S \subseteq N$, imposing the disjunction $(\sum_{e \in \delta(S)} x_e = 2) \vee (\sum_{e \in \delta(S)} x_e \geq 4)$. In both cases, the branching constraints for the formulation VRP2 can be translated into robust cuts in VRP4.

The BCP in Fukasawa et al. [19] adopted the branching over sets (already used in Lysgaard, Letchford, and Eglese [29]). In order to better choose the branching set, this BCP used strong branching. Each set in a collection containing from 5 to 10 candidate sets was heuristically evaluated by applying a small number of column generation iterations in its children nodes. Remark that the exact evaluation of each candidate, performing full column generation (perhaps also cut generation) in both children nodes, would be too expensive.

The recent work by Røpke [39] showed that is possible to obtain major improvements in the BCP performance by performing a more sophisticated and aggressive strong branching. His extensive experiments were performed both on CVRP and VRPTW instances.

- The simpler branching over individual edges (arcs in the VRPTW case) was used.

- The strong branching procedure starts by performing a quick evaluation of 30 candidate edges, producing a ranking. The best ranked candidate is then fully evaluated and becomes the incumbent winner. Then, other candidates with good ranking are better evaluated, but only while they have a reasonable chance of beating the incumbent winner.

- It is possible to collect statistics along the enumeration tree to help the strong branching. In particular, the previous evaluations of an edge (in different nodes) are good predictors of future evaluations.

This strong branching was the key algorithmic element that allowed his robust BCP to solve the hard instance M-n151-k12, with optimum 1015, starting from the rather modest root lower bound of 1001.5. Another robust BCP with the same strong branching was the first algorithm to solve all the 56 VRPTW Solomon instances with 100 vertices. We remark that a strong branching for a BCP, even with smart accelerating ideas, is a very "pricing-intensive" mechanism (i.e., it requires many calls to the pricing solver).

3.6.2 ▪ Route Enumeration

Baldacci, Christofides, and Mingozzi [3] use a route enumeration–based approach (already used in Baldacci, Bodin, and Mingozzi [2]) in order to close the duality gap after the root node. The key observation is that a route $r \in \Omega$ can only be part of a solution that improves the best known upper bound if (i) its reduced cost (the sum of its edge reduced costs with respect to the current dual solution) is smaller than the gap, and (ii) there is no other route visiting the same set of clients with smaller cost (with respect to the original edge costs). The enumeration of those routes may be performed by a label setting algorithm, similar to those used to price elementary routes, producing a set $R \subseteq \Omega$. A general MIP solver is then used to solve the formulation VRP4 restricted to the set R, using the known upper bound as a cutoff. If the restricted model is infeasible, it is proved that the solution that provided the upper bound is optimal. Otherwise, the optimal solution of the restricted model is the optimal solution of the complete problem.

The efficiency of this approach is related to the cardinality of the set R. In fact, if $|R|$ is too large, the label setting enumerating algorithm will run out of memory. Even assuming that it is possible to enumerate R, there is still the question of solving the restricted VRP4 model. The cardinality of R depends crucially on the duality gap, but there is a second important factor: again, the average number of customers per route. If it is small, say, $n/|K| \leq 8$, then R is not likely to be large, even with a significant gap. It is typical in those cases to generate sets with no more than a few tenths of thousands routes, the resulting restricted models being solved in minutes or even seconds. The speedup of the enumeration approach with respect to the traditional branching can be large. On the other hand, instances with many customers per route can be more problematic. Successful enumeration on those cases may require small gaps. Baldacci, Mingozzi, and Roberti [4] enhanced the enumeration procedure by considering two dual solutions π_1 and π_2, with associated lower bounds LB_1 and LB_2. A route r cannot belong to any improving solution either if $LB_1 + \bar{c}_1(r) \geq UB$ or if $LB_2 + \bar{c}_2(r) \geq UB$, where UB is the best known upper bound and $\bar{c}_1(r)$ and $\bar{c}_2(r)$ are the route reduced costs with respect to π_1 and π_2, respectively. In their algorithm, the second dual solution corresponds to the final solution at the root node, and the first dual solution is obtained at an earlier stage of the algorithm, so $LB_1 \leq LB_2$. In fact, after π_1 is obtained, the pricing of elementary routes used in their column and cut generation algorithm starts to test the condition $LB_1 + \bar{c}_1(r) < UB$. This restriction on the set of routes may improve the final bound LB_2.

Contardo [12] introduced a different strategy to better profit from the route enumeration. From time to time, it is tested whether the current solution would lead to a set R_1 with less than 5 million routes, still too large to allow the efficient resolution of the restricted VRP4 model by a MIP solver. Instead, the set R_1 is stored in a pool, and the column and cut generation proceeds. However, from this point, the pricing is not solved by a label setting algorithm anymore; it starts to be performed by a straightforward inspection of the routes in the pool. As a result, the separation of non-robust cuts, which was being done in a very controlled way, can now be aggressive. After each improvement of the current lower bound, reduced cost fixing is performed in the pool, decreasing its size. In most instances from the literature, the column and cut generation algorithm terminates with zero gap. If this is not the case, the final set of routes remaining in the pool defines the set R and the corresponding restricted VRP4 model is given to the MIP solver. For example, for solving instance M-n151-k12, a relaxation giving a lower bound of 1009.0 generates a set R_1 with about 4 million routes. In the end of the root node, the lower bound is increased to 1012.5 and the final set R only has about 13,000 routes. This strategy is also used in Contardo and Martinelli [14].

3.6.3 ▪ Hybridizing Branching and Route Enumeration

It is clear that route enumeration is an important element in some of the best performing algorithms for the CVRP, being capable of drastically reducing the running times of some instances. Nevertheless, it is disturbing to consider that route enumeration, no matter how cleverly done, is an inherently exponential space procedure (it would remain exponential even if P=NP) that is bound to fail on larger/harder instances.

However, one does not need to take a radical stance on completely avoiding branching. The hybrid strategy used in Pessoa, Poggi de Aragão, and Uchoa [33] and Pessoa, Uchoa, and Poggi de Aragão [35] performs route enumeration after solving each node. If the limit of 80,000 routes is reached, the enumeration is aborted and the BCP proceeds by traditional branching. Of course, since deeper nodes will have smaller gaps, at some point the enumeration will work. The overall effect may be a substantially reduced enumeration tree. For example, where the traditional branching would need to reach depth 10, that hybrid strategy would not go beyond depth 5.

The BCP in Pecin et al. [32] implements a more sophisticated hybrid strategy. Route enumeration (in Contardo's style) is tried in each node v, allowing quite large sets R_1, in some cases with up to 50 million routes:

- If the enumeration succeeds, node v continues to be solved using the pricing by inspection in the pool of routes. An unlimited number of SRCs and Clique Cuts (not used before the enumeration) are separated. Fixing by reduced costs eliminates routes from the pool. If the final set R has less than 20,000 routes, the node is finished by an MIP solver. However, there are situations where the final set R is still too large. When this happens, the BCP itself performs a branching. Each child node of v receives a copy of R; the pricing in those nodes will continue to be done by inspection.

- If the enumeration fails, the node performs an aggressive hierarchical strong branching. Actually, the branching effort is dimensioned according to the expected size of the subtree rooted in v.

3.7 ▪ Overview of Computational Results

In order to provide experimental information on the topics discussed in this chapter, we report results over the standard classes of instances used for testing exact methods for the CVRP. Classes A, B, and P were proposed in Augerat et al. [1]. Classes E, F, and M were proposed by Christofides and Eilon [8], by Fisher [18], and by Christofides, Mingozzi, and Toth [9], respectively. All those instances can currently be found at http://branchandcut.org/VRP/data.

We start this section with an analysis of the bounds that are obtained with formulation VRP4 without the addition of extra cuts. Tables 3.1 and 3.2 present the computing times and the percent gap from the proved optimal over a selected set of instances which includes those that, today, can be regarded as the most relevant when analyzing exact approaches for the CVRP. The following variants are considered: (i) q-routes with k-cycle elimination, for $k = 2, 3, 4, 5$, using the code from Fukasawa et al. [19]; and (ii) ng-routes with ng-sets of size 8, 16, 32, and 64, using the code from Martinelli, Pecin, and Poggi [30]. This last code is an unidirectional forward label setting algorithm that uses a kind of DSSR; the ng-sets are increased dynamically. Times correspond to runs on processor Intel Core 2 Duo E7400 2.8GHz. A time limit of 2 hours was set. The rows **Avg. values** are averages, always excluding instances F-n135-k7 and M-n121-k7, where some variants exceeded the time limit. For the columns corresponding to gaps, they are ordinary arithmetic averages. However, for the columns corresponding to times, we report geometrical averages (otherwise a few larger instances would determine the results). Some comments follow:

- Pricing routes without 2-cycles or 3-cycles is fast. This is expected from the theory; there is a proven upper bound of at most 2 and 6 labels per bucket, respectively. On the other hand, the resulting bounds are not close to those that could be obtained by pricing elementary routes. This suggests using k-cycle elimination for larger values of k. However, the approach does not scale well. Elimination of 5-cycles makes the pricing much slower and still does not reach the elementary bounds.

- The ng-route approach performed much better. Using ng-sets of size 8 already gives bounds similar to those obtained by 5-cycle elimination, taking a time comparable to 4-cycle elimination. The relative strength of ng-8 is somewhat counterintuitive, since it is not difficult to construct cycles with five (or even less) reasonably short edges that would not be forbidden by a memory of eight neighbors. Anyway, it seems that the ng-route approach (at least with the new DSSR implementation in [30]) scales very well. Increasing the size of the ng-sets to 16, 32, and 64 does not make the algorithm much slower. The bounds obtained by ng-64 over those instances are identical to the bounds obtained by elementary routes. In fact,

Bounds for the set partitioning formulation without extra cuts

Table 3.1. *Set partitioning formulation - no cuts - k-cycle elimination.*

Instance			2-cycle		3-cycle		4-cycle		5-cycle	
Name	n/k	OPT	T(s)	gap	T(s)	gap	T(s)	gap	T(s)	gap
A-n62-k8	7.6	1288	1.68	5.07	2.90	3.28	6.08	2.97	26.14	2.80
A-n80-k10	7.9	1763	4.37	2.88	4.52	2.28	12.16	2.02	30.34	1.95
B-n50-k8	6.1	1312	0.69	7.20	0.98	6.16	2.87	3.66	13.30	3.47
B-n78-k10	7.7	1221	1.62	7.90	3.05	6.21	4.36	6.00	32.66	5.63
E-n76-k7	10.7	682	3.95	2.74	4.37	2.71	11.88	2.67	57.72	2.66
E-n76-k8	9.4	735	2.71	2.49	3.72	2.40	7.47	2.34	36.26	2.34
E-n76-k10	7.5	830	1.65	2.24	2.04	2.22	3.92	2.20	15.70	2.20
E-n76-k14	5.4	1021	0.64	2.10	0.68	2.03	1.77	1.85	4.75	1.84
E-n101-k8	12.5	815	8.43	3.51	18.69	3.23	20.83	3.15	70.82	3.12
E-n101-k14	7.1	1067	2.39	2.05	3.25	1.88	6.00	1.86	21.76	1.86
F-n135-k7	19.1	1162	1400	6.04	3119	3.82	4757	2.64	-	-
M-n121-k7	17.1	1034	54.07	2.03	66.93	1.35	108.8	1.03	360.3	0.89
M-n151-k12	12.5	1015	32.92	2.32	47.55	2.05	74.93	2.01	194.6	1.94
M-n200-k16	12.5	1274	138.3	2.63	139.5	1.98	230.2	1.87	424.2	1.84
M-n200-k17	11.8	1275	76.62	2.50	94.46	1.87	159.5	1.78	420.4	1.75
P-n70-k10	6.9	827	0.81	2.26	1.25	2.15	2.68	2.05	6.30	2.01
Avg. values:			4.20	3.42	5.74	2.89	11.26	2.60	38.89	2.53

Table 3.2. *Set partitioning formulation - no cuts - n g-routes (ng-64 is equivalent to elementary route pricing in all those instances).*

Instance	ng-8		ng-16		ng-32		ng-64	
Name	T(s)	gap	T(s)	gap	T(s)	gap	T(s)	gap
A-n62-k8	3.26	2.93	3.5	2.58	4.24	2.58	5.09	2.58
A-n80-k10	6.03	1.88	6.12	1.79	7.57	1.78	10.75	1.78
B-n50-k8	1.50	3.47	1.56	3.46	1.82	3.46	1.85	3.46
B-n78-k10	5.50	4.44	5.57	4.38	7.07	4.38	9.24	4.38
E-n76-k7	9.86	2.61	10.95	2.52	11.20	2.41	15.17	2.41
E-n76-k8	6.66	2.34	7.72	2.21	7.25	2.21	10.14	2.21
E-n76-k10	3.80	2.20	4.63	2.18	4.91	2.11	5.56	2.11
E-n76-k14	2.65	1.88	2.56	1.79	2.47	1.79	3.03	1.79
E-n101-k8	23.90	3.14	23.31	2.98	27.57	2.95	35.14	2.95
E-n101-k14	8.29	1.86	9.01	1.74	9.10	1.55	10.39	1.55
F-n135-k7	6462	2.34	-	-	-	-	-	-
M-n121-k7	74.18	0.74	564.1	0.46	-	-	-	-
M-n151-k12	72.70	1.90	83.36	1.81	91.89	1.75	111.1	1.73
M-n200-k16	158.9	1.87	164.7	1.82	206.3	1.78	237.6	1.72
M-n200-k17	201.5	1.76	156.7	1.74	194.9	1.69	234.3	1.65
P-n70-k10	2.82	2.14	3.04	1.98	2.98	1.94	3.67	1.94
Avg. values:	10.24	2.46	10.70	2.35	11.89	2.31	14.60	2.30

setting sufficiently large ng-sets seems to be a very practical way of pricing elementary routes.

- The last comments are not valid for the instances with many customers per route. For example, it is not practical to use large ng-sets on instances F-n135-k7 $(n/|K| = 19.1)$ and M-n121-k7 $(n/|K| = 17.1)$.

As mentioned before, the bounds obtained by using VRP4, even with elementary routes, are not good enough to be the basis of an efficient CVRP algorithm. Tables 3.3 and 3.4 present times and bounds for VRP4 with the addition of rounded RCCs and comb cuts separated using the heuristics in the CVRPSEP library (see Lysgaard [28]). Those robust cuts typically eliminate more than half of the gaps, which are now small enough to

Bounds for the set partitioning formulation with robust cuts

Table 3.3. *Set partitioning formulation - RCCs + comb cuts - k-cycle elimination.*

Instance			2-cycle		3-cycle		4-cycle		5-cycle	
Name	n/k	OPT	T(s)	gap	T(s)	gap	T(s)	gap	T(s)	gap
A-n62-k8	7.6	1288	11.13	1.10	11.56	0.60	24.69	0.55	111.8	0.55
A-n80-k10	7.9	1763	19.27	0.75	16.68	0.52	47.79	0.43	157.8	0.42
B-n50-k8	6.1	1312	4.52	1.54	6.13	1.30	8.61	0.76	37.32	0.63
B-n78-k10	7.7	1221	17.94	0.69	16.71	0.56	26.02	0.43	171.4	0.35
E-n76-k7	10.7	682	13.52	1.77	17.68	1.73	44.87	1.69	500.2	1.68
E-n76-k8	9.4	735	17.08	1.21	21.96	1.16	39.06	1.15	263.2	1.13
E-n76-k10	7.5	830	7.20	1.56	6.43	1.59	14.00	1.53	104.7	1.52
E-n76-k14	5.4	1021	2.64	1.59	3.46	1.42	6.49	1.33	20.48	1.32
E-n101-k8	12.5	815	53.74	1.38	81.18	1.35	142.1	1.33	827.9	1.29
E-n101-k14	7.1	1067	7.12	1.44	13.19	1.26	24.50	1.24	112.3	1.23
F-n135-k7	19.1	1162	5203	0.21	-	-	-	-	-	-
M-n121-k7	17.1	1034	535.1	0.32	483.0	0.25	935.6	0.21	3302	0.21
M-n151-k12	12.5	1015	125.0	1.72	180.1	1.57	357.7	1.52	1878	1.45
M-n200-k16	12.5	1274	489.0	2.08	592.4	1.73	860.7	1.70	3369	1.67
M-n200-k17	11.8	1275	293.7	1.97	359.2	1.62	559.0	1.58	2358	1.53
P-n70-k10	6.9	827	3.76	1.67	4.26	1.53	11.55	1.48	83.31	1.44
Avg. values:			20.26	1.39	24.61	1.21	47.43	1.13	249.8	1.09

Table 3.4. *Set partitioning formulation - RCCs + comb cuts - ng-routes (ng-64 is equivalent to elementary route pricing in all those instances).*

Instance	ng-8		ng-16		ng-32		ng-64	
Name	T(s)	gap	T(s)	gap	T(s)	gap	T(s)	gap
A-n62-k8	6.74	0.56	7.63	0.46	9.41	0.46	8.21	0.47
A-n80-k10	10.42	0.37	10.24	0.36	10.76	0.35	15.78	0.34
B-n50-k8	3.01	0.65	2.38	0.64	2.67	0.63	3.61	0.63
B-n78-k10	9.54	0.32	9.92	0.30	12.82	0.31	16.01	0.30
E-n76-k7	14.18	1.60	15.05	1.57	15.51	1.48	21.33	1.48
E-n76-k8	12.59	1.15	12.69	1.08	14.53	1.07	16.29	1.05
E-n76-k10	5.22	1.53	6.36	1.52	6.55	1.45	7.61	1.44
E-n76-k14	3.23	1.38	3.73	1.32	3.37	1.32	3.63	1.32
E-n101-k8	47.37	1.29	50.39	1.24	49.31	1.22	61.53	1.23
E-n101-k14	10.01	1.25	9.35	1.21	11.71	1.12	12.30	1.12
F-n135-k7	-	-	-	-	-	-	-	-
M-n121-k7	108.2	0.15	1483	0.13	-	-	-	-
M-n151-k12	107.7	1.44	96.89	1.32	111.7	1.26	131.8	1.23
M-n200-k16	189.5	1.69	189.6	1.66	221.9	1.62	275.2	1.54
M-n200-k17	176.2	1.56	196.6	1.54	209.4	1.51	264.7	1.47
P-n70-k10	4.00	1.53	4.54	1.41	4.35	1.42	6.29	1.40
Avg. values:	15.38	1.10	15.88	1.05	17.35	1.02	21.02	1.01

provide efficient algorithms. Again, results with k-cycle elimination were obtained by the code from Fukasawa et al. [19], and results with ng-paths came from Martinelli, Pecin, and Poggi [30]. While the first algorithm only separates a round of cuts after column generation with exact pricing converges, the more recent algorithm found advantage in trying to separate cuts as soon as the heuristic pricing fails, reducing the number of calls to the more expensive exact pricing. This provides a partial explanation for the fact that the times for the 2-cycle elimination variant are comparable with the times for the much stronger ng-64 variant. Anyway, the previous comments for the formulation VRP4 without cuts are still valid. On most instances, ng-8 obtains bounds comparable to 5-cycle elimination in less time; ng-16, ng-32, and ng-64 are not much slower than ng-8. Moreover, ng-64 is still equivalent to elementary route pricing.

Bounds for the set partitioning formulation with SRCs

Table 3.5. *Set partitioning formulation · RCCs + lm-SRCs · ng-8.*

Instance			1SRC and 3SRCs						+4,5SRCs	
Name	n/k	OPT	T(s)	gap	1SRC	D1	3SRC	D3	T(s)	gap
A-n62-k8	7.6	1288	14.7	0.15	17	8.1	86	10.8	25.6	0.05
A-n80-k10	7.9	1763	16.0	0.09	23	7.3	109	10.3	36.6	0.01
B-n50-k8	6.1	1312	13.2	0.21	22	5.8	53	8.6	31.0	0.17
B-n78-k10	7.7	1221	6.9	0.00	3	5.7	62	6.9	7.5	0.00
E-n76-k7	10.7	682	102.9	0.16	33	11.4	193	13.5	224.0	0.05
E-n76-k8	9.4	735	38.5	0.12	39	9.9	169	11.3	45.3	0.04
E-n76-k10	7.5	830	22.9	0.43	19	5.8	143	8.8	68.6	0.19
E-n76-k14	5.4	1021	6.0	0.62	14	4.5	93	6.1	15.2	0.46
E-n101-k8	12.5	815	155.8	0.00	29	11.7	205	11.2	147.9	0.00
E-n101-k14	7.1	1067	40.1	0.35	35	6.7	175	7.9	120.8	0.27
F-n135-k7	19.1	1162	-	-	-	-	-	-	-	-
M-n121-k7	17.1	1034	82.9	0.05	25	15.4	47	26.0	228.2	0.05
M-n151-k12	12.5	1015	636.6	0.28	78	11.0	340	12.5	996.3	0.14
M-n200-k16	12.5	1274	2791	0.56	122	9.5	546	13.0	8400	0.42
M-n200-k17	11.8	1275	2405	0.45	92	9.6	537	13.0	7903	0.31
P-n70-k10	6.9	827	12.6	0.47	17	6.0	135	8.3	39.1	0.28
Avg. values:			57.1	0.28					118.3	0.17

Table 3.5 is intended to show the large supplementary gap reductions that can be obtained by non-robust SRCs. The results were obtained with the algorithm in Pecin et al. [32] on a single core of an i7-2620M 3.3GHz processor, pricing ng-routes with ng-sets of size 8. The first set of columns corresponds to the separation of RCCs, lm-1SRCs, and lm-3SRCs. As the separation is performed until convergence, the final lower bounds are the same that would be obtained by pricing elementary routes (the lm-1SRCs remove all cycles) and separating RCCs plus traditional 3SRCs. The table also shows additional information for that variant of the algorithm: the number of active lm-1SRCs, their densities (D1) (the average size of the corresponding M sets), the number of active lm-3SRCs, and also their densities (D3). Comparing with Table 3.4, it can be verified that the 3SRCs alone reduced the average gap from 1.01 to 0.28. The relatively small densities of the lm-SRCs explain how the pricing remains tractable even on the instances with 200 vertices. The last two columns in Table 3.5 show results for a variant that also separates lm-4SRCs, lm-5,1SRCs, and lm-5,2SRCs. The average gap decreased to 0.17. We remark that it is not a good BCP strategy to separate lm-SRCs until convergence; it is possible to obtain slightly larger gaps in a fraction of the times reported in Table 3.5.

The final set of results shown in this chapter is aimed at showing the evolution of the recent exact approaches for the CVRP. Tables 3.6, 3.7, and 3.8 summarize their performances over the set of 81 instances, divided by class, used in Fukasawa et al. [19]. As usual in the recent literature where similar tables appear, classes E and M are grouped together. Columns **Opt** indicate the number of instances solved to optimality. Columns **Gap** and **Time** are the average gap in the root node and average times in seconds (over the indicated machines), computed only over the solved instances. Some comments and additional information about those results follow:

- **LLE04** refers to the Branch-and-Cut in Lysgaard, Letchford, and Eglese [29]. That code represents the culmination of a sequence of Branch-and-Cut algorithms (started by Augerat et al. [1]), with increasingly better separation of valid cuts for the formulation VRP2. Analyzing its results in more detail, it can be noticed that the algorithm had particular difficulties solving instances with fewer customers per

Overall results of the more recent exact algorithms

Table 3.6. *Summary of results: LLE04, FLL+06, BCM08.*

Class	NP	LLE04			FLL+06			BCM08		
		Opt	Gap	Time	Opt	Gap	Time	Opt	Gap	Time
A	22	15	2.06	6638	22	0.81	1961	22	0.20	118
B	20	19	0.61	8178	20	0.47	4763	20	0.16	417
E-M	12	3	2.10	39592	9	1.19	126987	8	0.69	1025
F	3	3	0.06	1016	3	0.14	2398			
P	24	16	2.26	11219	24	0.76	2892	22	0.28	187
Total	81	56			78			72		
Machine		Intel Celeron 700MHz			Pentium 4 2.4GHz			Pentium 4 2.6GHz		

Table 3.7. *Summary of results: BMR11, Con12, CM14.*

Class	NP	BMR11			Con12			CM14		
		Opt	Gap	Time	Opt	Gap	Time	Opt	Gap	Time
A	22	22	0.13	30	22	0.07	59	22	0.09	59
B	20	20	0.06	67	20	0.05	89	20	0.08	34
E-M	12	9	0.49	303	10	0.30	2807	10	0.27	1548
F	3	2	0.11	164	2	0.06	3	3	0.03	27722
P	24	24	0.23	85	24	0.13	43	17	0.18	240
Total	81	77			78			79		
Machine		Xeon X7350 2.93GHz			Xeon E5462 2.8GHz			Xeon E5462 2.8GHz		

Table 3.8. *Summary of results: Rop12, PPPU14.*

Class	NP	Rop12			PPPU14		
		Opt	Gap	Time	Opt	Gap	Time
A	22	22	0.57	53	22	0.03	5.6
B	20	20	0.25	208	20	0.04	6.2
E-M	12	10	0.96	44295	12	0.19	3669
F	3	3	0.25	2163	3	0.00	3679
P	24	24	0.69	280	24	0.07	33
Total	81	79			81		
Machine		Core i7-2620M 2.7GHz			Core i7-2620M 3.3GHz		

vehicle; their root gaps were too large. For example, instances E-n76-k7 and E-n76-k8 could be solved, but E-n76-k10 and E-n76-k14 could not. On the other hand, the Branch-and-Cut had little trouble solving instance F-n135-k7 (already solved in [1]).

- **FLL+06** refers to the BCP in Fukasawa et al. [19]. As shown in Table 3.6, the combination of column and cut generation reduced a lot the root gaps with respect to those by LLE04. The improvement was more significant precisely in the instances with fewer customers per vehicle. Overall, only the three larger instances in the M series could not be solved to optimality. It should be remarked that in some instances (like F-n135-k7), when the algorithm realizes that the column generation in the root node is too slow, it switches to a Branch-and-Cut similar to LLE04.

- **BCM08** refers to the column and cut generation algorithm in Baldacci, Christofides, and Mingozzi [3]. It can be seen that the introduction of SCCs and Clique Cuts (the latter were particularly effective) led to significantly reduced gaps with respect to those in FLL+06. Moreover, the route enumeration dramatically reduced the overall running times for many instances. Those effects were especially acute on the instances with fewer customers per vehicle. For example, on instance E-n101-k14, the root node of FLL+06 yielded the lower bound of 1053.8, and the BCP completed after exploring 5848 nodes, taking 116,284 seconds. In the same instance, BCM08 obtained a bound of 1064.3 and proceeded to enumerate 76,756 routes (using an upper bound of 1071). The resulting restricted VRP4 model is then sent to a general MIP solver. The overall solving time was only 1230 seconds. On the other hand, the more recent algorithm had difficulties solving some instances with many customers per route. For example, instance E-n101-k8 could not be solved.

- **BMR11** refers to the column and cut generation algorithm in Baldacci, Mingozzi, and Roberti [4]. It can be seen that newly introduced elements improved the gaps and the running times significantly with respect to BCM08. More importantly, the algorithm became much more stable, being able to quickly solve some instances with many customers per route. For example, instance E-n101-k8 took 579 seconds. The only previously solved instance that could not be solved was F-n135-k7.

- **Con12** refers to the column and cut generation algorithm in Contardo [12]. The algorithm follows the path paved by BCM08 and BMR11, combining non-robust cut pricing and route enumeration. However, Con12 sticks to an easier base pricing problem, 2-cycle elimination, while non-robust cuts (including those to enforce partial elementarity) are carefully added. It seems that this approach leaves more room for introducing a larger number of 3SRCs; those cuts are particularly effective on reducing the gaps. The solution of instance M-n151-k12 in 19,699 seconds is an indication of its strength.

- **CM14** refers to the column and cut generation algorithm in Contardo and Martinelli [14]. The algorithm contains a number of improvements over Con12; ng-routes are priced instead of q-routes with 2-cycle elimination. Instance M-n151-k12 could be solved in less than 3 hours. Also, an improved pricing algorithm implementation allowed solving instance F-n135-k7 by column generation for the first time. It took just over 83,000 seconds. It should be remarked that the algorithm was run only on 17 of the 24 instances of series P. As the seven ignored instances are very small (up to 23 vertices) and very easy, we assumed zero gaps and zero running times for computing statistics in Table 3.7.

- **Rop12** refers to the BCP algorithm in Røpke [39]. This algorithm is closer to the BCP in Fukasawa et al. [19]; its root gaps (ng-10 pricing + CVRPSep cuts) are significantly worse than those by BCM08, BMR11, Con12, and CM14. On the other hand, the robust approach allows the fast solution of each node, which combined with an aggressive and effective strong branching yielded results that are comparable with the results of those algorithms. The run that solved instance M-n151-k12 used ng-15 pricing and took 5268 nodes and 417,146 seconds to be finished. The solution of instance F-n135-k7 also required an special parameterization; instead of ng-routes, faster 2-cycle elimination was used.

- **PPPU14** refers to the BCP algorithm in Pecin et al. [32]. This algorithm uses ng-8 pricing and separates RCCs and lm-SRCs (1SRCs, 3SRCs, 4SRCs, and 5SRCs). After enumeration it can also separate Clique Cuts. Its algorithmic engineering combines elements found in many previous works. Instance M-n151-k12 was solved in 212 seconds. Instance M-n200-k17 was solved in 3581 seconds, while M-n200-k16 took 39,869 seconds. The larger time spent in M-n200-k16 is explained by the relatively poor quality of the best known upper bound for that instance, 4 units above the optimal 1274. Instance F-n135-k7 was solved by the BCP in 5405 seconds. Four larger instances, ranging from 240 to 360 customers, from the set proposed in Golden et al. [21] were solved too.

Currently, LLE04 and PPPU14 are the non-dominated exact CVRP algorithms available, in the sense that there are instances from the literature that are better solved by one of them. Unhappily, the classical benchmark used in the CVRP has only three instances in the range from 150 to 200 vertices. However, extensive experiments with a new benchmark set (Uchoa et al. [42]) support the statement that current algorithms can indeed solve instances with up to 200 vertices in a consistent way. Furthermore, many instances, those with characteristics that are more favorable to BCP algorithms (like an $n/|K|$ ratio smaller than 8), are likely to be solved with up to 300 vertices.

3.8 ▪ Conclusions and Future Research Directions

The progress made in the last decade on exact CVRP algorithms was very expressive. Measuring it by the size where instances can be consistently solved, we observe an increase from 50 to 200 customers between Lysgaard, Letchford, and Eglese [29] and Pecin et al. [32]. The last algorithm was the result of a deliberate effort of testing, improving, and combining ideas proposed by several authors. Some of the newer elements found in CVRP algorithms are likely to be useful on algorithms for solving other VRP variants. For example, the lm-SRCs could be used in any variant that can be modeled by the set partitioning formulation VRP4.

Bibliography

[1] P. AUGERAT, J. BELENGUER, E. BENAVENT, A. CORBERÁN, D. NADDEF, AND G. RINALDI, *Computational results with a branch and cut code for the capacitated vehicle routing problem*, Technical Report 949-M, Université Joseph Fourier, Grenoble, France, 1995.

[2] R. BALDACCI, L. BODIN, AND A. MINGOZZI, *The multiple disposal facilities and multiple inventory locations rollon–rolloff vehicle routing problem*, Computers & Operations Research, 33 (2006), pp. 2667–2702.

[3] R. BALDACCI, N. CHRISTOFIDES, AND A. MINGOZZI, *An exact algorithm for the vehicle routing problem based on the set partitioning formulation with additional cuts*, Mathematical Programming, 115 (2008), pp. 351–385.

[4] R. BALDACCI, A. MINGOZZI, AND R. ROBERTI, *New route relaxation and pricing strategies for the vehicle routing problem*, Operations Research, 59 (2011), pp. 1269–1283.

[5] M. L. BALINSKI AND R. QUANDT, *On an integer program for a delivery problem*, Operations Research, 12 (1964), pp. 300–304.

[6] N. BOLAND, J. DETHRIDGE, AND I. DUMITRESCU, *Accelerated label setting algorithms for the elementary resource constrained shortest path problem*, Operations Research Letters, 34 (2006), pp. 58–68.

[7] A. CHABRIER, *Vehicle routing problem with elementary shortest path based column generation*, Computers & Operations Research, 33 (2006), pp. 2972–2990.

[8] N. CHRISTOFIDES AND S. EILON, *An algorithm for the vehicle-dispatching problem*, Operational Research Quarterly, 20 (1969), pp. 309–318.

[9] N. CHRISTOFIDES, A. MINGOZZI, AND P. TOTH, *The vehicle routing problem*, in Combinatorial Optimization, N. Christofides, A. Mingozzi, P. Toth, and C. Sandi, eds., vol. 1, Wiley Interscience, Chichester, 1979, pp. 315–338.

[10] ———, *Exact algorithms for the vehicle routing problem, based on spanning tree and shortest path relaxations*, Mathematical Programming, 20 (1981), pp. 255–282.

[11] ———, *State-space relaxation procedures for the computation of bounds to routing problems*, Networks, 11 (1981), pp. 145–164.

[12] C. CONTARDO, *A new exact algorithm for the multi-depot vehicle routing problem under capacity and route length constraints*, Technical Report, Archipel-UQAM 5078, Université du Québec à Montréal, Canada, 2012.

[13] C. CONTARDO, J.-F. CORDEAU, AND B. GENDRON, *A branch-and-cut-and-price algorithm for the capacitated location-routing problem*, Technical report, CIRRELT-2011-44, Université de Montréal, Canada, 2011.

[14] C. CONTARDO AND R. MARTINELLI, *A new exact algorithm for the multi-depot vehicle routing problem under capacity and route length constraints*, Discrete Optimization, 12 (2014), pp. 129–146.

[15] J. DESROSIERS, F. SOUMIS, AND M. DESROCHERS, *Routing with time windows by column generation*, Networks, 14 (1984), pp. 545–565.

[16] O. DU MERLE, D. VILLENEUVE, J. DESROSIERS, AND P. HANSEN, *Stabilized column generation*, Discrete Mathematics, 194 (1999), pp. 229–237.

[17] D. FEILLET, P. DEJAX, M. GENDREAU, AND C. GUEGUEN, *An exact algorithm for the elementary shortest path problem with resource constraints: Application to some vehicle routing problems*, Networks, 44 (2004), pp. 216–229.

[18] M. L. FISHER, *Optimal solution of vehicle routing problems using minimum k-trees*, Operations Research, 42 (1994), pp. 626–642.

[19] R. FUKASAWA, H. LONGO, J. LYSGAARD, M. POGGI DE ARAGÃO, M. REIS, E. UCHOA, AND R. F. WERNECK, *Robust branch-and-cut-and-price for the capacitated vehicle routing problem*, Mathematical Programming, 106 (2006), pp. 491–511.

[20] M. T. GODINHO, L. GOUVEIA, T. MAGNANTI, P. PESNEAU, AND J. PIRES, *On time-dependent models for unit demand vehicle routing problems*, in International Network Optimization Conference (INOC), 2007.

[21] B. GOLDEN, E. WASIL, J. KELLY, AND I.-M. CHAO, *The impact of metaheuristics on solving the vehicle routing problem: Algorithms, problem sets, and computational results*, in Fleet Management and Logistics, Springer, Berlin, 1998, pp. 33–56.

[22] E. HOSHINO AND C. DE SOUZA, *Column generation algorithms for the capacitated m-ring-star problem*, in Computing and Combinatorics, Springer, Berlin, 2008, pp. 631–641.

[23] S. IRNICH, G. DESAULNIERS, J. DESROSIERS, AND A. HADJAR, *Path-reduced costs for eliminating arcs in routing and scheduling*, INFORMS Journal on Computing, 22 (2010), pp. 297–313.

[24] S. IRNICH AND D. VILLENEUVE, *The shortest-path problem with resource constraints and k-cycle elimination for $k \geq 3$*, INFORMS Journal on Computing, 18 (2006), pp. 391–406.

[25] M. JEPSEN, B. PETERSEN, S. SPOORENDONK, AND D. PISINGER, *Subset-row inequalities applied to the vehicle-routing problem with time windows*, Operations Research, 56 (2008), pp. 497–511.

[26] G. LAPORTE AND Y. NOBERT, *A branch and bound algorithm for the capacitated vehicle routing problem*, OR Spectrum, 5 (1983), pp. 77–85.

[27] A. LETCHFORD AND J. J. SALAZAR GONZÁLEZ, *Projection results for vehicle routing*, Mathematical Programming, 105 (2006), pp. 251–274.

[28] J. LYSGAARD, *CVRPSEP: A package of separation routines for the capacitated vehicle routing problem*, Aarhus School of Business, Department of Management Science and Logistics, Aarhus, Denmark, 2003.

[29] J. LYSGAARD, A. LETCHFORD, AND R. EGLESE, *A new branch-and-cut algorithm for the capacitated vehicle routing problem*, Mathematical Programming, 100 (2004), pp. 423–445.

[30] R. MARTINELLI, D. PECIN, AND M. POGGI, *Efficient restricted non-elementary route pricing for routing problems*, Technical Report MCC 11/13, PUC-Rio, Departamento de Informática, Rio de Janeiro, Brasil, 2013.

[31] D. NADDEF AND G. RINALDI, *Branch-and-cut algorithms for the capacitated VRP*, in The Vehicle Routing Problem, P. Toth and D. Vigo, eds., SIAM, 2002, ch. 3, pp. 53–84.

[32] D. PECIN, A. PESSOA, M. POGGI, AND E. UCHOA, *Improved branch-and-cut-and-price for capacitated vehicle routing*, in Integer Programming and Combinatorial Optimization, vol. 8494 of Lecture Notes in Computer Science, Springer, Berlin, 2014, pp. 393–403.

[33] A. PESSOA, M. POGGI DE ARAGÃO, AND E. UCHOA, *Robust branch-cut-and-price algorithms for vehicle routing problems*, in the Vehicle Routing Problem: Latest Advances and New Challenges, B. Golden, S. Raghavan, and E. Wasil, eds., Springer, New York, 2008, pp. 297–325.

[34] A. PESSOA, R. SADYKOV, E. UCHOA, AND F. VANDERBECK, *In-out separation and column generation stabilization by dual price smoothing*, in Symposium on Experimental Algorithms, Springer, Berlin, 2013, pp. 354–365.

[35] A. PESSOA, E. UCHOA, AND M. POGGI DE ARAGÃO, *A robust branch-cut-and-price algorithm for the heterogeneous fleet vehicle routing problem*, Networks, 54 (2009), pp. 167–177.

[36] M. POGGI DE ARAGÃO AND E. UCHOA, *Integer program reformulation for robust branch-and-cut-and-price*, in Annals of Mathematical Programming in Rio, L. Wolsey, ed., Búzios, Brazil, 2003, pp. 56–61.

[37] G. RIGHINI AND M. SALANI, *Symmetry helps: Bounded bi-directional dynamic programming for the elementary shortest path problem with resource constraints*, Discrete Optimization, 3 (2006), pp. 255–273.

[38] ——, *New dynamic programming algorithms for the resource constrained elementary shortest path problem*, Networks, 51 (2008), pp. 155–170.

[39] S. RØPKE, *Branching decisions in branch-and-cut-and-price algorithms for vehicle routing problems*, Presentation in Column Generation 2012, 2012.

[40] A. SUBRAMANIAN, E. UCHOA, A. PESSOA, AND L. S. OCHI, *Branch-cut-and-price for the vehicle routing problem with simultaneous pickup and delivery*, Optimization Letters, 7 (2013), pp. 1569–1581.

[41] E. UCHOA, *Cuts over extended formulations by flow discretization*, in Progress in Combinatorial Optimization, A. Mahjoub, ed., ISTE-Wiley, London, 2011, pp. 255–282.

[42] E. UCHOA, D. PECIN, A. PESSOA, M. POGGI, A. SUBRAMANIAN, AND T. VIDAL, *New benchmark instances for the capacitated vehicle routing problem*, Presentation in Route 2014, 2014.

Chapter 4

Heuristics for the Vehicle Routing Problem

Gilbert Laporte
Stefan Ropke
Thibaut Vidal

4.1 ▪ Introduction

In recent years, several sophisticated mathematical programming decomposition algorithms have been put forward for the solution of the VRP. Yet, despite this effort, only relatively small instances involving around 100 customers can be solved optimally, and the variance of computing times is high. However, instances encountered in real-life settings are sometimes large and must be solved quickly within predictable times, which means that efficient heuristics are required in practice. Also, because the exact problem definition varies from one setting to another, it becomes necessary to develop heuristics that are sufficiently flexible to handle a variety of objectives and side constraints. These concerns are clearly reflected in the algorithms developed over the past few years. This chapter provides an overview of heuristics for the VRP, with an emphasis on recent results.

The history of VRP heuristics is as old as the problem itself. In their seminal paper, Dantzig and Ramser [19] sketched a simple heuristic based on successive matchings of vertices through the solution of linear programs and the elimination of fractional solutions by trial and error. The method was illustrated on an eight-vertex graph. It was not pursued, but may have inspired the developers of matching-based heuristics (see Altinkemer and Gavish [3], Desrochers and Verhoog [20], and Wark and Holt [91]). Since then, a wide variety of *constructive* and *improvement heuristics* have been proposed, culminating in recent years with the development of powerful *metaheuristics* capable of computing within a few seconds solutions whose value lies within less than one percent of the best known values.

The field of VRP heuristics is now so rich that it makes no sense to provide an exhaustive compilation of them in a book chapter such as this. Instead, we have decided to focus on methods and principles that have withstood the test of time or present some interesting distinctive features. For a more complete description of the classical heuristics and of the early metaheuristics, we refer the reader to the two chapters by Laporte and

Semet [46] and by Gendreau, Laporte, and Potvin [25] in the first edition of the Toth and Vigo [83] book.

The evolution of VRP heuristics over the past 10 years has taken place, almost exclusively, within the context of metaheuristics. The concept that best encapsulates this evolution is probably that of hybridization. First, we have witnessed the emergence of new heuristics combining several concepts initially developed independently from each other, often related to search principles such as simulated annealing, tabu search, variable neighborhood search, and genetic algorithms. Second, various other strategies, exotic large neighborhoods, exact mathematical techniques, decomposition, and cooperation schemes have found their way into the most successful methods. Third, there has been a hybridization of scope in the sense that researchers are now developing *flexible* methods (e.g., Cordeau, Laporte, and Mercier [15], Vidal et al. [86], and Subramanian, Uchoa, and Ochi [76]) which can be directly applied to the solution of a wide variety of VRP variants without any major structural change.

We will first summarize in Sections 4.2, 4.3, and 4.4 some of the relevant results pertaining to constructive heuristics, improvement heuristics, and metaheuristics for the classical VRP, i.e., the version of the problem containing capacity and route length restrictions only. We will then review and analyze a number of hybridization strategies in Section 4.5, and we will discuss unified algorithms in Section 4.6. Computational comparisons of some of the heuristics surveyed in this chapter are presented Section 4.7. Conclusions follow in Section 4.8.

4.2 ▪ Constructive Heuristics

Constructive heuristics are usually employed to provide a starting solution to an improvement heuristic. However, most metaheuristics are now so robust that they can be initialized from any random solution, feasible or not. Several such heuristics have been proposed over the years and are fully described in the first edition of this book (Toth and Vigo [83]). However, since most of these heuristics have now fallen into disuse, we will concentrate on two classical methods which still present a particular interest.

4.2.1 ▪ The Clarke and Wright Savings Heuristic

The Clarke and Wright heuristic [12] initially constructs back and forth routes $(0, i, 0)$ for $(i = 1, \ldots, n)$ and gradually merges them by applying a saving criterion. More specifically, merging the two routes $(0, \ldots, i, 0)$ and $(0, j, \ldots, 0)$ into a single route $(0, \ldots, i, j, \ldots, 0)$ generates a saving $s_{ij} = c_{i0} + c_{0j} - c_{ij}$. Since the savings remain the same throughout the algorithm, they can be computed a priori. In the so-called parallel version of the algorithm which appears to be the best (see Laporte and Semet [46]), the feasible route merger yielding the largest saving is implemented at each iteration, until no more merger is feasible. This simple algorithm possesses the advantages of being intuitive, easy to implement, and fast. It is often used to generate an initial solution in more sophisticated algorithms. Several enhancements and acceleration procedures have been proposed for this algorithm (see, e.g., Nelson et al. [59] and Paessens [62]), but given the speed of today's computers and the robustness of the latest metaheuristics, these no longer seem justified.

4.2.2 ▪ Petal Algorithms

Petal algorithms consist of generating a set S of feasible VRP routes and combining them through the solution of a set partitioning problem. More specifically, let d_k be the cost of route k, let a_{ik} be a binary coefficient equal to 1 if and only if customer i belongs to route k, and let x_k be a binary variable equal to 1 if and only if route k belongs to the solution. Then the problem is to

$$\text{minimize} \sum_{k \in S} d_k x_k$$

$$\text{s.t.} \sum_{k \in S} a_{ik} x_k = 1 \qquad \forall\, i = 1, \dots, n,$$

$$x_k \in \{0, 1\} \qquad \forall\, k \in S.$$

The idea of solving the VRP through a set partitioning formulation was first proposed as an exact algorithm by Balinski and Quandt [5], but proved impractical because the number of feasible routes is typically extremely large and computing the costs d_k usually requires the solution of an NP-hard problem. The problem is normally solved heuristically by generating a set of promising routes, as suggested by Foster and Ryan [23] and by Ryan, Hjorring, and Glover [72]. Renaud, Boctor, and Laporte [67] go one step further and also generate configurations called 2-petals, consisting of two embedded or intersecting routes. On the 14 benchmark instances of Christofides, Mingozzi, and Toth [11], these authors quickly obtained solution with costs lying within 2.38% of those of the best known solutions.

Petal algorithms are also particularly well suited for problems containing constraints, such as time windows, other than capacity and route duration constraints. Column generation then becomes a solution methodology of choice, especially for tightly constrained instances.

4.3 ▪ Classical Improvement Heuristics

Classical improvement heuristics perform intra-route and inter-route moves. In the first case, one can apply any improvement heuristic designed for the *Traveling Salesman Problem* (TSP), such as λ-OPT exchanges (Lin [49]), in which λ edges are removed and replaced by λ other edges. In the implementation of Lin and Kernighan [50], the value of λ is modified dynamically throughout the search. According to Johnson and McGeoch [43], who have conducted a detailed empirical analysis of several TSP heuristics, a careful implementation of the Lin and Kernighan algorithm yields the best results. One such implementation was proposed by Helsgaun [36] and is now incorporated in the Concorde TSP solver of Applegate et al. [4].

In practice, inter-route improvement moves are essential to achieve good results. These include classical operators such as removing k consecutive customers from their current route and reinserting them elsewhere (RELOCATE), swapping consecutive customers between different routes (SWAP), or removing two edges from different routes and reconnecting them differently (2-OPT*). These three types of moves are the most commonly used in recent state-of-the-art methods, along with 2-OPT. The number of exchanged customers usually remains small, no more than two in most cases.

A complete exploration of standard neighborhoods requires $O(n^2 k^2)$ operations. For large instances, it becomes necessary to reduce the list of considered moves. One such pruning technique, called granular search (see Johnson and McGeoch [42] and Toth and Vigo [84]), considers geographical restrictions to avoid moves between distant customers. Specific move evaluation orders and restrictions, considering the partial sums of length of exchanged edges as in Irnich, Funke, and Grünert [40], can also lead to important search effort reductions.

Most classical moves are special cases of so-called b-cyclic, k-transfer moves (Thompson and Psaraftis [81]) in which a circular permutation of b routes is considered and k customers from each route are shifted to the next route of the cyclic permutation. Interroute improvement moves can also be viewed as operators within destroy and repair schemes (see Shaw [75]) which alternate between moves that destroy part of the solution and moves that reconstruct it. In the *Adaptive Large Neighborhood Search* (ALNS) heuristic of Pisinger and Ropke [63], these moves are randomly selected by means of a roulette wheel mechanism, yielding excellent results.

4.4 ▪ Metaheuristics

Current metaheuristics for the VRP can broadly be classified into local search methods and population-based heuristics. Local search methods explore the solution space by moving at each iteration from a solution to another solution in its neighborhood. These include well-known schemes such as simulated annealing (see Kirkpatrick, Gelatt, and Vecchi [44] and Nikolaev and Jacobson [60]), deterministic annealing (see Dueck [21], Dueck and Scheuer [22], and Li, Golden, and Wasil [48]), tabu search (see Glover [29] and Gendreau and Potvin [27]), iterated local search (see Baxter [7] and Lourenço, Martin, and Stützle [51]), and variable neighborhood search (see Mladenović and Hansen [54]). Population-based heuristics evolve a population of solutions which may be combined together in the hope of generating better ones. This category includes ant colony optimization (see Reimann, Doerner, and Hartl [66]), genetic algorithms (see Holland [38] and Prins [64]), scatter search, and path relinking (see Glover [28] and Resende et al. [69]).

At the time of the publication of the first edition of this book (Toth and Vigo [83]), these metaheuristics tended to be rather distinct from each other. However, more than 10 years later, the frontiers between them have become more fuzzy and, more importantly, several powerful hybrids of these algorithms have emerged. Therefore, rather than providing a detailed account of every available implementation, a task which is now next to impossible given their number, we will concentrate on a few principles of general applicability. Readers interested in an overview of metaheuristic principles are referred to the handbook edited by Gendreau and Potvin [26].

4.4.1 ▪ Local Search Algorithms

Local search algorithms start from an initial solution x_1 and move at each iteration t from the current solution x_t to another solution x_{t+1} in its neighborhood $N(x_t)$. If $f(x)$ denotes the cost of x, then $f(x_{t+1})$ is not necessarily less than $f(x_t)$. Care must therefore be taken to avoid cycling.

Simulated Annealing (SA).   In SA, cycling is prevented by selecting a solution x randomly in $N(x_t)$. If $f(x) \leq f(x_t)$, then $x_{t+1} = x$. Otherwise,

$$x_{t+1} = \begin{cases} x & \text{with probability } p_t, \\ x_t & \text{with probability } 1 - p_t, \end{cases}$$

where p_t is usually a decreasing function of t and of $f(x) - f(x_t)$. It is common to define p_t as

$$p_t = \exp(-[f(x) - f(x_t)]/\theta_t),$$

where the *temperature* θ_t is a decreasing function of t. A well-known application of SA to the VRP is that of Osman [61].

Deterministic Annealing (DA). In DA, the rule for accepting x is deterministic. One of the best implementations of this scheme is the *Record-to-Record Travel* (RRT) algorithm of Li, Golden, and Wasil [48] which is based on the general idea put forward by Dueck [21]. In this algorithm, a *record* is the best known solution x^*. At iteration t, a solution x is drawn from $N(x_t)$ and $x_{t+1} = x$ if $f(x) \le \sigma f(x^*)$, where σ is a parameter usually slightly larger than 1 (e.g., $\sigma = 1.05$); if $f(x) > \sigma f(x^*)$, then $x_{t+1} = x_t$. The implementations proposed by Groër, Golden, and Wasil [31] and Li, Golden, and Wasil [48] are rather effective while being easy to reproduce.

Tabu Search (TS). TS moves from a solution t to the best non-tabu solution x_{t+1} in $N(x_t)$. In order to avoid cycling, solutions sharing some attributes with x_t are declared *tabu*, or forbidden, for a number of iterations. The tabu status of a potential solution is revoked whenever it corresponds to a new best known solution. Tens of implementations of TS have been proposed in the past 20 years. These usually contain rules to diversify the search or to intensify it in promising regions. One implementation feature stands out among all those that have been proposed. It consists of considering intermediate infeasible solutions during the search, as in Gendreau, Hertz, and Laporte [24]. This is achieved by minimizing a penalized function $f'(x) = f(x) + \sum_k \alpha_k V_k(x)$, where $V_k(x)$ is the total violation of constraint of type k in solution x (for example, $V_k(x)$ could be the total violation of the capacity constraint over all routes), and α_k is a self-adjusting parameter. Of course, $f'(x)$ and $f(x)$ coincide whenever x is feasible. The weights α_k are initially set at 1 and are adjusted throughout the search. At every iteration in the more recent implementations, if x is feasible with respect to constraint k, then α_k is divided by $1 + \delta$, where $\delta > 0$; otherwise, α_k is multiplied by $1 + \delta$ (see, e.g., Cordeau, Laporte, and Mercier [15]). A recent tabu search algorithm that provides good results is described in Zachariadis and Kiranoudis [92].

Iterated Local Search (ILS). ILS is a local search heuristic whose origins can be traced back to the 1980s (e.g., Baxter [7]) and which has been designated under a variety of names (see Lourenço, Martin, and Stützle [51] for a historical account). The algorithm is simple and can be applied on top of any local search procedure, be it a simple one like steepest descent based on a single neighborhood or a more complicated one like TS. The idea is to apply an embedded local search mechanism until it reaches a stopping criterion, perturb (or shake) the solution returned by the embedded local search to yield a new starting solution, and then reapply the embedded local search. This continues until a stopping criterion is met (e.g., a number of outer iterations, a time limit, or a number of consecutive iterations without solution improvement). The perturbation operation is application-specific and must be designed with care, so that the perturbation is not easily undone by the embedded local search, but it should not completely destroy the structure of the original solution. Chen, Huang, and Dong [10] have designed an iterated variable neighborhood search descent heuristic for the *Capacitated VRP* (CVRP), which combines several operators and a perturbation strategy based on an exchange of a few consecutive customers. Subramanian, Uchoa, and Ochi [76] have later designed a hybrid metaheuristic combining the design of vehicle routes by means of an ILS heuristic and the solution of a set partitioning problem. Shaking is performed by means of combined exchanges of customers.

Variable Neighborhood Search (VNS). VNS was proposed by Mladenović and Hansen [54] as a general search strategy. It works with several neighborhoods $N_1, \ldots, N_p$ which are often of increasing complexity and can even be embedded, for example, 2-OPT, 3-OPT, etc. Starting with an initial solution the method iteratively applies these neighborhoods in a descent fashion until no further improvement is possible. After the last neighborhood has been applied, a new cycle can be restarted. The algorithm stops after a preset number of cycles or when no further improvement is possible. Several variants of this basic mechanism have been proposed, and the method has been applied to a wide variety of contexts (see Hansen et al. [33] for a survey). VNS was successfully applied to the VRP by Kytöjoki et al. [45]. This search scheme was used to guide the application of several VRP improvement procedures. The method was designed with the solution of large-scale real-life instances in mind. It was capable of quickly identifying high quality solutions on such instances involving up to 20,000 customers.

4.4.2 ▪ Population-Based Algorithms

Whereas the previously described local search algorithms have been inspired by the necessity of escaping from local optima and avoiding cycling, population-based methods take their inspiration from natural concepts, e.g., the evolution of species and the behavior of social insects foraging. These methods implement a high-level guidance strategy based on different memory structures, such as neural networks, pools of solutions represented as chromosomes, or pheromone matrices. Furthermore, all known successful VRP heuristics of this type also rely on local search components to drive the search towards promising solutions. Most of the population-based methods in the VRP literature are thus inherently hybrid.

Ant Colony Optimization (ACO). The ACO algorithm of Reimann, Doerner, and Hartl [66] has been one of the most successful. New solutions are generated by means of a savings-based procedure and local search. Instead of using the standard saving definition $s_{ij} = c_{i0} + c_{0j} - c_{ij}$ of the Clarke and Wright algorithm [12], the authors use an attractiveness value $\chi_{ij} = \tau_{ij}^{\alpha} - s_{ij}^{\beta}$, where the *pheromone value* τ_{ij} measures how good combining i and j turned out to be in previous iterations, and α, β are user-controlled parameters. The combination of vertices i and j takes place with probability $p_{ij} = \chi_{ij}/(\sum_{(h,\ell)\in\Omega_k} \chi_{h\ell})$, where Ω_k is the set of the feasible (i, j) combinations yielding the k best savings.

Genetic Algorithms (GA). The first successful application of GAs (see Holland [38]) to the VRP is due to Prins [64]. The method combines genetic operators, selection, and crossover with an efficient local search which replaces the classic randomized mutation operator. This type of hybrid method is sometimes called memetic algorithm (see Moscato and Cotta [55]). Prins [64] represents a solution as a giant tour without trip delimiters. An optimal shortest path-based procedure (see [8]), called SPLIT, is used for inserting visits to the depots and delimiting the routes. Genetic operators are applied on giant-tour solutions, thus allowing for simple permutation-based crossover operations, while local search procedures are applied on the complete solution representation after SPLIT. Diversity management procedures are exploited to control the population. A new solution s_{NEW} with objective value $z(s_{\text{NEW}})$ is accepted in the population $\mathscr{P}$ if and only if there exists no solution $s \in \mathscr{P}$ such that $|z(s_{\text{NEW}}) - z(s)| \leq \Delta$, where Δ is an objective-spacing coefficient.

Other efficient GAs for the CVRP have later been proposed. The heuristic of Nagata and Bräysy [57] relies on an adaptation of the *Edge-Assembly Crossover* (EAX), which has been successful on the TSP (see Nagata and Kobayashi [58]). This crossover considers the graph associated with the merger of the edges from two parents and selects several cycles in the graph, alternating between edges of parents P1 and P2. A new offspring is then created by removing the selected edges of P1 and replacing them with those of P2. The resulting graph is not always a legit VRP solution, since it may contain cycles that are not connected to the depot or routes exceeding the vehicle capacity. Thus, a repair procedure is applied, including a greedy heuristic to merge disconnected cycles and routes and an efficient local search to reduce capacity infeasibility. Combining this crossover with well-designed local search procedures, including efficient neighborhood pruning methods, proved highly successful.

The hybrid GA of Vidal et al. [85] "with advanced diversity control" builds upon the solution representation of Prins [64], and also integrates a bicriteria evaluation of individuals, driven by solution quality as well as their contribution to the population diversity. The fitness $\phi_{\mathscr{P}}(S)$ of a solution S in population $\mathscr{P}$ is a weighted sum of its rank in the population in terms of contribution to the population's diversity $\phi^{\mathrm{DIV}}(S)$, evaluated as a Hamming distance to the other solutions, and its rank with respect to the solution cost $\phi^{\mathrm{COST}}(S)$. The parameter μ^{ELITE} governs the relative weight of each criterion. This fitness measure is used to select both the parents and the survivors:

$$(4.1) \qquad \phi_{\mathscr{P}}(S) = \phi^{\mathrm{COST}}(S) + \left(1 - \frac{\mu^{\mathrm{ELITE}}}{|\mathscr{P}|}\right) \phi^{\mathrm{DIV}}(S).$$

This efficient diversity management system allows the method to rely on an efficient granular local search, applied on all offspring, without risking a premature population convergence. Finally, penalized infeasible solutions are exploited during the solution process to further enhance the method exploration capabilities.

Scatter Search (SS) and Path Relinking (PR).   Solution recombinations and local improvement are very different and complementary methods. The former allows the algorithm to select and recombine large parts of high-quality solutions, while the latter is used for solution refinement. This may explain the success of recent hybrid GAs. Yet, in many current methods, the crossover operator often appends together sequences of visits from different solutions in a blind and randomized manner. As such, a solution immediately obtained after crossover may be of very low quality. Whereas some randomization and shaking is beneficial at the exploration level, too much of it can be detrimental. Thus, some research has focused on exploiting more purposeful and intelligent recombinations, e.g., in the context of SS or PR (see Glover [28] and Resende et al. [69]). The latter explores in a finer manner the intermediate solutions located on a path between an origin solution and a target solution. The recent PR algorithm of Tarantilis, Anagnostopoulou, and Repoussis [80] has proved very successful on the VRP with time windows.

Learning Mechanisms.   Early implementations of learning mechanisms to the VRP were based on neural networks. Some of these methods are described in Gendreau, Laporte, and Potvin [25]. However, their overall success on CVRP benchmark instances remains mitigated, and these methods are now mostly investigated in dynamic contexts (see Créput et al. [18]).

4.5 ▪ Hybridizations

Classifying the current state-of-the-art methods for the CVRP would have been easier a decade ago. However, as research progresses, we witness the emergence of a wider variety of hybrids methods which rely on concepts borrowed from various algorithmic paradigms such as local search, large neighborhoods, collective intelligence and population-based search, perturbations, integer programming, constraint programming, tree search, data mining, and parallel computing. The frontier between methods is thus becoming increasingly blurred. Some important families of hybridizations stand out and are now discussed.

Population-Based Search and Local Search. Population-based heuristics have been long complemented by local search, but the inverse process, i.e., complementing a local search method such as TS by population and recombination concepts, has also been considered and is sometimes called *Adaptive Memory Programming* (AMP) (see Taillard et al. [78]). During the search, recurrent fragments of visit sequences from elite solutions, e.g., local minima, are stored in an *adaptive memory*. Subsequences of visits from these routes are selected to create new solutions used as starting points. The subsequence selection process is driven by the number of occurrences in elite routes, as by some minimum length requirements. This technique has led to several successful heuristics for the CVRP (see Rochat and Taillard [70] and Tarantilis [79]). Extracting and using promising sequences of solutions is a complex and rewarding task, which is sometimes called *guidance* (see Le Bouthillier, Crainic, and Kropf [47]), and may even be performed by means of data mining techniques (see Santos et al. [73]).

Meta-Meta Hybridizations. A very common hybridization scheme is to combine several concepts borrowed from different metaheuristics. In some cases, the concepts and methods may be indissociable from each other and occur in a single phase, while otherwise diverse techniques can be applied, sequentially or concurrently, to the same or different starting points. The heuristic of Kytöjoki et al. [45] combines VNS with a technique known as *Guided Local Search* (GLS), in which some solution features are temporarily penalized in the objective in order to escape from a local optimum (see Voudouris and Tsang [89]). Performing restarts from different initial solutions is a mechanism generally associated with GRASP (see Resende and Ribeiro [68]). Yet, many other metaheuristics also take advantage of this strategy (e.g., Cordeau and Maischberger [16], Prins [65], and Subramanian, Uchoa, and Ochi [76]), and so do implicitly all authors who consider the best solution out of several runs when reporting results. In addition to restarts, [65] performs several randomized local searches on each incumbent solution of an ILS, leading to a population of solutions at each iteration, out of which the best new incumbent is extracted. Finally, [16] combines concepts of ILS with a perturbation performed as a ruin-and-recreate move on a cluster of customers, a tabu memory, and a GLS objective function within a parallel context and restarts from exchanged solutions.

Hybridizations with Large Neighborhoods. The variety of neighborhoods is the essence of VNS. It may arise as a result of a parametrization of classic neighborhoods, e.g., by relocating or exchanging sequences of visits of increasing size as in Hemmelmayr, Doerner, and Hartl [37]. One possible avenue of improvement is to rely on structurally different neighborhoods, such as the SWAP or RELOCATE operators, along with large neighborhoods based on ruin-and-recreate moves or ejection chains. The ALNS algorithm of Pisinger and Ropke [63] is a good example. Structurally different large neighborhoods are used and selected through a roulette wheel mechanism. The selection rate is adapted to the performance of the operator at hand. Mester and Bräysy [53] use a complex hybridization scheme, involving GLS, large neighborhoods based on ejection chains, decomposition phases, and $1 + 1$ evolution strategies where a single offspring replaces a single parent in case of improvement. The rationale behind metaheuristic hybridizations stems from the fact that each method leads to a different distribution of final solutions. A strongly attractive local optimum for a method, e.g., TS, may be easier to improve with a different technique, e.g., hybrid GA, and vice versa. Hybrids are therefore inspired by the same strategy as VNS, i.e., alternating between neighborhoods to further improve the solutions, albeit at a higher level.

Hybridizations with Mathematical Programming Solvers. Other forms of hybridizations combine metaheuristics with mathematical programming solvers or other exact algorithms. These methods are called *matheuristics* (see Maniezzo, Stützle, and Voss [52]). One successful strategy requires storing high-quality routes in a pool and applying an integer programming solver to a set covering formulation. The resulting method belongs to the family of petal algorithms (see Foster and Ryan [23] and Renaud, Boctor, and Laporte [67]), but it is more sophisticated than the basic scheme since it involves some cooperation between the MIP solver and the route generation heuristic. Some matheuristics of this kind are known to be particularly efficient for VRP instances containing a large number of small routes (see Groër, Golden, and Wasil [32], Muter, Birbil, and Sahin [56], and Subramanian, Uchoa, and Ochi [76]). The synergy between the exact method and the metaheuristic may also be enhanced by promptly communicating new upper bounds and using them as cut-offs, as well as applying a reactive mechanism that interactively controls the size of the set partitioning models (see Groër, Golden, and Wasil [32] and Subramanian, Uchoa, and Ochi [76]). A number of other matheuristics have been proposed for the CVRP. The use of constraint or integer programming algorithms for solution reconstructions, advocated by Shaw [75], is now infrequent for the CVRP. Furthermore, Toth and Tramontani [82] introduced a large-neighborhood improvement procedure derived from the TSP neighborhood of Sarvanov and Doroshko [74]. As illustrated by Figure 4.1, the method relies on a reassignment model to relocate at most $n/2$ groups of customers (white-filled circles) between fixed positions (gray-filled circles). Ahuja et al. [1] survey other very large neighborhoods based on linear or dynamic programming which may be combined with CVRP metaheuristics.

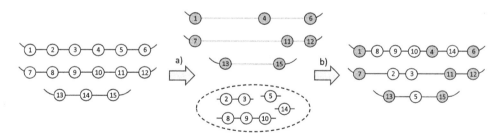

Figure 4.1. *Large Neighborhood of Toth and Tramontani [82]: (a) some fixed customers are selected, and the edges are short-circuited; (b) customer groups located between fixed points are optimally reassigned using integer programming.*

Parallel Algorithms. Parallel and cooperative search mechanisms have also led to well-performing hybrid heuristics. Most recent approaches consider a high-level parallelism (see Crainic and Toulouse [17]) in which different solutions and search trajectories are explored in different threads. At any point in time, the overall algorithm possesses a population of solutions, which can be exchanged, analyzed, and stored. Parallel methods therefore conveniently combine the exploration capacities of population-based search and the fast improvement mechanisms of local search. Homogeneous solvers are sometimes used. For example, Cordeau and Maischberger [16] combine several ILS with well-designed exchange mechanisms. Yet, most successful methods benefit from heterogeneous collaborating solvers, e.g., GA and TS cooperation in Le Bouthillier, Crainic, and Kropf [47], shaking and set covering solvers in Groër, Golden, and Wasil [32], or multiple TS with different neighborhood structures in Jin, Crainic, and Løkketangen [41].

Decompositions or Coarsening Phases. Several state-of-the-art metaheuristics are complemented by decomposition or coarsening techniques, thus allowing an efficient handling of larger-scale instances such as those of Golden et al. [30] and Li, Golden, and Wasil [48]. A common way to proceed is to rely on the customer-to-route assignments of an existing solution and combine subsets of routes and their associated customers to create subproblems. Different methods can be used to define the groups, e.g., (a) randomly, (b) relatively to a proximity criterion between routes such as the barycentric distance, (c) by increasing polar angle around the depot as in Vidal et al. [86], or (d) by dividing the geometrical space into sections, as in Groër, Golden, and Wasil [32]. These alternatives are depicted in Figure 4.2. Such decompositions, based on one elite solution, can

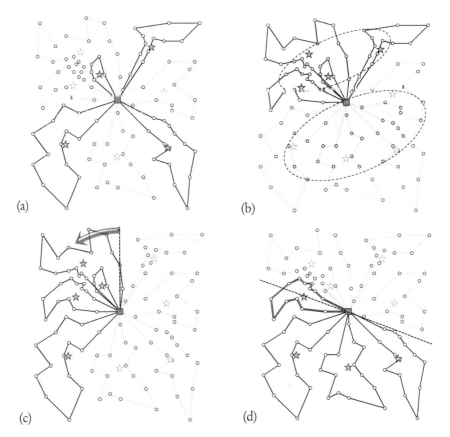

(a)   (b)

(c)   (d)

Figure 4.2. *Decomposition strategies on instance P03 of Christofides, Mingozzi, and Toth* [11] *with* 100 *customers. Route barycenters are represented by stars. Strategy* (a) *is a random selection of routes,* (b) *relies on a clustering of barycenters,* (c) *performs a circular sweeping of barycenters, and* (d) *divides the space into two sections (any route being part of two sections, e.g., the route displayed with double lines, being considered in both subproblems).*

be assimilated to projection methods, which fix some assignment decisions to obtain a separable problem. Another instance size reduction technique involves the identification of recurrent edges in good solutions, which can then be temporarily fixed by "merging" customers, as in multi-level heuristics, in order to focus on the remaining decision variables (see Walshaw [90]). It should be noted that, in this case, the reduced problem is asymmetric.

Diversification vs. Intensification. The balance between diversification, i.e., exploring unknown solution features, and intensification, i.e., focusing the search around known good solutions, has been the subject of considerable attention since the emergence of metaheuristics. Most hybrid heuristics for the CVRP seek to achieve a fine balance between these two search strategies. To illustrate, Table 4.1 summarizes some key intensification and diversification techniques. Numerous methodological elements and strategies have thus been suggested over the past years. In addition to finding new promising concepts, one main challenge of current research on metaheuristics is to better assess the scope of the application and the impact of some of these separate elements and combine them effectively.

4.6 ▪ Unified Algorithms

Recently, we have witnessed the emergence of a large number of new vehicle routing variants with additional constraints, objectives, and decision variables, called *attributes* in Vidal et al. [87]. Examples of recent attributes include fuel consumption optimization, turn penalties, time-dependent or flexible travel times, multiple compartments, and consistency. A very large number of metaheuristics dedicated to some attribute combinations have been proposed in the past years, for example for the solution of multi-period and multi-depot VRPs, open heterogeneous VRPs, or vehicle fleet mix problems with time windows. However, given the almost unlimited number of hybrid problems that can be formulated, it is becoming increasingly important to move from problem-specific methods to more flexible approaches which have the potential of being applied to a wider range of settings. This also leads to the issue of assessing the scope and the degree of generality of algorithm components.

A number of algorithms are sufficiently flexible to produce high-quality solutions on several multi-attribute VRPs through a single implementation. Thus, the *Unified Tabu Search* (UTS) heuristic of Cordeau, Laporte, and Mercier [15] can solve classical VRPs as well as *Pickup-and-Delivery Problems* (PDPs) with any combination of multiple depots, multiple planning periods, duration constraints, and time windows. The method was extended by Cordeau and Maischberger [16] into a parallel iterated TS heuristic. Similarly, the ALNS algorithm of Pisinger and Ropke [63] can efficiently solve VRPs and PDPs with multiple depots, vehicle-customer compatibility constraints, backhauls, and time windows. As a matter of fact, ALNS can be viewed as a metaheuristic framework in itself. It has been applied to a host of other multi-attribute VRPs, but often with separate implementations. A rich family of problems with time attributes was solved in Hashimoto et al. [34], Ibaraki et al. [39], and Hashimoto, Yagiura, and Ibaraki [35], who applied dynamic programming techniques for route evaluations. The list of covered attributes includes soft and multiple time windows, as well as time-dependent and flexible travel times. The hybrid ILS and set covering heuristic of Subramanian, Uchoa, and Ochi [76] efficiently solves a variety of problems with multiple depots, heterogeneous fleet, and simultaneous or mixed pickup and deliveries. Finally, the *Unified Hybrid Genetic Search* (UHGS) heuristic of Vidal et al. [88] was specifically designed to be highly flexible. Problem-specific

Table 4.1. *Key metaheuristic strategies.*

Diversification		Intensification	
Noise in edge costs	ALNS [63]	Incentives on elite solution features	[79]
Penalization of known solution features	GLS [89]	Pheromones	ACO
Incentive on under-explored solution features			
Penalized infeasible solutions: load, distance, or number of routes	[24], others		
Crossover and recombinations	GA, ES, SS	Enumerative neighborhoods: RELOCATE, SWAP, 2-OPT, among others	[50]
Variable neighborhoods	VNS [54]	Variable neighborhoods	VNS [54]
Large neighborhoods: ruin-and-recreate, ejection chains, set covering, among others	[75], [1], others	Large neighborhoods: ruin-and-recreate, ejection chains, set covering, among others	[75], [1], others
Randomized choice of neighbors: random first improvement, top X% savings	Various papers	Greedy choice of neighbors: best improvement, best saving	Various papers
Deteriorating moves	SA [61]	Deterioration thresholds	RRT [21]
Tabu memories	TS [15, 24]	Elite guiding solutions	PR [28]
Restarts and jumps on unexplored solutions	Various papers	Restart and jumps on elite solutions	Various papers [24], others
		False starts	[65]
		Multiple randomized LS on the same solution	
Populations of solutions	GA, ES	Exchange of elite solutions	Parallel [16, 41]
Parents and survivors selection with respect to diversity	[64, 85]	Parents and survivors selection with respect to solution cost	GA, ES
		Adaptation of method components relatively to performance measures	ALNS [63]
Fixing under-explored solution elements, uncoarsening	[32, 90]	Fixing variables from elite solutions, coarsening	[32, 90]
		Decomposition phases based on routes from an elite solution	Various papers

Acronyms: ACO – Ant Colony Optimization, ALNS – Adaptive Large Neighborhood Search, ES – Evolutionary Strategy, GA – Genetic Algorithm, GLS – Guided Local Search, PR – Path Relinking, RRT – Record-to-Record Travel, SA – Simulated Annealing, SS – Scatter Search, TS – Tabu Search, VNS – Variable Neighborhood Search.

features are relegated to small modular components, thus allowing the algorithm to be highly competitive on about 30 different problems.

Designing more unified solvers poses significant new challenges. As long as only a few attributes are considered, producing a unified method for a rich formulation merging all attributes is a feasible alternative. However, some attributes, such as soft time windows, hours of service regulations, or loading constraints, lead to complex and time-consuming solution evaluations and local search procedures. Attribute-related parameters may be set to default values when dealing with subproblems, e.g., time windows of $[-\infty, \infty]$, but heavy computations with dummy values may still be performed, with adverse consequences on computational performance. Finally, tailored strategies known to be efficient for specific attributes, e.g., efficient move evaluation techniques, may not apply for the general problem. For these reasons, it is sometimes necessary to design efficient methods that focus on the specific attributes of the problem at hand. This may require some form of adaptation relatively to the problem characteristics, as was attempted in Vidal et al. [88].

Finally, significant care should be paid to the calibration of parameters which may be highly correlated to the specific structure of the problems and instances, such as the distribution of customer locations, route size, and time window width. Thus, when moving towards more flexible methods, it is necessary to study not only the number of parameters but also their sensitivity to several problem features. This consideration applies in general to any choice of algorithm hybridization where some components are useful in some settings but detrimental in others. For this reason, adaptive heuristic frameworks such as ALNS (see Pisinger and Ropke [63]) and other hyper-heuristics (heuristics that select or build heuristics; see Burke et al. [9]) may prove very helpful.

4.7 ▪ Computational Comparison of Selected Metaheuristics

In this section we report on computational experiments from a selected set of successful metaheuristics. We compare the metaheuristics in terms of solution quality and speed, and we also touch upon issues such as simplicity, flexibility, and parameter sensitivity in the comparison. Table 4.2 lists the metaheuristics included in the computational

Table 4.2. *Metaheuristics included in the computational comparison.*

Name	Reference
CLM01	Cordeau, Laporte, and Mercier [15]
TV03	Toth and Vigo [84]
RDH04	Reimann, Doerner, and Hartl [66]
T05	Tarantilis [79]
MB07	Mester and Bräysy [53]
PR07	Pisinger and Ropke [63]
NB09	Nagata and Bräysy [57]
P09	Prins [65]
GGW10	Groër, Golden, and Wasil [31]
ZK10	Zachariadis and Kiranoudis [92]
GGW11	Groër, Golden, and Wasil [32]
CM12	Cordeau and Maischberger [16]
JCL12	Jin, Crainic, and Løkketangen [41]
VCGLR12	Vidal et al. [85]
SUO13	Subramanian, Uchoa, and Ochi [76]

comparison, along with the abbreviations we use to identify them. We have selected meta-heuristics proposed after the year 2000 and that perform well with respect to at least one of the criteria listed above.

4.7.1 ▪ Benchmark Data Sets

The two most widely used benchmark data sets for the VRP are those proposed by Christofides, Mingozzi, and Toth [11] and by Golden et al. [30]. We will refer to these as the *CMT* and the *GWKC* data sets, respectively. The first set contains 14 instances ranging from 50 to 200 customers, while the second one contains 20 instances involving from 240 to 483 customers. In this section we compare the selected metaheuristics using these data sets. Instances from Taillard [77] and Rochat and Taillard [70] (denoted *RT* instances in the following) and from Li, Golden, and Wasil [48] (denoted *LGW* instances) are also used but are not as widely accepted as the CMT and GWKC data sets. The RT set contains 13 instances ranging from 75 to 385 customers, and the LGW set contains 12 instances ranging from 560 to 1200 customers. The instances in the GWKC and the LGW data sets are all distributed spatially in symmetric patterns, while those from the CMT and RT sets have a more realistic appearance.

4.7.2 ▪ Comparing Computation Time and Solution Quality

We compare the metaheuristics in terms of solution quality and computation time. Performing such a comparison is difficult since we do not have access to all the metaheuristics and we must rely on the results reported in the respective papers. Typically, several sets of results are reported in the papers for each instance. A stochastic metaheuristic can, for example, have been run 10 times on the same instance, and the average solution quality along with the best solution value obtained are typically reported. Sometimes the best result "found during testing" is also reported. However, often little information is provided about how many test runs were performed. In order to make our comparison as fair as possible, we use the best result reported for which we know the computational effort. In the example above we would report the best result found in the 10 runs, as well as the time needed to perform all runs. We would not report the best result found during testing since we do not know how much effort was put into obtaining this result. Typically, we report results from several configurations of each metaheuristic in order to show what results can be obtained with different running times. Even with this approach, it can be difficult to compare results since authors often have quite different measures for what constitutes an acceptable running time. Also, the fact that computational tests are performed on different computers makes it difficult to compare running times. To partly remedy this problem we normalize the running times to match an Intel Core i7 CPU running at 2.93 GHz. We use the CINT2006, CINT2000, and CINT95 benchmarks (see the Standard Performance Evaluation Corporation web page, http://www.spec.org) to perform the normalization. However, the result of the normalization should only be taken as a rough estimate. Comparing sequential and parallel metaheuristics poses another problem. Should one simply compare the "wall clock" time spent by each metaheuristic, thereby giving a huge advantage to the parallel metaheuristics, or should the wall clock time be multiplied by the number of threads that were run in parallel? We have chosen the latter approach, but we realize that it places parallel metaheuristics at a disadvantage because of the communication overhead. The parallel metaheuristics included in the comparison are CGW11 and JCL12. The CM12 metaheuristic is also parallel, but the results we report are from a sequential run.

Tables 4.3 and 4.4 contain aggregated results for the CMT and GWKC instances, respectively. The tables contain five columns each. The first column identifies the paper using the abbreviations of Table 4.2, the second column shows the configuration used (we refer to the original papers for further details), and the third column shows the average gap to the best known solution. For each algorithm and each instance in the data set we calculate the percentage gap as $100(z - z^b)/z^b$, where z is the solution value obtained by the metaheuristic and z^b is the best known solution value for the instance. The gaps are averaged over all instances in the data set. Columns 4 and 5 show the normalized and original computation time per instance, respectively. We refer to the original papers for information about the computers used in the experiments. Both tables are sorted according to solution quality. We note that not all metaheuristics are present in Table 4.3, either because results were omitted from the original paper or because the paper did not provide enough information.

Table 4.3. *Computational results for the CMT instances. Note that the result for the CLM01 metaheuristic is from the survey by Cordeau et al.* [13].

Heuristic	Configuration	Gap (%)	Time (s)	Time* (s)
NB09	Best of 10 runs	0.00	506	831
SUO13	Best of 10 runs	0.00	1008	1008
GGW11	129 threads, best of 5 runs	0.00	138899	193500
VCGLR12	Average of 10 runs – 50,000 it.	0.02	356	585
NB09	Average of 10 runs	0.03	51	83
MB07	Best configuration	0.03	87	163
GGW11	8 threads, best of 5 runs	0.03	4102	5714
VCGLR12	Average of 10 runs – 10,000 it.	0.05	81	133
GGW11	4 threads, best of 5 runs	0.05	2051	2857
P09	-	0.07	9	16
MB07	Fast configuration	0.08	2	3
SUO13	Average of 10 runs	0.08	101	101
PR07	Best of 10 runs – 50,000 it.	0.11	593	1046
T05	Standard configuration	0.18	21	338
GGW10	Set partitioning	0.28	7	9
GGW10	Ejection - random	0.31	17	22
PR07	Average of 10 runs – 50,000 it.	0.31	59	105
CLM01	-	0.56	529	1477
TV03	-	0.64	5	230

One sees that all metaheuristics in the comparison achieve good results on the CMT instances, but their running times vary significantly. From a practical point of view, the solution quality obtained is more than adequate for typical applications. For the larger instances of the GWKC data set, one observes a larger spread in solution quality. The increased difficulty of these instances is also illustrated by the fact that we have witnessed a steady stream of new best solutions for the GWKC data set in the last decade, while only one improved solution has been found for the CMT data set. The difference in computational effort spent is striking, ranging from seven seconds to more than 200,000 seconds per instance (using normalized times). The trade-off between time and solution quality for the GWKC data set is depicted in Figure 4.3. The normalized computation time is shown on the x-axis using a logarithmic scale, while solution quality, measured as the average gap, is displayed along the y-axis. Most configurations from Table 4.4 are shown

Table **4.4.** *Computational results for the GWKC data set. We note that the time spent for the CM12 metaheuristic probably is underestimated since the paper only reports time averaged over both the GWKC and the CMT data sets and the CMT instances typically are solved faster than the GWKC instances. We also note that the results for the CLM01 and RDH04 metaheuristics are from the survey by Cordeau et al. [13].*

Heuristic	Configuration	Gap (%)	Time (s)	Time* (s)
GGW11	129 threads, best of 5 runs	0.12	138899	193500
VCGLR12	Average of 10 runs, 50,000 iterations	0.16	3387	5563
NB09	Best of 10 runs	0.17	13006	21359
ZK10	Best of 10 runs	0.22	14128	24300
VCGLR12	Average of 10 runs, 10,000 iterations	0.27	1042	1712
NB09	Average of 10 runs	0.27	1301	2136
GGW11	8 threads, best of 5 runs	0.30	8470	11800
MB07	Best configuration	0.33	782	1461
JCL12	8 threads, best of 10 runs	0.35	200978	200978
SUO13	Best of 10 runs	0.40	39382	39382
ZK10	Average of 10 runs	0.43	1413	2430
GGW11	4 threads, best of 5 runs	0.44	4235	5900
CM12	Best of 10, 10^6 iterations	0.56	18770	18770
SUO13	Average of 10 runs	0.56	3938	3938
P09	-	0.63	233	436
JCL12	8 threads, average of 10 runs	0.60	20098	20098
PR07	Best of 10 runs, 50,000 iterations	0.82	3662	6457
T05	Standard	0.93	169	2729
RDH04	-	0.93	599	2960
CM12	Average of 10 runs, 10^6 iterations	0.94	1877	1877
GGW10	Ejection - random	1.19	43	55
MB07	Fast configuration	1.23	7	13
GGW10	Set partitioning	1.27	10	13
PR07	Average of 10 runs, 50,000 iterations	1.35	366	646
CM12	Average of 10 runs, 10^5 iterations	1.46	188	188
CLM01	-	1.79	1207	3366.6
TV03	-	3.21	21	1053

as data points. However, we have left out the configurations with gaps of more than 2% to make the figure less cluttered. One can see that configurations MB07-B, CGW10-A, T05, P09, MB07-A, VCGLR12-A, VCGLR12-B, and CGW11-A define the Pareto front (meaning that any other configuration is dominated by one of these configurations). We acknowledge that a different benchmark for normalizing time may yield a slightly different Pareto front. Configurations CGW10-B and NB09-B are, for example, very close to being part of the Pareto front. We therefore highlight the metaheuristics MB07, CGW10, T05, P09, VCGLR12, NB09, and CGW11 as outstanding when it comes to either solution quality, computing time, or a combination of these two statistics.

For completeness we include detailed results for the CMT and GWKC instances in Tables 4.5–4.6 and Tables 4.7–4.9, respectively. These tables only include results for one

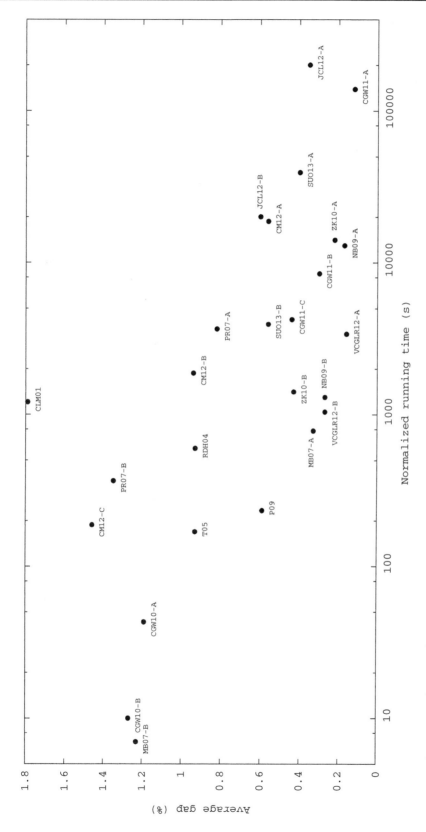

Figure 4.3. *Solution quality vs. running time for the GWKC instances. The x-axis shows the average computation time per instance in seconds (logarithmic scale), while the y-axis shows the solution quality measured as the average percent deviation from best known solution. Each label identifies a heuristic using the identifiers defined in Table 4.2. If several configurations of a heuristic are used, then each configuration is indicated by a letter "A", "B", or "C" after the heuristic identifier, with A being the most powerful configuration. The data for the table can be found in Table 4.4.*

Table 4.5. *CMT data set, detailed computational results part* I. *Results for the CLM*01 *meta-heuristic are from the survey by Cordeau et al.* [13].

#	n	Best known	CLM01 Value	%	TV03 Value	%	T05 Value	%	MB07 Value	%
1	50C	**524.61**	**524.61**	0.00	**524.61**	0.00	**524.61**	0.00	**524.61**	0.00
2	75C	**835.26**	835.28	0.00	838.6	0.40	**835.26**	0.00	**835.26**	0.00
3	100C	**826.14**	**826.14**	0.00	828.56	0.29	**826.14**	0.00	**826.14**	0.00
4	150C	**1028.42**	1032.68	0.41	1033.21	0.47	1029.64	0.12	**1028.42**	0.00
5	199C	**1291.29**	1315.76	1.90	1318.25	2.09	1311.48	1.56	**1291.29**	0.00
6	50CD	**555.43**	**555.43**	0.00	**555.43**	0.00	**555.43**	0.00	**555.43**	0.00
7	75CD	**909.68**	**909.68**	0.00	920.72	1.21	**909.68**	0.00	**909.68**	0.00
8	100CD	**865.94**	865.95	0.00	869.48	0.41	**865.94**	0.00	**865.94**	0.00
9	150CD	**1162.55**	1167.85	0.46	1173.12	0.91	1163.19	0.06	**1162.55**	0.00
10	199CD	**1395.85**	1416.84	1.50	1435.74	2.86	1407.21	0.81	1401.12	0.38
11	120C	**1042.11**	1073.47	3.01	1042.87	0.07	**1042.11**	0.00	**1042.11**	0.00
12	100C	**819.56**	**819.56**	0.00	**819.56**	0.00	**819.56**	0.00	**819.56**	0.00
13	120CD	**1541.14**	1549.25	0.53	1545.51	0.28	1544.01	0.19	**1541.14**	0.00
14	100CD	**866.37**	**866.37**	0.00	**866.37**	0.00	**866.37**	0.00	**866.37**	0.00
Average				0.56		0.64		0.20		0.03
Normalized time (s)			529.5		4.5		20.8		87.0	

#	n	Best known	PR07 Value	%	P09 Value	%	NB09 Value	%	GGW10 Value	%
1	50C	**524.61**	**524.61**	0.00	**524.61**	0.00	**524.61**	0.00	**524.61**	0.00
2	75C	**835.26**	**835.26**	0.00	**835.26**	0.00	**835.26**	0.00	842.37	0.85
3	100C	**826.14**	**826.14**	0.00	**826.14**	0.00	**826.14**	0.00	827.39	0.15
4	150C	**1028.42**	1029.56	0.11	1029.48	0.10	**1028.42**	0.00	1032.12	0.36
5	199C	**1291.29**	1297.12	0.45	1294.09	0.22	1291.45	0.01	1307.36	1.24
6	50CD	**555.43**	**555.43**	0.00	**555.43**	0.00	**555.43**	0.00	**555.43**	0.00
7	75CD	**909.68**	**909.68**	0.00	**909.68**	0.00	**909.68**	0.00	**909.68**	0.00
8	100CD	**865.94**	**865.94**	0.00	**865.94**	0.00	**865.94**	0.00	**865.94**	0.00
9	150CD	**1162.55**	1163.68	0.10	**1162.55**	0.00	**1162.55**	0.00	1164.12	0.14
10	199CD	**1395.85**	1405.88	0.72	1401.46	0.40	**1395.85**	0.00	1410.97	1.08
11	120C	**1042.11**	1042.12	0.00	**1042.11**	0.00	**1042.11**	0.00	**1042.11**	0.00
12	100C	**819.56**	**819.56**	0.00	**819.56**	0.00	**819.56**	0.00	**819.56**	0.00
13	120CD	**1541.14**	1542.86	0.11	1545.43	0.28	**1541.14**	0.00	1542.86	0.11
14	100CD	**866.37**	**866.37**	0.00	**866.37**	0.00	**866.37**	0.00	**866.37**	0.00
Average				0.11		0.07		0.00		0.28
Normalized time (s)			593.1		8.5		506.2		7.3	

configuration per paper. We include the configuration that obtains the best solution quality within a set time limit. Each table starts with three columns describing the instances. The first column contains the instance name, and the next column shows the number of customers in the instance and indicates whether the instance is constrained by capacity only (C) or by both capacity and distance (or duration) (CD). The third column provides the best known solution value for the instance. For each metaheuristic we provide the solution value obtained by the chosen configuration and the percentage gap. When a metaheuristic obtains the best known solution value, it is highlighted in bold. For each metaheuristic the table also displays the average gap and the average time spent per instance (using normalized times) after reporting results on the individual instances.

Table 4.6. *CMT data set, detailed computational results part* II.

#	n	Best known	GGW11 Value	%	VCGLR12 Value	%	SUO13 Value	%
1	50C	**524.61**	**524.61**	0.00	**524.61**	0.00	**524.61**	0.00
2	75C	**835.26**	**835.26**	0.00	**835.26**	0.00	**835.26**	0.00
3	100C	**826.14**	**826.14**	0.00	**826.14**	0.00	**826.14**	0.00
4	150C	**1028.42**	**1028.42**	0.00	**1028.42**	0.00	**1028.42**	0.00
5	199C	**1291.29**	1291.45	0.01	1291.74	0.03	1291.45	0.01
6	50CD	**555.43**	**555.43**	0.00	**555.43**	0.00	**555.43**	0.00
7	75CD	**909.68**	**909.68**	0.00	**909.68**	0.00	**909.68**	0.00
8	100CD	**865.94**	**865.94**	0.00	**865.94**	0.00	**865.94**	0.00
9	150CD	**1162.55**	**1162.55**	0.00	**1162.55**	0.00	**1162.55**	0.00
10	199CD	**1395.85**	**1395.85**	0.00	1397.7	0.13	**1395.85**	0.00
11	120C	**1042.11**	**1042.11**	0.00	**1042.11**	0.00	**1042.11**	0.00
12	100C	**819.56**	**819.56**	0.00	**819.56**	0.00	**819.56**	0.00
13	120CD	**1541.14**	**1541.14**	0.00	1542.86	0.11	**1541.14**	0.00
14	100CD	**866.37**	**866.37**	0.00	**866.37**	0.00	**866.37**	0.00
Average				0.00		0.02		0.00
Normalized time (s)			138898.5		356.4		1008.2	

4.7.3 ▪ Other Measures of Quality

The preceding section focused on solution quality and running time. These may be the two most obvious and the easiest criteria used to assess the quality of an algorithm, but other criteria are also relevant when comparing metaheuristics (see, e.g., Cordeau et al. [14]). Here we discuss such criteria and point out a few metaheuristics that perform especially well for one or more of these. Desirable features of a VRP metaheuristics are the following.

Simplicity. Ideally, a metaheuristic should be simple in the sense that it should be easy to understand, implement, and fine-tune. Focusing on improving solution quality may to some extent be detrimental to the objective of producing simple metaheuristics since there exists a tendency to take successful components from many past metaheuristics and combine them into new metaheuristics that perform slightly better than the previous generation but are more complex. Simplicity may come at the expense of speed or accuracy, but a successful metaheuristic should strike a reasonable balance between the criteria. Examples of relatively simple metaheuristics in recent years are those of Cordeau, Laporte, and Mercier [15], Toth and Vigo [84], Prins [64], and Groër, Golden, and Wasil [31].

Number of Parameters and Sensitivity to Parameters. This feature is clearly related to the previous one. We prefer metaheuristics that are controlled by few parameters, and ideally the metaheuristic should not be too sensitive to the parameters in the sense that the same set of parameters should be used for a wide range of instances. The metaheuristics of Groër, Golden, and Wasil [31] are controlled by relatively few parameters, and so are the TS algorithms of Cordeau, Laporte, and Mercier [15] and Toth and Vigo [84]. Another way of avoiding the curse of parameters is to let the metaheuristic select some of the parameters itself. Examples of such metaheuristics are the reactive TS of Battiti and Tecchiolli [6] and the ALNS metaheuristic of Ropke and Pisinger [63, 71].

Table 4.7. *GWKC data set, detailed computational results part I. Results for the CLM01 and RDH04 metaheuristics are from the survey by Cordeau et al.* [13].

#	n	Best known	CLM01 Value	%	TV03 Value	%	RDH04 Value	%	T05 standard value	%	MB07 value	%
1	240C	5623.47	5681.97	1.04	5736.15	2.00	5644.02	0.37	5676.97	0.95	5627.54	0.07
2	320C	8404.61	8657.36	3.01	8553.03	1.77	8449.12	0.53	8459.91	0.66	8447.92	0.52
3	400C	11030.80	11037.40	0.06	11402.75	3.37	11036.22	0.05	11036.22	0.05	11036.22	0.05
4	480C	13592.88	13740.60	1.09	14910.62	9.69	13699.11	0.78	13637.53	0.33	13624.52	0.23
5	200C	6460.98	6756.44	4.57	6697.53	3.66	6460.98	0.00	6460.98	0.00	6460.98	0.00
6	280C	8403.25	8537.17	1.59	8963.32	6.66	8412.90	0.11	8414.28	0.13	8412.88	0.11
7	360C	10102.68	10267.40	1.63	10547.44	4.40	10195.59	0.92	10216.50	1.13	10195.56	0.92
8	440C	11635.34	11869.50	2.01	12036.24	3.45	11828.78	1.66	11936.16	2.59	11663.55	0.24
9	255CD	579.71	587.39	1.32	593.35	2.35	586.87	1.24	585.43	0.99	583.39	0.63
10	323CD	736.26	752.76	2.24	751.66	2.09	750.77	1.97	746.56	1.40	741.56	0.72
11	399CD	912.84	929.07	1.78	936.04	2.54	927.27	1.58	923.17	1.13	918.45	0.61
12	483CD	1102.69	1119.52	1.53	1147.14	4.03	1140.87	3.46	1130.40	2.51	1107.19	0.41
13	252CD	857.19	875.88	2.18	868.80	1.35	865.07	0.92	865.01	0.91	859.11	0.22
14	320CD	1080.55	1102.03	1.99	1096.18	1.45	1093.77	1.22	1086.07	0.51	1081.31	0.07
15	396CD	1337.92	1363.76	1.93	1369.44	2.36	1358.21	1.52	1353.91	1.20	1345.23	0.55
16	480CD	1612.50	1647.06	2.14	1652.32	2.47	1635.16	1.41	1634.74	1.38	1622.69	0.63
17	240CD	707.76	710.93	0.45	711.07	0.47	708.76	0.14	708.74	0.14	707.79	0.00
18	300CD	995.13	1014.62	1.96	1016.83	2.18	998.83	0.37	1006.90	1.18	998.73	0.36
19	360CD	1365.60	1383.79	1.33	1400.96	2.59	1367.20	0.12	1371.01	0.40	1366.86	0.09
20	420CD	1818.25	1854.24	1.98	1915.83	5.37	1822.94	0.26	1837.67	1.07	1820.09	0.10
Average				1.79		3.21		0.93		0.93		0.33
Normalized time (s)			1206.7		20.6		598.9		168.5		782.0	

Table 4.8. GWKC data set, detailed computational results part II.

#	n	Best known	PR07 Value	%	P09 Value	%	NB09 Value	%	GGW10 Value	%	ZK10 Value	%
1	240C	**5623.47**	5650.91	0.49	5644.52	0.37	5626.81	0.06	5662.23	0.69	5626.81	0.06
2	320C	**8404.61**	8469.32	0.77	8447.92	0.52	8431.66	0.32	8466.92	0.74	8447.92	0.52
3	400C	**11030.80**	11047.01	0.15	11036.22	0.05	11036.22	0.05	11078.85	0.44	11036.22	0.05
4	480C	**13592.88**	13635.31	0.31	13624.52	0.23	**13592.88**	0.00	13634.11	0.30	13624.53	0.23
5	200C	**6460.98**	6466.68	0.09	**6460.98**	0.00	**6460.98**	0.00	6472.38	0.18	**6460.98**	0.00
6	280C	**8403.25**	8416.13	0.15	8412.90	0.11	8404.26	0.01	8415.21	0.14	8412.90	0.11
7	360C	**10102.68**	10181.75	0.78	10195.59	0.92	10156.58	0.53	10195.59	0.92	10169.26	0.66
8	440C	**11635.34**	11713.62	0.67	11643.90	0.07	11691.06	0.48	11869.61	2.01	11651.67	0.14
9	255CD	**579.71**	585.14	0.94	586.23	1.12	580.42	0.12	588.31	1.48	581.28	0.27
10	323CD	**736.26**	748.89	1.72	744.36	1.10	738.49	0.30	747.76	1.56	738.57	0.31
11	399CD	**912.84**	922.70	1.08	922.40	1.05	914.72	0.21	934.85	2.41	916.99	0.45
12	483CD	**1102.69**	1119.06	1.48	1116.12	1.22	1106.76	0.37	1133.01	2.75	1105.93	0.29
13	252CD	**857.19**	864.68	0.87	862.32	0.60	**857.19**	0.00	867.11	1.16	858.45	0.15
14	320CD	**1080.55**	1095.40	1.37	1089.35	0.81	**1080.55**	0.00	1093.69	1.22	1081.05	0.05
15	396CD	**1337.92**	1359.94	1.65	1352.39	1.08	1342.53	0.34	1360.68	1.70	1341.46	0.26
16	480CD	**1612.50**	1639.11	1.65	1634.27	1.35	1620.85	0.52	1639.24	1.66	1617.48	0.31
17	240CD	**707.76**	708.90	0.16	708.85	0.15	**707.76**	0.00	709.56	0.25	**707.76**	0.00
18	300CD	**995.13**	1002.42	0.73	1002.15	0.71	995.13	0.00	1012.66	1.76	996.55	0.14
19	360CD	**1365.60**	1374.24	0.63	1371.67	0.44	1365.97	0.03	1381.01	1.13	1366.75	0.08
20	420CD	**1818.25**	1830.80	0.69	1830.98	0.70	1820.02	0.10	1842.57	1.34	1824.46	0.34
Average				0.82		0.63		0.17		1.19		0.22
Normalized time (s)			3662.0		233.4		13005.5		43.4		14127.9	

Table 4.9. *GWKC data set, detailed computational results part III.*

#	n	Best known	GGW11 Value	%	CM12 Value	%	JCL12 Value	%	VCGLR12 Value	%	SUO13 Value	%
1	240C	**5623.47**	**5623.47**	0.00	5643.97	0.36	**5623.47**	0.00	5625.10	0.03	5657.74	0.61
2	320C	**8404.61**	8447.92	0.52	8445.70	0.49	8419.50	0.18	8419.25	0.17	8447.92	0.52
3	400C	**11030.80**	11036.22	0.05	11036.20	0.05	**11030.80**	0.00	11036.22	0.05	11036.22	0.05
4	480C	**13592.88**	13624.52	0.23	13629.30	0.27	13615.20	0.16	13624.53	0.23	13624.53	0.23
5	200C	**6460.98**	**6460.98**	0.00	**6460.98**	0.00	**6460.98**	0.00	**6460.98**	0.00	**6460.98**	0.00
6	280C	**8403.25**	8412.90	0.11	8412.90	0.11	**8403.25**	0.00	8412.90	0.11	8412.90	0.11
7	360C	**10102.68**	10195.59	0.92	10183.50	0.80	10184.40	0.81	10134.90	0.32	10195.58	0.92
8	440C	**11635.34**	11663.55	0.24	11673.70	0.33	11671.00	0.31	**11635.34**	0.00	11710.47	0.65
9	255CD	**579.71**	**579.71**	0.00	582.95	0.56	581.73	0.35	581.08	0.24	583.24	0.61
10	323CD	**736.26**	737.28	0.14	739.95	0.50	738.50	0.30	738.92	0.36	741.96	0.77
11	399CD	**912.84**	913.35	0.06	920.54	0.84	914.98	0.23	914.37	0.17	921.46	0.94
12	483CD	**1102.69**	1102.76	0.01	1111.76	0.82	1109.93	0.66	1105.97	0.30	1113.30	0.96
13	252CD	**857.19**	**857.19**	0.00	865.27	0.94	861.92	0.55	859.08	0.22	**857.19**	0.00
14	320CD	**1080.55**	**1080.55**	0.00	1085.76	0.48	1082.52	0.18	1081.99	0.13	**1080.55**	0.00
15	396CD	**1337.92**	1338.19	0.02	1355.86	1.34	1351.13	0.99	1341.95	0.30	1347.13	0.69
16	480CD	**1612.50**	1613.66	0.07	1633.84	1.32	1629.78	1.07	1616.92	0.27	1624.55	0.75
17	240CD	**707.76**	**707.76**	0.00	708.17	0.06	707.83	0.01	707.84	0.01	**707.76**	0.00
18	300CD	**995.13**	**995.13**	0.00	1000.86	0.58	1000.27	0.52	996.95	0.18	995.65	0.05
19	360CD	**1365.60**	**1365.60**	0.00	1373.07	0.55	1367.31	0.13	1366.39	0.06	1366.29	0.05
20	420CD	**1818.25**	**1818.25**	0.00	1832.65	0.79	1827.39	0.50	1819.75	0.08	1821.16	0.16
Average				0.12		0.56		0.35		0.16		0.40
Normalized time (s)			138898.5		18770.0		200978.4		3387.5		39382.3	

It is hard to single out metaheuristics that are robust with respect to parameter settings since it is not common to publish detailed parameter tuning results. However, it is safe to say that the metaheuristics we characterize as flexible in the next paragraph are relatively insensitive to parameter settings since they often are able to solve different variants of the VRP using the same set of parameters.

Flexibility. It is desirable for metaheuristics to be able to solve other variants of the VRP with little or no changes. Such variants could, for example, include multiple depots, pickups and deliveries, or time windows. Such a feature is especially important for software developers since they often face the challenge of producing vehicle routing solvers for several users who rarely have exactly the same VRP. We have surveyed such flexible metaheuristics in Section 4.6, and we highlight those of Cordeau, Laporte, and Mercier [15], Pisinger and Ropke [63], Cordeau and Maischberger [16], Vidal et al. [85], and Subramanian, Uchoa, and Ochi [76] as metaheuristics that have proven their ability to solve many VRP variants. The genetic algorithm of Prins [64] is also rather flexible since the method has been extended to many other VRP types even though the original paper was focused on the VRP.

Robustness. One would like metaheuristics to return a good solution consistently since in real life an occasional poor solution can lead to financial losses or a loss of faith in the metaheuristic. The robustness of a heuristics can to some extent be judged by inspecting the detailed results of Tables 4.5–4.9. It can be seen that there exists a high correlation between robust results and the good overall solution quality measured in Tables 4.3 and 4.4.

Taking Advantage of Parallel Processing. Compared with what we have witnessed between 1960 and 2000, the speed of a single processing unit has improved relatively slowly in recent years. At the same time, CPUs have been equipped with more and more cores, thus enabling parallel processing on standard desktop computers. It looks as if this trend will continue in the near future, and it therefore seems advisable to parallelize algorithms in order to take full advantage of the current and next generation of computers. It is possible to use parallel processing for almost any existing metaheuristic, simply by launching several copies of the metaheuristic on several processors, each starting from a different solution. However, more sophisticated approaches are of course possible, as we have pointed out in Section 4.5. We note that among the metaheuristics mentioned in this section, those of Groër, Golden, and Wasil [32], Jin, Crainic, and Løkketangen [41] and Cordeau and Maischberger [16] can already take advantage of parallel processing, and population-based methods are typically easy to parallelize since the work involved in combining solutions from the population or in improving individual solutions can often be performed in parallel. Some authors have suggested parallelizing other classical metaheuristics (see, for example, Alba [2] and Crainic and Toulouse [17]).

4.8 ▪ Conclusions and Future Research Directions

The past 10 years have seen the emergence of several powerful metaheuristics for the VRP, mostly derived from the hybridization of concepts initially developed independently from each other. The best known algorithms typically combine several search principles rooted in early local search heuristics and genetic algorithms. The best metaheuristics make use of sophisticated neighborhoods, exact optimization methods, decomposition strategies, and cooperation principles. Heuristics are also becoming more flexible in the sense that they can sometimes be applied, without any structural change, to a wide range of VRP extensions and variations.

Even though we do not yet know the optimal solutions for most of the instances used in the computational comparison, it is safe to say that current metaheuristics are capable of producing high quality solutions for instances with up to 500 customers, and solution quality has improved steadily over the last decade. However, the gains in solution quality are now becoming marginal, an indication that the current solution methodologies may have reached a plateau.

It may now be worthwhile considering new more realistic and varied large-scale benchmark instances and expending more efforts toward conceptual simplicity, parsimony of parameters, robustness, and flexibility. We also believe that the research effort in the coming years will be aimed at developing a new generation of metaheuristics that can yield the same solution quality as those of today but within significantly less time. One promising avenue of research toward this goal lies in the hybridization of heuristic concepts with MIP solvers, which are becoming increasingly powerful.

We have witnessed the emergence of new research streams in the field of vehicle routing, namely in the areas of green logistics, city logistics, and humanitarian logistics. We expect that these emerging fields will continue to grow in the coming years. Some other problem characteristics, such as synchronization, service consistency, balanced work allocation, congestion, and speed optimization, have emerged, giving rise to a new generation of difficult problems which require methodological developments.

Bibliography

[1] R. K. AHUJA, Ö. ERGUN, J. B. ORLIN, AND A. P. PUNNEN, *A survey of very large-scale neighborhood search techniques*, Discrete Applied Mathematics, 123 (2002), pp. 75–102.

[2] E. ALBA, ed., *Parallel Metaheuristics: A New Class of Algorithms*, Wiley-Interscience, Hoboken, NJ, 2005.

[3] K. ALTINKEMER AND B. GAVISH, *Parallel savings based heuristics for the delivery problem*, Operations Research, 39 (1991), pp. 456–469.

[4] D. L. APPLEGATE, R. E. BIXBY, V. CHVÁTAL, AND W. J. COOK, *The Traveling Salesman Problem: A Computational Study*, Princeton University Press, Princeton, NJ, 2006.

[5] M. L. BALINSKI AND R. E. QUANDT, *On an integer program for a delivery problem*, Operations Research, 12 (1964), pp. 300–304.

[6] R. BATTITI AND G. TECCHIOLLI, *The reactive tabu search*, ORSA Journal on Computing, 6 (1994), pp. 128–140.

[7] J. BAXTER, *Depot location: A technique for the avoidance of local optima*, European Journal of Operational Research, 18 (1984), pp. 208–214.

[8] J. E. BEASLEY, *Route first-cluster second methods for vehicle routing*, Omega, 11 (1983), pp. 403–408.

[9] E. K. BURKE, M. HYDE, G. KENDALL, G. OCHOA, E. ÖZCAN, AND J. R. WOOD-WARD, *A classification of hyper-heuristic approaches*, in Handbook of Metaheuristics, M. Gendreau and J.-Y. Potvin, eds., Springer, New York, 2010, pp. 449–468.

[10] P. CHEN, H.-K. HUANG, AND X.-Y. DONG, *Iterated variable neighborhood descent algorithm for the capacitated vehicle routing problem*, Expert Systems with Applications, 37 (2010), pp. 1620–1627.

[11] N. CHRISTOFIDES, A. MINGOZZI, AND P. TOTH, *The vehicle routing problem*, in Combinatorial Optimization, N. Christofides, A. Mingozzi, P. Toth, and C. Sandi, eds., Wiley, Chichester, UK, 1979, pp. 315–338.

[12] G. CLARKE AND J. W. WRIGHT, *Scheduling of vehicles from a central depot to a number of delivery points*, Operations Research, 12 (1964), pp. 568–581.

[13] J.-F. CORDEAU, M. GENDREAU, A. HERTZ, G. LAPORTE, AND J.-S. SORMANY, *New heuristics for the vehicle routing problem*, in Logistics Systems: Design and Optimization, A. Langevin and D. Riopel, eds., Springer, New York, 2005, pp. 279–297.

[14] J.-F. CORDEAU, M. GENDREAU, G. LAPORTE, J.-Y. POTVIN, AND F. SEMET, *A guide to vehicle routing heuristics*, Journal of the Operational Research Society, 53 (2002), pp. 512–522.

[15] J.-F. CORDEAU, G. LAPORTE, AND A. MERCIER, *A unified tabu search heuristic for vehicle routing problems with time windows*, Journal of the Operational Research Society, 52 (2001), pp. 928–936.

[16] J.-F. CORDEAU AND M. MAISCHBERGER, *A parallel iterated tabu search heuristic for vehicle routing problems*, Computers & Operations Research, 39 (2012), pp. 2033–2050.

[17] T. G. CRAINIC AND M. TOULOUSE, *Parallel meta-heuristics*, in Handbook of Metaheuristics, M. Gendreau and J.-Y. Potvin, eds., Springer, New York, 2010, pp. 497–541.

[18] J.-C. CRÉPUT, A. HAJJAM, A. KOUKAM, AND O. KUHN, *Self-organizing maps in population based metaheuristic to the dynamic vehicle routing problem*, Journal of Combinatorial Optimization, 24 (2012), pp. 437–458.

[19] G. B. DANTZIG AND J. H. RAMSER, *The truck dispatching problem*, Management Science, 6 (1959), pp. 80–91.

[20] M. DESROCHERS AND T. W. VERHOOG, *A matching based savings algorithm for the vehicle routing problem*, Techical Report, Les Cahiers du GERAD, G-89-04, HEC, Montréal, Canada, 1989.

[21] G. DUECK, *New optimization heuristics: The great deluge algorithm and the record-to-record travel*, Journal of Computational Physics, 104 (1993), pp. 86–92.

[22] G. DUECK AND T. SCHEUER, *Threshold accepting: A general purpose optimization algorithm appearing superior to simulated annealing*, Journal of Computational Physics, 90 (1990), pp. 161–175.

[23] B. A. FOSTER AND D. M. RYAN, *An integer programming approach to the vehicle scheduling problem*, Operational Research Quarterly, 27 (1976), pp. 367–384.

[24] M. GENDREAU, A. HERTZ, AND G. LAPORTE, *A tabu search heuristic for the vehicle routing problem*, Management Science, 40 (1994), pp. 1276–1290.

[25] M. GENDREAU, G. LAPORTE, AND J.-Y. POTVIN, *Metaheuristics for the capacitated VRP*, in The Vehicle Routing Problem, P. Toth and D. Vigo, eds., SIAM, Philadelphia, 2002, pp. 129–154.

[26] M. GENDREAU AND J.-Y. POTVIN, eds., *Handbook of Metaheuristics*, Springer, New York, 2010.

[27] ———, *Tabu search*, in Handbook of Metaheuristics, M. Gendreau and J.-Y. Potvin, eds., vol. 146, Springer, New York, 2010, pp. 41–59.

[28] F. GLOVER, *Heuristics for integer programming using surrogate constraints*, Decision Sciences, 8 (1977), pp. 156–166.

[29] ———, *Future paths for integer programming and links to artificial intelligence*, Computers & Operations Research, 13 (1986), pp. 533–549.

[30] B. L. GOLDEN, E. A. WASIL, J. P. KELLY, AND I.-M. CHAO, *The impact of metaheuristics on solving the vehicle routing problem: Algorithms, problem sets, and computational results*, in Fleet Management and Logistics, T. G. Crainic and G. Laporte, eds., Kluwer, Boston, 1998, pp. 33–56.

[31] C. GROËR, B. L. GOLDEN, AND E. A. WASIL, *A library of local search heuristics for the vehicle routing problem*, Mathematical Programming Computation, 2 (2010), pp. 79–101.

[32] ———, *A parallel algorithm for the vehicle routing problem*, INFORMS Journal on Computing, 23 (2011), pp. 315–330.

[33] P. HANSEN, N. MLADENOVIĆ, J. BRIMBERG, AND J. A. MORENO PÉREZ, *Variable neighborhood search*, in Handbook of Metaheuristics, M. Gendreau and J.-Y. Potvin, eds., Springer, New York, 2010, pp. 61–86.

[34] H. HASHIMOTO, T. IBARAKI, S. IMAHORI, AND M. YAGIURA, *The vehicle routing problem with flexible time windows and traveling times*, Discrete Applied Mathematics, 154 (2006), pp. 2271–2290.

[35] H. HASHIMOTO, M. YAGIURA, AND T. IBARAKI, *An iterated local search algorithm for the time-dependent vehicle routing problem with time windows*, Discrete Optimization, 5 (2008), pp. 434–456.

[36] K. HELSGAUN, *An effective implementation of the Lin-Kernighan traveling salesman heuristic*, European Journal of Operational Research, 126 (2000), pp. 106–130.

[37] V. C. HEMMELMAYR, K. F. DOERNER, AND R. F. HARTL, *A variable neighborhood search heuristic for periodic routing problems*, European Journal of Operational Research, 195 (2009), pp. 791–802.

[38] J. H. HOLLAND, *Adaptation in Natural and Artificial systems. An Introductory Analysis with Applications to Biology, Control and Artificial Intelligence*, The University of Michigan Press, Ann Arbor, MI, 1975.

[39] T. IBARAKI, S. IMAHORI, M. KUBO, T. MASUDA, T. UNO, AND M. YAGIURA, *Effective local search algorithms for routing and scheduling problems with general time-window constraints*, Transportation Science, 39 (2005), pp. 206–232.

[40] S. IRNICH, B. FUNKE, AND T. GRÜNERT, *Sequential search and its application to vehicle-routing problems*, Computers & Operations Research, 33 (2006), pp. 2405–2429.

[41] J. JIN, T. G. CRAINIC, AND A. LØKKETANGEN, *A parallel multi-neighborhood cooperative tabu search for capacitated vehicle routing problems*, European Journal of Operational Research, 222 (2012), pp. 441–451.

[42] D. S. JOHNSON AND L. A. MCGEOCH, *The traveling salesman problem: A case study in local optimization*, in Local Search in Combinatorial Optimization, E. H. L. Aarts and J. K. Lenstra, eds., Princeton University Press, Princeton, NJ, 1997, pp. 215–310.

[43] ———, *Experimental analysis of heuristics for the STSP*, in The Traveling Salesman Problem and Its Variations, G. Gutin and A. P. Punnen, eds., Kluwer, Dordrecht, 2002, pp. 369–443.

[44] S. KIRKPATRICK, C. D. GELATT, AND M. P. VECCHI, *Optimization by simulated annealing*, Science, 220 (1983), pp. 671–680.

[45] J. KYTÖJOKI, T. NUORTIO, O. BRÄYSY, AND M. GENDREAU, *An efficient variable neighborhood search heuristic for very large scale vehicle routing problems*, Computers & Operations Research, 34 (2007), pp. 2743–2757.

[46] G. LAPORTE AND F. SEMET, *Classical heuristics for the capacitated VRP*, in The Vehicle Routing Problem, P. Toth and D. Vigo, eds., SIAM, Philadelphia, 2002, ch. 5, pp. 109–128.

[47] A. LE BOUTHILLIER, T. G. CRAINIC, AND P. KROPF, *A guided cooperative search for the vehicle routing problem with time windows*, Intelligent Systems, IEEE, 20 (2005), pp. 36–42.

[48] F. LI, B. L. GOLDEN, AND E. A. WASIL, *Very large-scale vehicle routing: new test problems, algorithms, and results*, Computers & Operations Research, 32 (2005), pp. 1165–1179.

[49] S. LIN, *Computer solutions of the traveling salesman problem*, Bell System Technical Journal, 44 (1965), pp. 2245–2269.

[50] S. LIN AND B. W. KERNIGHAN, *An effective heuristic algorithm for the traveling-salesman problem*, Operations Research, 21 (1973), pp. 498–516.

[51] H. R. LOURENÇO, O. C. MARTIN, AND T. STÜTZLE, *Iterated local search: Framework and applications*, in Handbook of Metaheuristics, M. Gendreau and J.-Y. Potvin, eds., Springer, New York, 2010, pp. 363–397.

[52] V. MANIEZZO, T. STÜTZLE, AND S. VOSS, eds., *Matheuristics: Hybridizing Metaheuristics and Mathematical Programming*, Springer, New York, 2009.

[53] D. MESTER AND O. BRÄYSY, *Active-guided evolution strategies for large-scale capacitated vehicle routing problems*, Computers & Operations Research, 34 (2007), pp. 2964–2975.

[54] N. MLADENOVIĆ AND P. HANSEN, *Variable neighborhood search*, Computers & Operations Research, 24 (1997), pp. 1097–1100.

[55] P. MOSCATO AND C. COTTA, *A modern introduction to memetic algorithms*, in Handbook of Metaheuristics, M. Gendreau and J.-Y. Potvin, eds., Springer, New York, 2010, pp. 141–183.

[56] I. MUTER, S. I. BIRBIL, AND G. SAHIN, *Combination of metaheuristic and exact algorithms for solving set covering-type optimization problems*, INFORMS Journal on Computing, 22 (2010), pp. 603–619.

[57] Y. NAGATA AND O. BRÄYSY, *Edge assembly-based memetic algorithm for the capacitated vehicle routing problem*, Networks, 54 (2009), pp. 205–215.

[58] Y. NAGATA AND S. KOBAYASHI, *Edge assembly crossover: A high-power genetic algorithm for the travelling salesman problem*, in Proceedings of the 7th International Conference on Genetic Algorithms, East Lansing, MI, Morgan Kaufmann, San Francisco, 1997, pp. 450–457.

[59] M. D. NELSON, K. E. NYGARD, J. H. GRIFFIN, AND W. E. SHREVE, *Implementation techniques for the vehicle routing problem*, Computers & Operations Research, 12 (1985), pp. 273–283.

[60] A. G. NIKOLAEV AND S. H. JACOBSON, *Simulated annealing*, in Handbook of Metaheuristics, M. Gendreau and J.-Y. Potvin, eds., Springer, New York, 2010, pp. 1–39.

[61] I. H. OSMAN, *Metastrategy simulated annealing and tabu search algorithms for the vehicle routing problem*, Annals of Operations Research, 41 (1993), pp. 421–451.

[62] H. PAESSENS, *The savings algorithm for the vehicle routing problem*, European Journal of Operational Research, 34 (1988), pp. 336–344.

[63] D. PISINGER AND S. ROPKE, *A general heuristic for vehicle routing problems*, Computers & Operations Research, 34 (2007), pp. 2403–2435.

[64] C. PRINS, *A simple and effective evolutionary algorithm for the vehicle routing problem*, Computers & Operations Research, 31 (2004), pp. 1985–2002.

[65] ——, *A GRASP × evolutionary local search hybrid for the vehicle routing problem*, in Bio-Inspired Algorithms for the Vehicle Routing Problem, F. Pereira and J. Tavares, eds., Springer, Berlin, Heidelberg, 2009, pp. 35–53.

[66] M. REIMANN, K. F. DOERNER, AND R. F. HARTL, *D-Ants: Savings based ants divide and conquer the vehicle routing problem*, Computers & Operations Research, 31 (2004), pp. 563–591.

[67] J. RENAUD, F. F. BOCTOR, AND G. LAPORTE, *An improved petal heuristic for the vehicle routing problem*, Journal of the Operational Research Society, 47 (1996), pp. 329–336.

[68] M. G. C. RESENDE AND C. C. RIBEIRO, *Greedy randomized adaptive search procedures: Advances, hybridizations, and applications*, in Handbook of Metaheuristics, M. Gendreau and J.-Y. Potvin, eds., Springer, New York, 2010, pp. 283–319.

[69] M. G. C. RESENDE, C. C. RIBEIRO, F. GLOVER, AND R. MARTÍ, *Scatter search and path-relinking: Fundamentals, advances, and applications*, in Handbook of Metaheuristics, M. Gendreau and J.-Y. Potvin, eds., Springer, New York, 2010, pp. 87–107.

[70] Y. ROCHAT AND É. D. TAILLARD, *Probabilistic diversification and intensification in local search for vehicle routing*, Journal of Heuristics, 1 (1995), pp. 147–167.

[71] S. ROPKE AND D. PISINGER, *An adaptive large neighborhood search heuristic for the pickup and delivery problem with time windows*, Transportation Science, 40 (2006), pp. 455–472.

[72] D. M. RYAN, C. HJORRING, AND F. GLOVER, *Extensions of the petal method for vehicle routeing*, Journal of the Operational Research Society, 44 (1993), pp. 289–296.

[73] H. SANTOS, L. OCHI, E. MARINHO, AND L. DRUMMOND, *Combining an evolutionary algorithm with data mining to solve a single-vehicle routing problem*, Neurocomputing, 70 (2006), pp. 70–77.

[74] V. I. SARVANOV AND N. N. DOROSHKO, *The approximate solution of the travelling salesman problem by a local algorithm with scanning neighborhoods of factorial cardinality in cubic time (in Russian)*, in Software: Algorithms and Programs 31, Mathematical Institute of the Belarusian Academy of Sciences, Minsk, 1981, pp. 11–13.

[75] P. SHAW, *A new local search algorithm providing high quality solutions to vehicle routing problems*, Technical report, University of Strathclyde, Glasgow, 1997.

[76] A. SUBRAMANIAN, E. UCHOA, AND L. S. OCHI, *A hybrid algorithm for a class of vehicle routing problems*, Computers & Operations Research, 40 (2013), pp. 2519–2531.

[77] É. D. TAILLARD, *Parallel iterative search methods for vehicle routing problems*, Networks, 23 (1993), pp. 661–673.

[78] É. D. TAILLARD, L. M. GAMBARDELLA, M. GENDREAU, AND J.-Y. POTVIN, *Adaptive memory programming: A unified view of metaheuristics*, European Journal of Operational Research, 135 (2001), pp. 1–16.

[79] C. D. TARANTILIS, *Solving the vehicle routing problem with adaptive memory programming methodology*, Computers & Operations Research, 32 (2005), pp. 2309–2327.

[80] C. D. TARANTILIS, A. K. ANAGNOSTOPOULOU, AND P. P. REPOUSSIS, *Adaptive path relinking for vehicle routing and scheduling problems with product returns*, Transportation Science, 47 (2013), pp. 356–379.

[81] P. M. THOMPSON AND H. N. PSARAFTIS, *Cyclic transfer algorithms for multivehicle routing and scheduling problems*, Operations Research, 41 (1993), pp. 935–946.

[82] P. TOTH AND A. TRAMONTANI, *An integer linear programming local search for capacitated vehicle routing problems*, in The Vehicle Routing Problem: Latest Advances and New Challenges, B. L. Golden, S. Raghavan, and E. A. Wasil, eds., Springer, New York, 2008, pp. 275–295.

[83] P. TOTH AND D. VIGO, eds., *The Vehicle Routing Problem*, SIAM, Philadelphia, 2002.

[84] ———, *The granular tabu search and its application to the vehicle-routing problem*, INFORMS Journal on Computing, 15 (2003), pp. 333–346.

[85] T. VIDAL, T. G. CRAINIC, M. GENDREAU, N. LAHRICHI, AND W. REI, *A hybrid genetic algorithm for multidepot and periodic vehicle routing problems*, Operations Research, 60 (2012), pp. 611–624.

[86] T. VIDAL, T. G. CRAINIC, M. GENDREAU, AND C. PRINS, *A hybrid genetic algorithm with adaptive diversity management for a large class of vehicle routing problems with time-windows*, Computers & Operations Research, 40 (2013), pp. 475–489.

[87] ———, *Heuristics for multi-attribute vehicle routing problems: A survey and synthesis*, European Journal of Operational Research, 231 (2013), pp. 1–21.

[88] ———, *A unified solution framework for multi-attribute vehicle routing problems*, European Journal of Operational Research, 234 (2014), pp. 658–673.

[89] C. VOUDOURIS AND E. TSANG, *Guided local search and its application to the traveling salesman problem*, European Journal of Operational Research, 113 (1999), pp. 469–499.

[90] C. WALSHAW, *A multilevel approach to the travelling salesman problem*, Operations Research, 50 (2002), pp. 862–877.

[91] P. WARK AND J. HOLT, *A repeated matching heuristic for the vehicle routeing problem*, Journal of the Operational Research Society, 45 (1994), pp. 1156–1167.

[92] E. E. ZACHARIADIS AND C. T. KIRANOUDIS, *A strategy for reducing the computational complexity of local search-based methods for the vehicle routing problem*, Computers & Operations Research, 37 (2010), pp. 2089–2105.

Part II

Important Variants of the Vehicle Routing Problem

Chapter 5

The Vehicle Routing Problem with Time Windows

Guy Desaulniers
Oli B.G. Madsen
Stefan Ropke

5.1 ▪ Introduction

The *Vehicle Routing Problem with Time Windows* (VRPTW) is the extension of the *Capacitated Vehicle Routing Problem* (CVRP) where the service at each customer must start within an associated time interval, called a time window. Time windows may be hard or soft. In case of hard time windows, a vehicle that arrives too early at a customer must wait until the customer is ready to begin service. In general, waiting before the start of a time window incurs no cost. In the case of soft time windows, every time window can be violated barring a penalty cost. The time windows may be one-sided, e.g., stated as the latest time for delivery.

Time windows arise naturally in problems faced by business organizations which work on flexible time schedules. Specific problems with hard time windows include security patrol service, bank deliveries, postal deliveries, industrial refuse collection, grocery delivery, school bus routing, and urban newspaper distribution. Among the soft time window problems, dial-a-ride problems constitute an important example. In this chapter, we focus mainly on the hard time-window variant that has been the mostly studied one.

In the existing literature on the VRPTW, the number of vehicles available to serve the customers is usually considered unlimited and the objective function depends on the nature of the chosen solution method. For exact methods the objective is to minimize the total distance traveled. For heuristics the primary objective is to minimize the number of vehicles used and the secondary to minimize the total distance traveled. There may be exceptions to this general statement.

Since the CVRP is NP-hard, by restriction, the VRPTW is also NP-hard. In fact even finding a feasible solution to the VRPTW for a fixed number of vehicles is itself an NP-complete problem (Savelsbergh [109]). From a computational point of view one may distinguish between tight and loose time windows, and between narrow and wide time windows. A tight time window is a window which influences the solution; i.e., the window is an active constraint. A time window is considered as narrow if it is narrow relative to the planning horizon, e.g., 10 minutes compared to 12 hours. A narrow time window is not necessarily tight at the same time. If all the time windows are very wide, the VRPTW turns almost into a CVRP.

In the CVRP the geography is usually the important factor determining the shape of the routes. If the number of customers is small, say 20, a trained dispatcher can do very well in planning the routes only by looking at a map showing the location of the customers. However, if the capacity constraint is binding on some of the routes, it is much more difficult to overlook the planning situation. For the VRPTW when some of the routes are constrained by the capacity and others by the time windows, it is even more difficult to plan the routes manually. The interplay between the spatial and the temporal elements of the routes may result in optimal routes which are far from the classical picture of routes formed as clover leafs or petals. If the number of customers is increased to a realistic figure, say at least 100–200, it then becomes really difficult to make routes by hand. It is here that the computerized solution methods show their advantages.

The early works on the VRPTW were case study oriented (see Pullen and Webb [100], Knight and Hofer [69], and Madsen [79]). The solution methods were based on relatively simple heuristics. The first exact Branch-and-Bound algorithms appeared in the beginning of the 1980s (see Christofides, Mingozzi, and Toth [19], Baker and Rushinek [4], and Trienekens [118]). In 1987, Solomon [113] introduced benchmark instances involving 100 customers that were accepted as standard benchmark problems by most researchers working on the VRPTW and served as a catalyst to increase research on the VRPTW. In the following decade, many heuristics were developed, mostly local search ones, but also the first metaheuristics (tabu search and genetic algorithms). Several exact algorithms based on complex methodologies such as Lagrangian relaxation and column generation were also designed.

In the previous book on the VRP by Toth and Vigo [117], the most recent references on the VRPTW chapter by Cordeau et al. [21] dates back to 2000. The following sections will deal with the scientific progress made on the VRPTW since 2000. The reader can find additional information in the earlier surveys by Bräysy and Gendreau [16, 17] (heuristics and metaheuristics), Kallehauge [65] (exact algorithms), Potvin [96] (evolutionary algorithms), Gendreau and Tarantilis [47] (metaheuristics), Desaulniers, Desrosiers, and Spoorendonk [29] (exact algorithms), and Baldacci, Mingozzi, and Roberti [7] (exact algorithms).

This chapter is organized as follows. Section 5.2 presents mathematical formulations for the VRPTW. Sections 5.3 and 5.4 are devoted to the recent exact and the heuristic solution algorithms for the VRPTW, respectively. Each of these sections includes summary computational results. Section 5.5 discusses extensions of the VRPTW. Finally, conclusions are drawn in Section 5.6.

5.2 ▪ Mathematical Formulations

Starting from the notation given in Chapter 1, the VRPTW is defined on the directed graph $G = (V, A)$, where the depot is represented by the two vertices 0 and $n + 1$, referred to as the *source* and *sink* vertices, respectively. Let $N = V \setminus \{0, n + 1\}$ be the set of customer vertices. All feasible vehicle routes correspond to source-to-sink elementary paths in G. The converse is, however, not necessarily true; that is, some source-to-sink elementary paths in G may not represent feasible routes because they violate the time windows or the vehicle capacity. To simplify notation, zero demands and zero service times are defined for vertices 0 and $n + 1$, i.e., $q_0 = q_{n+1} = s_0 = s_{n+1} = 0$. Furthermore, a time window is associated with them, i.e., $[a_0, b_0] = [a_{n+1}, b_{n+1}]$, where a_0 and b_0 are the earliest possible departure time from the depot and the latest possible arrival time at the depot, respectively. Assuming that the travel time matrix satisfies the triangle inequality, feasible solutions exist only if $a_0 \leq \min_{i \in V \setminus \{0\}} \{b_i - t_{0i}\}$ and

$$b_0 \geq \max_{i \in V \setminus \{0\}} \{\max\{a_0 + t_{0i}, a_i\} + s_i + t_{i,n+1}\}.$$

Note that an arc $(i, j) \in A$ can be omitted due to temporal considerations, if $a_i + s_i + t_{ij} > b_j$, or capacity limitations, if $q_i + q_j > Q$, or by other factors. Finally, let us mention that when vehicles are allowed to remain at the depot, especially in the case where the primary objective consists of minimizing the number of vehicles used, the arc $(0, n + 1)$ with $c_{0,n+1} = t_{0,n+1} = 0$ must be added to the arc set A.

We first present a *Mixed-Integer Programming* (MIP) formulation for the VRPTW involving two types of variables: for each arc $(i, j) \in A$ and each vehicle $k \in K$, there is a binary arc-flow variable x_{ijk} that is equal to 1 if arc (i, j) is used by vehicle k, and 0 otherwise; and, for each vertex $i \in V$ and vehicle $k \in K$, there is a time variable T_{ik} specifying the start of service time at vertex i when serviced by vehicle k.

The VRPTW can be formulated as the following multi-commodity network flow model with time-window and capacity constraints:

$$(5.1) \quad \text{(VRPTW1) minimize} \sum_{k \in K} \sum_{(i,j) \in A} c_{ij} x_{ijk}$$

$$(5.2) \quad \text{s.t.} \sum_{k \in K} \sum_{j \in \delta^+(i)} x_{ijk} = 1 \qquad \forall\, i \in N,$$

$$(5.3) \quad \sum_{j \in \delta^+(0)} x_{0jk} = 1 \qquad \forall\, k \in K,$$

$$(5.4) \quad \sum_{i \in \delta^-(j)} x_{ijk} - \sum_{i \in \delta^+(j)} x_{jik} = 0 \qquad \forall\, k \in K,\, j \in N,$$

$$(5.5) \quad \sum_{i \in \delta^-(n+1)} x_{i,n+1,k} = 1 \qquad \forall\, k \in K,$$

$$(5.6) \quad x_{ijk}(T_{ik} + s_i + t_{ij} - T_{jk}) \leq 0 \qquad \forall\, k \in K,\, (i,j) \in A,$$

$$(5.7) \quad a_i \leq T_{ik} \leq b_i \qquad \forall\, k \in K,\, i \in V,$$

$$(5.8) \quad \sum_{i \in N} q_i \sum_{j \in \delta^+(i)} x_{ijk} \leq Q \qquad \forall\, k \in K,$$

$$(5.9) \quad x_{ijk} \in \{0, 1\} \qquad \forall\, k \in K,\, (i,j) \in A.$$

Objective function (5.1) aims at minimizing the total cost. Constraints (5.2) ensure that each customer is assigned to exactly one route. Next, constraints (5.3)–(5.5) define a source-to-sink path in G for each vehicle k. Additionally, constraints (5.6)–(5.7) and (5.8) guarantee schedule feasibility with respect to time windows and vehicle capacity, respectively. Note that, for a given k, the value of T_{ik} is meaningless whenever customer i is not visited by vehicle k. Finally, the arc-flow variables are subject to binary requirements (5.9).

Model (5.1)–(5.9) is nonlinear due to constraints (5.6) that can, however, be linearized as

$$(5.6a) \quad T_{ik} + s_i + t_{ij} - T_{jk} \leq (1 - x_{ijk}) M_{ij} \qquad \forall\, k \in K,\, (i,j) \in A,$$

where M_{ij}, $(i, j) \in A$, are large constants that can be set to $\max\{b_i + s_i + t_{ij} - a_j, 0\}$.

The linear relaxation of model (5.1)–(5.5), (5.6a), (5.7)–(5.9) provides, in general, very weak lower bounds. This model has, however, a block-angular structure that can be exploited, where each block is composed of the constraints (5.3)–(5.5), (5.6a), (5.7)–(5.9) for a specific vehicle $k \in K$ and defines an *Elementary Shortest Path Problem with Resource Constraints* (ESPPRC). Applying the Dantzig–Wolfe decomposition principle [25] to this model yields the following set partitioning model once the (identical) vehicles and their corresponding variables are aggregated (see, e.g., Desrosiers et al. [33]). In this model, Ω denotes the set of all feasible routes, c_r the cost of route $r \in \Omega$, and a_{ir} the number of visits to customer $i \in N$ in route $r \in \Omega$ ($a_{ir} \in \{0,1\}$ when r is an elementary route. With each route $r \in \Omega$ is associated a binary path-flow variable y_r that takes value 1 if route r is selected in the solution and 0 otherwise. The set partitioning model is

(5.10)   (VRPTW2) minimize $\sum_{r \in \Omega} c_r y_r$

(5.11)   s.t. $\sum_{r \in \Omega} a_{ir} y_r = 1$ $\forall \, i \in N,$

(5.12) $y_r \in \{0,1\}$ $\forall \, r \in \Omega.$

Objective function (5.10) seeks to minimize total cost. Set partitioning constraints (5.11) impose that each customer be visited exactly once by a vehicle. The binary requirements on the path-flow variables are expressed by (5.12). Notice that, as in the VRPTW literature, the above model assumes that the number of vehicles available to service the customers is unlimited, that is, $|K|$ is as large as needed. If this was not the case, the constraint

(5.13) $\sum_{r \in \Omega} y_r \leq |K|$

enforcing the selection of at most one route per available vehicle would be added to the model.

The linear relaxation of model (5.10)–(5.12) produces better lower bounds than that of model (5.1)–(5.5), (5.6a), (5.7)–(5.9). On the other hand, the set partitioning model contains a huge number of variables, one per feasible route. In Section 5.3.1, we shall see how this difficulty can be overcome using column gene ration.

Several other models were proposed for the VRPTW. In particular, two-index formulations were used in conjunction with Branch-and-Cut algorithms (see Section 5.3.2). We present such a formulation below that involves one type of variables: for each arc $(i,j) \in A$, there is a binary variable x_{ij} that is equal to 1 if arc (i,j) is used in the solution and 0 otherwise. Denote by P the set of paths (not necessarily from the source-to-sink vertex) in G that do not respect the time-window constraints and by $A(p)$ the set of arcs in path $p \in P$. Let $r(S)$ be the minimum number of vehicles required to serve the customers in subset $S \subseteq N$ according to their demands. This number is, in general, replaced by $\lceil q(S)/Q \rceil$, where $q(S) = \sum_{i \in S} q_i$. The two-index formulation corresponds to

(5.14)   (VRPTW3) minimize $\sum_{(i,j) \in A} c_{ij} x_{ij}$

(5.15)   s.t. $\sum_{j \in \delta^-(i)} x_{ji} = 1$ $\forall \, i \in N,$

(5.16) $\sum_{j \in \delta^+(i)} x_{ij} = 1$ $\forall \, i \in N,$

$$(5.17) \qquad \sum_{i \notin S} \sum_{j \in S} x_{ij} \geq r(S) \qquad \forall \, S \subseteq N, \, S \neq \emptyset,$$

$$(5.18) \qquad \sum_{(i,j) \in A(p)} x_{ij} \leq |A(p)| - 1 \qquad \forall \, p \in P,$$

$$(5.19) \qquad x_{ij} \in \{0, 1\} \qquad \forall \, (i,j) \in A.$$

The objective function (5.14) minimizes the total cost for serving the customers. Constraints (5.15)–(5.16) ensure that a vehicle arrives at and departs from each customer, respectively. Capacity inequalities (5.17) ensure that vehicle capacity is satisfied on all selected routes and also tighten the linear relaxation by imposing a minimum number of vehicles to serve every subset of customers S. Furthermore, they act as subtour elimination constraints. Infeasible path inequalities (5.18) forbid the selection of paths that do not respect the time windows. Finally, the flow variables x_{ij} are subject to binary requirements (5.19).

The two-index formulation (5.14)–(5.19) contains an exponential number of constraints (5.17) and (5.18). For practical-sized instances, they need to be generated dynamically as in a cutting plane algorithm. Additional valid inequalities can also be considered to tighten the linear relaxation of this model (see Section 5.3.2).

5.3 ▪ Exact Solution Methods

Numerous exact solution methods have been proposed for the VRPTW. The reader is referred to Cordeau et al. [21] for a survey on the methods that were developed before 2000. Since then, several algorithms were also designed. They can be classified into the following three families: Branch-and-Cut-and-Price, Branch-and-Cut, and reduced set partitioning.

5.3.1 ▪ Branch-and-Cut-and-Price

Branch-and-Price is a leading methodology for solving a wide variety of constrained vehicle routing and crew scheduling problems (see Barnhart et al. [9], Desaulniers, Desrosiers, and Solomon [28], and Lübbecke and Desrosiers [77]). For the VRPTW, it was introduced by Desrochers, Desrosiers, and Solomon [32]. It corresponds to a Branch-and-Bound in which the linear relaxations are solved by column generation. When cutting planes are generated to strengthen the linear relaxations, the method is called Branch-and-Cut-and-Price. The first Branch-and-Cut-and-Price for the VRPTW is due to Kohl et al. [70].

Column generation (Dantzig and Wolfe [25] and Gilmore and Gomory [49]) is an iterative process that can be applied for solving certain linear programs involving a huge number of variables such as the linear relaxation of the set partitioning model (5.10)–(5.12). In this context, the linear program is called the *Master Problem* (MP). At each iteration, a *Restricted Master Problem* (RMP) and an auxiliary problem, called the pricing problem or the subproblem, are solved sequentially. The RMP is the MP restricted to a subset of its variables. It is solved by a linear programming solver to provide a primal and a dual solution. Given this dual solution, the subproblem is problem-specific and consists of finding negative reduced cost columns (variables) for the RMP when some exist. When such columns are found, they are added to the RMP before starting a new iteration. If no negative reduced cost columns can be identified, the solution process stops and the current RMP primal solution is declared optimal for the MP.

For the VRPTW, the subproblem is an ESPPRC (see, e.g., Irnich and Desaulniers [61]) defined on network G. The resource constraints impose the time window and vehicle capacity constraints. Let α_i, $i \in N$, be the dual variables associated with constraints (5.11), and let $\alpha_0 = 0$. The subproblem can be stated as

$$(5.20) \qquad \min_{r \in \Omega} \ c_r - \sum_{i \in N} a_{ir} \alpha_i$$

or, equivalently, as

$$(5.21) \qquad \text{minimize} \ \sum_{(i,j) \in A} (c_{ij} - \alpha_i) x_{ij}$$

$$(5.22) \qquad \text{s.t.} \quad \sum_{j \in \delta^+(0)} x_{0j} = 1,$$

$$(5.23) \qquad \sum_{i \in \delta^-(j)} x_{ij} - \sum_{i \in \delta^+(j)} x_{ji} = 0 \qquad \forall\, j \in N,$$

$$(5.24) \qquad \sum_{i \in \delta^-(n+1)} x_{i,n+1} = 1,$$

$$(5.25) \qquad x_{ij}(T_i + s_i + t_{ij} - T_j) \leq 0 \qquad \forall\, (i,j) \in A,$$

$$(5.26) \qquad a_i \leq T_i \leq b_i \qquad \forall\, i \in V,$$

$$(5.27) \qquad \sum_{i \in N} q_i \sum_{j \in \delta^+(i)} x_{ij} \leq Q,$$

$$(5.28) \qquad x_{ij} \in \{0,1\} \qquad \forall\, (i,j) \in A,$$

where constraints (5.22)–(5.28) correspond to the constraints (5.3)–(5.9) of any specific vehicle k and index k is omitted. The objective function (5.21) aims at minimizing the path reduced cost, ensuring that a negative reduced cost column is found when one exists. Note that, in this formulation, path elementarity is imposed through the uniqueness of the time variable at each vertex and the positivity of the travel and service times.

In a Branch-and-Cut-and-Price algorithm, cuts and branching decisions are added to derive integer solutions when the computed solution of the MP is fractional. These cuts and branching decisions impact the MP or the subproblem definition. In the following paragraphs, we review the recent contributions on Branch-and-Cut-and-Price algorithms for the VRPTW.

Labeling Algorithms for the ESPPRC. The ESPPRC is known to be NP-hard (Dror [36]). It is usually solved by dynamic programming, namely, by a labeling algorithm (Irnich and Desaulniers [61]). In such an algorithm, a label E is a vector of $3 + |N|$ components representing a partial path from the source vertex to any vertex i of G. Its components indicate the path reduced cost ($Z(E)$); its (earliest) start of service time at vertex i ($T(E)$); its cumulated load up to vertex i ($L(E)$); and whether vertex $\ell \in N$ can still be visited in an extension of this path ($W_\ell(E)$). A vertex can still be visited if it has not been visited yet and if it can be reached without violating its time window or the vehicle capacity. Starting from an initial label $E_0 = (0, a_0, \ldots, 0)$ at the source vertex, a labeling algorithm propagates the labels forward using extension functions towards the sink vertex. Given a label $E_i = (Z(E_i), T(E_i), L(E_i), (W_\ell(E_i))_{\ell \in N})$ representing a path ending at node i, its extension along an arc $(i,j) \in A$ yields a new label $E_j = (Z(E_j), T(E_j), L(E_j), (W_\ell(E_j))_{\ell \in N})$

at node j with

(5.29) $\qquad Z(E_j) = Z(E_i) + c_{ij} - \alpha_i,$

(5.30) $\qquad T(E_j) = \max\{T(E_i) + s_i + t_{ij}, a_j\},$

(5.31) $\qquad L(E_j) = L(E_i) + q_j,$

(5.32) $\qquad W_\ell(E_j) = \begin{cases} W_\ell(E_i) + 1 & \text{if } j = \ell, \\ \max\{W_\ell(E_i), U_\ell(T(E_j), L(E_j))\} & \text{otherwise,} \end{cases}$

where $U_\ell(T(E_j), L(E_j))$ is equal to 1 if it is not possible to feasibly reach vertex $\ell \in N$ from vertex j with a start of service time at j equal to $T(E_j)$ and a load of $L(E_j)$. Label E_j corresponds to an infeasible partial path if $T(E_j) > b_j$, $L(E_j) > Q$, or $W_\ell(E_j) > 1$ for at least one customer $\ell \in N$. When one of these conditions is met, E_j is discarded.

To avoid enumerating all feasible partial paths, a labeling algorithm applies a dominance rule. This rule aims at identifying partial paths that cannot yield an optimal source-to-sink path. Given that the extension functions (5.29)–(5.32) are nondecreasing, the following proposition states the dominance rule used for the ESPPRC in the context of the VRPTW.

Proposition 1 (Desaulniers et al. [27]). *Let E and E' be two labels representing partial paths ending at the same vertex. Label E dominates label E' (which can be discarded) if*

(5.33) $\qquad Z(E) \leq Z(E'),$

(5.34) $\qquad T(E) \leq T(E'),$

(5.35) $\qquad L(E) \leq L(E'),$

(5.36) $\qquad W_\ell(E) \leq W_\ell(E') \qquad \forall\, \ell \in N.$

When equality holds for all components, one of the two labels must be kept.

For solving the linear relaxation of the VRPTW, Feillet et al. [40] were the first authors to propose a column generation algorithm that relies on an ESPPRC subproblem and the above labeling algorithm. They demonstrated empirically that, for the instances with relatively wide time windows, their algorithm produces much better lower bounds than a column generation algorithm based on the generation of paths allowing cycles (see the paragraphs on path relaxations below). It also finds more optimal integer solutions at the root node.

Different acceleration strategies for the labeling algorithm were developed. Noticing that the conditions of the dominance rule in Proposition 1 are only sufficient to identify dominated labels, Chabrier [18] proposed a modified dominance rule that permits the identification of dominated labels even if some of the conditions (5.36) do not hold: condition (5.33) must be replaced by

(5.37) $\qquad\qquad Z(E) \leq Z(E') - \sum_{\ell \in N(E,E')} \alpha_\ell,$

where $N(E, E')$ is the subset of customers $\ell \in N$ for which $W_\ell(E) > W_\ell(E')$. In their computational experiments, they considered this dominance rule only when $|N(E, E')| \leq 2$. Otherwise, the rule of Proposition 1 is used.

To speed up the labeling algorithm, Righini and Salani [103] designed a bidirectional search that consists of extending labels forwardly from the source vertex and backwardly from the sink vertex before joining forward labels with backward labels to yield complete source-to-sink feasible paths. Forward and backward extensions are forbidden when a predetermined resource (time or load) reaches the midpoint of its possible values over the whole network G. In this way, the total number of labels generated is significantly reduced. Bounds on the reduced cost of any extension (computed through binary knapsack problems) can also be used to prune labels prematurely. Such bounds were also proposed by Feillet, Gendreau, and Rousseau [41]. These authors also introduced the concept of label loading and meta-extensions. Before starting the labeling algorithm, labels representing all partial paths (originating from the source vertex) of the paths associated with the basic variables in the current RMP are attached to the vertices of the network. Furthermore, meta-vertices corresponding to the ends of these paths are appended to the network. When (forward) labeling reaches one of these meta-vertices, a so called meta-extension is performed to extend the label directly to the sink vertex in the hope of finding rapidly negative reduced cost paths.

Another efficient acceleration strategy, called decremental state space relaxation, was developed independently by Boland, Dethridge, and Dumitrescu [14] and Righini and Salani [104]. This technique starts by solving the ESPRCC considering labels with no customer components $W_\ell()$, $\ell \in N$, that is, allowing all cycles. If no negative reduced cost paths (elementary or not) are found in this way, then column generation stops. Otherwise the paths with a negative reduced cost are tested for elementarity. If elementary ones are found, they are added to the RMP before starting a new column generation iteration. If none are identified, then a customer component $W_{\bar{\ell}}()$ is added to the labels, where $\bar{\ell}$ is chosen as a vertex visited more than once in the non-elementary path with the least reduced cost (other criteria for selecting $\bar{\ell}$ can also be considered). The labeling algorithm is then reapplied using enlarged labels. This process repeats until no negative reduced paths are found or at least one negative reduced cost elementary path is identified. With this strategy, the number of customer components to consider is relatively small (that is, not many iterations of the above search are required) unless the time windows and travel times allow many cycles.

For the difficult instances, solving the NP-hard ESPPRC subproblem is often the bottleneck of the Branch-and-Price algorithm: a large part of the total computational time is devoted to solving it. As is well known for various applications, heuristics can be used to solve the subproblem as long as they succeed to generate negative reduced cost columns. When this is not the case, an exact algorithm must be invoked to ensure the exactness of the overall column generation algorithm. Examples of heuristics that were used for the ESPPRC subproblem in the VRPTW context can be found in Chabrier [18] and Desaulniers, Lessard, and Hadjar [31]. Chabrier [18] proposes omitting conditions (5.36) in the dominance rule or to heuristically eliminate arcs from network G. On top of these ideas, Desaulniers, Lessard, and Hadjar [31] developed an efficient multi-start tabu search column generator that applies a tabu search procedure starting from the path associated with each variable in the current basis of the RMP. They present computational results which show that, with such heuristics, the exact labeling algorithm needs to be invoked only once or twice per linear relaxation most of the times. Another heuristic strategy, called limited search discrepancy, that comes from constraint programming was proposed by Feillet, Gendreau, and Rousseau [41]. For each vertex, a set of "good" outgoing arcs is determined based on their reduced cost at each column generation iteration. The other arcs are qualified as "bad" arcs. When solving the subproblem, a maximum number of bad arcs can be traversed in a path. When no negative reduced cost paths are found, this maximum is increased before executing heuristic labeling again.

Finally, Irnich et al. [62] proposed an exact procedure to eliminate arcs in the subproblem network that cannot be part of an optimal solution. Given an upper bound UB on the optimal value of an integer program and an optimal dual solution α to its linear relaxation providing a lower bound LB, it is known that any integer variable whose reduced cost with respect to α is greater than $UB - LB$ can be fixed to zero (see, e.g., Nemhauser and Wolsey [91, p. 389]). To eliminate arcs in the context of a Branch-and-Price algorithm for the VRPTW, this criterion can be applied as follows. After solving an MP in a Branch-and-Bound node, one obtains a dual solution α and a lower bound LB. If one can prove that the reduced costs (with respect to this dual solution) of all variables y_r associated with a path traversing arc $(i, j) \in A$ are greater than $UB - LB$, then arc (i, j) can be eliminated. Irnich et al. [62] showed how to compute a lower bound on the reduced cost of all paths containing an arc (i, j) simultaneously for all arcs in A. Their procedure consists of using forward and backward labeling to compute feasible shortest paths from the source vertex to all other vertices in V and from the sink vertex to all other vertices in V, respectively. Their approach ensures a maximum value for these lower bounds and, thus, the maximum number of eliminated arcs. Because the ESPPRC can be highly difficult to solve, a relaxed shortest path problem, such as the ones discussed below, can also be used in this arc elimination procedure, yielding, however, fewer eliminated arcs.

Labeling algorithms are widely used for solving the ESPPRC arising as a subproblem of the VRPTW. Alternative solution methods are rare. A constraint programming algorithm was designed by Rousseau, Gendreau, and Pesant [108]. Attempts at developing Branch-and-Cut algorithms were presented in recent conferences, but, to our knowledge, they were not successful enough to appear in refereed publications.

Path Relaxations. Given the difficulty of solving the ESPPRC subproblem, Branch-and-Price algorithms can rely on certain relaxed subproblems to generate columns. These subproblems allow the generation of paths containing cycles, but they are easier to solve. On the other hand, with such paths, the MP yields, in general, a weaker lower bound. Paths with cycles are eliminated through branching decisions.

In the first Branch-and-Price algorithm for the VRPTW, Desrochers, Desrosiers, and Solomon [32] used as a subproblem a *Shortest Path Problem with Resource Constraints* (SPPRC) and 2-cycle elimination; that is, all cycles are allowed except those of the form $i - j - i$ with $i, j \in V$. In this case, the labels contain no customer components $W_\ell()$, $\ell \in N$, and a maximum of two labels is kept at each vertex for each pair of admissible time and load values (see Irnich and Desaulniers [61]). More recently, Irnich and Villeneuve [63] considered the SPPRC with k-cycle elimination for any $k \geq 2$; that is, all cycles of length k or less are forbidden. Handling the case for $k \geq 3$ needs, however, sophisticated data structures. Test results showed that, compared to the 2-cycle elimination case, using 3-cycle and 4-cycle elimination can yield higher lower bounds for instances with wide time windows and much faster computational times. On the other hand, eliminating k-cycles for $k > 4$ does not bring a sufficient increase in the lower bound to compensate for the additional time spent solving the subproblem.

Desaulniers, Lessard, and Hadjar [31] proposed another path relaxation called partial elementarity. It consists of considering a subset of the customer components $W_\ell()$, $\ell \in N$, that is selected dynamically: after solving the MP, components that would forbid cycles present in the solution are added to this subset before solving the MP again. This iterative process continues until no cycles appear in the solution or the subset contains a prespecified maximum number of components. Note that this relaxation is akin to the decremental state space relaxation discussed above. With this approach, Desaulniers, Lessard, and Hadjar [31] showed that, for most tested 100-customer instances, a maximum of 40 customer components is sufficient to obtain only elementary paths in an MP solution.

Baldacci et al. [5] and Baldacci, Mingozzi, and Roberti [6] introduced the ng-path relaxation. With every customer $i \in N$ is associated a neighborhood $N_i \subset N$ that contains i and its "closest" neighbors. An ng-path can contain a cycle $i - \cdots - j - \cdots - i$ only if it contains a vertex j such that $i \notin N_j$. The rationale behind accepting such cycles is that because i is not a neighbor of j, then returning to i after visiting j is considered a long detour and, therefore, a route containing such a cycle should not be part of an optimal MP solution. The labeling algorithm described above can easily be modified to find ng-paths. Indeed, the sole modification concerns the extension functions of the customer components $W_\ell()$, $\ell \in N$, that must set to 0 all components such that $\ell \notin N_j$ when creating a label at vertex $j \in V$. Consequently, a maximum of $|N_j|$ customer components can take value 1, increasing the chance of dominating labels when $|N_j|$ is relatively small and accelerating the solution of the subproblem. In practice, neighborhoods of size 15 seem sufficient to ensure that the paths in the computed MP solution are elementary for instances with 100 customers. Note that the neighborhoods can also be built dynamically as for the partial elementarity relaxation that can be seen as a special case of the ng-path relaxation in which all neighborhoods N_i, $i \in N$, are identical.

Dual Variable Stabilization. Column generation is well known to be subject to dual value oscillations from one iteration to the next throughout the solution process (see Vanderbeck [119]). In the first half of the iterations, these oscillations are due to poor dual information that allows the generation of irrelevant columns (this phenomenon is called the *heading-in effect*). In the second half, primal degeneracy is often present, yielding multiple dual solutions and many generated columns that are not useful. Slow convergence (called the *tailing-off effect*) is thus observed; that is, the RMP objective value converges very slowly to the MP optimal value, while searching for a complementary dual optimal solution. Clearly, dual value oscillations increase the number of iterations required to achieve optimality.

To avoid such oscillations and reduce the number of iterations, a dual variable stabilization strategy can be applied. For the VRPTW, Rousseau, Gendreau, and Feillet [107] proposed computing at each iteration a dual solution corresponding to an interior point of the RMP optimal dual facet. To do so, they first solve modified RMPs (with various right-hand sides) to generate several optimal dual solutions. Then, the interior point is obtained as a convex combination of these dual solutions computed. This simple method allows a substantial reduction of the number of iterations required to solve an MP.

Another stabilization method was developed by Kallehauge, Larsen, and Madsen [67]. Instead of solving directly the RMP at each column generation iteration, they solve its dual. This approach corresponds to a Lagrangian relaxation method in which the Lagrangian dual problem is solved by a cutting-plane algorithm. To stabilize the Lagrangian multipliers (the dual variables), each of them is restricted to taking a value within a trust region centered around the value taken in the previous iteration. With this method, the number of iterations is not necessarily reduced (and may even increase substantially), but the duals of the RMP seem to be much easier to solve than their primal counterparts, yielding an overall reduction in computational time. This method is a sequel to the Lagrangian relaxation method of Kohl and Madsen [71] that used a bundle method to stabilize the dual variables.

Valid Inequalities. As discussed by Desaulniers, Desrosiers, and Spoorendonk [30], valid inequalities in a column generation context can be defined on the variables of the compact formulation or solely on the variables of the extended formulation. In our case, the compact and extended formulations correspond to (5.1)–(5.9) and (5.10)–(5.12), respectively. Cutting planes of the first type can easily be rewritten in terms of the MP variables using the variable substitution of the Dantzig–Wolfe reformulation process (see Desaulniers et al. [27]). When involving only the arc-flow variables, they have the advantage of not modifying the structure of the subproblem. An example of such inequalities that can be applied to the VRPTW are the 2-path inequalities (Kohl et al. [70]). Let $S \subseteq N$ be a subset of customers and $v(S)$ the minimum number of vehicles required to service the customers in S while respecting vehicle capacity and time windows. Given a subset of customers S that cannot be serviced by a single vehicle (that is, with $v(S) \geq 2$), the corresponding 2-path cut forces the traversal of at least two arcs entering subset S in any feasible solution. The 2-path cuts are expressed by

$$(5.38) \qquad \sum_{k \in K} \sum_{\substack{(i,j) \in A: \\ i \in V \setminus S, j \in S}} x_{ijk} \geq 2 \qquad \forall\, S \subseteq N \mid v(S) \geq 2$$

or, equivalently, by

$$(5.39) \qquad \sum_{r \in \Omega} \beta_r^S y_r \geq 2 \qquad \forall\, S \subseteq N \mid v(S) \geq 2,$$

where β_r^S is equal to the number of arcs in route r entering subset S. The separation of these cuts consists of finding a subset S such that $v(S) \geq 2$. Given a subset S (obtained by enumeration or by a heuristic), two sufficient conditions are typically checked to determine whether $v(S) \geq 2$. First, $r(S) = \lceil q(S)/Q \rceil \geq 2$ implies $v(S) \geq 2$. Second, if the traveling salesman problem with time windows defined over S is infeasible, then we deduce that $v(S) \geq 2$. To handle these inequalities in a column generation context, one simply needs to subtract the associated dual values from the reduced cost of the arcs involved in the cuts. The 2-path cuts can be generalized to ℓ-path cuts for $\ell \geq 2$ by replacing the right-hand side of (5.38) or (5.39) by ℓ whenever $v(S) = \ell$. If $v(S) = r(S)$, these inequalities are capacity inequalities. Otherwise, to prove that $v(S) > 2$ when $v(S) > r(S)$ is rather difficult, as it amounts to solving a VRPTW with the objective of minimizing the number of vehicles used to cover the customers in S.

Other families of valid inequalities of the first type (that is, defined on the arc-flow variables) can also be applied to the VRPTW, in particular, several that were developed for the CVRP or the traveling salesman problem, such as the comb inequalities and the odd-cat inequalities.

The valid inequalities of the second type cannot be expressed as linear combinations of arc-flow variables. They are more difficult to handle, as they require modifying the subproblem. For the VRPTW, Jepsen et al. [64] were the first to propose such inequalities, which are called the *Subset Row* (SR) inequalities and correspond to a subset of the Chvátal–Gomory rank 1 cuts for the set partitioning polytope. They are defined over subsets S of the constraints (5.11) as follows:

$$(5.40) \qquad \sum_{r \in \Omega} \left\lfloor \frac{1}{k} \sum_{i \in S} a_{ir} \right\rfloor y_r \leq \left\lfloor \frac{|S|}{k} \right\rfloor \qquad \forall\, S \subseteq N,\, 0 < k < |S|.$$

In their computational experiments, Jepsen et al. considered only the SR inequalities for $|S| = 3$ and $k = 2$, which can be separated by enumeration. In this case, the SR inequalities correspond to a subset of the clique inequalities stipulating that, for any 3-customer subset S, at most one route covering at least two of these customers can be part of a feasible solution. Handling these 3-customer SR inequalities increases the complexity of the

ESPPRC subproblem, as each SR cut requires one additional resource, that is, one additional component in each label. More precisely, let $\mathscr{SR}$ be the subset of SR inequalities generated (with $|S| = 3$ and $k = 2$). Denote by S_g the subset S associated with inequality $g \in \mathscr{SR}$ and by ζ_g its nonpositive dual variable. This dual value must be subtracted from the reduced cost of a route that visits at least two customers in S_g, but it cannot be included in the arc reduced costs. In the labeling algorithm used for solving the subproblem, an additional component $B_g(\cdot)$ is added to the labels for each SR inequality $g \in \mathscr{SR}$ to count the number of customers visited in S_g. When extending a label E_i along an arc $(i, j) \in A$ to create a new label E_j, the dual values ζ_g are subtracted in the reduced cost extension function (5.29) for all cuts $g \in \mathscr{SR}$ such that $B_g(E_i) = 1$ and $B_g(E_j) = 2$. Furthermore, the dominance condition (5.33) is replaced by

$$(5.41) \qquad\qquad Z(E) - \sum_{g \in \mathscr{SR}(E,E')} \zeta_g \leq Z(E'),$$

where $\mathscr{SR}(E, E')$ is the subset of SR cuts g such that $B_g(E) = 1$ and $B_g(E') = 0$. No direct dominance conditions are required on the $B_g(\cdot)$ components of a label. Finally, note that these modifications to the labeling algorithm need further adjustments when a path relaxation allowing cycles is used to generate columns.

The computational results obtained by Jepsen et al. [64] and Desaulniers, Lessard, and Hadjar [31] show that using the SR inequalities can significantly improve the lower bound computed at the root node of the search tree, completely closing the integrality gap for most tested instances. These cuts also help in reducing the computational time for most instances. For the most difficult ones, their impact on the subproblem complexity results in instances of the subproblem that are too hard to be solved in acceptable computational times.

Other valid inequalities defined directly on the y_r variables of the VRPTW extended formulation (5.10)–(5.12) and handled in a column generation framework were also studied recently: Chvátal–Gomory rank 1 cuts by Petersen, Pisinger, and Spoorendonk [93] and clique inequalities by Spoorendonk and Desaulniers [114]. In both cases, these inequalities can help raise the lower bounds achieved at the root node, but the difficulty in treating them does not necessarily allow a reduction of the computational time.

Branching Decisions. To derive integer solutions, the column and cut generation method is embedded into a Branch-and-Bound algorithm to yield a Branch-and-Cut-and-Price method. It is well known that it is difficult to take branching decisions directly on the MP variables y_r, $r \in \Omega$. Indeed, setting $y_r = 0$ implies adding a difficult constraint to the subproblem, namely, to forbid the (re)generation of route r. The branching decisions are rather defined on the arc-flow variables and can be imposed either in the subproblem by modifying the underlying network or in the MP by adding constraints.

The most popular strategy, used by Desrochers, Desrosiers, and Solomon [32], Irnich and Villeneuve [63], Kallehauge, Larsen, and Madsen [67], Kohl et al. [70], and by Desaulniers, Lessard, and Hadjar [31], among others, consists of branching on the aggregated flow on an arc $(i, j) \in A$, that is, on the value of $x_{ij} = \sum_{k \in K} x_{ijk}$. To impose $x_{ij} = 0$, the arc (i, j) is removed from the subproblem network. To impose $x_{ij} = 1$, all arcs $(i', j) \in A$ such that $i' \neq i$ and $i' \neq 0$, and all arcs $(i, j') \in A$ such that $j \neq j'$ and $j \neq n + 1$, are removed from the network. In this case, all variables y_r in the current RMP associated with a route r that contains a removed arc must be deleted from the RMP.

Another popular strategy (see Desrochers, Desrosiers, and Solomon [32], Kohl et al. [70], and Irnich and Villeneuve [63]) that is not sufficient in itself to guarantee integrality is to branch on the total number of vehicles used $\sum_{r\in\Omega} y_r$. Such decisions are imposed by adding constraints in the MP.

Alternative branching strategies were also developed in specific works. In these strategies, the decisions were made on the time at which the service starts at a customer, on the possible successors of a customer along a route, or on the flow on the arcs entering or exiting a subset of customers.

5.3.2 ▪ Branch-and-Cut

Branch-and-Cut (see, e.g., Padberg and Rinaldi [92]) is another popular method for solving combinatorial optimization problems. It consists of a cutting plane method embedded into a Branch-and-Bound procedure. Cutting planes are added in each node of the search tree to tighten the linear relaxation as much as possible. When a linear relaxation is feasible, its computed optimal solution is fractional, and no more cutting planes can be identified by the implemented separation procedures; branching decisions are imposed to derive integer solutions.

Bard, Kontoravdis, and Yu [8] were the first to propose a Branch-and-Cut algorithm for the VRPTW in the special case where the objective function aims at minimizing the number of vehicles used (but it can easily be adapted for minimizing total traveling cost or distance). Their algorithm relies on the two-index model (5.14)–(5.19) to which the following relaxed time-window and capacity constraints are added dynamically:

$$(5.42) \qquad T_i - T_j + (s_i + t_{ij})x_{ij} \leq (1 - x_{ij})(b_i - a_j) \qquad \forall\, (i,j) \in A,$$

$$(5.43) \qquad a_i \leq T_i \leq b_i \qquad \forall\, i \in V,$$

$$(5.44) \qquad h_i - h_j + q_j \leq (1 - x_{ij})(Q - q_j) \qquad \forall\, (i,j) \in A,$$

$$(5.45) \qquad 0 \leq h_i \leq Q \qquad \forall\, i \in V,$$

where T_i and h_i, $i \in V$, are resource variables indicating the start of service time and the accumulated load at vertex i, respectively. Furthermore, Bard, Kontoravdis, and Yu suggest lifting the infeasible path inequalities (5.18) by replacing their right-hand-side member with $|A(p)| - \max\{1, r(N(p))\}$, where $N(p)$ corresponds to the set of customer vertices in path $p \in P$. Their Branch-and-Cut algorithm also involves other valid inequalities, namely, comb, incompatible pair, D_k^-, and D_k^+ inequalities. To identify violated inequalities, they developed ad hoc heuristic separation algorithms.

Lysgaard [78] proposed another Branch-and-Cut algorithm for the VRPTW that relies on the two-index model (5.14)–(5.19), where the infeasible path inequalities (5.18) are replaced by the stronger tournament inequalities. Given an infeasible path $p = (v_1, v_2, \ldots, v_\ell) \in P$, the corresponding tournament inequality is

$$(5.46) \qquad \sum_{i=1}^{\ell-1} \sum_{j=i+1}^{\ell} x_{v_i v_j} \leq |A(p)| - 1.$$

Lysgaard also introduces a new family of valid inequalities, called the reachability inequalities, that can be viewed as a strengthening of the ℓ-path inequalities. In fact, he develops different variants of these cuts. Let us present one. For a customer vertex $i \in N$, let $A_i^- \subset A$ be its reachable arc set, that is, the arc set containing all the arcs (j, j') that are part of a feasible path from the source vertex 0 to vertex i. For a subset of customers $U \subseteq N$, its reachable arc set is given by $A_U^- = \cup_{i \in U} A_i^-$. Furthermore, we say that a customer subset U (that may be a singleton) is conflicting if and only if the customers in U must be serviced on $|U|$ distinct routes in any feasible solution. A subset of the reachability inequalities is expressed as follows:

$$(5.47) \qquad \sum_{\substack{(i,j) \in A \cap A_U^-: \\ i \in V \setminus S, j \in S}} x_{ij} \geq |U| \qquad \forall\, U \subseteq S \subseteq N \text{ such that } U \text{ is conflicting.}$$

The inequality for subsets U and S stipulates that at least $|U|$ vehicles must enter S through arcs allowing one to reach the customers in U. Symmetric inequalities involving the arcs exiting subset S (and reaching arc sets) were defined as well as inequalities involving semi-conflicting pairs of customers, that is, customers that can be visited on the same route but only consecutively. The separation of the reachability inequalities requires first computing the reachable and reaching arc sets for every customer vertex $i \in N$. Then, conflicting subsets and semi-conflicting customer pairs are enumerated. Finally, for each conflicting subset and each semi-conflicting pair, an exact polynomial-time procedure permits finding violated inequalities.

A third Branch-and-Cut algorithm was designed by Kallehauge, Boland, and Madsen [66]. Again, a modified version of the two-index formulation (5.14)–(5.19) serves as the basis of this algorithm. In the version used, the capacity constraints (5.17) are replaced by subtour elimination constraints; that is, $r(S)$ is replaced by 1. Furthermore, the infeasible path inequalities are defined not only for paths that violate time windows but also for those that do not respect vehicle capacity. As in Lysgaard [78], Kallehauge, Boland, and Madsen suggest using tournament inequalities (5.46) instead of infeasible path inequalities (5.18). Furthermore, based on a previous work on the asymmetric traveling salesman problem with replenishment arcs, they develop the following lifted path inequalities:

$$(5.48) \qquad \sum_{i=1}^{\ell-1} \sum_{(v_i,j) \in \tilde{\delta}_{v_i}^+(p)} x_{v_i,j} \geq 1 \qquad \forall\, p = (v_1, v_2, \ldots, v_\ell) \in P,$$

where $\tilde{\delta}_{v_i}^+(p) = \{(v_i,j) \in A \mid j \neq v_1, \ldots, v_{i+1},\ T_{v_i} + s_{v_i} + t_{v_i,j} \leq b_j,\ \sum_{h=1}^{i} q_h + q_j \leq Q\} \cup \{(v_i, n+1) \in A\}$ for $i = 1, 2, \ldots, \ell - 1$ contains all the arcs leaving vertex v_i that allow escaping path p while respecting time windows and vehicle capacity. They show that, under certain assumptions, these lifted path inequalities are facet-defining. They also transfer to the VRPTW precedence inequalities that were introduced for the precedence constrained asymmetric traveling salesman problem. Indeed, the time windows allow one to define predecessor and successor relationships between the customers. Let $\pi(j) = \{i \in V \mid a_j + s_j + t_{ji} > b_i\}$ and $\sigma(j) = \{i \in V \mid a_i + s_i + t_{ij} > b_j\}$ be the predecessor and successor sets of customer $j \in N$. The precedence inequalities are given by

$$(5.49) \qquad \sum_{\substack{(i,\ell) \in A: \\ i \in S \setminus \pi(j), j \in \bar{S} \setminus \pi(j)}} x_{ij} \geq 1 \qquad\qquad \forall\, S \subseteq N, j \in S,$$

$$(5.50) \qquad \sum_{\substack{(i,\ell) \in A: \\ i \in \bar{S} \setminus \sigma(j), j \in S \setminus \sigma(j)}} x_{ij} \geq 1 \qquad\qquad \forall\, S \subseteq N, j \in S,$$

where $\bar{S} = V \setminus S$. The inequalities (5.49) and (5.50) are, more precisely, called the predecessor and successor inequalities (or π- and σ-inequalities), respectively. The Branch-and-Cut algorithm of Kallehauge, Boland, and Madsen [66] also involves odd-cat, D_k^-, and D_k^+ inequalities.

Letchford and Salazar González [74] compare the strength of the lower bounds provided by different formulations of the CVRP. In particular, they show that several types of valid inequalities that can be used to tighten a two-index formulation are implied by the set partitioning formulation when elementary routes are considered, whereas a subset of them are also implied when non-elementary routes are allowed. These results can be transferred to the VRPTW. These authors also introduce a two-commodity flow formulation for the VRPTW that allows them to derive a new family of valid inequalities. These inequalities are called projection inequalities because they are obtained by a projection onto the subspace of the arc-flow variables x_{ij}. They can thus be used with the two-index model (5.14)–(5.19). Intuitively, these inequalities state that, for any subset $S \subseteq N$ of customers, the sum of the upper time limits of the vehicles leaving S must be greater than or equal to the sum of the lower time limits of vehicles entering S plus the time spent by the vehicles inside S.

5.3.3 ▪ Reduced Set Partitioning

Baldacci, Mingozzi, and Roberti [6] (see also Baldacci et al. [5]) developed an efficient procedure to reduce the size of the VRPTW set partitioning model (5.10)–(5.12) by fixing to 0 the value of a very large number of its variables. The reduced set partitioning model is then solved using a MIP commercial solver such as CPLEX. The following criterion, which is valid for any integer linear program, is used to fix variables to 0: Given an integer linear program and an upper bound U on its optimal value, a non-negative integer variable takes value zero in every optimal integer solution if its reduced cost with respect to a feasible dual solution of a linear programming relaxation exceeds $U - L$, where L is the cost of the dual solution that provides a lower bound on the optimal value. The variable can therefore be fixed to 0 or, equivalently, eliminated from the problem formulation.

The upper bound U can be set to the cost of a good heuristic solution. To compute a lower bound L and a feasible dual solution $(\pi_i)_{i \in N \cup \{0\}}$ of the linear relaxation of (5.10)–(5.12), Baldacci, Mingozzi, and Roberti [6] proposed a dual ascent method that combines Lagrangian relaxation, subgradient optimization, and column generation. In this method, up to four heuristics can be used sequentially. In the first heuristic, routes can contain cycles and may not respect either vehicle capacity or time windows. In the second heuristic, ng-routes are generated to yield a better dual solution. In the third heuristic, elementary paths are considered. Finally, the fourth heuristic integrates subset row inequalities defined on subsets of three customers. The dual solution obtained with a heuristic serves as the starting point of the next heuristic.

Given the final computed dual solution $(\pi_i)_{i \in N \cup \{0\}}$ and its cost L, a dynamic algorithm is applied to find all feasible routes in network G whose reduced cost with respect to $(\pi_i)_{i \in N \cup \{0\}}$ is less than or equal to $U - L$. Only the variables associated with these routes are considered in the reduced set partitioning model that also includes the subset row inequalities identified by the fourth heuristic. When there are too many variables, only a subset of the variables (those with the least reduced costs) is enumerated by dynamic programming, yielding an overall heuristic solution method.

5.3.4 ▪ Summary Computational Results for Exact Methods

For more than two decades, the exact solution methods have been tested on the VRPTW benchmark instances proposed by Solomon [113]. These 100-customer instances are

Table 5.1. *Results of the most recent exact methods.*

Instances		JPSP08		DLH08		BMR11	
		No.	Avg.	No.	Avg.	No.	Avg.
Series	No.	Solved	Time	Solved	Time	Solved	Time
C1	9	9	468	9	18	9	25
RC1	8	8	11,005	8	2,150	8	276
R1	12	12	27,412	12	2,327	12	251
C2	8	7	2,795	8	2,093	8	40
RC2	8	5	3,204	6	15,394	8	3,767
R2	11	4	35,292	8	63,068	10	28,680
Total	56	45		51		55	

divided into two classes of 29 (class 1) and 27 (class 2) instances, each containing three series (C, RC, and R). In general, the time windows are relatively narrow in class 1 and wide in class 2. The locations of the customers are clustered in the C instances, random in the R instances, and mixed in the RC series.

For each series of 100-customer instances, Table 5.1 reports computational results obtained by the three most recent exact algorithms, namely, those of Jepsen et al. [64], Desaulniers, Lessard, and Hadjar [31], and Baldacci, Mingozzi, and Roberti [6], abbreviated by JSPS08, DLH08, and BMR11, respectively. In this table, the first two columns indicate the instance series and the number of instances it contains. For each series and each algorithm, we report the number of instances that were solved to optimality (no time limit imposed) and the average computational time in seconds to solve them. These times are those reported by the authors and were obtained on computers with different characteristics: P4 at 3.0 GHz for JSPS08, AMD Opteron at 2.6 GHz for DLH08, and IBM Intel Xeon X7350 at 2.93 GHz for BMR11. The last line of Table 5.1 provides the total number of instances solved by each algorithm.

These results clearly show that the instances in class 2 are much more difficult to solve than those in class 1 because wide time windows increase the number of feasible routes and the number of customers per feasible route, yielding harder to solve instances. Furthermore, we observe that, on average, the clustered instances are the easiest to solve and the totally random instances the hardest. The results also show the rapid evolution of the recent algorithms. Before the paper of Jepsen et al. [64], only 35 of the 56 instances were solved to optimality. With the introduction of the SR inequalities in a Branch-and-Cut-and-Price algorithm, Jepsen et al. raised this number to 45. Using an efficient tabu search column generator, the Branch-and-Cut-and-Price algorithm of Desaulniers, Lessard, and Hadjar [31] solved six additional instances. With their approach based on a reduced set partitioning model and relying on ng-paths, Baldacci, Mingozzi, and Roberti [6] solved all instances except one due to a lack of memory. Their computational times are much smaller than the previous ones, but it should be noted that their method requires an upper bound and, for their experiments, they used the best known upper bounds from the literature. Note that, in a recent conference presentation, Ropke [105] reported solving all Solomon instances using a Branch-and-Cut-and-Price algorithm relying on a strong branching strategy. Because no technical paper describing the algorithm is available yet, this approach was not covered above.

In the future, one can expect to see more computational results for the Gehring and Homberger benchmark instances [45], which extend the Solomon data set to instances involving 200, 400, 600, 800, and 1000 customers. To the best of our knowledge, only

Larsen [73], Cook and Rich [20], and Kallehauge, Larsen, and Madsen [67] report solving some of these instances.

5.4 ▪ Heuristics

Heuristics are solution methods that can often find good-quality feasible solutions relatively quickly. However, there is no guarantee regarding solution quality. Heuristics are, thus, tested empirically, and their performance is judged by their computational results. Currently, most VRPs encountered in the industry are solved using heuristics because of their speed and their ability to handle large instances. Tradition dictates that a hierarchical objective function is used when heuristics are applied: the first priority is to minimize the number of vehicles used and the second to minimize the cost of the traversed arcs. This differs from the exact algorithms that do not consider the number of vehicles in the objective function.

In the past decade research on heuristic methods for the VRPTW has focused on metaheuristics and little research has been conducted on construction heuristics. An exception is the work of Ioannou, Kritikos, and Prastacos [59], which proposes a relatively fast construction heuristic which can produce solutions of good quality considering the devoted computational time. Construction heuristics for the VRPTW were surveyed by Bräysy and Gendreau [16].

5.4.1 ▪ Basics

A central concept in most successful heuristics for the VRPTW is that of *local search*. Local search algorithms are based on neighborhoods. Let $\mathscr{S}$ be the set of feasible solutions to a given VRPTW instance, and let $c : \mathscr{S} \to \mathbb{R}$ be a function that maps from a solution to the *cost* to this solution. The set $\mathscr{S}$ is finite, but it is often extremely large. Since the VRPTW is a minimization problem, our goal is to find a solution s^* for which $c(s^*) \leq c(s)$ for all $s \in \mathscr{S}$. However, with heuristics, we are willing to settle for a solution that might be slightly inferior to s^*.

Let $\mathscr{P}(\mathscr{S})$ be the set of subsets of solutions in $\mathscr{S}$. We define a *neighborhood function* as a function $N : \mathscr{S} \to \mathscr{P}(\mathscr{S})$ that maps from a solution s to a subset of solutions $N(s)$. This subset is called the *neighborhood* of s. A solution s is said to be *locally optimal* or a *local optimum* with respect to a neighborhood $N(s)$ if $c(s) \leq c(s')$ for all $s' \in N(s)$. With these definitions, we can describe a steepest descent algorithm (see Algorithm 5.1). The algorithm takes an initial solution s as input. At each iteration, it finds the best solution s' in the neighborhood $N(s)$ of the current solution s (line 4). If s' is better than s (line

---

Algorithm 5.1 Steepest descent

1: input: Initial solution $s \in \mathscr{S}$
2: done = false
3: **while** done $\neq$ true **do**
4:      $s' \in \arg\min_{s'' \in N(s)} \{c(s'')\}$
5:      **if** $c(s') < c(s)$ **then**
6:          $s = s'$
7:      **else**
8:          done = true
9: return s

---

5), then s' replaces s as the current solution (line 6). Lines 3–8 are repeated as long as s' is an improved solution. When the loop stops, the algorithm returns s as the best solution found. The algorithm is called a steepest descent algorithm because it always chooses the best solution in the current neighborhood. As we will see in the following, there are other ways to use neighborhoods to explore the solution space.

In general, larger neighborhoods lead to solutions of better quality when the neighborhoods are used, for example, in a steepest descent algorithm. The drawback of larger neighborhoods is that the evaluation of all solutions is more time consuming in a large neighborhood compared to a smaller one unless clever algorithmic ideas can be used to speed up the search. In the following subsections, we review a number of neighborhoods that have been used in the most successful VRPTW heuristics. We classify these neighborhoods into two categories: *traditional* and *large* neighborhoods. The first category contains the neighborhoods whose size is growing polynomially with n in a controlled manner such that all solutions in the neighborhood can be evaluated by explicit enumeration. Polynomial-sized neighborhoods whose growth is so rapid that they cannot be searched explicitly and neighborhoods whose size is growing exponentially with n fall in the second category (as in Ahuja et al. [1]).

5.4.1.1 ▪ Traditional Neighborhoods

Traditional neighborhoods can be divided into two categories: *intra-route* and *inter-route*. The neighborhoods in the former category contain solutions in which a single route is changed with respect to the reference solution, whereas, in the second category, they contain solutions obtained by moving customers between two or more routes. A thorough survey of VRPTW neighborhoods can be found in Bräysy and Gendreau [16]. Below we review the traditional neighborhoods that have been used in the most important metaheuristics of the past decade.

2-opt Neighborhood (intra-route).   A 2-opt neighborhood (see Figure 5.1) contains solutions obtained by removing two arcs from a route and replacing them by two other arcs to reconnect the route, while changing the orientation of the subpath that does not contain the depot. In the figure the square represents the depot and the circles are customers. The dashed arcs correspond to subpaths in network G (see Section 5.2), involving one or several arcs, while each solid arc corresponds to a single arc in G. Notice that changing the orientation of the subpath $(i+1,\ldots,j+1)$ can be problematic because of the time windows.

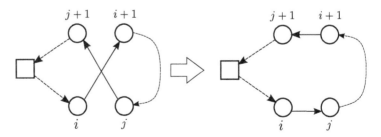

Figure 5.1. *Illustration of the 2-opt neighborhood.*

Or-opt Neighborhood (intra-route).   Each solution in the Or-opt neighborhood (see Figure 5.2) is defined by relocating a subpath $(i+1,\ldots,j)$ to a different position in a route. The orientation of the relocated subpath is preserved. The move is carried out by

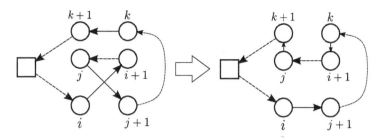

Figure 5.2. *Illustration of the Or-opt neighborhood.*

removing arcs $(i, i+1)$, $(j, j+1)$, and $(k, k+1)$ and replacing them by arcs $(i, j+1)$, $(k, i+1)$, and $(j, k+1)$. Typically the neighborhood is reduced by considering only subpaths containing a limited number of customers or by trying only insertions that are close to the original position. Bräysy [15] introduced a slight variant of the Or-opt neighborhood, denoted IOPT, by allowing the relocated subpath to be reversed when it is reinserted.

2-opt* Neighborhood (inter-route). The 2-opt* neighborhood (see Figure 5.3) is defined similarly to the 2-opt neighborhood, but its solutions are derived by modifying two routes instead of one. Two arcs, $(i, i+1)$ and $(j, j+1)$, from two distinct routes are removed, and the routes are reconnected by inserting the arcs $(i, j+1)$ and $(j, i+1)$. The effect is that the tails of the two routes are exchanged. The 2-opt* neighborhood does not change the orientation of the subpaths, as opposed to the 2-opt.

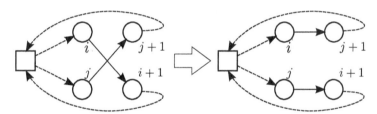

Figure 5.3. *Illustration of the 2-opt* neighborhood.*

Cross Exchange Neighborhood (inter-route). In the cross exchange neighborhood (see Figure 5.4), two subpaths are selected and their positions are exchanged. This is done by removing four arcs $(i, i+1)$, $(j, j+1)$, $(k, k+1)$, and $(l, l+1)$ and replacing them by the arcs $(i, k+1)$, $(l, j+1)$, $(k, i+1)$, and $(j, l+1)$. The size of the cross exchange neighborhood is typically reduced by considering only subpaths containing a limited number of customers. We note that the cross exchange neighborhood can be used in an intra-route fashion where the subpaths to be exchanged belong to the same route.

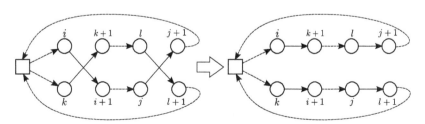

Figure 5.4. *Illustration of the cross exchange neighborhood.*

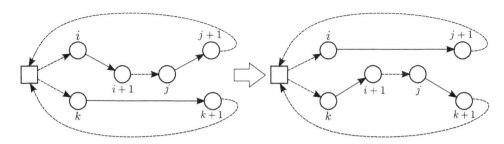

Figure 5.5. *Illustration of the path relocation neighborhood.*

Path Relocation Neighborhood (inter-route). In the path relocation neighborhood (see Figure 5.5), the solutions are obtained by relocating a subpath from one route to another one. This is done by removing three arcs $(i, i + 1)$, $(j, j + 1)$, and $(k, k + 1)$ and replacing them by the arcs $(i, j + 1)$, $(j, k + 1)$, and $(k, i + 1)$. This neighborhood can be seen as a special case of the cross exchange neighborhood if we allow one empty subpath in the latter neighborhood. The length of the relocated subpath is typically limited, and some implementations allow the subpath to be reversed when it is reinserted.

Speedup Techniques. In local search, it is important to implement neighborhood evaluation efficiently in order to search as many solutions as possible within the time allocated for the search. Each move in the neighborhoods mentioned above can be evaluated in constant time using the techniques described in Kindervater and Savelsbergh [68], but for large instances this might not be enough. It is therefore common to truncate the neighborhood search in a heuristic way such that some moves that do not seem promising are not evaluated. This can be done by considering only moves that connect nearby customers; time-window considerations can also be taken into account. Two examples of such *filtering* can be found in Hashimoto and Yagiura [53] and Nagata, Bräysy, and Dullaert [88].

Irnich [60] introduced *sequential search* for the VRPTW. Sequential search is able to speed up local search by using exact filtering where large subsets of the possible moves can be skipped based on cost considerations. Computational results show impressive speedups for large-scale VRPTW instances. Sequential search has, to the best of our knowledge, not yet been used in the most successful VRPTW heuristics.

5.4.1.2 ▪ Large Neighborhoods

The use of large neighborhoods in heuristics has been surveyed by Ahuja et al. [1] and Pisinger and Ropke [95]. Methods for searching large neighborhoods can be divided in two categories: exact and heuristic evaluation methods. An exact evaluation should always identify the best solution within the neighborhood, while a heuristic evaluation may well miss it. Some exponential-sized neighborhoods can be searched exactly in polynomial time. An example is the assignment neighborhood for the *Traveling Salesman Problem* (TSP), which is described by Ahuja et al. [1]. Other exponential-sized neighborhoods are searched exactly by solving NP-hard optimization problems. For example, the cyclic exchange neighborhood of Thompson and Psaraftis [116] can be searched exactly using the algorithms presented in Chapter 4 of Dumitrescu [38].

Exact evaluation of large neighborhoods have not been used in the most successful heuristics for the VRPTW, but large neighborhood heuristic evaluation is an important component in several recent heuristics (for example, see Lim and Zhang [75]). In the

VRPTW literature, the most common large neighborhoods are encountered in the *Large Neighborhood Search* (LNS) framework put forward by Shaw [111] (and the closely related ruin and recreate algorithm of Schrimpf et al. [110]). The key elements in LNS are the *destroy* and *repair* operators. A destroy operator partially disintegrates a solution, while a repair operator reconstructs a complete solution starting from a partial solution. The two operators are used repeatedly, as illustrated in the pseudo-code of Algorithm 5.2. In this pseudo-code x is the current solution, x^* is the best solution observed during the search, and x' is a temporary solution. The function $d(\cdot)$ destroys a solution, $r(\cdot)$ rebuilds a complete solution, and $c(\cdot)$ calculates the cost of a solution. The algorithm repeats lines 4 to 8 until a stopping criterion is met (e.g., a maximum number of iterations). In line 4 the destroy and repair operators are applied. In line 5 a test determines whether the new solution x' should be accepted as the current solution. In the most simple case only improving solutions are accepted.

Algorithm 5.2 Large neighborhood search

1: Input: initial solution x
2: $x^* = x$
3: **while** stop criterion not met **do**
4:    $x' = r(d(x))$
5:    **if** accept(x, x') **then**
6:       $x = x'$
7:    **if** $c(x') < c(x^*)$ **then**
8:       $x^* = x'$
9: return x^*

An illustration of how the destroy/repair operators can be used to improve a VRPTW solution is shown in Figure 5.6. In this example the destroy operator removes a subset of customers from their routes and the repair operator reinserts unassigned customers into the routes.

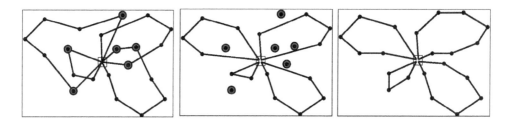

Figure 5.6. *Example of destroy/repair operation. Left figure shows the initial solution with customers to be removed marked. The middle figure illustrates the partial solution after the destroy operation. The right figure shows the solution after the repair operation where the unassigned customers are reinserted.*

5.4.1.3 ▪ Allowing Infeasible Solutions

When designing metaheuristics for the VRPTW, one must decide whether infeasible solutions should be allowed during the search. When allowed, they are typically penalized in the objective function by one or more terms that aim at reducing infeasibility. Allowing infeasible solutions makes it easier to maneuver in the solution space, providing shortcuts between areas of high-quality solutions. On the other hand, it adds complexity to the algorithm: evaluation of the objective may be more difficult, penalty parameters must be adjusted (potentially in a adaptive way), and one must ensure that feasible solutions are visited at least occasionally. Despite these drawbacks, allowing infeasible solutions appears to be an important tool for finding high-quality solutions.

In the current metaheuristics three types of infeasibilities are allowed: time window, vehicle capacity, and customer service violations. The last type occurs when some customers are not visited. It is the most common type in the algorithms that minimize the number of vehicles used (see the next subsection). Time window and capacity violations are typically allowed together and lead to a penalized objective function $\bar{c}(s)$:

$$\bar{c}(s) = c(s) + \alpha q(s) + \beta w(s),$$

where $c(s)$ is the standard objective function and $q(s)$ (resp., $w(s)$) measures the total capacity (resp., time-window) violation over all routes. α and β are parameters that are usually adjusted during the search, reacting to the performance of the algorithm. A good example of how the α and β parameters can managed is given in Cordeau, Laporte, and Mercier [23].

Time-window violation along a route $r = (v_0 = 0, v_1, v_2, \ldots, v_k, v_{k+1} = n+1)$ has traditionally been calculated straightforwardly: the start of service time T_{v_i} at vertex v_i is computed as

$$T_{v_i} = \begin{cases} a_0 & \text{if } i = 0, \\ \max\{a_{v_i}, T_{v_{i-1}} + s_{v_{i-1}} + t_{v_{i-1}v_i}\} & \text{otherwise,} \end{cases}$$

and the total violation is $\sum_{i=1}^{k+1} \max\{0, T_{v_i} - b_{v_i}\}$. The drawback of this procedure is that the evaluation of a move in the traditional neighborhoods can no longer be performed in constant time. Instead, Nagata [85] proposed an artificial start of service time evaluation

$$T'_{v_i} = \begin{cases} a_0 & \text{if } i = 0, \\ \max\{a_{v_i}, \min\{b_{v_i}, T'_{v_{i-1}}\} + s_{v_{i-1}} + t_{v_{i-1}v_i}\} & \text{otherwise} \end{cases}$$

that moves back the start of service time to the end of the time window whenever it is violated. Time-window violation is given by $\sum_{i=1}^{k+1} \max\{0, T'_{v_i} - b_{v_i}\}$, a formula that still captures time-window violations and makes it possible to evaluate in constant time a single move in the traditional neighborhoods (see Vidal et al. [120]).

5.4.1.4 ▪ Minimizing the Number of Vehicles Used

The vehicle minimization/traveling cost minimization is typically carried out in a two-phase manner where one first seeks to find the least number of vehicles necessary and then keeps this number fixed while minimizing the total traveling cost. If the heuristic allows visiting infeasible solutions, then one can select a number of vehicles m and create a (possibly infeasible) solution with that number of vehicles. Then a search is conducted with the aim of finding a feasible solution. If the search is successful, the number of vehicles is reduced by one by deleting a complete route before repeating the search. Otherwise the number of vehicles is increased by one and the search is repeated if no feasible solutions have been found yet. Using a computed lower bound on the number of vehicles, it is possible to stop the search for a fewer number of vehicles when the lower bound is met. The search for the minimum number of vehicles can also be aborted when a sufficient effort has been invested. Nagata and Bräysy [87] have devised one of the most effective route minimization procedure.

5.4.2 ▪ Single Trajectory Search

Starting from a single solution, single trajectory search algorithms generate a sequence of solutions that can be seen as a trajectory through the solution space. At every iteration, only the current solution is used to determine the next one. In the following paragraphs, we review two families of single trajectory search algorithms that were applied to solve the VRPTW, namely, iterated local search and LNS algorithms. Note that these algorithms can rely on tabu search and simulated annealing schemes.

5.4.2.1 ▪ Iterated Local Search

Iterated local search (Lourenço, Martin, and Stützle [76]) is a simple metaheuristic that is based on a series of steepest descent local searches. A pseudo-code of this algorithm is given in Algorithm 5.3. Starting from an initial solution x, the algorithm iterates (lines 4 to 9) between perturbing the current solution x to obtain a temporary solution x' and applying local search to improve this solution until a stopping criterion is met (e.g., a maximum number of iterations). In line 4, x is perturbed, for example, by applying one or more random moves defining a chosen neighborhood function. In line 5, the local search algorithm is typically a descent algorithm, but it could also be a more advanced metaheuristic like tabu search (see, e.g., Cordeau and Maischberger [24]). In lines 6 and 7, the current solution x is updated if the solution x' obtained from the local search is accepted. In lines 8 and 9, the best known solution x^* is updated, if necessary.

Algorithm 5.3 Iterated local search

1: Input: initial solution x
2: $x^* = x$
3: **while** stop criterion not met **do**
4:     $x' = \text{perturb}(x)$
5:     $x' = \text{localsearch}(x')$
6:     **if** $\text{accept}(x, x')$ **then**
7:         $x = x'$
8:     **if** $c(x') < c(x^*)$ **then**
9:         $x^* = x'$
10: return x^*

Ibaraki et al. [57] developed an iterated local search algorithm for a variant of the VRPTW where the time windows are represented by convex, piecewise linear penalty functions. The VRPTW is a special case of this problem when the penalty function is set appropriately (infinite penalty outside the feasible region). Because of the penalty functions, it is non-trivial to determine the best possible starting time for a given route. This problem is called the *Optimal Start Time Problem* (OSTP). The authors present a dynamic programming algorithm for the OSTP and describe how the dynamic programming algorithm can be sped up in a local search algorithm. Note that the OSTP was previously studied by Dumas, Soumis, and Desrosiers [37], who considered arbitrary convex penalty functions.

Their iterated local search algorithm uses a number of neighborhoods: Or-opt (or IOPT), 2-opt, 2-opt*, path relocation, and cross exchange. Visiting infeasible solutions with respect to time windows or capacity is allowed, but penalized with the penalization parameters set adaptively. When a local minimum is encountered, a perturbation that consists of performing one to three random cross exchange moves is applied.

Hashimoto, Yagiura, and Ibaraki [54] proposed an iterated local search algorithm for a time-dependent VRPTW with soft time windows by iterated local search. As before, it is a non-trivial task to determine the optimal starting time of a route, and a dynamic programming algorithm is presented. The iterated local search algorithm uses three neighborhoods: the 2-opt* and cross exchange inter-route neighborhoods and the Or-opt intra-route neighborhood. When a local optimum is encountered, the current solution is perturbed by applying a single, random cross exchange move to the solution. The algorithm is used to solve VRPTW instances by allowing time window violations and adjusting penalties appropriately.

Cordeau and Maischberger [24] introduced an iterated local search algorithm that relies on an updated version of the tabu search algorithm of Cordeau, Laporte, and Mercier [22, 23] as the local search algorithm. The tabu search algorithm allows solutions that are infeasible with respect to both the time windows and vehicle capacity. It uses a simple neighborhood based on customer relocation. The perturbation step is inspired by LNS and consists of removing a cluster of customers and reinserting them in a random order. Each customer is reinserted in a route using a cheapest-insertion policy.

Another iterated local search algorithm was proposed by Lim and Zhang [75]. Unlike the above iterated local search algorithms, their algorithm does not allow infeasible solutions but yields, nevertheless, good results. In our opinion, the Lim and Zhang algorithm is, however, more complex than the others.

5.4.2.2 ▪ Large Neighborhood Search

The original LNS heuristic by Shaw [111] already solved the VRPTW with good results on a reduced set of instances, but it was only with the work of Bent and Van Hentenryck [11] that the method's ability for obtaining good solutions to the VRPTW was fully revealed. Bent and Van Hentenryck used a two-phase approach. In the first phase a simulated annealing heuristic based on traditional neighborhoods is applied to minimize the number of vehicles, while, in the second phase, LNS is invoked to minimize travel costs. In the LNS algorithm, solutions are destroyed by removing groups of related customers, while the repair mechanism uses a Branch-and-Bound algorithm to reinsert the customers. The Branch-and-Bound algorithm is truncated using *limited discrepancy search* (Harvey and Ginsberg [52]) in order to keep the computational times low. Only improving solutions are accepted.

Pisinger and Ropke [94] solved the VRPTW (and other VRP variants) by transforming each instance into a pickup-and-delivery problem with time windows that is then solved using the LNS heuristic developed in Ropke and Pisinger [106]. Several destroy and repair operators are implemented, and an adaptive mechanism is used to favor the application of the most successful destroy/repair operators in the previous iterations. Because of this feature the heuristic is termed *Adaptive Large Neighborhood Search* (ALNS). The repair operators are based on quick, greedy heuristics. Simulated annealing is used as an outer metaheuristic framework to decide whether deteriorating solutions should be accepted.

Another paper that proposes an LNS heuristic relying on quick (greedy) repair methods is due to Mester and Bräysy [80]. They embedded the LNS moves within an evolution strategy metaheuristic (see Beyer and Schwefel [12]) and also exploited traditional neighborhoods during the search.

Prescott-Gagnon, Desaulniers, and Rousseau [98] developed an ALNS heuristic based on a much stronger repair operator. Solution destruction is performed using four removal heuristics, while the repair step uses heuristic Branch-and-Price, adapted from the

Branch-and-Cut-and-Price algorithm of Desaulniers, Lessard, and Hadjar [31]. In order to make each repair step sufficiently fast the original algorithm is modified in several ways: (1) the column generation subproblem is only solved heuristically using a tabu search algorithm; (2) for large instances, the column generation process is stopped prematurely if the master problem objective value has not improved for a given number of iterations; (3) no cuts are generated; and (4) when branching is required, the value of the largest fractional variable in the master problem is permanently fixed to one without any backtracking possibility. To diversify the search, all feasible solutions generated by the repair step are accepted, even if the new solution is worse than the current solution.

5.4.3 ▪ Population-Based Search

A number of metaheuristics are based on the idea of maintaining a pool of solutions, called a *population*, that evolves at each iteration of the solution process. Unlike in single trajectory search, new solutions are derived from a population of solutions that offers diversity in itself. Below, we survey the recent works on the VRPTW based on evolutionary and path relinking algorithms, two families of population-based search heuristics.

5.4.3.1 ▪ Evolutionary Algorithms

Evolutionary algorithms combine solutions from the current population to produce offsprings. The most fit of them are then retained to update the population. Examples of this type of algorithms are *genetic algorithms* (Holland [56] and Reeves [101]) and *memetic algorithms* (Moscato [81] and Moscato and Cotta [82]).

Memetic algorithms hybridize genetic algorithms with local search and potentially problem-specific algorithms. The pseudo-code of a simple memetic algorithm is sketched in Algorithm 5.4. The size of the initial population created in line 1 is obviously an important parameter. Lines 2 to 6 constitute the main loop of the algorithm which can be stopped according to various criteria based on the number of iterations performed, the elapsed time, or a measure of convergence. In line 3, new solutions forming set S are constructed by combining existing solutions from P. This operation is called a *crossover*. In line 4, mutation is applied to a subset of the solutions in S to create a new subset of offsprings S', perturbing in this way the offspring set. The mutation step is not included in all memetic algorithms. In line 5, local search is executed on each offspring solution in S' to generate a set S'' of improved solutions. Sometimes this local search step is seen as a mutation operation. The solutions in S'' are then used to update the population P (line 6). A simple way to perform this update consists of keeping the $|P|$ best solutions from the set $P \cup S''$, but a more advanced procedure is typically used to favor diversity in population P. A classical genetic algorithm is obtained by removing the local search step (line 5) in Algorithm 5.4.

Algorithm 5.4 Memetic algorithm

1: Create an initial pool P of solutions
2: **while** stopping criterion is not met **do**
3:    $S = \text{crossover}(P)$
4:    $S' = \text{mutation}(S)$
5:    $S'' = \text{localsearch}(S')$
6:    $P = \text{updatepool}(P, S'')$
7: return best solution observed during the search

A vital component in memetic algorithms is the crossover algorithm. Below we give an overview of the crossover methods used in the most successful evolutionary algorithms for the VRPTW.

EAX Crossover. The EAX crossover algorithm is based on a strong crossover algorithm for the TSP that was first introduced by Nagata and Kobayashi [89] and has been at the foundation of the best evolutionary algorithms for the TSP since then (see, e.g., Nagata and Kobayashi [90]). The EAX algorithm was later adapted to the CVRP (Nagata [84] and Nagata and Bräysy [86]) and, finally, to the VRPTW by Nagata, Bräysy, and Dullaert [88]. In five steps, the EAX algorithm combines two solutions α and β defined by the arc sets A_α and A_β, respectively. Step 1 constructs a graph $G_{\alpha\beta}$ containing the arcs $A_{\alpha\beta} = (A_\alpha \cup A_\beta) \setminus (A_\alpha \cap A_\beta)$. Step 2 partitions $A_{\alpha\beta}$ into a number of $\alpha\beta$-cycles. An $\alpha\beta$-cycle is a cycle in $G_{\alpha\beta}$ alternating between arcs in A_α and arcs in A_β and such that all arcs in A_α have an opposite orientation to that of the arcs in A_β. Step 3 builds an *E-set* by combining $\alpha\beta$-cycles, the simplest strategy being to select a single $\alpha\beta$-cycle as an E-set. Step 4 creates an offspring solution defined by the arcs in $(A_\alpha \setminus (E \cap A_\alpha)) \cup (E \cap A_\beta)$, where E are the arcs in the selected E-set. Because this offspring solution may contain subtours, step 5 eliminates subtours one by one by deleting some arcs in the subtours and adding others to connect the resulting pieces. Notice that the built offspring is dependent on the choices made in steps 2, 3, and 5. Furthermore, it does not necessarily satisfy the time-window and capacity constraints. To obtain a feasible solution, a repair procedure performing 2-opt*, Or-exchange, node relocation, and node exchange moves is applied. Solutions that remain infeasible after the repair step are discarded, while feasible solutions are improved using local search. The EAX crossover was later used by Błocho and Czech [13], who embedded it in a parallel algorithm.

OX Crossover and Giant Tour Representation. The OX crossover is a classic crossover for permutation-based solution representations (see, e.g., Falkenauer and Bouffouix [39]). Figure 5.7 illustrates this crossover for permutations of the numbers from 1 to 9. One chooses two random numbers i and j and creates an offspring permutation by copying the piece of parent 1 from position i to j into position i to j of the offspring. The remaining part of the offspring is found by copying elements of parent 2, not already in the offspring, into position $j + 1$ and forward (in a circular fashion), respecting the order of parent 2.

position	1	2	3	4	5	6	7	8	9
Parent 1	4	3	8	2	9	6	5	7	1
Parent 2	5	7	3	2	8	4	9	1	6
offspring	3	4	8	2	9	6	1	5	7

Figure 5.7. *OX crossover example with $i = 3$ and $j = 6$.*

In order to apply the OX crossover for the VRPTW, one needs to define a mapping from a solution to a permutation and back again. To do so for the CVRP, Prins [99] proposed a mapping from a solution to a permutation based on a *giant tour representation* of a solution. This representation selects an ordering of the routes in the solution and then lists the customers of the first route (in order of appearance), the customers of the second route, and so on. To transform a permutation back into a CVRP solution, Prins suggested a *split* procedure (introduced by Beasley [10]) that cuts the permutation into a number of segments, each constituting a route. The cutting points are determined optimally (that is, they must minimize the total cost of the resulting solution) in polynomial time using a shortest path calculation in an auxiliary network. Recently, Vidal et al. [120] applied the same crossover operator to the VRPTW with great success. Their algorithm allows the split procedure to generate infeasible solutions by adding penalized arcs in the auxiliary network corresponding to infeasible routes. A local search heuristic based on traditional neighborhoods can be called with a certain probability to try to repair the infeasible solutions. Both feasible and infeasible solutions are kept separately in the solution pool.

Multiparent Recombination Repoussis, Tarantilis, and Ioannou [102] took a different approach to obtaining offspring solutions. Their operator produces an offspring from a single parent by taking the entire population into account. In each generation the algorithm counts the number of occurrences m_{ij} of each arc (i, j) in the entire population. An arc is considered promising if it occurs in more than θ solutions in the population. The threshold θ is calculated in each generation based on the current diversity of the population. If the population is diverse, θ is lower than if the population contains similar solutions. Offsprings are generated from a parent solution using an LNS algorithm. The destroy operator removes customers adjacent to arcs (i, j) for which $m_{ij} < \theta$. The customers are reinserted by a greedy randomized heuristic. The generated offsprings are then improved using traditional neighborhood searches in two phases: tabu search is applied in the vehicle number minimization phase, whereas guided local search aims at distance minimization.

5.4.3.2 • Path Relinking

Hashimoto and Yagiura [53] developed a path relinking algorithm for the VRPTW that works with a pool of solutions and allows infeasible solutions which are penalized. The penalty weights are controlled adaptively depending on whether feasibility is easy or hard to attain. In each outer iteration of the algorithm, two solutions σ_A and σ_B from the pool are selected randomly. The algorithm transforms σ_A into σ_B by a series of 2-opt* and Or-opt moves. This generates a number of solutions in between the σ_A and σ_B solutions. In the inner loop, some of these solutions are improved using 2-opt*, cross exchange, and Or-opt neighborhoods. Certain generated solutions are added to the solution pool if they improve the quality of the pool, taking into account total infeasibility, time infeasibility, and capacity infeasibility.

5.4.4 • Metaheuristic Overview

This section provides an overview of the features of the heuristics cited above. The following abbreviations identify them: BC12: Błocho and Czech [13], BVH04: Bent and Van Hentenryck [11], CM12: Cordeau and Maischberger [24], HY08: Hashimoto and Yagiura [53], HYI08: Hashimoto, Yagiura, and Ibaraki [54], IINSUY08: Ibaraki et al. [57], LZ07: Lim and Zhang [75], MB05: Mester and Bräysy [80], NB09: Nagata and Bräysy [86], NBD10: Nagata, Bräysy, and Dullaert [88], PDR09: Prescott-Gagnon, Desaulniers, and Rousseau [98], PR07: Pisinger and Ropke [94], RTI09: Repoussis, Tarantilis, and Ioannou [102], and VCGP13: Vidal et al. [120].

Table 5.2 compares the features of these heuristics. The columns in this table indicate the neighborhoods used as presented in Sections 5.4.1.1 and 5.4.1.2, including the maximum string size considered in the cross-exchange and path relocation neighborhoods

Table 5.2. *Metaheuristic comparison.*

	Intra			Inter		Metaheuristic									Recomb.			Infeas.				
	Or-Opt	2-opt	2-opt*	Cross-exchange	Path relocation	LNS	Neighborhood filtering	Parallel metaheuristic	ILS	SA	TS	GLS	PR	EA	EAX	OX+split	Multiparent	TW	Capacity	Customers	Richer than VRPTW	Vehicle minimization
BVH04	X	X	X	1	1	X				X												X
MB05		X		1	1	X						X		X								X
PR07	X	X	X	X		X				X										X	X	X
LZ07	X		X	3		CE			X		X									X	X	X
HYI08	X	X	X	3			X		X				X					A			X	
HY08	X		X	3	3		X		X				X					A	X		X	
IINSUY08	X	X	X	4	3				X		X	X		X				A	X		X	X
RTI09		X		1	1	(X)					X						X					X
PDR09						X											X					X
NB09			X	1	1		X					X		X	X			B	X	X		X
NBD10	X		X	1	1		X							X	X			B	X	X		X
BC12					1			X			X							A	X			X
CM12			X	2	2			X								X		A	X		X	
VCGP13	X	X	X	2	2	X	X							X				B	X		X	X

(columns 2 to 7); if a heuristic filtering of the neighborhoods was used (column 8); if parallel processing can be exploited (column 9); which metaheuristic frameworks are used among iterated local search (ILS), simulated annealing (SA), tabu search (TS), guided local search (GLS), path relinking (PR), evolutionary algorithms or strategies (EA) (columns 10 to 15); the crossover operator used in evolutionary algorithms (columns 16 to 18); if time windows can be violated during the search (column 19), where an "A" means that the traditional way of measuring time-window violation is used, while a "B" rather refers to the method of Nagata [85]; if the capacity and the all-customers-served constraints can be violated (columns 20 and 21); if the algorithm is able to solve VRP variants other than the VRPTW (column 22); and if the algorithm can minimize the number of vehicles used (column 23). Algorithms without this last feature are dependent on being initialized with the desired number of vehicles. Notice that BC12 uses local search, but the authors have not specified which neighborhoods are used. LZ07 does not specify the maximum string size for cross exchanges: this is indicated by an "X" in the corresponding column. All heuristics with "X" in column 7 apply the destroy/repair idea from Shaw's LNS [111]. The parentheses around the "X" for RTI09 highlight a quite different usage of this destroy/repair idea. Finally, LZ07 use large neighborhoods, but of the cyclic exchange type, which explains the "CE" in column 7.

Some immediate observations are that local search is playing a part in all the chosen heuristics and that most of the algorithms allow visiting infeasible solutions.

5.4.5 ▪ Summary Computational Results for Heuristics

In this section we compare the most important heuristics from the past decade based on the results they achieved on the data sets of Solomon [113] and Gehring and Homberger [45]. Tables 5.3 and 5.4 provide computational results for the Solomon and Gehring–Homberger instances, respectively. For each pair of heuristic and instance sets, we report three numbers: the *Cumulative Number of Vehicles* (CNV) summed over all instances in the set, the *Cumulative Total Distance* (CTD), and the average time spent per instance in minutes. Computational times have been normalized to match the performance of an Intel Xeon 2.93 GHz computer. Normalization factors were found using the CINT2006 and CINT2000 benchmarks (see the Standard Performance Evaluation Corporation [115]). We acknowledge that such normalization is far from being exact. Note that some papers do not contain enough information to report average computational times, which explains the blank entries in the two tables. To report the computational time for the BC12 parallel heuristic we multiplied the number of cores used by the computational time. This is of course not completely fair because the heuristic will not be able to take advantage of all processing units at the same time, but it is the best we can do.

Table 5.3. *Results on the Solomon instances (sorted according to CNV first and CTD second).*

	NBD10	VCGP13	RTI09	PDR09	BVH04	HYI08
CNV	405	405	405	405	405	405
CTD	57187	57196	57216	57240	57273	57282
Time (min)	16.9	13.4	33.7	97.0		29.7
	PR07	LZ07	HY08	CM12	IINSUY08	
CNV	405	405	405	406	407	
CTD	57332	57368	57484	57199	57545	
Time (min)	15.3	15.6	16.2	392.8	9.9	

Table 5.4. *Gehring–Homberger results (sorted according to total CNV first and total CTD second). Columns* 200, 400, . . . , 1000 *show results for each size class. The last column shows total CNV and CTD and average time.*

		200	400	600	800	1000	Overall
	CNV	694	1380	2065	2734	3417	10290
NB09	CTD	-	-	-	-	-	-
	Time (min)	40.5	80.9	121.4	161.9	202.4	121.4
	CNV	694	1381	2066	2736	3419	10296
BC12	CTD	168088	388939	792278	1354477	2056153	4759935
	Time (min)	933.1	3740.7	6175.2	7473.0	8244.9	5313.4
	CNV	694	1381	2068	2739	3420	10302
VCGP13	CTD	168092	388013	786373	1334963	2036700	4714141
	Time (min)	42.0	170.5	497.0	1075.0	1745.0	705.9
	CNV	694	1383	2068	2737	3420	10302
HY08	CTD	169070	392507	800982	1367971	2085125	4815655
	Time (min)	31.5	64.0	95.6	127.1	159.6	95.6
	CNV	694	1381	2067	2738	3424	10304
NBD10	CTD	168067	388466	789592	1357695	2045720	4749540
	Time (min)	13.8	54.6	85.3	93.1	119.1	73.2
	CNV	694	1381	2066	2739	3428	10308
RTI09	CTD	169163	395936	816326	1424321	2144830	4950576
	Time (min)	56.5	113.1	169.6	226.2	282.7	169.6
	CNV	694	1382	2068	2742	3429	10315
LZ07	CTD	169296	393695	802681	1372427	2071643	4809742
	Time (min)	55.3	175.4	383.5	752.6	1105.9	494.5
	CNV	694	1383	2069	2747	3430	10323
HYI08	CTD	171018	406109	847470	1442957	2204728	5072282
	Time (min)	59.2	118.5	177.9	237.1	296.5	177.8
	CNV	694	1385	2071	2745	3432	10327
PDR09	CTD	168556	389011	800561	1391344	2096823	4846295
	Time (min)	171.3	287.7	339.4	417.0	523.6	347.8
	CNV	694	1387	2070	2750	3431	10332
IINSUY08	CTD	170484	398938	825172	1421225	2155374	4971193
	Time (min)	19.7	39.5	59.3	79.0	98.8	59.3
	CNV	694	1385	2071	2758	3438	10346
PR07	CTD	169042	393210	807470	1358291	2110925	4838938
	Time (min)	48.4	49.6	57.5	71.3	83.6	62.1
	CNV	694	1389	2082	2765	3446	10376
MB05	CTD	168573	390386	796172	1361586	2078110	4794827
	Time (min)	2.9	6.1	14.4	52.0	215.3	58.1
	CNV	697	1394	2091	2778	3468	10428
BVH04	CTD	168502.6	410111.6	858040	1469790	2266959	5173403
	Time (min)	-	-	-	-	-	-

Heuristics were often tested using several parameter configurations. In Tables 5.3 and 5.4 we report the results from the best configuration (and, typically, the most time-consuming configuration) in order to make the comparison consistent. When the heuristic was executed multiple times, e.g., 10 times, on each instance, we report the best solution found in the 10 tests and the time for performing all 10 runs. Note that for BVH04 we have obtained the Gehring–Homberger results from the technical report version of Bent and Van Hentenryck [11]. For IINSUY08, we chose the ILS-1 configuration and, for PR07, we chose the results produced by letting the algorithm run for 50,000 iterations.

The results on the Solomon instances show that the best heuristics proposed are all very close to one another in terms of solution quality. Most of them produce a CNV of 405. Many of the heuristics show similar computational times, but a few methods have significantly higher running times.

Results on the Gehring and Homberger instances show more variations in CNV and average computational time. We can conclude that Nagata and Bräysy [87] obtain the best results in terms of the primary objective and that the race is very close among the best algorithms. Matching the results in Table 5.4 with the algorithm features from Table 5.2, one can observe that allowing infeasible solutions, especially time and capacity violations, are key factors to a successful metaheuristic. We also note that the population-based search methods yield high-quality solutions. Furthermore, we conclude that the Solomon instances are no longer sufficient for comparing state-of-the-art heuristics and that large-sized instances are necessary to highlight differences.

5.4.6 ▪ Other Measures of Quality

Running time and solution quality achieved are obvious performance indicators for a metaheuristic. However, they are not the only ones. As mentioned in Chapter 4 it also makes sense to look at simplicity, number of parameters and parameter sensitivity, flexibility, and robustness. Several of the heuristics that were highlighted in Chapter 4 as especially promising regarding one or more of these indicators are also able to solve the VRPTW. Among the heuristics only mentioned in this chapter we would like to highlight the iterated local search algorithms of Hashimoto, Yagiura, and Ibaraki [54] and Ibaraki et al. [57]. The core metaheuristics in these papers are very simple, but good results are obtained nevertheless. Note also that the heuristic of Prescott-Gagnon, Desaulniers, and Rousseau [98] appears to be flexible in the sense that it potentially can handle many VRP variants by making small adjustments to the subproblem that generates routes. A drawback of this heuristic is that it is rather complex to implement since it embeds a well-tuned Branch-and-Price framework.

5.5 ▪ Extensions

This section surveys the scientific literature on extensions of the VRPTW. It contains three parts. The first two are dedicated to the split-delivery VRPTW and the time-dependent VRPTW that have been studied for more than a decade now. The last part briefly comments on three extensions that were recently addressed.

5.5.1 ▪ Split Deliveries

The *Split-Delivery Vehicle Routing Problem with Time Windows* (SDVRPTW) consists of a VRPTW where the demand of each customer can be fulfilled by several vehicles and where demands greater than the vehicle capacity is allowed. The literature on the SDVRPTW is quite limited. Frizzell and Giffin [44] and Mullaseril, Dror, and Trudeau [83] developed simple construction and improvement heuristics. In the first paper the proposed heuristics are especially tailored for the problem with a special network structure. In the second paper the proposed heuristic is applied to a real-life problem of managing the trucks for distributing feed in a cattle ranch. In Ho and Haugland [55], a solution method based on tabu search is proposed. The well-known Solomon test instances are applied with modified customer demands.

Gendreau et al. [46] introduced an exact Branch-and-Cut-and-Price method. As in many VRPTW algorithms the lower bounds in the search tree are computed by a column generation algorithm. However, for the SDVRPTW, an additional difficulty is that the quantities delivered are also decision variables. These decisions on the delivered quantities are taken in the MP leading to an exponential number of constraints and a complexified ESPPRC subproblem. The algorithm succeeded at solving modified Solomon instances with up to 50 customers. Desaulniers [26] designed an exact Branch-and-Cut-and-Price method for the SDVRPTW. To avoid an exponential number of constraints in the MP due to the quantities delivered, the delivered quantities are handled in the column generation subproblem, which is a combined ESPPRC and an LP-relaxed bounded knapsack problem. Enhancements to this algorithm were proposed in Archetti, Bouchard, and Desaulniers [2]. Within one hour of computational time, the resulting algorithm succeeded at solving to optimality 262 of the 504 modified Solomon instances (namely, 168, 86, and 8 instances, with 25, 50, and 100 customers, respectively).

5.5.2 ▪ Time-Dependent Costs or Travel Times

The *Time-Dependent Vehicle Routing Problem with Time Windows* (TDVRPTW) consists of a VRPTW where the travel time or travel cost between two locations depends on the time of the day. This can occur, for example, when congestion during rush hours result in longer travel times. An important, desirable, and reasonable property for this type of problems is the *First-In-First-Out* (FIFO) property, which stipulates that a vehicle leaving a location i at any time t to go to a location j cannot arrive earlier at j than another identical vehicle that left location i before t and went directly to j.

The TDVRPTW has received very little attention in the scientific literature compared to the CVRP or the VRPTW. The first authors to deal with the TDVRP after 2000 were Ichoua, Gendreau, and Potvin [58], who developed a parallel tabu search heuristic for the case with soft time windows (that is, barring a penalty, service can start later than a time-window upper bound). Fleischmann, Gietz, and Gnutzmann [43] present a general heuristic-based framework for the implementation of time-varying travel times in various vehicle routing algorithms, including TDVRPTW. Haghani and Jung [51] propose a genetic algorithm. Hashimoto, Yagiura, and Ibaraki [54] design an iterated local search heuristic that takes into account time-dependent service times, traveling times, and traveling costs in the case with soft time windows. Donati et al. [35] propose a solution based on an ant colony heuristic. Soler, Albiach, and Martínez [112] transform the TDVRPTW through several steps into an asymmetric CVRP. They solve very small instances to optimality, but none of realistic size. Figliozzi [42] develops an iterative route construction and improvement heuristic and introduces new replicable test problems. Ghiani and Guerriero [48] exploit some properties of the model of Ichoua, Gendreau, and Potvin [58] and show that this model is quite general.

To conclude, a few heuristic methods have been designed for variants of the TDVRPTW. Only one exact method suitable for very small-sized instances was also proposed.

5.5.3 ▪ Driver Considerations, Multiple Use of Vehicles, and Multiple Time Windows

The VRPTW with driver regulations takes into account various regulations concerning the working hours, breaks, and rests of the drivers. These regulations are appearing more and more often all around the world and may differ from area to area. For this problem, Goel [50] introduced an LNS heuristic based on local search operators, Prescott-Gagnon et al. [97] developed an LNS method based on a column generation heuristic, and Kok et al. [72] proposed a dynamic programming heuristic. According to the reported computational results obtained on instances involving 100 customers over a one-week horizon, the algorithm of Prescott-Gagnon et al. outperforms the other two algorithms.

The CVRP with multiple use of vehicles is a variant of the classical CVRP that allows one to assign several routes to the same vehicle over a finite planning horizon (e.g., one work day) in a context where vehicle availability, vehicle fixed costs, or maximum working time per vehicle must be considered. Azi, Gendreau, and Potvin [3] addressed the case with time windows and designed an exact Branch-and-Price solution method. They solved instances with up to 40 customers.

Doerner et al. [34] studied the VRP with multiple interdependent time windows in which each customer may be required to be visited several times and the elapsed time between two consecutive visits must not exceed a maximum time. This VRPTW variant is inspired by a blood transportation application. The authors developed an exact algorithm and heuristic algorithms. Their results show that the heuristic algorithms find solutions reasonably close to optimal solutions in a fraction of a second.

5.6 ▪ Conclusions and Future Research Directions

The previous sections have reported a considerable progress in VRPTW methodologies in the last decade.

Concerning the exact methods the research has concentrated on stabilization of the dual variables, effective solutions of the resource constrained shortest path subproblems, introduction of new valid inequalities, and smarter Branch-and-Bound techniques. The result is that it is now possible to solve to optimality all the Solomon instances. The computational time needed has also been reduced considerably during the last decade but may, however, be still quite long.

Concerning the heuristic methods the research has focused almost solely on metaheuristics. It appears that population-based search methods, large neighborhoods, and allowing infeasible solutions during the search are important ingredients in current state-of-the-art heuristics.

We envision that research for the VRPTW will proceed. On the exact methods, research will focus on the development of better solution algorithms for solving the subproblem and on the introduction of new cuts and smarter Branch-and-Bound procedures. An important challenge is the design of an efficient separation algorithm for the 2-path cuts which are currently separated by enumerating a large number of customer subsets and solving a TSP with time windows for each subset. Even more challenging is the development of a separation algorithm for the 3-path cuts that may greatly help to close the integrality gap. With such research endeavors, we can foresee that more of the Gehring and Homberger instances will be solved to optimality.

We expect that research on heuristics will continue but at a slightly slower pace compared to what was witnessed in the last decade. This prediction is based on the observation that most research currently aims at solving CVRP variants that are richer than the classical VRPTW. In terms of future research directions we believe that we still have not seen the full potential of methods combining mathematical programming and metaheuristics. We also think that the development of heuristics that can match the solution quality of the best methods of today in a fraction of the time would be of interest for practical applications where interactive planning seems more widespread than batch planning.

As in the past, the future exact and heuristic methods for the VRPTW will also serve as a basis for solving more complex VRPs. Finally, we think that the VRPTW will remain a privileged platform for validating the usefulness of new research ideas for exact and heuristic methods due to the high-level maturity of the existing VRPTW solution methods.

Bibliography

[1] R. K. AHUJA, Ö. ERGUN, J. ORLIN, AND A. PUNNEN, *A survey of very large-scale neighborhood search techniques*, Discrete Applied Mathematics, 123 (2002), pp. 75–102.

[2] C. ARCHETTI, M. BOUCHARD, AND G. DESAULNIERS, *Enhanced branch-and-price-and-cut for vehicle routing with split deliveries and time windows*, Transportation Science, 45 (2011), pp. 285–298.

[3] N. AZI, M. GENDREAU, AND J.-Y. POTVIN, *An exact algorithm for a vehicle routing problem with time windows and multiple use of vehicles*, European Journal of Operational Research, 202 (2010), pp. 756–763.

[4] E. K. BAKER AND S. RUSHINEK, *Large scale implementation of a time oriented vehicle scheduling model*, Technical Report, US Department of Transportation, Urban Mass Transit Administration, Washington, D.C., 1982.

[5] R. BALDACCI, E. BARTOLINI, A. MINGOZZI, AND R. ROBERTI, *An exact solution framework for a broad class of vehicle routing problems*, Computational Management Science, 7 (2010), pp. 229–268.

[6] R. BALDACCI, A. MINGOZZI, AND R. ROBERTI, *New route relaxation and pricing strategies for the vehicle routing problem*, Operations Research, 59 (2011), pp. 1269–1283.

[7] ——, *Recent exact algorithms for solving the vehicle routing problem under capacity and time window constraints*, European Journal of Operational Research, 218 (2012), pp. 1–6.

[8] J. BARD, G. KONTORAVDIS, AND G. YU, *A branch-and-cut procedure for the vehicle routing problem with time windows*, Transportation Science, 36 (2002), pp. 250–269.

[9] C. BARNHART, E. L. JOHNSON, G. L. NEMHAUSER, M. W. P. SAVELSBERGH, AND P. H. VANCE, *Branch-and-price: Column generation for solving huge integer programs*, Operations Research, 46 (1998), pp. 316–329.

[10] J. E. BEASLEY, *Route first — cluster second methods for vehicle routing*, Omega, 11 (1983), pp. 403–408.

[11] R. BENT AND P. VAN HENTENRYCK, *A two-stage hybrid local search for the vehicle routing problem with time windows*, Transportation Science, 38 (2004), pp. 515–530.

[12] H.-G. BEYER AND H.-P. SCHWEFEL, *Evolution strategies—A comprehensive introduction*, Natural Computing, 1 (2002), pp. 3–52.

[13] M. BŁOCHO AND Z. J. CZECH, *A parallel memetic algorithm for the vehicle routing problem with time windows*, Parallel Computing, submitted, 2012.

[14] N. BOLAND, J. DETHRIDGE, AND I. DUMITRESCU, *Accelerated label setting algorithms for the elementary resource constrained shortest path problem*, Operations Research Letters, 34 (2006), pp. 58–68.

[15] O. BRÄYSY, *A reactive variable neighborhood search for the vehicle-routing problem with time windows*, INFORMS Journal on Computing, 15 (2003), pp. 347–368.

[16] O. BRÄYSY AND M. GENDREAU, *Vehicle routing problem with time windows, part I: Route construction and local search algorithms*, Transportation Science, 39 (2005), pp. 104–118.

[17] ———, *Vehicle routing problem with time windows, part II: Metaheuristics*, Transportation Science, 39 (2005), pp. 119–139.

[18] A. CHABRIER, *Vehicle routing problem with elementary shortest path based column generation*, Computers & Operations Research, 33 (2006), pp. 2972–2990.

[19] N. CHRISTOFIDES, A. MINGOZZI, AND P. TOTH, *State-space relaxation procedures for the computation of bounds to routing problems*, Networks, 11 (1981), pp. 145–164.

[20] W. COOK AND J. L. RICH, *A parallel cutting plane algorithm for the vehicle routing problem with time windows*, Technical Report, Computational and Applied Mathematics, Rice University, Houston, TX, 1999.

[21] J.-F. CORDEAU, G. DESAULNIERS, J. DESROSIERS, M. M. SOLOMON, AND F. SOUMIS, *VRP with time windows*, in The Vehicle Routing Problem, P. Toth and D. Vigo, eds., SIAM, Philadelphia, 2002, ch. 7, pp. 157–193.

[22] J.-F. CORDEAU, G. LAPORTE, AND A. MERCIER, *A unified tabu search heuristic for vehicle routing problems with time windows*, Journal of the Operational Research Society, 52 (2001), pp. 928–936.

[23] ———, *Improved tabu search algorithm for the handling of route duration constraints in vehicle routing problems with time windows*, Journal of the Operational Research Society, 55 (2004), pp. 542–546.

[24] J.-F. CORDEAU AND M. MAISCHBERGER, *A parallel iterated tabu search heuristic for vehicle routing problems*, Computers & Operations Research, 39 (2012), pp. 2033–2050.

[25] G. B. DANTZIG AND P. WOLFE, *The decomposition algorithm for linear programming*, Operations Research, 8 (1960), pp. 101–111.

[26] G. DESAULNIERS, *Branch-and-price-and-cut for the split delivery vehicle routing problem with time windows*, Operations Research, 58 (2010), pp. 179–192.

[27] G. DESAULNIERS, J. DESROSIERS, I. IOACHIM, M. M. SOLOMON, F. SOUMIS, AND D. VILLENEUVE, *A unified framework for deterministic time constrained vehicle routing and crew scheduling problems*, in Fleet Management and Logistics, T. Crainic and G. Laporte, eds., Kluwer, Norwell, MA, 1998, pp. 57–93.

[28] G. DESAULNIERS, J. DESROSIERS, AND M. M. SOLOMON, *Column generation*, Springer, New York, 2005.

[29] G. DESAULNIERS, J. DESROSIERS, AND S. SPOORENDONK, *The vehicle routing problem with time windows: State-of-the-art exact solution methods*, in Wiley Encyclopedia of Operations Research and Management Science, Vol. 8, J. Cochrane, ed., Wiley, New York, 2010, pp. 5742–5749.

[30] ———, *Cutting planes for branch-and-price algorithms*, Networks, 58 (2011), pp. 301–310.

[31] G. DESAULNIERS, F. LESSARD, AND A. HADJAR, *Tabu search, generalized k-path inequalities, and partial elementarity for the vehicle routing problem with time windows*, Transportation Science, 42 (2008), pp. 387–404.

[32] M. DESROCHERS, J. DESROSIERS, AND M. M. SOLOMON, *A new optimization algorithm for the vehicle routing problem with time windows*, Operations Research, 40 (1992), pp. 342–354.

[33] J. DESROSIERS, Y. DUMAS, F. SOUMIS, AND M. M. SOLOMON, *Time constrained routing and scheduling*, in Network Routing, vol. 8 of Handbooks in Operations Research and Management Science, M. O. Ball, T. L. Magnanti, C. L. Monma, and G. L. Nemhauser, ed., Elsevier Science, Amsterdam, 1995, pp. 35–139.

[34] K. DOERNER, M. GRONALT, R. HARTL, G. KIECHLE, AND M. REIMANN, *Exact and heuristic algorithms for the vehicle routing problem with multiple interdependent time windows*, Computers & Operations Research, 35 (2008), pp. 3034–3048.

[35] A. V. DONATI, R. MONTEMANNI, N. CASAGRANDE, A. E. RIZZOLI, AND L. M. GAMBARDELLA, *Time dependent vehicle routing problem with a multi ant colony system*, European Journal of Operational Research, 185 (2008), pp. 1174–1191.

[36] M. DROR, *Note on the complexity of the shortest path models for column generation in VRPTW*, Operations Research, 42 (1994), pp. 977–979.

[37] Y. DUMAS, F. SOUMIS, AND J. DESROSIERS, *Optimizing the schedule for a fixed vehicle path with convex inconvenience costs*, Transportation Science, 24 (1990), pp. 145–152.

[38] I. DUMITRESCU, *Constrained Path and Cycle Problems*, PhD thesis, Department of Mathematics and Statistics, The University of Melbourne, Australia, 2002.

[39] E. FALKENAUER AND S. BOUFFOUIX, *A genetic algorithm for job shop*, in Proceedings of the 1991 IEEE International Conference on Robotics and Automation, 1991, pp. 824–829.

[40] D. FEILLET, P. DEJAX, M. GENDREAU, AND C. GUEGUEN, *An exact algorithm for the elementary shortest path problem with resource constraints: Application to some vehicle routing problems*, Networks, 44 (2004), pp. 216–229.

[41] D. FEILLET, M. GENDREAU, AND L.-M. ROUSSEAU, *New refinements for the solution of vehicle routing problems with branch and price*, INFOR, 45 (2007), pp. 239–256.

[42] M. A. FIGLIOZZI, *The time dependent vehicle routing problem with time windows: Benchmark problems, an efficient solution algorithm, and solution characteristics*, Transportation Research Part E: Logistics and Transportation Review, 48 (2012), pp. 616–636.

[43] B. FLEISCHMANN, M. GIETZ, AND S. GNUTZMANN, *Time-varying travel times in vehicle routing*, Transportation Science, 38 (2004), pp. 160–173.

[44] P. W. FRIZZELL AND J. W. GIFFIN, *The split delivery vehicle scheduling problem with time windows and grid network distances*, Computers & Operations Research, 22 (1995), pp. 655–667.

[45] H. GEHRING AND J. HOMBERGER, *A parallel two-phase metaheuristic for routing problems with time windows*, Asia-Pacific Journal of Operational Research, 18 (2001), pp. 35–47.

[46] M. GENDREAU, P. DEJAX, D. FEILLET, AND C. GUEGUEN, *Vehicle routing with time windows and split deliveries*, Technical Report 2006-851, Laboratoire Informatique d'Avignon, Avignon, France, 2006.

[47] M. GENDREAU AND C. D. TARANTILIS, *Solving large-scale vehicle routing problems with time windows: The state of the art*, Technical Report 2010-04, CIRRELT, Montréal, Canada, 2010.

[48] G. GHIANI AND E. GUERRIERO, *A note on the Ichoua et al. (2003) travel time model*, Technical Report, Optimization Online, January 2012.

[49] P. C. GILMORE AND R. E. GOMORY, *A linear programming approach to the cutting stock problem*, Operations Research, 9 (1961), pp. 849–859.

[50] A. GOEL, *Vehicle routing and scheduling with drivers' working hours*, Transportation Science, 43 (2009), pp. 17–26.

[51] A. HAGHANI AND S. JUNG, *A dynamic vehicle routing problem with time-dependent travel times*, Computers & Operations Research, 32 (2005), pp. 2959–2986.

[52] W. D. HARVEY AND M. L. GINSBERG, *Limited discrepancy search*, in Proceedings of the Fourteenth International Joint Conference on Artificial Intelligence, Montréal, Canada, 1995.

[53] H. HASHIMOTO AND M. YAGIURA, *A path relinking approach with an adaptive mechanism to control parameters for the vehicle routing problem with time windows*, Lecture Notes in Computer Science, 4972 (2008), pp. 254–265.

[54] H. HASHIMOTO, M. YAGIURA, AND T. IBARAKI, *An iterated local search algorithm for the time-dependent vehicle routing problem with time windows*, Discrete Optimization, 5 (2008), pp. 434–456.

[55] S. C. HO AND D. HAUGLAND, *A tabu search heuristic for the vehicle routing problem with time windows and split deliveries*, Computers & Operations Research, 31 (2004), pp. 1947–1964.

[56] J. H. HOLLAND, *Adaptation in Natural and Artificial Systems: An Introductory Analysis with Applications to Biology, Control and Artificial Intelligence*, MIT Press, Cambridge, MA, 1992.

[57] T. IBARAKI, S. IMAHORI, K. NONOBE, K. SOBUE, T. UNO, AND M. YAGIURA, *An iterated local search algorithm for the vehicle routing problem with convex time penalty functions*, Discrete Applied Mathematics, 156 (2008), pp. 2050–2069.

[58] S. ICHOUA, M. GENDREAU, AND J.-Y. POTVIN, *Vehicle dispatching with time-dependent travel times*, European Journal of Operational Research, 144 (2003), pp. 379–396.

[59] G. IOANNOU, M. KRITIKOS, AND G. PRASTACOS, *A greedy look-ahead heuristic for the vehicle routing problem with time windows*, Journal of the Operational Research Society, 52 (2001), pp. 523–537.

[60] S. IRNICH, *A unified modeling and solution framework for vehicle routing and local search-based metaheuristics*, INFORMS Journal on Computing, 20 (2008), pp. 270–287.

[61] S. IRNICH AND G. DESAULNIERS, *Shortest path problems with resource constraints*, in Column Generation, G. Desaulniers, J. Desrosiers, and M. Solomon, eds., Springer, New York, 2005, ch. 2, pp. 33–65.

[62] S. IRNICH, G. DESAULNIERS, J. DESROSIERS, AND A. HADJAR, *Path reduced costs for eliminating arcs*, INFORMS Journal on Computing, 22 (2010), pp. 297–313.

[63] S. IRNICH AND D. VILLENEUVE, *The shortest path problem with resource constraints and k-cycle elimination for $k \geq 3$*, INFORMS Journal on Computing, 18 (2006), pp. 391–406.

[64] M. JEPSEN, B. PETERSEN, S. SPOORENDONK, AND D. PISINGER, *Subset-row inequalities applied to the vehicle-routing problem with time windows*, Operations Research, 56 (2008), pp. 497–511.

[65] B. KALLEHAUGE, *Formulations and exact algorithms for the vehicle routing problem with time windows*, Computers & Operations Research, 35 (2008), pp. 2307–2330.

[66] B. KALLEHAUGE, N. BOLAND, AND O. B. G. MADSEN, *Path inequalities for the vehicle routing problem with time windows*, Networks, 49 (2007), pp. 273–293.

[67] B. KALLEHAUGE, J. LARSEN, AND O. B. G. MADSEN, *Lagrangian duality applied to the vehicle routing problem with time windows*, Computers & Operations Research, 33 (2006), pp. 1464–1487.

[68] G. A. P. KINDERVATER AND M. W. P. SAVELSBERGH, *Vehicle routing: Handling edge exchanges*, in Local Search in Combinatorial Optimization, E. Aarts and J. Lenstra, eds., Wiley, Chichester, UK, 1997, pp. 337–360.

[69] K. KNIGHT AND J. HOFER, *Vehicle scheduling with timed and connected calls: A case study*, Operational Research Quarterly, 19 (1968), pp. 299–310.

[70] N. KOHL, J. DESROSIERS, O. B. G. MADSEN, M. M. SOLOMON, AND F. SOUMIS, *2-Path cuts for the vehicle routing problem with time windows*, Transportation Science, 33 (1999), pp. 101–116.

[71] N. KOHL AND O. B. G. MADSEN, *An optimization algorithm for the vehicle routing problem with time windows based on Lagrangean relaxation*, Operations Research, 45 (1997), pp. 395–406.

[72] A. L. KOK, C. M. MEYER, H. KOPFER, AND J. M. J. SCHUTTEN, *A dynamic programming heuristic for the vehicle routing problem with time windows and European community social legislation*, Transportation Science, 44 (2010), pp. 442–454.

[73] J. LARSEN, *Parallellization of the vehicle routing problem with time windows*, PhD thesis IMM-PHD-1999-62, Department of Mathematical Modelling, Technical University of Denmark, Lyngby, Denmark, 1999.

[74] A. N. LETCHFORD AND J. J. SALAZAR GONZÁLEZ, *Projection results for vehicle routing*, Mathematical Programming, Ser. B, 105 (2006), pp. 251–274.

[75] A. LIM AND X. ZHANG, *A two-stage heuristic with ejection pools and generalized ejection chains for the vehicle routing problem with time windows*, INFORMS Journal on Computing, 19 (2007), pp. 443–457.

[76] H. R. LOURENÇO, O. C. MARTIN, AND T. STÜTZLE, *Iterated local search: Framework and applications*, in Handbook of Metaheuristics, M. Gendreau and J.-Y. Potvin, eds., Springer, New York, 2nd ed., 2010, pp. 363–397.

[77] M. E. LÜBBECKE AND J. DESROSIERS, *Selected topics in column generation*, Operations Research, 53 (2005), pp. 1007–1023.

[78] J. LYSGAARD, *Reachability cuts for the vehicle routing problem with time windows*, European Journal of Operational Research, 175 (2006), pp. 210–223.

[79] O. B. G. MADSEN, *Optimal scheduling of trucks - A routing problem with tight due times for delivery*, in Optimization Applied to Transportation Systems, H. Strobel, R. Genser and M. Etschmaier, eds., IIASA, International Institute for Applied System Analysis, Laxenburgh, 1976, pp. 126–136.

[80] D. MESTER AND O. BRÄYSY, *Active guided evolution strategies for large-scale vehicle routing problems with time windows*, Computers & Operations Research, 32 (2005), pp. 1593–1614.

[81] P. MOSCATO, *On evolution, search, optimization, genetic algorithms and martial arts: Toward memetic algorithms*, Caltech Concurrent Computation Program 826, California Institute of Technology, 1989.

[82] P. MOSCATO AND C. COTTA, *A modern introduction to memetic algorithms*, in Handbook of Metaheuristics, M. Gendreau and J.-Y. Potvin, eds., Springer, New York, 2nd ed., 2010, pp. 141–183.

[83] P. A. MULLASERIL, M. DROR, AND P. TRUDEAU, *Split-delivery routing heuristics in livestock feed distribution*, Journal of the Operational Research Society, 48 (1997), pp. 107–116.

[84] Y. NAGATA, *Edge assembly crossover for the capacitated vehicle routing problem*, Lecture Notes in Computer Science, 4446 (2007), pp. 142–152.

[85] ——, *Efficient evolutionary algorithm for the vehicle routing problem with time windows: edge assembly crossover for the VRPTW*, in Proceedings of the 2007 IEEE Congress on Evolutionary Computation, 2007. CD-ROM.

[86] Y. NAGATA AND O. BRÄYSY, *Edge assembly based memetic algorithm for the capacitated vehicle routing problem*, Networks, 54 (2009), pp. 205–215.

[87] ——, *A powerful route minimization heuristic for the vehicle routing problem with time windows*, Operations Research Letters, 37 (2009), pp. 333–338.

[88] Y. NAGATA, O. BRÄYSY, AND W. DULLAERT, *A penalty-based edge assembly memetic algorithm for the vehicle routing problem with time windows*, Computers & Operations Research, 37 (2010), pp. 724–737.

[89] Y. NAGATA AND S. KOBAYASHI, *Edge assembly crossover: A high-power genetic algorithm for the travelling salesman problem*, in Proceedings of the 7th International Conference on Genetic Algorithms, 1997, pp. 450–457.

[90] ———, *A powerful genetic algorithm using edge assembly crossover for the traveling salesman problem*, INFORMS Journal on Computing, 25 (2013), pp. 346–363.

[91] G. L. NEMHAUSER AND L. A. WOLSEY, *Integer and Combinatorial Optimization*, John Wiley & Sons, New York, 1988.

[92] M. W. PADBERG AND G. RINALDI, *Optimization of a 532-city symmetric traveling salesman problem by branch-and-cut*, Operations Research Letters, 6 (1987), pp. 1–7.

[93] B. PETERSEN, D. PISINGER, AND S. SPOORENDONK, *Chvátal-Gomory rank-1 cuts used in a Dantzig-Wolfe decomposition of the vehicle routing problem with time windows*, in The Vehicle Routing Problem: Latest Advances and New Challenges, B. L. Golden, S. Raghavan, and E. A. Wasil, eds., vol. 43 of Operations Research/Computer Science Interfaces Series, Springer, New York, 2008, pp. 397–419.

[94] D. PISINGER AND S. ROPKE, *A general heuristic for vehicle routing problems*, Computers & Operations Research, 34 (2007), pp. 2403–2435.

[95] ———, *Large neighborhood search*, in Handbook of Metaheuristics, M. Gendreau and J.-Y. Potvin, eds., Springer-Verlag, Berlin, 2nd ed., 2010, ch. 13, pp. 399–419.

[96] J.-Y. POTVIN, *State-of-the art review: Evolutionary algorithms for vehicle routing*, INFORMS Journal on Computing, 21 (2009), pp. 518–548.

[97] E. PRESCOTT-GAGNON, G. DESAULNIERS, M. DREXL, AND L.-M. ROUSSEAU, *European driver rules in vehicle routing with time windows*, Transportation Science, 44 (2010), pp. 455–473.

[98] E. PRESCOTT-GAGNON, G. DESAULNIERS, AND L.-M. ROUSSEAU, *A branch-and-price-based large neighborhood search algorithm for the vehicle routing problem with time windows*, Networks, 54 (2009), pp. 190–204.

[99] C. PRINS, *A simple and effective evolutionary algorithm for the vehicle routing problem*, Computers & Operations Research, 31 (2004), pp. 1985–2002.

[100] H. PULLEN AND M. WEBB, *A computer application to a transport scheduling problem*, Computer Journal, 10 (1967), pp. 10–13.

[101] C. R. REEVES, *Genetic algorithms*, in Handbook of Metaheuristics, M. Gendreau and J.-Y. Potvin, eds., Springer, New York, 2nd ed., 2010, pp. 109–139.

[102] P. P. REPOUSSIS, C. D. TARANTILIS, AND G. IOANNOU, *Arc-guided evolutionary algorithm for the vehicle routing problem with time windows*, IEEE Transactions on Evolutionary Computation, 13 (2009), pp. 624–647.

[103] G. RIGHINI AND M. SALANI, *Symmetry helps: bounded bi-directional dynamic programming for the elementary shortest path problem with resource constraints*, Discrete Optimization, 3 (2006), pp. 255–273.

[104] ———, *New dynamic programming algorithms for the resource constrained elementary shortest path problem*, Networks, 51 (2008), pp. 155–209.

[105] S. ROPKE, *Branching decisions in branch-and-cut-and-price algorithms for vehicle routing problems*, International Workshop on Column Generation, June 10–13, 2012, Bromont, Canada.

[106] S. ROPKE AND D. PISINGER, *An adaptive large neighborhood search heuristic for the pickup and delivery problem with time windows*, Transportation Science, 40 (2006), pp. 455–472.

[107] L.-M. ROUSSEAU, M. GENDREAU, AND D. FEILLET, *Interior point stabilization for column generation*, Operations Research Letters, 35 (2007), pp. 660–668.

[108] L.-M. ROUSSEAU, M. GENDREAU, AND G. PESANT, *Solving VRPTWs with constraint programming based column generation*, Annals of Operations Research, 130 (2004), pp. 119–216.

[109] M. W. P. SAVELSBERGH, *Local search in routing problems with time windows*, Annals of Operations Research, 4 (1985), pp. 285–305.

[110] G. SCHRIMPF, J. SCHNEIDER, H. STAMM-WILBRANDT, AND G. DUECK, *Record breaking optimization results using the ruin and recreate principle*, Journal of Computational Physics, 159 (2000), pp. 139–171.

[111] P. SHAW, *Using constraint programming and local search methods to solve vehicle routing problems*, in CP-98 (Fourth International Conference on Principles and Practice of Constraint Programming), vol. 1520 of Lecture Notes in Computer Science, Springer, Berlin, 1998, pp. 417–431.

[112] D. SOLER, J. ALBIACH, AND E. MARTÍNEZ, *A way to optimally solve a time-dependent vehicle routing problem with time windows*, Operations Research Letters, 37 (2009), pp. 37–42.

[113] M. M. SOLOMON, *Algorithms for the vehicle routing and scheduling problem with time window constraints*, Operations Research, 35 (1987), pp. 254–265.

[114] S. SPOORENDONK AND G. DESAULNIERS, *Clique inequalities applied to the vehicle routing problem with time windows*, INFOR, 48 (2010), pp. 53–67.

[115] STANDARD PERFORMANCE EVALUATION CORPORATION. *SPEC CPU2006 benchmark suite*. http://www.spec.org, 2006.

[116] P. M. THOMPSON AND H. N. PSARAFTIS, *Cyclic transfer algorithms for multivehicle routing and scheduling problems*, Operations Research, 41 (1993), pp. 935–946.

[117] P. TOTH AND D. VIGO, EDS., *The Vehicle Routing Problem*, SIAM, Philadelphia, 2002.

[118] H. W. J. M. TRIENEKENS, *The time constrained vehicle routing problem*, Technical Report, Erasmus University, Rotterdam, The Netherlands, 1982.

[119] F. VANDERBECK, *Implementing mixed integer column generation*, in Column Generation, G. Desaulniers, J. Desrosiers, and M. M. Solomon, eds., Springer, New York, 2005, ch. 12, pp. 331–358.

[120] T. VIDAL, T. G. CRAINIC, M. GENDREAU, AND C. PRINS, *A hybrid genetic algorithm with adaptive diversity management for a large class of vehicle routing problems with time windows*, Computers & Operations Research, 40 (2013), pp. 475–489.

Chapter 6

Pickup-and-Delivery Problems for Goods Transportation

Maria Battarra
Jean-François Cordeau
Manuel Iori

6.1 ▪ Introduction

Pickup-and-Delivery Problems (PDPs) constitute an important family of routing problems in which goods or passengers have to be transported from different origins to different destinations. These problems are usually defined on a graph in which vertices represent origins or destinations for the different entities (or *commodities*) to be transported. PDPs can be classified into three main categories according to the type of demand and route structure being considered. In *many-to-many* (M-M) problems, each commodity may have multiple origins and destinations and any location may be the origin or destination of multiple commodities. These problems arise, for example, in the repositioning of inventory between retail stores or in the management of bicycle or car sharing systems. *One-to-many-to-one* (1-M-1) problems are characterized by the presence of some commodities to be delivered from a depot to many customers and of other commodities to be collected at the customers and transported back to the depot. These have applications, for example, in the distribution of beverages and the collection of empty cans and bottles. They also arise in forward and reverse logistics systems where, in addition to delivering new products, one must plan the collection of used, defective, or obsolete products. Finally, in *one-to-one* (1-1) problems, each commodity has a single origin and a single destination between which it must be transported. Typical applications of these problems are less-than-truckload transportation and urban courier operations. The three types of PDPs are illustrated in Figure 6.1, where the square represents the depot and the other vertices are customers.

Like other VRPs, PDPs may also be classified according to the decision framework being considered and to the availability of information. In *static* problems, one assumes that all problem parameters are deterministic and known a priori before vehicle routes are constructed. *Dynamic* problems are characterized by the fact that some of the information required to make decisions is gradually revealed over time and requires the solution

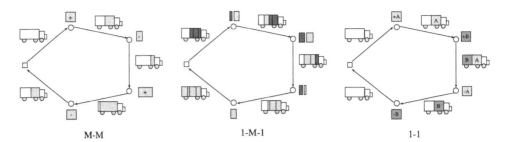

Figure 6.1. *The three types of PDPs.*

to be updated. Dynamic problems may also be *stochastic* when some information about the uncertain parameters is known in the form of probability distributions.

Most variants of PDPs have been studied intensively in the last three decades and have been the object of several literature surveys. In particular, we refer the reader to Parragh, Doerner, and Hartl [103, 104] for a general survey of PDPs and to Berbeglia et al. [15] and Berbeglia, Cordeau, and Laporte [16] for more focused reviews of static and dynamic problems, respectively.

This chapter is concerned with PDPs arising in the transportation of goods. Variants related to passenger transportation are covered in Chapter 7. The next three sections review models and algorithms for many-to-many, one-to-many-to-one, and one-to-one problems, respectively. Problems incorporating vehicle loading constraints are reviewed in Section 6.5.

6.2 ▪ Many-to-Many Problems

Most of the literature on *many-to-many* (M-M) problems focuses on the single-commodity case where a unique commodity must be transported between multiple origins and destinations. The *One-Commodity M-M Pickup and Delivery Vehicle Routing Problem* (1-PDVRP) can be defined as follows. We are given a complete graph $G = (V, A)$ and a fleet K of identical vehicles, each having capacity Q. Vertex 0 is the depot, where vehicles are initially located, and the other vertices of $V \setminus \{0\}$ represent customers. Each customer i has a demand q_i, where $q_i > 0$ means that the customer requires a pickup and $q_i < 0$ that it requires a delivery. Vehicles can leave the depot either empty or with some load. The cost of traveling along an arc $(i, j) \in A$ is c_{ij}. The 1-PDVRP is to find a set of routes that start and finish at the depot, serve all requests without violating the vehicle capacity, and have minimum total cost.

The 1-PDVRP can be modeled as a *Mixed-Integer Linear Program* (MILP) by using binary variables x_{ijk} taking value 1 if arc $(i, j) \in A$ is traversed by vehicle $k \in K$, and 0 otherwise, and non-negative variables f_{ij} indicating the load transported on arc $(i, j) \in A$. The model is the following:

(6.1)   (1-PDVRP)   minimize $\displaystyle\sum_{i \in V}\sum_{j \in V}\sum_{k \in K} c_{ij} x_{ijk}$

(6.2)   s.t. $\displaystyle\sum_{j \in V}\sum_{k \in K} x_{ijk} = 1$ $\forall\, i \in V \setminus \{0\}$,

(6.3) $\displaystyle\sum_{j \in V} x_{ijk} - \sum_{j \in V} x_{jik} = 0$ $\forall\, i \in V,\ k \in K$,

(6.4) $\qquad 0 \leq f_{ij} \leq Q \sum_{k \in K} x_{ijk} \qquad \forall\, i, j \in V,$

(6.5) $\qquad \sum_{j \in V} f_{ji} - \sum_{j \in V} f_{ij} = q_i \qquad \forall\, i \in V \setminus \{0\},$

(6.6) $\qquad \sum_{i \in S} \sum_{j \in S} x_{ijk} \leq |S| - 1 \qquad \forall\, S \subseteq V \setminus \{0\},\, S \neq \emptyset,\, k \in K,$

(6.7) $\qquad x_{ijk} \in \{0, 1\} \qquad \forall\, i, j \in V,\, k \in K.$

The objective function requires the minimization of the cost. Constraints (6.2) impose that each customer be visited once. Constraints (6.3) enforce that a vehicle does not end its route at a vertex other than the depot, while constraints (6.4) restrict the flow on each arc to be at most Q. Constraints (6.5) set the flow conservation, and constraints (6.6) avoid the presence of subtours in the solution. Note that in the presence of negative demands constraints (6.6) are necessary because flow conservation is not sufficient to ensure that all routes start and finish at the depot. Note also that it is easy to enrich the above model by considering different weight capacities, or by including time windows or a maximum duration constraint for each route.

As mentioned above, model (6.1)–(6.7) does not impose any restriction on the load that leaves and enters the depot. Vehicles are thus allowed to start their routes carrying some goods, thus possibly visiting delivery customers first. Imposing zero initial and final load may result in a high increase in solution cost. Note also that in the 1-PDVRP description we disregarded customers with zero demand, but these can be easily included in the above model with no further modification in the constraints.

We now describe the main algorithms introduced for the single-vehicle and multiple-vehicle variants of the 1-PDVRP.

6.2.1 ▪ Single Vehicle

The single-vehicle case of the 1-PDVRP was formally introduced by Hernández-Pérez and Salazar-González [59], under the name *One-commodity Pickup and Delivery Traveling Salesman Problem* (1-PDTSP). The authors presented mathematical formulations for the symmetric and asymmetric cases, both based on the use of binary variables x_{ij} taking value one if the edge or arc (i, j) is used, and non-negative variables f_{ij} giving the flow of commodity passing on edge or arc (i, j). They then focused on the solution of the symmetric formulation by Branch-and-Cut. By Benders' decomposition they projected out the flow variables and obtained a pure 0-1 model including an exponential number of Benders' cuts. They proposed an exact procedure to separate these cuts using a maximum flow algorithm on a supporting graph. The Benders' cuts are then lifted to stronger rounded inequalities, which are separated with a four-step heuristic algorithm using in turn two max-flow computations, a constructive greedy procedure and a local search phase. The Branch-and-Cut was enriched by the additional separation of three-edge clique inequalities and a simple primal heuristic invoked at the root node. The proposed formulation was also adapted to solve the 1-M-1 TSP with pickups and deliveries (see Section 6.3.2.1). To test their algorithm they created a set of instances that has become the standard benchmark for comparing algorithms for the 1-PDTSP. These instances have Euclidean distances and a number of requests between 20 and 50. Their Branch-and-Cut could solve to optimality all such instances within two hours of computing time. Computational evidence showed that (i) the 1-PDTSP is harder than the 1-M-1 TSP with pickups and deliveries (see Section 6.3.2.1) as well as the TSP with the same number of vertices;

(ii) the difficulty of the 1-PDTSP increases when the capacity decreases because the vehicle is forced to perform long detours to achieve feasibility and the lower bound is far from the optimum.

Hernández-Pérez and Salazar-González [61] later improved their results with a new Branch-and-Cut algorithm. They again considered the symmetric formulation involving only the x_{ij} binary variables, but enriched it with several families of valid inequalities, namely rank inequalities (including the simpler clique and clique-cluster inequalities), generalized inhomogeneous multistar inequalities (including the simpler positive and negative multistar inequalities), and generalized homogeneous partial multistar inequalities. Most of the separation procedures were based on heuristics and were provided in the form of pseudo-codes. New and existing instances with up to 200 requests were solved within two hours.

To tackle larger instances, Hernández-Pérez and Salazar-González [60] proposed two simple heuristics. The first one is based on a greedy algorithm improved with 2-opt and 3-opt moves, and the second one is based on Branch-and-Cut. The algorithms were tested on small instances ($n \le 60$) for which optimal solutions are known and on larger instances with $n \le 500$. On the small instances the optimality gap was always below 2%. The second heuristic provided better solutions than the first. Improved results were obtained by Hernández-Pérez, Rodríguez-Martín, and Salazar-González [58], who introduced a *Variable Neighborhood Descent* (VND) heuristic based on 2-opt and 3-opt local search operators. To diversify the search, the VND was embedded into a GRASP framework. The best solution obtained was post-optimized using a second VND based on vertex exchange operators. On the small instances this heuristic obtained the optimal solution for 145 instances out of 150. For the large instances, by running their algorithm 25 times on each instance they could improve the results of Hernández-Pérez and Salazar-González [60] on 113 of 150 instances. The average CPU time was under 10 minutes.

Zhao et al. [151] obtained good quality solutions by means of a genetic algorithm making use of pheromone information and local search. Their computing times (reported only for the large instances) were comparable to those of Hernández-Pérez, Rodríguez-Martín, and Salazar-González [58]. Hosny and Mumford [64] obtained further improvements by using a *Variable Neighborhood Search* (VNS) algorithm, but at the expense of very high computing times (up to 42 hours on the large instances). To our knowledge, the best computational results have been obtained by Mladenović et al. [92]. They proposed a VNS algorithm that makes use of a collection of neighborhood structures to solve the classical TSP and a binary indexed tree to efficiently check the feasibility of the routes generated during the neighborhoods exploration.

To the best of our knowledge, only one study has addressed the case of multiple commodities: Hernández-Pérez and Salazar-González [62] considered the problem in which a single vehicle is used to transport a set of multiple commodities. They solved the problem with a Branch-and-Cut algorithm similar to that of Hernández-Pérez and Salazar-González [61] for the single-commodity case.

6.2.2 ▪ Multiple Vehicles

The multiple-vehicle case has attracted far less attention than its single-vehicle counterpart. Shi, Zhao, and Gong [125] have presented the mathematical formulation (6.1)–(6.7) and a genetic algorithm derived from that of Zhao et al. [151]. Computational experiments were performed only for the genetic algorithm, on a set of instances derived from those of Hernández-Pérez, Rodríguez-Martín, and Salazar-González [58], by introducing a maximum length constraint for each route. Dror, Fortin, and Roucairol [40] studied the distribution of electric cars in France. Cars are parked in some slots and, after being used, may be returned to a different location. Redistribution is then performed by means of auto-carriers to restore the best distribution for the users. The authors studied the split delivery generalization of the 1-PDVRP, obtained by splitting a vertex of demand q_i into q_i vertices of unit demand. They also considered a heterogeneous fleet of auto-carriers. They modeled the problem with a three-index formulation, similar to (6.1)–(6.7), and also presented heuristic algorithms. Note that this work may be considered as an antecedent of the 1-PDTSP presented by Hernández-Pérez and Salazar-González [59], although it does not report extensive computational results.

A problem that recently attracted the interest of many researchers is the bike sharing rebalancing problem. Bike sharing systems offer public bikes located in several stations. A user can take a bike from a station, use it for a journey, and then leave it at a possibly different station. To keep the system operational, it is often necessary to redistribute the bikes among the stations. Chemla, Meunier, and Wolfler Calvo [27] studied the case of a single vehicle, allowing temporary drop off and split deliveries. Raviv, Tzur, and Forma [111] focused instead on the multiple-vehicle case, proposing a three-index mathematical formulation and testing it on the systems in use in Paris and Washington DC.

6.3 ▪ One-to-Many-to-One Problems

One-to-many-to-one (1-M-1) problems refer to PDPs in which deliveries and pickups concern two distinct sets of commodities: some are shipped from the depot to the customers, and others are picked up at the customers and delivered to the depot. In the last two decades, 1-M-1 problems have become increasingly popular, mainly because of the trend toward recycling and product reuse. For example, in Europe the delivery of new appliances should be accompanied by the pickup of old ones in order to comply with Waste Electrical and Electronic Equipment (WEEE) regulations (see, e.g., Beullens, Van Oudheusden, and Van Wassenhove [17]). The beverage industry provides another popular area of application with full bottles being delivered and empty ones being returned to the depot (see, e.g., Privé et al. [107]). There are also applications in which loaded pallets or containers have to be delivered to the customers while empty ones are collected from the customers (see, e.g., Crainic, Gendreau, and Dejax [35]).

The literature on 1-M-1 PDPs is vast and varied. It is therefore convenient to further classify its variants with respect to additional features concerning the customer ordering, the demand types, and the presence of time windows or maximum route duration constraints. Similar classifications have been proposed for the single-vehicle variants in the survey of Gribkovskaia and Laporte [55]. In particular, problems can be classified with respect to the type of service considered: customers may ask for a *simultaneous* (or combined) service of pickup and delivery, when at least one customer requires both a pickup and a delivery. *Single-demand* service occurs when every customer requires either a pickup or a delivery, but not both.

1-M-1 problems are also intrinsically connected with handling issues: pickups on board may obstruct deliveries. Situations in which handling operations are not allowed may force the decision maker to serve all deliveries first, followed by all pickups. In such a case, if a customer requires both services, it will have to be visited twice. These problems are referred to as *backhaul* problems. On the contrary, problems in which handling is acceptable are referred to as *mixed*, because routes allow for mixed sequences of pickups and deliveries. Intermediate situations are those in which a subset of customers (or all of them) are allowed to be visited more than once. The resulting routes are not Hamiltonian tours, but may allow one to avoid handling operations at the expense of an increase in routing cost. Variants in which handling operations are avoided by altering the Hamiltonian tour shape are presented in Section 6.5.

In Figure 6.2, single-vehicle problems with simultaneous and single demand, backhauls, and mixed services are depicted. Vertex 0 is the depot, and each customer i requires d_i units of delivery from the depot and provides p_i units to be picked up (denoted by the pair (d_i, p_i)). Cases (c) and (d) differ from (a) and (b) because customer 1 requires a combined pickup and delivery service of 2 delivery units and 1 pickup. Tour (a) differs from tour (b) with respect to the backhaul constraint: in tour (b) the delivery customer 2 has to be served before the others (requiring pickup), resulting in a higher routing cost. In tour (d), customer 1 has to be visited twice: first to perform the delivery and second to perform the pickup, thus respecting the backhaul constraint.

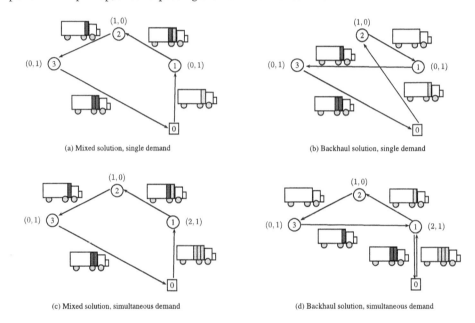

(a) Mixed solution, single demand   (b) Backhaul solution, single demand

(c) Mixed solution, simultaneous demand   (d) Backhaul solution, simultaneous demand

Figure 6.2. *Different* 1-*M*-1 *problems.*

In the remainder of this section, problems with combined and single demands as well as problems with mixed routes are presented, in that order.

6.3.1 ▪ Simultaneous Demands

The *Vehicle Routing Problem with Simultaneous Pickup and Delivery* (VRPSPD) is probably the most studied and most general variant of the 1-M-1 problems, and it is also known as the *Multiple-Vehicle Hamiltonian 1-M-1-PDP with Combined Demands*. It can be formally defined as follows. Let $G = (V, A)$ be a directed graph where V is the vertex set and A is the arc set. Vertex 0 represents the depot, which acts as the origin of the delivery commodities and destination of the pickup commodities, and the other vertices of V are the customers. We assume that a homogeneous fleet K of vehicles with capacity Q is available. As for the 1-PDVRP, the cost of traversing the arc (i, j) is c_{ij}. Each customer i has non-negative demands d_i for the delivery commodity and p_i for the pickup commodity. The problem is to construct $|K|$ routes of minimum total cost, satisfying the pickup and delivery requests of each customer in a single visit and not exceeding the capacity of the vehicles.

The following is an adaptation of the three-index MIP formulation for the VRPSPD of Montané and Galvão [93]. In this formulation, x_{ijk} takes value 1 if the arc (i,j) is traversed by vehicle k, and 0 otherwise. The flow variables y_{ij} and z_{ij} are the amounts of pickup and delivery commodities traveling on arc (i,j), respectively:

$$(6.8) \qquad \text{(VRPSPD)} \quad \text{minimize} \sum_{i \in V} \sum_{j \in V} \sum_{k \in K} c_{ij} x_{ijk}$$

$$(6.9) \qquad \text{s.t.} \quad \sum_{j \in V} \sum_{k \in K} x_{ijk} = 1 \qquad \forall\, i \in V \setminus \{0\},$$

$$(6.10) \qquad \sum_{j \in V} x_{ijk} - \sum_{j \in V} x_{jik} = 0 \qquad \forall\, i \in V,\, k \in K,$$

$$(6.11) \qquad \sum_{i \in V} x_{0ik} \leq 1 \qquad \forall\, k \in K,$$

$$(6.12) \qquad y_{ij} + z_{ij} \leq Q \sum_{k \in K} x_{ijk} \qquad \forall\, i,j \in V,$$

$$(6.13) \qquad \sum_{j \in V} y_{ij} - \sum_{j \in V} y_{ji} = p_i \qquad \forall\, i \in V \setminus \{0\},$$

$$(6.14) \qquad \sum_{j \in V} z_{ji} - \sum_{j \in V} z_{ij} = d_i \qquad \forall\, i \in V \setminus \{0\},$$

$$(6.15) \qquad y_{ij}, z_{ij} \geq 0 \qquad \forall\, i,j \in V,$$

$$(6.16) \qquad x_{ijk} \in \{0,1\} \qquad \forall\, i,j \in V,\, k \in K.$$

The objective function (6.8) aims at minimizing the routing cost. Constraints (6.9) and (6.10) force each customer to be visited exactly once. Constraints (6.11) impose that each vehicle in the fleet be used at most once, while (6.12) bound the flow of goods traveling on each arc to be at most equal to the vehicle capacity. Constraints (6.13) and (6.14) are the flow conservation constraints for pickups and deliveries, respectively. Note that this formulation does not consider additional constraints such as time windows (see Angelelli and Mansini [4]) or maximum route duration constraints (see Montané and Galvão [93], Ai and Kachitvichyanukul [1], and Mingyong and Erbao [89]).

We now present a summary of the main algorithms for the single-vehicle and multiple-vehicle cases, respectively.

6.3.1.1 ▪ Single Vehicle

The single-vehicle problem, which is usually called the *TSP with Simultaneous Pickup and Delivery* (TSPSPD) or *Single-Vehicle Hamiltonian 1-M-1-PDP with Combined Demands*, asks for the minimum-cost Hamiltonian tour serving each customer once without violating the capacity of the vehicle. The TSPSPD can be transformed into a problem with single demands (see, e.g., Berbeglia et al. [15]) by setting the vehicle capacity equal to $Q' = Q - \sum_{i \in V \setminus \{0\}} d_i$ and defining the demand of customer i as $\delta_i = p_i - d_i$ for all $i \in V \setminus \{0\}$. Obviously, the single-demand variant is also a special case of the TSPSPD. Therefore, algorithms for the combined and single-demand problems are interchangeable (when the tour is Hamiltonian). Most of the research has focused on the case with single demands. We thus refer the reader to Section 6.3.2.1 for algorithms that can be adapted to the TSPSPD. We finally note that it is also possible to solve the TSPSPD as a 1-PDTSP, as mentioned by Hernández-Pérez and Salazar-González [59].

6.3.1.2 ▪ Multiple Vehicles

The multiple-vehicle Hamiltonian problem is often called the *VRP with Simultaneous Pickup and Delivery* (VRPSPD) and has attracted a lot of attention in the scientific community. The VRPSPD was introduced by Min [87], inspired by the problem of distributing and collecting books in a library. He proposed a three-stage constructive heuristic in which customers are first clustered, then vehicles are assigned to clusters, and finally routes are assigned to the vehicles. Anily [5] proposed a lower bound and a heuristic algorithm that is asymptotically optimal, under mild probabilistic conditions on the customer locations.

Among early heuristics, one may mention the insertion-based heuristic of Salhi and Nagy [120], able to solve problems with single and multiple depots, the cheapest insertion algorithm with quadratic complexity insertion criteria of Dethloff [39], the cluster-first-route-second algorithm of Halse [56], and the local search heuristic of Nagy and Salhi [96], in which different degrees of solution feasibility are accepted during the execution.

Several of the metaheuristics proposed to solve the VRPSPD are based on *Tabu Search* (TS) or are hybrid algorithms borrowing the concept of tabu list from TS. Crispim and Brandão [36] proposed a TS algorithm in which the neighborhood exploration is performed according to a VND procedure and a single tabu list keeps track of forbidden moves for all neighborhoods. Chen and Wu [28] proposed another hybrid metaheuristic, where a record-to-record algorithm is enhanced by adopting a tabu list and classical local search operators. Montané and Galvão [93] presented a TS algorithm, as well as a mathematical formulation for the VRPSPD with maximum route duration constraints. Bianchessi and Righini [18] enhanced the TS approach by a VND scheme for the neighborhood selection. We also mention the reactive TS algorithms of Wassan, Nagy, and Ahmadi [143] and Wassan, Wassan, and Nagy [144] and the hybrid TS and guided local search algorithm of Zachariadis, Tarantilis, and Kiranoudis [148].

Among successful population-based metaheuristics, Çatay [26] introduced an effective ant colony algorithm. Zachariadis, Tarantilis, and Kiranoudis [149] proposed an adaptive memory approach, in which genetic algorithms and TS are hybridized. More precisely, promising solution components, or *bones*, are extracted from a pool of routes. Bones are selected according to a utility function, combining the bone's length, the number of times the bone was selected in previous iterations, and the cost of the best solution in which the bone was present. Moreover, bones are selected so that each customer appears only once, allowing for an easy recombination of the selected bones into a feasible VRPSPD solution. A granular TS algorithm is applied to improve the resulting solution, and the pool of routes is updated before the whole algorithm is iterated. Finally, Zachariadis and Kiranoudis [147] proposed a local search metaheuristic exploring 2-opt and variable length bone exchange neighborhoods. The latter aims at exchanging the positions of sequences of customers of length shorter than μ. A static move descriptor strategy allows for a faster exploration of these rich neighborhoods, and the "promises" concept enhances the algorithm by preventing cycling and by inducing diversification.

Successful and flexible single-solution-based metaheuristics for the VRPSPD are the *Adaptive Large Neighborhood Search* (ALNS) of Ropke and Pisinger [117], the *Parallel Iterated Local Search* (PILS) algorithm of Subramanian et al. [131], and the Unified local search and Hybrid Genetic Search (UHGS) of Vidal et al. [140]. ALNS iteratively perturbs and improves a solution by means of many diversification and intensification neighborhoods. The neighborhoods are chosen at each iteration by a roulette wheel selection principle, in which weights are adaptively updated according to their performance in previous iterations. The PILS algorithm consists of an intensification stage, in which neighborhoods with quadratic complexity are exhaustively explored in random order, and a diversification stage, in which a randomly selected rich neighborhood perturbs the current solution. The candidate diversification neighborhoods are double-bridge, double swap, and ejection chains. The algorithm is initialized by a randomized constructive heuristic and implemented in a master-slave parallel architecture. A similar *Iterated Local Search* (ILS) hybrid algorithm was presented in a non-parallel implementation by Subramanian, Uchoa, and Ochi [132]. This algorithm consists of a cooperative framework, in which an ILS algorithm similar to that of Subramanian et al. [131] and a *Set Partitioning* (SP) formulation share a pool of routes and upper bounds. The ILS algorithm provides high quality upper bounds and columns in reasonable computing time to the SP formulation, whereas the ILS algorithm benefits from high-quality initial solutions yielded by the SP formulation. The resulting framework, denoted by H-ILS-SP, provides high quality results not only when applied to the VRPSPD, but to many routing variants. The UHGS of Vidal et al. [140] is based on a hybrid genetic algorithm, in which local search operators improve individuals in the population, benefiting from a solution representation without trip delimiters. The individual evaluation preserves (but penalizes) infeasible solutions and diverse individuals; the selection strategy promotes diversity and cost efficiency by keeping track of subpopulations (each updated with respect to the survivor selection mechanism, when the size is too large).

The most common benchmark instances for the VRPSPD are the real-world instance of Min [87] (with 22 customers), the instances of Dethloff [39] (40 problems with 50 customers), the set by Salhi and Nagy [120] (14 problems with 50 to 199 customers), and the instances of Montané and Galvão [93] (18 problems with 100 to 400 customers). The latter two sets are probably the most challenging arena for metaheuristics: the best results seem to be achieved by the adaptive memory algorithm of Zachariadis, Tarantilis, and Kiranoudis [149], the PILS, the H-ILS-SP, and the UHGS. More precisely, UHGS and PILS algorithms obtain a 0.00% average percentage deviation from the best known solutions for the instances of Salhi and Nagy [120], whereas the average gap for Zachariadis, Tarantilis, and Kiranoudis [149] is 0.11%. Note that PILS is considerably faster than UHGS (given the parallel implementation) but provides a slightly worse average performance. The best results on the benchmark set of Montané and Galvão [93] are achieved by H-ILS-SP, with 0.00% and 0.08% average gaps for the best and the average solution values among the 10 executions being performed. UHGS produces slightly worse results (0.07% and 0.20% average gaps for the best and the average of 10 replications).

To the best of our knowledge, the only approximation algorithm for the VRPSPD is that of Katoh and Yano [69], where the networks being analyzed are trees.

Exact algorithms include the Branch-and-Cut-and-Price algorithm for the VRPSPD with time windows by Angelelli and Mansini [4], capable of solving to optimality instances with up to 20 customers, and the algorithm of Dell'Amico, Righini, and Salani [37]. This algorithm is based on a set covering formulation where exact dynamic programming and state space relaxation techniques are used to solve the pricing subproblem. Bi-directional search, an upper bound on the number of customers visited in a route, as well as problem-specific branching strategies allowed the algorithm to solve to optimality problems with up to 40 customers.

Recently, Subramanian et al. [133] proposed a Branch-and-Cut algorithm based on the lazy separation of the capacity inequalities, improving the best known lower bounds and finding some new optimal solutions. This approach seems to outperform the Branch-and-Cut algorithms based on single-commodity, undirected two-commodity, and directed two-commodity formulations (capable of solving instances with up to 100 customers) from Subramanian [129]. Subramanian et al. [134] also proposed a Branch-and-Cut-and-

Price method based on the *pd*-routes idea: *pd*-routes are non-elementary paths in which both pickups and deliveries are taken into account to identify capacity violations. The pricing problem is solved by dynamic programming, generating *pd*-routes, but scaling and sparsification are adopted at early stages of the algorithm to speed up computation. This algorithm has improved many best known lower bounds and seems to outperform Branch-and-Cut algorithms when the average number of customers per route is small.

6.3.2 ▪ Single Demands

Single-demand 1-M-1 problems represent a special case of simultaneous demand problems where each customer can either require a pickup or a delivery, but not both. These problems can be further classified depending on the constraints imposed on the customer order: mixed routes allow pickup and delivery customers to be mixed (see Section 6.3.2.1 for single-vehicle problems and Section 6.3.2.2 for multiple-vehicle problems), whereas backhaul routes force all delivery customers to precede the pickups. Backhaul problems will not be presented extensively in this chapter but are covered in Chapter 9.

6.3.2.1 ▪ Single Vehicle

In problems with mixed routes, deliveries and pickups can be performed in any order. As mentioned by Gribkovskaia and Laporte [55], the single-vehicle problem is usually referred to as the *TSP with Pickups and Deliveries* (TSPPD), the *TSP with Delivery and Backhauls*, and the *Mixed TSP*. In the classification of Berbeglia et al. [15], the problem is the *1-M-1-PDP with Single Demands and Mixed Solutions*. In the following, we will adopt the acronym TSPPD.

The TSPPD was introduced by Mosheiov [94], who proved that any TSP solution can be converted into a feasible TSPPD solution by inserting the depot in an appropriate location along the tour. Assuming that the TSP tour is obtained by an algorithm with worst-case performance α, the overall approach guarantees a worst-case ratio of $1 + \alpha$. Mosheiov [94] also described a constructive heuristic based on a cheapest insertion criterion. Anily and Mosheiov [6] proposed an approximation algorithm based on the construction of a minimum spanning tree: the tree is traversed in a depth-first fashion, prioritizing the subtrees with positive net demand and delivery customers in the same subtree. Finally, a Hamiltonian tour is obtained by means of shortcut operations. Assuming symmetric distances satisfying the triangular inequality, the worst-case performance of this algorithm is 2.

Gendreau, Laporte, and Vigo [51] considered the special case of a TSPPD defined on a cycle and designed a linear-time exact algorithm. The algorithm was then adapted to general graphs and provides a worst-case performance equal to 3. The authors also proposed a TS metaheuristic, based on 2-opt moves, and a set of instances that now constitute the standard benchmark for the TSPPD. These instances were adapted from VRP instances, and the number of customers varies between 6 and 261.

Some effective algorithms for the M-M problem have also been extended to the TSPPD, such as the Hernández-Pérez and Salazar-González truncated Branch-and-Cut algorithm for the 1-PDTSP [60] (see Section 6.2.1). Another adaptation to the TSPPD is the hybrid genetic algorithm of Zhao et al. [152] proposed for the 1-PDTSP: the algorithm is based on a pheromone-driven crossover operator and local search operators. Recently, Çinar, Öncan, and Haldun [29] proposed another genetic algorithm using a depot removal-insertion-based tour improvement procedure.

The ILS of Subramanian and Battarra [130] combines a VND intensification phase (in which Or-opt, 2-opt, exchange, and relocate operators are applied in random order) with an effective diversification phase (in which the double-bridge operator is used), similar to the ILS metaheuristic for the VRPSPD described in Section 6.3.1.2.

Finally, Baldacci, Hadjiconstantinou, and Mingozzi [10] proposed a Branch-and-Cut algorithm, based on a two-commodity flow formulation, with initial upper bounds obtained by a genetic algorithm. This approach could solve to optimality some instances with up to 200 customers. The state-of-the-art exact algorithm for the TSPPD is an adaptation of the Branch-and-Cut of Hernández-Pérez and Salazar-González [59, 61], presented in Section 6.2.1. This algorithm is able to solve instances with up to 260 customers.

Classical benchmark problems for the TSPPD are the Euclidean instances of Mosheiov [94] and the instances of Gendreau, Laporte, and Vigo [51]. Both sets are available at http://webpages.ull.es/users/hhperez/PDsite/. The Gendreau, Laporte, and Vigo set [51] consists of instances adapted from the VRP (6 to 261 customers), instances in which the coordinates of the customers are randomly generated and Euclidean instances are considered (25 to 200 customers), and instances with randomly generated arc costs (25 to 200 customers). Gendreau, Laporte, and Vigo [51], Hernández-Pérez and Salazar-González [60], Zhao et al. [152], and Çinar, Öncan, and Haldun [29] report high-quality average performance for the Gendreau, Laporte, and Vigo problems [51]. Note that the genetic algorithm used by Baldacci, Hadjiconstantinou, and Mingozzi [10] to compute upper bounds had also found optimal solutions to all instances in the set but seven. The randomly generated instances of Hernández-Pérez and Salazar-González [60] provide a more challenging benchmark set involving up to 500 customers. The algorithms of Hernández-Pérez and Salazar-González [60] and of Subramanian and Battarra [130] produce the best known results: both are capable of solving to optimality all small instances (up to 60 customers) in fractions of a second of computing time. The former is slightly faster than the latter on large instances (100 to 500 customers), but the latter, despite its simplicity, was capable of finding 44 new best known solutions and provides a better performance than the former, even by considering its average performance over 10 executions.

6.3.2.2 ▪ Multiple Vehicles

The *Vehicle Routing Problem with Mixed Pickup and Delivery* (VRPMPD) or *Multiple-Vehicle 1-M-1 PDP with Single Demands and Mixed Solutions* was first studied by Golden et al. [53] under the name *Vehicle Routing with Backhauling*. The authors introduced a stop-based backhaul procedure according to which delivery customers are scheduled first, while pickups are added to the routes in a second stage by an insertion algorithm. Casco, Golden, and Wasil [25] enhanced this idea by incorporating a penalty term in the insertion procedure that is applied whenever deliveries are performed after pickups.

Mosheiov [95] proposed a two-index formulation, a lower bounding procedure, and iterative tour partitioning algorithms with related worst-case performance for the VRPMPD. These heuristics iteratively break a giant tour into routes, trying to optimize the utilization of the vehicles. The constructive heuristics of Salhi and Nagy [120] and of Dethloff [39] (presented in Section 6.3.1.2) are capable of solving the VRPMPD, and so are the ALNS of Ropke and Pisinger [117], the local search heuristic of Nagy and Salhi [96], and the hybrid TS and VND algorithm by Crispim and Brandão [36]. Wade and Salhi [142] proposed an ant colony algorithm dedicated to the VRPMPD, while Wassan, Nagy, and Ahmadi [143] discussed the relationship between the simultaneous and mixed variants and developed another ant colony algorithm. We refer the interested reader to Wassan, Nagy, and Ahmadi [143] for a description of five sets of benchmark instances.

Subramanian [129] presented both exact and heuristic algorithms for the VRPMPD. The exact algorithms are Branch-and-Cut algorithms based on lazy separation of the capacity constraints (see also Subramanian et al. [133]), on single-commodity, undirected two-commodity, directed two-commodity, and two-index formulations, as well as on the Branch-and-Cut-and-Price algorithm presented in Section 6.3.1.2 (see also Subramanian et al. [134]). These algorithms provide the best lower bounds on the Salhi and Nagy benchmark set [120] (21 instances without route duration constraint and 21 with route duration constraint, considering between 50 and 199 customers). The Branch-and-Cut-and-Price algorithm improves some of the best known lower bounds and outperforms the Branch-and-Cut algorithms, given that the average number of customers per route in the Salhi and Nagy problems [120] is relatively small. The ILS of Subramanian [129] and the hybrid algorithm of Subramanian, Uchoa, and Ochi [132] (presented in Section 6.3.1.2) are also capable of solving the VRPMPD: both algorithms outperform the ALNS of Ropke and Pisinger [116]. The hybrid algorithm is capable of finding or improving the best known solutions on all instances but a few. The ILS is not as accurate as the hybrid, but about six times faster.

Kontoravdis and Bard [70] presented a GRASP algorithm for the VRP with time windows and adapted the algorithm for the VRPMDP with time windows. Zhong and Cole [153] proposed a guided local search heuristic for the same variant. Wade and Salhi [141] studied a VRPMDP variant in which pickups may be performed only once a percentage of the deliveries has been delivered. This problem is solved by means of an insertion heuristic. Finally, good computational results were recently obtained by Tarantilis, Anagnostopoulou, and Repoussis [136]. They studied the effect of different backhauling strategies, varying from the pure mixed strategy to the one allowing only backhaul routes (see the next section). They proposed an adaptive path relinking method that evolves a set of reference solutions and combines them using path relinking. Their algorithm obtained good improvements with respect to existing metaheuristics for the VRPMDP with time windows.

6.3.2.3 ▪ Backhaul Routes

In problems with backhaul routes, all deliveries must be performed before any of the pickups. Therefore, handling issues are always avoided at the expense of extra routing costs. We refer the reader to the survey of Gribkovskaia and Laporte [55] and to Chapter 9 for the single-vehicle and multiple-vehicle problems, respectively.

Another family of related problems arises in the context of transportation with *cross-docking* (see, e.g., Wen et al. [145] and Tarantilis [135]), where vehicles performing pickups stop at a consolidation facility to reorganize the loads before making deliveries. Recently, Santos, Mateus, and da Cunha [121] explicitly considered cross-docking in pickup and delivery routing, allowing for both routes that visit pickup locations followed by delivery locations and routes that stop at an intermediate cross-dock.

6.4 ▪ One-to-One Problems

In *one-to-one* (1-1) PDPs, each customer request consists of transporting a load from one pickup vertex to one destination vertex. This problem arises in urban courier services, less-than-truckload transportation systems and, maritime shipping (see Andersson, Christiansen, and Fagerholt [3] and Korsvik, Fagerholt, and Laporte [71]), among others. In the context of passenger transportation, it is often called the dial-a-ride problem (see Chapter 7).

We first provide a formulation for the general case involving multiple capacitated vehicles and time windows. We then review the literature on the single-vehicle case and on the multiple-vehicle case, respectively.

Following the notation introduced by Cordeau [30], 1-1 problems can be formulated on a graph $G = (V, A)$ where the vertex set V is the union of the set of pickup nodes

$P = \{1, \ldots, n\}$, the set of delivery nodes $D = \{n+1, \ldots, 2n\}$, and the depot nodes $\{0, 2n + 1\}$. Each request i then consists of transporting a load of size q_i from the pickup node $i \in P$ to the delivery node $n+i \in D$. A non-negative service duration s_i is also associated with every node $i \in V$. We assume that $s_0 = s_{2n+1} = 0$, $q_{n+i} = -q_i$, and $q_0 = q_{2n+1} = 0$. A time window $[a_i, b_i]$ is also associated with node $i \in V$, where a_i and b_i represent the earliest and latest times, respectively, at which service may begin at node i. Finally, with each arc $(i,j) \in A$ are associated a routing cost c_{ij} and a travel time t_{ij}. Let L_k be the maximum duration of the route performed by vehicle k.

For each arc $(i,j) \in A$ and each vehicle $k \in K$, let x_{ijk} be a binary variable equal to 1 if vehicle k travels from node i to node j, and 0 otherwise. For each node $i \in V$ and each vehicle $k \in K$, let T_{ik} be the time at which vehicle k begins service at node i, and let Q_{ik} be the load of vehicle k after visiting node i. The *Pickup-and-Delivery Vehicle Routing Problem with Time Windows* (PDVRPTW) can be formulated as the following mixed-integer program:

$$(6.17) \qquad \text{(PDVRPTW)} \qquad \text{minimize} \sum_{k \in K} \sum_{i \in V} \sum_{j \in V} c_{ij} x_{ijk}$$

$$(6.18) \qquad \text{s.t.} \qquad \sum_{k \in K} \sum_{j \in V} x_{ijk} = 1 \qquad\qquad \forall\, i \in P,$$

$$(6.19) \qquad\qquad \sum_{j \in V} x_{ijk} - \sum_{j \in V} x_{n+i,jk} = 0 \qquad \forall\, i \in P,\, k \in K,$$

$$(6.20) \qquad\qquad \sum_{j \in V} x_{0jk} = 1 \qquad\qquad \forall\, k \in K,$$

$$(6.21) \qquad\qquad \sum_{j \in V} x_{jik} - \sum_{j \in V} x_{ijk} = 0 \qquad \forall\, i \in P \cup D,\, k \in K,$$

$$(6.22) \qquad\qquad \sum_{i \in V} x_{i,2n+1,k} = 1 \qquad\qquad \forall\, k \in K,$$

$$(6.23) \qquad\qquad T_{jk} \geq (T_{ik} + s_i + t_{ij}) x_{ijk} \qquad \forall\, i \in V,\, j \in V,\, k \in K,$$

$$(6.24) \qquad\qquad Q_{jk} \geq (Q_{ik} + q_j) x_{ijk} \qquad \forall\, i \in V,\, j \in V,\, k \in K,$$

$$(6.25) \qquad\qquad T_{n+i,k} - T_{ik} - s_i - t_{i,n+i} \geq 0 \qquad \forall\, i \in P,$$

$$(6.26) \qquad\qquad T_{2n+1,k} - T_{0k} \leq L_k \qquad\qquad \forall\, k \in K,$$

$$(6.27) \qquad\qquad a_i \leq T_{ik} \leq b_i \qquad\qquad \forall\, i \in V,\, k \in K,$$

$$(6.28) \qquad\qquad \max\{0, q_i\} \leq Q_{ik} \leq \min\{Q_k, Q_k + q_i\} \qquad \forall\, i \in V,\, k \in K,$$

$$(6.29) \qquad\qquad x_{ijk} \in \{0,1\} \qquad\qquad \forall\, i \in V,\, j \in V,\, k \in K.$$

Constraints (6.18) and (6.19) ensure that each request is served exactly once and that the origin and destination nodes of a request are visited by the same vehicle. Constraints (6.20)–(6.22) guarantee that all routes start at the origin depot and end at the destination depot. Consistency of the time and load variables is ensured by constraints (6.23) and (6.24). Precedence constraints are imposed through inequalities (6.25). Finally, inequalities (6.26) bound the duration of each route, while (6.27) and (6.28) impose time-window and capacity constraints, respectively.

6.4.1 ▪ Single Vehicle

Most of the research on single-vehicle problems has focused on variants without time windows. The special case of the problem where the load of each request is equal to 1 and the vehicle itself has a capacity of 1 is called the *stacker crane problem*. This problem, which has applications in port operations and in various industrial contexts, was studied, among others, by Frederickson, Hecht, and Kim [49].

The more general *Pickup-and-Delivery TSP* (PDTSP) was first studied formally in the context of passenger transportation by Stein [127, 128], who performed a probabilistic analysis of the problem and introduced construction heuristics based on region-partitioning principles. Several dynamic programming approaches were then developed for the PDTSP. In particular, Psaraftis [108] and Desrosiers, Dumas, and Soumis [38] addressed problems arising in dial-a-ride systems.

The incorporation of precedence constraints in models and algorithms for the TSP was first discussed by Lokin [80]. Later, Kalantari, Hill, and Arora [68] extended the TSP Branch-and-Bound algorithm of Little et al. [79] to solve the PDTSP with a single or multiple vehicles and with or without capacity constraints. A polyhedral study of the PDTSP was done by Ruland and Rodin [118], who introduced a Branch-and-Cut algorithm based on four families of valid inequalities. This algorithm was capable of solving instances with up to 15 requests. This work was later extended by Dumitrescu et al. [42], who generalized some of these inequalities and introduced several other families. Their Branch-and-Cut algorithm was able to solve some instances with up to 35 requests.

The development of exact algorithms for the PDTSP has largely benefited from the work done on the closely related TSP with precedence constraints in which a vertex may have multiple predecessors. The polyhedron of this problem was analyzed by Balas, Fischetti, and Pulleyblank [8], whereas Fischetti and Toth [48] and Ascheuer, Jünger, and Reinelt [7] solved the problem by additive Branch-and-Bound and by Branch-and-Cut, respectively. Dynamic programming approaches were also proposed by Bianco et al. [19] and Mingozzi, Bianco, and Ricciardelli [88]. Finally, additional valid inequalities were described by Gouveia and Pesneau [54] and by Mak and Ernst [85].

Local search heuristics for the PDTSP were first investigated by Psaraftis [109], who considered k-interchanges as in the Lin and Kernighan heuristic [78] for the TSP. The author proposed a procedure to identify the best such move in $O(n^k)$ despite the presence of the precedence constraints. The idea consists of separating the improvement checks, i.e., finding the best move, from the feasibility checks, i.e., verifying its feasibility. Feasibility checks can be performed in a preprocessing step, and the information can be stored in a matrix which is updated every time an improved solution is found. Later, Kubo and Kasugai [72] introduced several insertion heuristics along with two local search algorithms extending the Or-opt procedure for the TSP by Or [99]. The efficient implementation of local search operators for the TSP with precedence constraints was also addressed by Savelsbergh [122]. A variable-depth search procedure relying on seven types of arc exchange mechanisms was developed by Van der Bruggen, Lenstra, and Schuur [138] to solve the PDTSP with time windows. An improvement to local search heuristics for the PDTSP was also introduced by Healy and Moll [57], who suggested alternating between moves aiming to improve a solution in the traditional sense and moves bringing the search toward solutions that have larger feasible neighborhoods. This principle, called "sacrificing", can be seen as a type of algorithmic perturbation. Several other heuristics for the PDTSP were introduced by Renaud, Boctor, and Ouenniche [113] and Renaud, Boctor, and Laporte [112]. The most effective of these heuristics are based on solution perturbation ideas. Vertices can be randomly removed and reinserted in the solution, or two solutions can be combined into one as in crossover operations performed in genetic algorithms. Finally, a genetic algorithm, a simulated annealing approach, and a hill climbing heuristic were developed by Hosny and Mumford [65] to solve the PDTSP with time windows.

6.4.2 • Multiple Vehicles

A large number of exact and heuristic algorithms were developed for the multiple-vehicle 1-to-1 problem, which is usually referred to as the *Pickup-and-Delivery Vehicle Routing Problem* (PDVRP). The vast majority of these algorithms handle the more general problem with time windows.

A Branch-and-Cut algorithm for both the PDVRP and the PDVRPTW was introduced by Lu and Dessouky [81], who formulated the problem with a polynomial number of variables and constraints through the use of binary variables b_{ij} indicating whether node i comes before another node j on the tour. The formulation was also strengthened by the addition of four families of valid inequalities relying on these variables. The authors were able to solve instances with up to five vehicles and 25 requests. Another Branch-and-Cut algorithm was developed by Ropke, Cordeau, and Laporte [115], who focused on the PDVRPTW and introduced two formulations with an exponential number of constraints based on the precedence constraints of Ruland and Rodin [118]. The most compact formulation in terms of the number of variables uses only binary routing variables, and time-window constraints are imposed by forbidding infeasible paths. These formulations were improved by several families of valid inequalities, including strengthened capacity constraints that take the pickup-and-delivery structure into account and fork constraints that combine this structure with time-window restrictions. The Branch-and-Cut algorithm was able to solve some instances with up to 75 requests.

Several algorithms based on column generation were also proposed for the PDVRPTW. The first such algorithm was introduced by Dumas, Desrosiers, and Soumis [41], who formulated the problem as an SP problem with an extra constraint on the number of vehicles. Columns are generated by solving through forward dynamic programming a non-elementary shortest path problem with pickup, delivery, and time-window constraints. In their implementation of the shortest path algorithm, labels contain the set of open requests, i.e., requests for which the pickup node has been visited, but not the delivery. This algorithm was capable of solving tightly constrained instances with up to 55 requests. Later, Ropke and Cordeau [114] presented a Branch-and-Cut-and-Price algorithm incorporating some of the inequalities introduced by Ropke, Cordeau, and Laporte [115] in a Branch-and-Price framework. The authors have compared solving the elementary and non-elementary shortest path problems to generate columns. This algorithm outperformed the Branch-and-Cut algorithm of Ropke, Cordeau, and Laporte [115] and was able to solve some tightly constrained instances with up to 500 requests. Very recently, another exact algorithm based on column generation was introduced by Baldacci, Bartolini, and Mingozzi [9]. This algorithm relies on a bounding procedure that finds a near-optimal dual solution of the LP relaxation of the SP formulation by combining two dual ascent heuristics and a cut-and-column generation procedure. This solution is then used to enumerate the routes with a reduced cost smaller than the gap between the lower and upper bounds. If the number of columns is small enough, an optimal solution can be found by solving an integer program. Otherwise, Branch-and-Cut-and-Price is used to close the gap. This algorithm is much faster than that of Ropke and Cordeau [114] and was able to solve 15 more instances to optimality.

One of the first heuristics for the PDVRPTW was introduced by Nanry and Barnes [97], who developed a reactive TS relying on three move operators: moving a pickup-delivery pair from its current route to a different route, swapping two pickup-delivery pairs between two different routes, and moving an individual pickup or delivery node within its route. An improved implementation of this heuristic was later described by Lau and Liang [74], who also proposed a hybrid partitioned insertion heuristic to generate initial solutions. At approximately the same time, Li and Lim [76] proposed another heuristic based on very similar move operators. This heuristic combines TS and simulated annealing and forces the search to return to the incumbent solution after a number of iterations without improvement. Pankratz [102] proposed a grouping genetic algorithm in which each gene represents a group of requests instead of single requests. An insertion heuristic is used to update the routing corresponding to a group of requests whenever an individual is modified by a genetic operator. Lu and Dessouky [82] focused on the creation of initial solutions and introduced an insertion-based construction heuristic that differs from classical procedures by taking into consideration the increase in travel time as well as the reduction in time window slack due to the insertion. In addition, this heuristic tries to reduce the number of crossings in the constructed routes so as to improve the visual attractiveness of the solution.

The two best performing metaheuristics for the PDVRPTW are those of Bent and Van Hentenryck [14] and of Ropke and Pisinger [116], both based on large neighborhood search. The first is a two-stage hybrid algorithm that consists of a first stage aiming to decrease the number of routes by using simulated annealing and a second stage trying to reduce the total travel cost by using large neighborhood search. This algorithm produced very good solutions on a large number of test instances. However, improved solutions were obtained slightly later by the ALNS of Ropke and Pisinger [116]. This heuristic relies on multiple simple heuristics both to remove and to reinsert customers in the solution. To remove nodes from the solution, the authors consider an adaptation of the relatedness measure of Shaw [124], as well as additional rules such as worst removal, which removes the requests leading to the largest cost savings. Reinsertions are performed either via a greedy or a regret heuristic. On test problems involving up to 1000 requests, this heuristic has produced better results on average than the two-stage approach of Bent and Van Hentenryck [14], both in terms of the number of vehicles used and of the total distance traveled.

While most algorithms developed for the PDVRP consider the static variant of the problem where all requests are known in advance, several authors have also studied the dynamic case in which some requests arrive dynamically as other requests are already being served. Savelsbergh and Sol [123] developed a Branch-and-Price algorithm to address a problem arising at a road transportation company operating in the Benelux region. The algorithm was used heuristically to solve large-scale dynamic instances over a multiple-day planning horizon by decomposing them into a number of smaller subproblems containing known requests within a rolling horizon framework. Mitrović-Minić and Laporte [91] defined and compared four waiting strategies to decide where vehicles should wait when requests arrive dynamically in the presence of time windows on pickups and deliveries. The authors considered a rolling horizon in which static problems are solved by a cheapest insertion procedure followed by a TS. This approach was then extended by Mitrović-Minić, Krishnamurti, and Laporte [90] by considering a double horizon objective: a short term objective of minimizing distance and a long term objective of maximizing slack time to accommodate new requests. A different TS heuristic for this problem was developed by Gendreau et al. [50]. This heuristic relies on ejection chain neighborhoods and an adaptive memory that maintains a pool of routes extracted from the best visited solutions. A hybrid adaptive and predictive control algorithm combining genetic algorithms and fuzzy clustering was introduced by Sáez, Cortés, and Núñez [119] for a dynamic PDVRP. The algorithm takes into account future demand and prediction of expected waiting time and travel times for the requests. Finally, Ghiani et al. [52] have described and assessed anticipatory heuristics for the problem that take advantage of stochastic information about future demand in the hope of making better vehicle assignment, routing, and waiting decisions. The algorithms work by sampling the short-term future every time a new request arrives so as to take into account the inconvenience cost of future requests when evaluating potential solutions. The authors have devised both an insertion procedure and a local search algorithm, which they applied to instances containing up to 600 requests. Both algorithms significantly outperformed reactive algorithms that do not incorporate sampling.

Several other variants of the PDVRP have also been studied in recent years. One such variant is the PDVRP with split loads that was addressed by Nowak, Ergun, and White [98]. The authors quantified the benefits of allowing split loads in the context of the PDP and conjectured that load splitting can lead to cost reductions of at most 50%. They have also proposed a simple heuristic to generate an initial solution and then select loads for splitting according to some guidelines. On a related note, Venkateshan and Mathur [139] designed a specialized column generation algorithm for a variant of the PDVRP where multiple visits are required by the same or different vehicles to satisfy each request. Another interesting variant is the PDVRP with transshipment that was recently introduced by Qu and Bard [110] in the context of air freight transportation. A GRASP incorporating a reactive ALNS heuristic was proposed to solve the problem. This heuristic makes use of specialized insertion and removal procedures to accommodate transshipments.

Several industrial applications of the PDVRP have also been described. For example, Xu et al. [146] addressed a rich PDP with many side constraints and features: multiple carriers and vehicle types, multiple time windows, compatibility constraints between requests and vehicles as well as between requests, last-in-first-out vehicle loading policy, working hour rules for drivers, and a complex objective function. This model aims to capture most practical situations arising at the client firms of an optimization software provider. The authors have proposed solving the problem with a column generation-based approach that solves the LP relaxation to optimality and then obtains an integer solution by considering only the set of columns generated when solving this relaxation. An application concerning the transportation of live animals was studied by Sigurd, Pisinger, and Sig [126]. In this context, precedence constraints must be enforced to avoid the spread of diseases, with healthier animals being transported first. The authors formulated the problem as a PDVRPTW with precedence constraints and developed a column generation approach to solve it to optimality.

6.5 ▪ Problems with Loading Constraints

In several pickup-and-delivery applications, the nature of the freight to be transported and the characteristics of the vehicle and of the handling equipment may cause some loading or unloading operations to be costly or even impossible. This happens, for example, when the vehicle is rear-loaded (i.e., it has a single door at the back for loading and unloading), and items are heavy or fragile and hence cannot be rearranged at the customer locations. In this situation an item being picked up is placed at the end of the cargo, and the next item to be delivered must be taken only from those at the back of the vehicle. Consider, for example, the 1-1 problem depicted in Figure 6.1. After the two pickups of A and B, the item available at the end of the cargo is B, and hence the first delivery to be performed can only be that of B, followed by A. This typically leads to an increase in solution cost with respect to problems where loading issues need not be taken into account.

Loading considerations are usually modeled by imposing additional constraints in the model, by including handling costs in the objective function to represent the time lost (or risk incurred) for a cargo rearrangement, or by reducing the loading space available on the vehicle so as to allow any possible loading or unloading operation. Several real-world problems involving combined routing (not necessarily with pickup and delivery) and loading have been studied in the literature. We refer the reader to Iori and Martello [66] for a recent survey. In this section we focus on 1-1 and 1-M-1 PDPs involving loading constraints. To the best of our knowledge, these constraints have not been considered in the context of M-M problems.

6.5.1 ▪ One-to-One Problems

An obvious situation where pickup and delivery operations are affected by loading considerations is when the shapes of the items complicate their arrangement in the cargo and must thus be explicitly taken into account when planning vehicle routes.

Malapert et al. [86] studied the generalization of the PDVRP (see Section 6.4.2), in which items are two-dimensional rectangles and vehicles are rear-loaded and have a two-dimensional rectangular loading surface. This problem also generalizes the *Capacitated Vehicle Routing Problem* (CVRP) with two-dimensional loading constraints (see, e.g., Iori, Salazar-González, and Vigo [67]) by introducing pickup and delivery of the items. A pickup of an item may be performed only if there is a feasible non-overlapping placement for the item in the vehicle surface. A delivery may be performed only if the unloading of the item can be obtained with a single straight movement, as would happen when using an automatic fork. No reshuffling of the cargo is allowed. The resulting problem was modeled using constraint programming, but no results were given.

Bartók and Imreh [11] studied a generalization of the above problem, where vehicles have a three-dimensional loading space and items are three-dimensional boxes. They solved the problem using heuristics based on local search operators. Zachariadis, Tarantilis, and Kiranoudis [150] investigated a further generalization in which time windows are imposed and the three-dimensional items are first loaded on pallets, which are then loaded on vehicles.

6.5.1.1 ▪ TSP with Pickup and Delivery and LIFO or FIFO Loading

The natural extension of the PDTSP (see Section 6.4.1) in which the vehicle is rear-loaded and the loading or unloading operations must be performed according to a *Last-In-First-Out* (LIFO) policy is known in the literature as the *Traveling Salesman Problem with Pickup and Delivery and LIFO Loading* (TSPPDL). The first work related to this problem dates back to Ladany and Mehrez [73], who studied a real-life delivery problem. Then, Pacheco [100, 101] proposed a Branch-and-Bound and a heuristic algorithm based on Or-opt exchanges.

More recently, heuristic algorithms were developed by Carrabs, Cordeau, and Laporte [23], who proposed exchange operators and a variable neighborhood search heuristic, and by Li et al. [77], who presented a metaheuristic based on a tree representation of the problem. Carrabs, Cerulli, and Cordeau [21] implemented an exact Branch-and-Bound using additive lower bounds and consistently solved instances with up to 17 requests. Cordeau et al. [32] solved the problem with a Branch-and-Cut based on the classical two-index vehicle flow formulation and consistently solved instances with up to 21 requests. By looking at the results, it is obvious that the TSPPDL is computationally more difficult than the original PDTSP.

Generalizations of the TSPPDL involving multiple vehicles were studied by Levitin and Abezgaouz [75], who focused on the routing of multiple-load LIFO AGVs in industrial plants, and by Xu et al. [146], who solved a real-world delivery problem with several complicating constraints (see Section 6.4.2 for more details).

A variant of the TSPPDL that considers the *First-In-First-Out* (FIFO) rule, instead of the LIFO, has been studied under the name of *Traveling Salesman Problem with Pickup and Delivery and FIFO Loading* (TSPPDF). Erdoğan, Cordeau, and Laporte [43] proposed a multi-commodity ILP formulation and several heuristics, including TS and iterated local search. Carrabs, Cerulli, and Cordeau [21] solved the problem by Branch-and-Bound, and Cordeau, Dell'Amico, and Iori [31] used Branch-and-Cut. As for the TSPPDL, the TSPPDF is hard and some instances with just 15 requests are still not solved to proven optimality.

6.5.1.2 ▪ Double TSP with Multiple Stacks

A variant that recently attracted the interest of many researchers is the *Double Traveling Salesman Problem with Multiple Stacks* (DTSPMS), another generalization of the PDTSP discussed in Section 6.4.1. The main difference between the two problems is the fact that in the DTSPMS each customer request consists of one pallet, and the single vehicle is equipped with multiple stacks, each of which accommodates pallets by obeying the LIFO policy. The vehicle must perform a first Hamiltonian tour on the pickup vertices and then a second Hamiltonian tour on the delivery vertices. During the pickup phase each pallet is loaded at the top of a stack, and during the delivery phase only the pallets located at the top of a stack can be unloaded. The aim is to find a pair of cycles of minimum cost for which there exists a feasible loading and unloading plan.

The DTSPMS was motivated by a real-world distribution case and was first studied by Petersen and Madsen [106], who proposed an ILP formulation and a simulated annealing metaheuristic. In terms of exact algorithms, Lusby et al. [84] presented an algorithm based on the iterated execution of two phases: in the first phase, the k best solutions of the two separate TSPs are generated; in the second phase, attempts are made to find feasible loading plans for each pair of TSP solutions. The approach fails in solving some instances with 12 requests. It was later improved by Lusby and Larsen [83], who proposed pre-processing techniques and increased the size of the smallest unsolved instance to 14 requests. An enumerative Branch-and-Bound for the case in which the vehicle has exactly two stacks was developed by Carrabs, Cerulli, and Speranza [22].

The best results on the DTSPMS have been obtained by Branch-and-Cut. Petersen, Archetti, and Speranza [105] studied mathematical formulations with exponentially many constraints and solved them with Branch-and-Cut algorithms. Alba Martínez et al. [2] considered one of the formulations of Petersen, Archetti, and Speranza [105] and enriched it with several families of valid inequalities. Their Branch-and-Cut could solve to optimality all instances with 18 requests. The fact that instances with about 20 requests remain unsolved after several attempts is an indication of the combinatorial difficulty of the DTSPMS.

To address larger instances, heuristic algorithms have been proposed. Felipe, Ortuño, and Tirado [45, 46] presented neighborhood structures and derived a variable neighborhood search algorithm. This last algorithm was extended by Felipe, Ortuño, and Tirado [47] with the use of algorithms that also search in the space of infeasible solutions. The complexity of the DTSPMS and of several subproblems that may arise from it is discussed by Toulouse and Wolfler Calvo [137], Casazza, Ceselli, and Nunkesser [24], and Bonomo, Mattia, and Oriolo [20].

The generalization of the DTSPMS in which pickups and deliveries may be performed in mixed order has been studied by Côté, Gendreau, and Potvin [34], who solved it with a large neighborhood search procedure, and by Côté et al. [33], who proposed mathematical formulations and Branch-and-Cut algorithms. Batista-Galván, Riera-Ledesma, and

Salazar-González [12] proposed a Branch-and-Cut algorithm and several valid inequalities for a further relaxation of the DTSPMS in which delivery vertices represent customers requiring a given product, whereas pickup vertices are markets where products are available at different costs. The problem is to find the minimum cost tour that visits a subset of markets, loads the products into the stacks, and then delivers the products to the customers.

6.5.2 ▪ One-to-Many-to-One Problems

In some cases the loading of the vehicle is organized in such a way that all loading and unloading operations are possible at any moment along the route. This, however, is obtained at the expense of a reduction in the overall loading capacity of the vehicle. This practice is applied, for example, to overcome linehaul/backhaul distribution and save routing costs. For example, Hoff et al. [63] studied a real-world beverage distribution problem arising in Norway. This is a typical situation of the 1-M-1 transportation discussed in Section 6.3, where full bottles must be delivered and empty bottles collected. In their study, each vehicle is rear-loaded, and its initial cargo is organized so that all full bottles are packed along the sides of the vehicle. An empty corridor is left at the center of the cargo, making all bottles accessible for unloading. Empty bottles, when loaded, take the place of the full ones that have been unloaded.

In other cases the full loading space of the vehicle can be used, but in order to perform unloading operations some items may have to be reshuffled at the customer sites. In this case the additional effort required for the reshuffling is usually modeled as a handling cost. A typical example is given by the *Traveling Salesman Problem with Pickups, Deliveries, and Handling Costs* (TSPPD-H). The TSPPD-H generalizes the TSP with Simultaneous Pickup and Delivery (see Section 6.3.1.1) by adding handling costs for all those unloading operations that cannot be performed directly from the bottom of a single rear-loaded vehicle. The problem was introduced by Battarra et al. [13], who modeled it as an ILP. Since several possible rearrangements of a cargo may be performed after a delivery, the authors introduced some simplified policies to make the problem more tractable and proposed heuristics and a Branch-and-Cut algorithm. These algorithms were tested on instances involving up to 25 customers and hundreds of items. Larger instances were solved by Erdoğan et al. [44] using metaheuristic algorithms. They first considered the subproblem of computing the optimal rearrangement of a cargo once the vehicle route is fixed. They then proposed an exact and an approximate method to solve this subproblem, and used them within TS and iterated local search algorithms.

6.6 ▪ Conclusions and Future Research Directions

This chapter has described the main variants of PDPs that have been studied in the last two decades. It has also reviewed the main exact and heuristic algorithms developed to solve these problems. One can notice that there is an abundant literature concerning each of the three main problem variants: many-to-many, one-to-many-to-one, and one-to-one transportation. There is also a growing interest in problems that combine pickup and delivery routing with vehicle loading considerations.

While the focus has for a long time been on heuristics to handle large-scale instances, one now witnesses an increasing number of exact algorithms capable of solving instances of realistic size. Nevertheless, most PDP instances remain much harder to solve than classical VRP instances of the same size. This is largely due to the presence of the precedence constraints which typically lead to poor linear programming relaxations. More research is thus needed on the polyhedral study of PDPs. In the current state of the art, it seems fair to say that methods based on Branch-and-Price appear to be the most promising for problems with multiple vehicles.

When looking at the pickup and delivery literature as a whole, one cannot fail to notice that there exists a very large number of problem variants which differ in their structure but nevertheless share many similarities. One can thus hope to see the development of general modeling and solution techniques capable of handling multiple variants with a unified framework. This would also parallel the evolution observed in other fields of vehicle routing.

This survey has also highlighted the fact that the literature for multiple-vehicle many-to-many problems is relatively scarce with respect to the single-vehicle counterparts. Loading features have also mainly been considered in single-vehicle PDPs. Therefore, multiple-vehicle many-to-many problems and multiple-vehicle problems with loading features still leave ample scope for further research opportunities

Finally, as with other routing problems, one observes a gradual trend toward the inclusion of dynamic and stochastic aspects. Most applications of PDPs are dynamic in nature and call for the frequent update of solutions as new information becomes available. With the current availability of information technologies that allow real-time communication between drivers and dispatchers, one can expect that there will be a growing need for algorithms capable of reacting to incoming requests and other changing conditions that require quick decision making.

Bibliography

[1] T. J. AI AND V. KACHITVICHYANUKUL, *A particle swarm optimization for the vehicle routing problem with simultaneous pickup and delivery*, Computers & Operations Research, 36 (2009), pp. 1693–1702.

[2] M. A. ALBA MARTÍNEZ, J.-F. CORDEAU, M. DELL'AMICO, AND M. IORI, *A branch-and-cut algorithm for the double traveling salesman problem with multiple stacks*, INFORMS Journal on Computing, 25 (2013), pp. 41–55.

[3] H. ANDERSSON, M. CHRISTIANSEN, AND K. FAGERHOLT, *The maritime pickup and delivery problem with time windows and split loads*, INFOR, 49 (2011), pp. 79–91.

[4] E. ANGELELLI AND R. MANSINI, *The vehicle routing problem with time windows and simultaneous pick-up and delivery*, in Quantitative Approaches to Distribution Logistics and Supply Chain Management, Lecture Notes in Economics and Mathematical Systems, A. Klose, M. G. Speranza, and L. N. Van Wassenhove, eds., Springer-Verlag, Berlin, 2001, pp. 249–267.

[5] S. ANILY, *The vehicle-routing problem with delivery and back-haul options*, Naval Research Logistics, 43 (1996), pp. 415–434.

[6] S. ANILY AND G. MOSHEIOV, *The traveling salesman problem with delivery and backhauls*, Operations Research Letters, 16 (1994), pp. 11–18.

[7] N. ASCHEUER, M. JÜNGER, AND G. REINELT, *A branch & cut algorithm for the asymmetric traveling salesman problem with precedence constraints*, Computational Optimization and Applications, 17 (2000), pp. 61–84.

[8] E. BALAS, M. FISCHETTI, AND W. R. PULLEYBLANK, *The precedence-constrained asymmetric traveling salesman polytope*, Mathematical Programming, 68 (1995), pp. 241–265.

[9] R. BALDACCI, E. BARTOLINI, AND A. MINGOZZI, *An exact algorithm for the pickup and delivery problem with time windows*, Operations Research, 59 (2011), pp. 414–426.

[10] R. BALDACCI, E. HADJICONSTANTINOU, AND A. MINGOZZI, *An exact algorithm for the traveling salesman problem with deliveries and collections*, Networks, 42 (2003), pp. 26–41.

[11] T. BARTÓK AND C. IMREH, *Pickup and delivery vehicle routing with multidimensional loading constraints*, Acta Cybernetica, 20 (2011), pp. 17–33.

[12] M. BATISTA-GALVÁN, J.-J. RIERA-LEDESMA, AND J.-J. SALAZAR-GONZÁLEZ, *The traveling purchaser problem, with multiple stacks and deliveries: A branch-and-cut approach*, Computers & Operations Research, 40 (2013), pp. 2103–2115.

[13] M. BATTARRA, G. ERDOĞAN, G. LAPORTE, AND D. VIGO, *The traveling salesman problem with pickups, deliveries, and handling costs*, Transportation Science, 44 (2010), pp. 383–399.

[14] R. BENT AND P. V. HENTENRYCK, *A two-stage hybrid algorithm for pickup and delivery vehicle routing problems with time windows*, Computers & Operations Research, 33 (2006), pp. 875–893.

[15] G. BERBEGLIA, J.-F. CORDEAU, I. GRIBKOVSKAIA, AND G. LAPORTE, *Static pickup and delivery problems: A classification scheme and survey*, TOP, 15 (2007), pp. 1–31.

[16] G. BERBEGLIA, J.-F. CORDEAU, AND G. LAPORTE, *Dynamic pickup and delivery problems*, European Journal of Operational Research, 202 (2010), pp. 8–15.

[17] P. BEULLENS, D. VAN OUDHEUSDEN, AND L. N. VAN WASSENHOVE, *Collection and vehicle routing issues in reverse logistics*, in Reverse Logistics: Quantitative Models for Closed-Loop Supply Chains, R. Dekker, M. Fleischmann, K. Inderfurth, and L. Van Wassenhove, eds., Springer-Verlag, Heidelberg, 2004, pp. 95–134.

[18] N. BIANCHESSI AND G. RIGHINI, *Heuristic algorithms for the vehicle routing problem with simultaneous pick-up and delivery*, Computers & Operations Research, 34 (2007), pp. 578–594.

[19] L. BIANCO, A. MINGOZZI, S. RICCIARDELLI, AND M. SPADONI, *Exact and heuristic procedures for the traveling salesman problem with precedence constraints, based on dynamic programming*, INFOR, 32 (1994), pp. 19–31.

[20] F. BONOMO, S. MATTIA, AND G. ORIOLO, *Bounded coloring of co-comparability graphs and the pickup and delivery tour combination problem*, Theoretical Computer Science, 412 (2011), pp. 6261–6268.

[21] F. CARRABS, R. CERULLI, AND J.-F. CORDEAU, *An additive branch-and-bound algorithm for the pickup and delivery traveling salesman problem with LIFO or FIFO loading*, INFOR, 45 (2007), pp. 223–238.

[22] F. CARRABS, R. CERULLI, AND M. G. SPERANZA, *A branch-and-bound algorithm for the double TSP with two stacks*, Networks, 61 (2013), pp. 58–65.

[23] F. CARRABS, J.-F. CORDEAU, AND G. LAPORTE, *Variable neighbourhood search for the pickup and delivery traveling salesman problem with LIFO loading*, INFORMS Journal on Computing, 19 (2007), pp. 618–632.

[24] M. CASAZZA, A. CESELLI, AND M. NUNKESSER, *Efficient algorithms for the double travelling salesman problem with multiple stacks*, Computers & Operations Research, 39 (2012), pp. 1044–1053.

[25] D. O. CASCO, B. L. GOLDEN, AND E. A. WASIL, *Vehicle routing with backhauls: Models, algorithms, and case studies*, in Vehicle Routing: Methods and Studies, B. L. Golden, ed., North–Holland, Amsterdam, 1988, pp. 127–147.

[26] B. ÇATAY, *A new saving-based ant algorithm for the vehicle routing problem with simultaneous pickup and delivery*, Expert Systems with Applications, 37 (2010), pp. 6809–6817.

[27] D. CHEMLA, F. MEUNIER, AND R. WOLFLER CALVO, *Bike sharing systems: Solving the static rebalancing problem*, Discrete Optimization, 10 (2013), pp. 120–146.

[28] J.-F. CHEN AND T.-H. WU, *Vehicle routing problem with simultaneous deliveries and pickups*, Journal of the Operational Research Society, 57 (2006), pp. 579–587.

[29] V. ÇINAR, T. ÖNCAN, AND S. HALDUN, *A genetic algorithm for the traveling salesman problem with pickup and delivery using depot removal and insertion moves*, in Applications of Evolutionary Computation, C. Di Chio, A. Brabazon, G. Di Caro, M. Ebner, M. Farooq, A. Fink, J. Grahl, G. Greenfield, P. Machado, M. O'Neill, E. Tarantino, and N. Urquhart, eds., vol. 6025 of Lecture Notes in Computer Science, Springer-Verlag, Berlin, 2010, pp. 431–440.

[30] J.-F. CORDEAU, *A branch-and-cut algorithm for the dial-a-ride problem*, Operations Research, 54 (2006), pp. 573–586.

[31] J.-F. CORDEAU, M. DELL'AMICO, AND M. IORI, *Branch-and-cut for the pickup and delivery traveling salesman problem with FIFO loading*, Computers & Operations Research, 37 (2010), pp. 970–980.

[32] J.-F. CORDEAU, M. IORI, G. LAPORTE, AND J.-J. SALAZAR-GONZÁLEZ, *A branch-and-cut algorithm for the pickup and delivery traveling salesman problem with LIFO loading*, Networks, 55 (2010), pp. 46–59.

[33] J.-F. CÔTÉ, C. ARCHETTI, M. G. SPERANZA, M. GENDREAU, AND J.-Y. POTVIN, *A branch-and-cut algorithm for the pickup and delivery traveling salesman problem with multiple stacks*, Networks, 60 (2012), pp. 212–226.

[34] J.-F. CÔTÉ, M. GENDREAU, AND J.-Y. POTVIN, *Large neighborhood search for the single vehicle pickup and delivery problem with multiple loading stacks*, Networks, 60 (2012), pp. 19–30.

[35] T. G. CRAINIC, M. GENDREAU, AND P. DEJAX, *Dynamic and stochastic models for the allocation of empty containers*, Operations Research, 41 (1993), pp. 102–126.

[36] J. CRISPIM AND J. BRANDÃO, *Metaheuristics applied to mixed and simultaneous extensions of vehicle routing problems with backhauls*, Journal of the Operational Research Society, 56 (2005), pp. 1296–1302.

[37] M. DELL'AMICO, G. RIGHINI, AND M. SALANI, *A branch-and-price approach to the vehicle routing problem with simultaneous distribution and collection*, Transportation Science, 40 (2006), pp. 235–247.

[38] J. DESROSIERS, Y. DUMAS, AND F. SOUMIS, *A dynamic programming solution of the large-scale single-vehicle dial-a-ride problem with time windows*, American Journal of Mathematical and Management Sciences, 6 (1986), pp. 301–325.

[39] J. DETHLOFF, *Vehicle routing and reverse logistics: The vehicle routing problem with simultaneous delivery and pick-up*, OR Spectrum, 23 (2001), pp. 79–96.

[40] M. DROR, D. FORTIN, AND C. ROUCAIROL, *Redistribution of self-service electric cars: A case of pickup and delivery*, Technical Report 3543, INRIA, Institut National de Rercherche en Informatique et en Automatique, Rocquencourt, France, 1998.

[41] Y. DUMAS, J. DESROSIERS AND F. SOUMIS, *The pickup and delivery problem with time windows*, European Journal of Operational Research, 54 (1991), pp. 7–22.

[42] I. DUMITRESCU, S. ROPKE, J.-F. CORDEAU, AND G. LAPORTE, *The traveling salesman problem with pickup and delivery: Polyhedral results and a branch-and-cut algorithm*, Mathematical Programming, 121 (2010), pp. 269–305.

[43] G. ERDOĞAN, J.-F. CORDEAU, AND G. LAPORTE, *The pickup and delivery traveling salesman problem with first-in-first-out loading*, Computers & Operations Research, 36 (2009), pp. 1800–1808.

[44] G. ERDOĞAN, M. BATTARRA, G. LAPORTE, AND D. VIGO, *Metaheuristics for the traveling salesman problem with pickups, deliveries and handling costs*, Computers & Operations Research, 39 (2012), pp. 1074–1086.

[45] A. FELIPE, M. T. ORTUÑO, AND G. TIRADO, *The double traveling salesman problem with multiple stacks: A variable neighborhood search approach*, Computers & Operations Research, 36 (2009), pp. 2983–2993.

[46] ――――, *New neighborhood structures for the double traveling salesman problem with multiple stacks*, TOP, 17 (2009), pp. 190–213.

[47] ――――, *Using intermediate non-feasible solutions to approach vehicle routing problems with precedence and loading constraints*, European Journal of Operational Research, 211 (2011), pp. 66–75.

[48] M. FISCHETTI AND P. TOTH, *An additive bounding procedure for combinatorial optimization problems*, Operations Research, 37 (1989), pp. 319–328.

[49] G. N. FREDERICKSON, M. S. HECHT, AND C. E. KIM, *Approximation algorithms for some routing problems*, SIAM Journal on Computing, 7 (1978), pp. 178–193.

[50] M. GENDREAU, F. GUERTIN, J.-Y. POTVIN, AND R. SÉGUIN, *Neighborhood search heuristics for a dynamic vehicle dispatching problem with pick-ups and deliveries*, Transportation Research Part C: Emerging Technologies, 14 (1998), pp. 157–174.

[51] M. GENDREAU, G. LAPORTE, AND D. VIGO, *Heuristics for the traveling salesman problem with pickup and delivery*, Computers & Operations Research, 26 (1999), pp. 699–714.

[52] G. GHIANI, E. MANNI, A. QUARANTA, AND C. TRIKI, *Anticipatory algorithms for same-day courier dispatching*, Transportation Research Part E: Logistics and Transportation Review, 45 (2009), pp. 96–106.

[53] B. L. GOLDEN, E. K. BAKER, J. L. ALFARO, AND J. R. SCHAFFER, *The vehicle routing problem with backhauling*, in Proceedings of the Twenty-First Annual Meeting of the S. E. TIMS, Myrtle Beach, SC, 1985, pp. 90–92.

[54] L. GOUVEIA AND P. PESNEAU, *On extended formulations for the precedence constrained asymmetric traveling salesman problem*, Networks, 48 (2006), pp. 77–89.

[55] I. GRIBKOVSKAIA AND G. LAPORTE, *One-to-many-to-one single vehicle pickup and delivery problems*, in The Vehicle Routing Problem: Latest Advances and New Challenges, B. L. Golden, S. Raghavan, E. A. Wasil, R. Sharda, and S. Voss, eds., vol. 43 of Operations Research/Computer Science Interface Series, Springer-Verlag, Berlin, 2008, pp. 359–377.

[56] K. HALSE, *Modeling and solving complex vehicle routing problems*, PhD thesis, Institute of Mathematical Statistics and Operations Research, Technical University of Denmark, Lyngby, Denmark, 1992.

[57] P. HEALY AND R. MOLL, *A new extension of local search applied to the dial-a-ride problem*, European Journal of Operational Research, 83 (1995), pp. 83–104.

[58] H. HERNÁNDEZ-PÉREZ, I. RODRÍGUEZ-MARTÍN, AND J.-J. SALAZAR-GONZÁLEZ, *A hybrid GRASP/VND heuristic for the one-commodity pickup-and-delivery traveling salesman problem*, Computers & Operations Research, 36 (2009), pp. 1639–1645.

[59] H. HERNÁNDEZ-PÉREZ AND J.-J. SALAZAR-GONZÁLEZ, *A branch-and-cut algorithm for a traveling salesman problem with pickup and delivery*, Discrete Applied Mathematics, 145 (2004), pp. 126–139.

[60] ——, *Heuristics for the one-commodity pickup-and-delivery traveling salesman problem*, Transportation Science, 38 (2004), pp. 245–255.

[61] ——, *The one-commodity pickup-and-delivery traveling salesman problem: Inequalities and algorithms*, Networks, 50 (2007), pp. 258–272.

[62] ——, *The multi-commodity pickup-and-delivery traveling salesman problem*, Networks, 63 (2014), pp. 46–59.

[63] A. HOFF, I. GRIBKOVSKAIA, G. LAPORTE, AND A. LØKKETANGEN, *Lasso solution strategies for the vehicle routing problem with pickups and deliveries*, European Journal of Operational Research, 192 (2009), pp. 755–766.

[64] M. HOSNY AND C. MUMFORD, *Solving the one-commodity pickup and delivery problem using an adaptive hybrid VNS/SA approach*, in Parallel Problem Solving from Nature, PPSN XI, R. Schaefer, C. Cotta, J. Kolodziej, and G. Rudolph, eds., vol. 6239 of Lecture Notes in Computer Science, Springer-Verlag, Berlin, 2010, pp. 189–198.

[65] ———, *The single vehicle pickup and delivery problem with time windows: Intelligent operators for heuristic and metaheuristic algorithms*, Journal of Heuristics, 16 (2010), pp. 417–439.

[66] M. IORI AND S. MARTELLO, *Routing problems with loading constraints*, TOP, 18 (2010), pp. 4–27.

[67] M. IORI, J.-J. SALAZAR-GONZÁLEZ, AND D. VIGO, *An exact approach for the vehicle routing problem with two-dimensional loading constraints*, Transportation Science, 41 (2007), pp. 253–264.

[68] B. KALANTARI, A. V. HILL, AND S. R. ARORA, *An algorithm for the traveling salesman problem with pickup and delivery customers*, European Journal of Operational Research, 22 (1985), pp. 377–386.

[69] N. KATOH AND T. YANO, *An approximation algorithm for the pickup and delivery vehicle routing problem on trees*, Discrete Applied Mathematics, 154 (2006), pp. 2335–2349.

[70] G. KONTORAVDIS AND J. F. BARD, *A GRASP for the vehicle routing problem with time windows*, ORSA Journal on Computing, 7 (1995), pp. 10–23.

[71] J. E. KORSVIK, K. FAGERHOLT, AND G. LAPORTE, *A large neighbourhood search heuristic for ship routing and scheduling with split loads*, Computers & Operations Research, 38 (2011), pp. 474–483.

[72] M. KUBO AND H. KASUGAI, *Heuristic algorithms for the single vehicle dial-a-ride problem*, Journal of the Operations Research Society of Japan, 33 (1990), pp. 354–365.

[73] S. P. LADANY AND A. MEHREZ, *Optimal routing of a single vehicle with loading constraints*, Transportation Planning and Technology, 8 (1984), pp. 301–306.

[74] H. C. LAU AND Z. LIANG, *Pickup and delivery with time windows: Algorithms and test case generation*, International Journal on Artificial Intelligence Tools, 11 (2002), pp. 455–472.

[75] K. LEVITIN AND R. ABEZGAOUZ, *Optimal routing of multiple-load AGV subject to LIFO loading constraints*, Computers & Operations Research, 30 (2003), pp. 397–410.

[76] H. LI AND A. LIM, *A metaheuristic for the pickup and delivery problem with time windows*, International Journal on Artificial Intelligence Tools, 12 (2003), pp. 173–186.

[77] Y. LI, A. LIM, W.-C. OON, H. QIN, AND D. TU, *The tree representation for the pickup and delivery traveling salesman problem with LIFO loading*, European Journal of Operational Research, 212 (2011), pp. 482–496.

[78] S. LIN AND B. W. KERNIGHAN, *An effective heuristic algorithm for the traveling-salesman problem*, Operations Research, 21 (1973), pp. 498–516.

[79] J. D. C. LITTLE, K. G. MURTY, D. W. SWEENEY, AND C. KAREL, *An algorithm for the traveling salesman problem*, Operations Research, 11 (1963), pp. 972–989.

[80] F. C. J. LOKIN, *Procedures for travelling salesman problems with additional constraints*, European Journal of Operational Research, 3 (1979), pp. 135–141.

[81] Q. LU AND M. M. DESSOUKY, *An exact algorithm for the multiple vehicle pickup and delivery problem*, Transportation Science, 38 (2004), pp. 503–514.

[82] ———, *A new insertion-based construction heuristic for solving the pickup and delivery problem with time windows*, Discrete Optimization, 175 (2006), pp. 672–87.

[83] R. M. LUSBY AND J. LARSEN, *An exact method for the double TSP with multiple stacks*, Networks, 58 (2011), pp. 290–300.

[84] R. M. LUSBY, J. LARSEN, M. EHRGOTT, AND D. RYAN, *An exact method for the double TSP with multiple stacks*, International Transactions on Operations Research, 17 (2010), pp. 637–652.

[85] V. MAK AND A. ERNST, *New cutting-planes for the time- and/or precedence-constrained ATSP and directed VRP*, Mathematical Methods of Operations Research, 66 (2007), pp. 69–98.

[86] A. MALAPERT, C. GUERÉT, N. JUSSIEN, A. LANGEVIN, AND L.-M. ROUSSEAU, *Two-dimensional pickup and delivery routing problem with loading constraints*, in Proceedings of the First CPAIOR Workshop on Bin Packing and Placement Constraints (BPPC'08), Paris, France, 2008.

[87] H. MIN, *The multiple vehicle routing problem with simultaneous delivery and pick-up points*, Transportation Research Part A: Policy and Practice, 23 (1989), pp. 377–386.

[88] A. MINGOZZI, L. BIANCO, AND S. RICCIARDELLI, *Dynamic programming strategies for the traveling salesman problem with time window and precedence constraints*, Operations Research, 45 (1997), pp. 365–377.

[89] L. MINGYONG AND C. ERBAO, *An improved differential evolution algorithm for vehicle routing problem with simultaneous pickups and deliveries and time windows*, Engineering Applications of Artificial Intelligence, 23 (2010), pp. 188–195.

[90] S. MITROVIĆ-MINIĆ, R. KRISHNAMURTI, AND G. LAPORTE, *Double-horizon based heuristics for the dynamic pickup and delivery problem with time windows*, Transportation Research Part B: Methodological, 38 (2004), pp. 669–685.

[91] S. MITROVIĆ-MINIĆ AND G. LAPORTE, *Waiting strategies for the dynamic pickup and delivery problem with time windows*, Transportation Research Part B: Methodological, 38 (2004), pp. 635–655.

[92] N. MLADENOVIĆ, D. UROŠEVIĆ, S. HANAFI, AND A. ILIĆ, *A general variable neighborhood search for the one-commodity pickup-and-delivery travelling salesman problem*, European Journal of Operational Research, 220 (2012), pp. 270–285.

[93] F. A. T. MONTANÉ AND R. D. GALVÃO, *A tabu search algorithm for the vehicle routing problem with simultaneous pick-up and delivery service*, Computers & Operations Research, 33 (2006), pp. 595–619.

[94] G. MOSHEIOV, *The travelling salesman problem with pick-up and delivery*, European Journal of Operational Research, 79 (1994), pp. 299–310.

[95] ——, *Vehicle routing with pick-up and delivery: Tour-partitioning heuristics*, Computers & Industrial Engineering, 34 (1998), pp. 669–684.

[96] G. NAGY AND S. SALHI, *Heuristic algorithms for single and multiple depot vehicle routing problems with pickups and deliveries*, European Journal of Operational Research, 162 (2005), pp. 126–141.

[97] W. P. NANRY AND J. W. BARNES, *Solving the pickup and delivery problem with time windows using reactive tabu search*, Transportation Research Part B: Methodological, 34 (2000), pp. 107–121.

[98] M. NOWAK, O. ERGUN, AND C. C. WHITE, *Pickup and delivery with split loads*, Transportation Science, 42 (2008), pp. 32–43.

[99] I. OR, *Traveling Salesman-type Combinatorial Problems and Their Relation to the Logistics of Blood Banking*, PhD thesis, Department of Industrial Engineering and Management Science, Northwestern University, Evanston, IL, 1976.

[100] J. A. PACHECO, *Problemas de rutas con carga y descarga en sistemas LIFO: Soluciones exactas*, Estudios de Economía Aplicada, 3 (1995), pp. 69–86.

[101] ——, *Heuristico para los problemas de ruta con carga y descarga en sistemas LIFO*, SORT, Statistics and Operations Research Transactions, 21 (1997), pp. 153–175.

[102] G. PANKRATZ, *A grouping genetic algorithm for the pickup and delivery problem with time windows*, OR Spectrum, 27 (2005), pp. 21–41.

[103] S. PARRAGH, K. DOERNER, AND R. HARTL, *A survey on pickup and delivery problems part I: Transportation between customers and depot*, Journal für Betriebswirtschaft, 58 (2008), pp. 21–51.

[104] ——, *A survey on pickup and delivery problems part II: Transportation between pickup and delivery locations*, Journal für Betriebswirtschaft, 58 (2008), pp. 81–117.

[105] H. L. PETERSEN, C. ARCHETTI, AND M. G. SPERANZA, *Exact solutions to the double travelling salesman problem with multiple stacks*, Networks, 56 (2010), pp. 229–243.

[106] H. L. PETERSEN AND O. B. G. MADSEN, *The double travelling salesman problem with multiple stacks*, European Journal of Operational Research, 198 (2009), pp. 139–147.

[107] J. PRIVÉ, J. RENAUD, F. BOCTOR, AND G. LAPORTE, *Solving a vehicle-routing problem arising in soft-drink distribution*, Journal of the Operational Research Society, 57 (2006), pp. 1045–1052.

[108] H. N. PSARAFTIS, *A dynamic programming approach to the single-vehicle, many-to-many immediate request dial-a-ride problem*, Transportation Science, 14 (1980), pp. 130–154.

[109] ——, *k-interchange procedures for local search in a precedence-constrained routing problem*, European Journal of Operational Research, 13 (1983), pp. 391–402.

[110] Y. QU AND J. F. BARD, *A GRASP with adaptive large neighborhood search for pickup and delivery problems with transshipment*, Computers & Operations Research, 39 (2012), pp. 2439–2456.

[111] T. RAVIV, M. TZUR, AND I. FORMA, *Static repositioning in a bike-sharing system: Models and solution approaches*, EURO Journal on Transportation and Logistics, 2 (2013), pp. 187–229.

[112] J. RENAUD, F. F. BOCTOR, AND G. LAPORTE, *Perturbation heuristics for the pickup and delivery traveling salesman problem*, Computers & Operations Research, 29 (2002), pp. 1129–1141.

[113] J. RENAUD, F. F. BOCTOR, AND J. OUENNICHE, *A heuristic for the pickup and delivery traveling salesman problem*, Computers & Operations Research, 27 (2000), pp. 905–916.

[114] S. ROPKE AND J.-F. CORDEAU, *Branch and cut and price for the pickup and delivery problem with time windows*, Transportation Science, 43 (2009), pp. 267–286.

[115] S. ROPKE, J.-F. CORDEAU, AND G. LAPORTE, *Models and branch-and-cut algorithms for pickup and delivery problems with time windows*, Networks, 49 (2007), pp. 258–272.

[116] S. ROPKE AND D. PISINGER, *An adaptive large neighborhood search heuristic for the pickup and delivery problem with time windows*, Transportation Science, 40 (2006), pp. 455–472.

[117] ———, *A unified heuristic for a large class of vehicle routing problems with backhauls*, European Journal of Operational Research, 171 (2006), pp. 750–775.

[118] K. S. RULAND AND E. Y. RODIN, *The pickup and delivery problem: Faces and branch-and-cut algorithm*, Computers & Mathematics with Applications, 33 (1997), pp. 1–13.

[119] D. SÁEZ, C. E. CORTÉS, AND A. NÚÑEZ, *Hybrid adaptive predictive control for the multi-vehicle dynamic pick-up and delivery problem based on genetic algorithms and fuzzy clustering*, Computers & Operations Research, 35 (2008), pp. 3412–3438.

[120] S. SALHI AND G. NAGY, *A cluster insertion heuristic for single and multiple depot vehicle routing problems with backhauling*, Journal of the Operational Research Society, 50 (1999), pp. 1034–1042.

[121] F. A. SANTOS, G. R. MATEUS, AND A. S. DA CUNHA, *The pickup and delivery problem with cross-docking*, Computers & Operations Research, 40 (2013), pp. 1085–1093.

[122] M. W. P. SAVELSBERGH, *An efficient implementation of local search algorithms for constrained routing problems*, European Journal of Operational Research, 47 (1990), pp. 75–85.

[123] M. W. P. SAVELSBERGH AND M. SOL, *DRIVE: Dynamic routing of independent vehicles*, Operations Research, 46 (1998), pp. 474–490.

[124] P. SHAW, *Using constraint programming and local search methods to solve vehicle routing problems*, in Principles and Practice of Constraint Programming, M. Maher and J.-F. Puget, eds., Springer-Verlag, New York, 1998, pp. 417–431.

[125] X. SHI, F. ZHAO, AND Y. GONG, *Genetic algorithm for the one-commodity pickup-and-delivery vehicle routing problem*, in IEEE International Conference on Intelligent Computing and Intelligent Systems, 2009, vol. 1, 2009, pp. 175–179.

[126] M. SIGURD, D. PISINGER, AND M. SIG, *Scheduling transportation of live animals to avoid the spread of diseases*, Transportation Science, 38 (2004), pp. 197–209.

[127] D. M. STEIN, *An asymptotic, probabilistic analysis of a routing problem*, Mathematics of Operations Research, 3 (1978), pp. 89–101.

[128] D. M. STEIN, *Scheduling dial-a-ride transportation systems*, Transportation Science, 12 (1978), pp. 232–249.

[129] A. SUBRAMANIAN, *Heuristic, Exact and Hybrid Approaches for Vehicle Routing Problems*, PhD thesis, Universidade Federal Fluminense, Rio de Janeiro, Brazil, 2012.

[130] A. SUBRAMANIAN AND M. BATTARRA, *An iterated local search algorithm for the travelling salesman problem with pickups and deliveries*, Journal of the Operational Research Society, 64 (2013), pp. 402–409.

[131] A. SUBRAMANIAN, L. M. A. DRUMMOND, C. BENTES, L. S. OCHI, AND R. FARIAS, *A parallel heuristic for the vehicle routing problem with simultaneous pickup and delivery*, Computers & Operations Research, 37 (2010), pp. 1899–1911.

[132] A. SUBRAMANIAN, E. UCHOA, AND L. S. OCHI, *A hybrid algorithm for a class of vehicle routing problems*, Computers & Operations Research, 40 (2013), pp. 2519–2531.

[133] A. SUBRAMANIAN, E. UCHOA, A. A. PESSOA, AND L. S. OCHI, *Branch-and-cut with lazy separation for the vehicle routing problem with simultaneous pickup and delivery*, Operations Research Letters, 39 (2011), pp. 338–341.

[134] A. SUBRAMANIAN, E. UCHOA, A. A. PESSOA, AND L. S. OCHI, *Branch-cut-and-price for the vehicle routing problem with simultaneous pickup and delivery*, Optimization Letters, 7 (2013), pp. 1569–1581.

[135] C. D. TARANTILIS, *Adaptive multi-restart tabu search algorithm for the vehicle routing problem with cross-docking*, Optimization Letters, 7 (2013), pp. 1583–1596.

[136] C. D. TARANTILIS, A. K. ANAGNOSTOPOULOU, AND P. P. REPOUSSIS, *Adaptive path relinking for vehicle routing and scheduling problems with product returns*, Transportation Science, 47 (2013), pp. 356–379.

[137] S. TOULOUSE AND R. WOLFLER CALVO, *On the complexity of the multiple stack TSP, kSTSP*, Lecture Notes in Computer Science, Theory and Applications of Models of Computation, 5532 (2009), pp. 360–369.

[138] L. J. J. VAN DER BRUGGEN, J. K. LENSTRA, AND P. C. SCHUUR, *Variable-depth search for the single-vehicle pickup and delivery problem with time windows*, Transportation Science, 27 (1993), pp. 298–311.

[139] P. VENKATESHAN AND K. MATHUR, *An efficient column-generation-based algorithm for solving a pickup-and-delivery problem*, Computers & Operations Research, 38 (2011), pp. 1647–1655.

[140] T. VIDAL, T. G. CRAINIC, M. GENDREAU, AND C. PRINS, *A unified solution framework for multi-attribute vehicle routing problems*, European Journal of Operational Research, 234 (2014), pp. 658–673.

[141] A. C. WADE AND S. SALHI, *An investigation into a new class of vehicle routing problem with backhauls*, Omega, 30 (2002), pp. 479–487.

[142] ——, *An ant system algorithm for the mixed vehicle routing problem with backhauls*, in Metaheuristics, M. Resende, J. de Sousa, and A. Viana, eds., Kluwer Academic Publishers, Norwell, MA, 2004, pp. 699–719.

[143] N. A. WASSAN, G. NAGY, AND S. AHMADI, *A heuristic method for the vehicle routing problem with mixed deliveries and pickups*, Journal of Scheduling, 11 (2008), pp. 149–161.

[144] N. A. WASSAN, A. H. WASSAN, AND G. NAGY, *A reactive tabu search algorithm for the vehicle routing problem with simultaneous pickups and deliveries*, Journal of Combinatorial Optimization, 15 (2008), pp. 368–386.

[145] M. WEN, J. LARSEN, J. CLAUSEN, J.-F. CORDEAU, AND G. LAPORTE, *Vehicle routing with cross-docking*, Journal of the Operational Research Society, 60 (2009), pp. 1708–1718.

[146] H. XU, Z.-L. CHEN, S. RAJAGOPAL, AND S. ARUNAPURAM, *Solving a practical pickup and delivery problem*, Transportation Science, 37 (2003), pp. 347–364.

[147] E. E. ZACHARIADIS AND C. T. KIRANOUDIS, *A local search metaheuristic algorithm for the vehicle routing problem with simultaneous pick-ups and deliveries*, Expert Systems with Applications, 38 (2011), pp. 2717–2726.

[148] E. E. ZACHARIADIS, C. D. TARANTILIS, AND C. T. KIRANOUDIS, *A hybrid metaheuristic algorithm for the vehicle routing problem with simultaneous delivery and pick-up service*, Expert Systems with Applications, 36 (2009), pp. 1070–1081.

[149] ——, *An adaptive memory methodology for the vehicle routing problem with simultaneous pick-ups and deliveries*, European Journal of Operational Research, 202 (2010), pp. 401–411.

[150] ——, *The pallet-packing vehicle routing problem*, Transportation Science, 46 (2012), pp. 341–358.

[151] F. ZHAO, S. LI, J. SUN, AND D. MEI, *Genetic algorithm for the one-commodity pickup-and-delivery traveling salesman problem*, Computers & Industrial Engineering, 56 (2009), pp. 1642–1648.

[152] F.-G. ZHAO, J.-S. SUN, S.-J. LI, AND W.-M. LIU, *A hybrid genetic algorithm for the traveling salesman problem with pickup and delivery*, International Journal of Automation and Computing, 6 (2009), pp. 97–102.

[153] Y. ZHONG AND M. H. COLE, *A vehicle routing problem with backhauls and time windows: A guided local search solution*, Transportation Research Part E: Logistics and Transportation Review, 41 (2005), pp. 131–144.

Chapter 7

Pickup-and-Delivery Problems for People Transportation

Karl F. Doerner
Juan-José Salazar-González

7.1 ▪ Introduction

Modern societies demand transportation systems that offer high efficiency and comfort but also minimal noise, pollution, cost, and delays. These requirements apply to the transportation of both goods and people. The previous chapter was devoted to vehicle routing problems related to the transportation of goods; this chapter focuses on vehicle routing problems associated with the transportation of people. The wide variety of different applications in this context creates a range of different optimization problems. This chapter starts dealing with one of these problems, the so-called *dial-a-ride problem*, as presented in Section 7.2. This service, provided by public authorities, primarily serves elderly and handicapped people who cannot use regular transit systems. In this context, cost minimization is important, but the quality of the service is the main aim. A mathematical formulation is described in Section 7.3, and prior research approaches in the literature are summarized in Section 7.4. Section 7.5 details other articles dealing with related vehicle routing problems with regard to transporting people.

7.2 ▪ Dial-a-Ride Problems

Dial-a-Ride problem is a name used to define a transportation system providing multi-occupancy, door-to-door transport service for people. In a sense this transportation mode covers the gap between a rigid bus system and a flexible taxicab system. Ideally a vehicle moves a large number of passengers with personalized service. Although in some applications a customer may be transported by two vehicles, one after the other joined by a transfer point, in most cases only one vehicle is assigned to serve each customer's request. An example applies to elderly or handicapped people. Including customer satisfaction in the planning for this service is the major difference between transporting people (this chapter) and transporting goods (the previous chapter).

Dial-a-Ride Problem (DARP) implies that a set of customers makes a request for service. Each one wants to be collected from a given location (origin), wants to be delivered in a different location (destination), and requires some time constraints. Typical time constraints are the so-called *time-window constraints*, which consist of imposing a desired pickup time within an interval and a desired delivery time within an interval for each customer. Extreme values defining these intervals may create infeasible instances, especially in the special cases of time-of-delivery and time-of-pickup constraints, for which a customer specifies a desired maximum delivery time and a desired minimum pickup time. These special cases are also associated with the aim of minimizing the difference between actual and desired times.

The optimization problem consists of assigning a vehicle (or sequence of vehicles) to each customer's request and designing the routes for the vehicles to perform all services within the time constraints, with minimum cost. This cost is a combination of the total travel distance and the inconvenience to customers. Several routes must be determined, and each route must satisfy the following constraints:

(i) Visiting: each customer's request must be served once.

(ii) Depot: each vehicle returns to the depot from which it departed.

(iii) Pairing (coupling): the pickup and the delivery services of a customer's request are served by the same vehicle.

(iv) Precedence: each customer's request must be picked up before being delivered.

(v) Vehicle capacity: the number of customers in a vehicle cannot exceed a given value.

(vi) Time windows: each customer must be picked up and delivered within given time intervals.

(vii) Maximum waiting time: the vehicle serving a customer cannot be waiting in an intermediate stop more than a given value.

(viii) Maximum ride time: the time a customer spends in a vehicle cannot exceed a given value.

This DARP is a multi-vehicle routing problem and is also categorized as a *Vehicle Routing Problem with Pickup and Delivery and Time Windows*. However, the DARP focuses specifically on the operational constraints associated with transporting people, such that high-quality service is mandatory. For example, vehicles should not idle when carrying passengers, which motivates the maximum waiting time constraints (vii). Indeed, the violation of constraints (vi)–(viii) creates the so-called *customer inconveniences*, and the main aim of DARP is to minimize them.

Depending on when the customer's request arrives in the system, the DARP may be static or dynamic. The *static* (also called *advance request*) case occurs when all the customer's requests are precisely known before planning the routes. It does not consider any probabilistic information describing the spatial or time distribution of other further requests. The *dynamic* case arises in two situations: (1) customer's requests are included in the problem when they occur, so a reoptimization procedure is needed, or (2) probabilistic information about customer's requests is considered. The first situation suggests online programming, whereas the second situation is typical in stochastic programming. Within the online situation appears the *immediate request* case, which occurs when customers desire to be collected as soon as possible.

The simplest DARP variant appears in the static situation, with one vehicle and with only precedence constraints (iv). The DARP is then a generalization of the *Precedence-Constraints Traveling Salesman Problem*, also known as *sequential ordering problem*, where each customer's origin must precede that customer's destination on the route. When there is also a vehicle capacity, i.e., constraint (v), then the DARP is called *Capacitated Traveling Salesman Problem with Pickups and Deliveries*. The literature on this problem concerns good transportation and allows a customer request to be temporarily stored in intermediate locations. If the vehicle capacity equals 1 (i.e., the vehicle can only move one customer at a time), then the problem is known as a *stacker-crane problem*. This problem finds most of the application in good transportation, and several studies exist also on the *preemptive* version (where the transported load can be temporally stored in an intermediate location). By adding the time-window constraints (vi), the problem becomes a *vehicle routing problem with pickup and delivery and time windows*. As mentioned before, in all these particular problems, the objective function must include customer inconvenience to properly remain in the DARP area. Therefore, if the time constraints are not included as *hard constraints*, some of them should be considered at least as *soft constraints* through coefficients in the objective function. The issue of whether customer service guarantees (or the corresponding time constraints) should be hard or soft is a debatable one; each approach has advantages and disadvantages. A general DARP may contain all the time constraints in a hard or a soft way. Clearly, other extensions are possible by also considering transit points, heterogeneous fleets of vehicles, different types of customers, probabilistic information, multiple objective functions, and so forth.

7.3 ▪ Problem Formulation

The DARP is a combination of a vehicle routing problem and a scheduling problem. The first consists of designing the routes for the vehicles, assigning each customer's request to a vehicle, and sequencing the locations to be visited. The second consists of deciding the time at which each vehicle starts the service at each location. Although both problems must be integrated in a single one when searching for an optimal DARP solution, some heuristic approaches support the scheduling subproblem for given vehicle routes. In these approaches, the scheduling subproblem aims to minimize the time duration of each route.

We provide mathematical formulations for each of these problems. First of all we need some notation. We adopt notation previously used in, e.g., Cordeau [12]. The number of customers is represented by n. For each customer $i \in \{1,\dots,n\}$, the pickup location (origin) is represented by i, and the delivery location (destination) is indicated by $i + n$. The set of all origins is denoted by P and the set of all destinations by D. The depot is represented by two locations: 0 represents the location from where each route starts, and $2n + 1$ represents the location where each route ends. The set $P \cup D \cup \{0, 2n + 1\}$ is the vertex set V of a directed graph $G = (V, A)$. The arc set A contains all possible pairs of different locations (i, j) except

- $i = 0$ and $j = 2n + 1$;

- $i = 0$ and $j \in D$;

- $i \in P$ and $j = 2n + 1$.

Then, given a set of vertices $S \subset V$, the set $A(S)$ denotes all the arcs with tail and head in S, and the set $\delta^+(S)$ denotes all the arcs with tail in S and head in $V \setminus S$. When $S \subseteq P \cup D$, let S' be $(P \cup D) \setminus S$. The time window at a location $i \in P \cup D$ is denoted by $[a_i, b_i]$.

The maximum ride time acceptable by customer i is R_i, and the maximum waiting time is denoted by W_i. These values are assumed to be given. We also assume that all customer requests consume the same amount of capacity in a vehicle. The vehicle capacity is known and represented by Q.

7.3.1 ▪ The Whole Problem

Because the DARP is a generalization of the classical VRP, a natural attempt to attain a whole formulation is to adapt a VRP formulation, which is not always easy. Under some conditions it can be done. To exemplify this procedure, consider a static DARP with one depot, a homogeneous fleet of vehicles, and the (hard) constraints (i)–(viii). When there are no soft constraints, the DARP objective functions may coincide with the VRP, which already includes the visiting constraint (i), the depot constraint (ii), and the vehicle capacity constraint (v). Therefore, to ensure the DARP feasibility of a VRP solution we need to eliminate infeasible routes, i.e., routes violating the pairing constraint (iii), the precedence constraint (iv), or the time constraints (vi)–(viii). To this end, suppose one has a model for the VRP on the binary variable x_a ($a \in A$) assuming value 1 if a vehicle travels through arc a, and 0 otherwise. The literature has several models based on these variables. One of them is, e.g., the so-called *two-index vehicle flow model* introduced by Laporte and Nobert [41]. As is standard, when a is the arc with tail i and head j, we write $x_{i,j}$ instead of x_a. Then a valid DARP model can be obtained by extending the VRP model with the inequalities

$$(7.1) \qquad\qquad \sum_{a \in R} x_a \le |R| - 1$$

for each set of arcs R defining a (piece of the) route from 0 to $2n+1$ violating the pairing, precedence, or time constraints of some customer's requests. These *infeasible path inequalities* have been used extensively in prior literature to impose different side constraints on routing problems (see, e.g., Ascheuer, Fischetti, and Grötschel [2, 3]). Although they are linear inequalities on the x_a variables, they are not nice in a mathematical formulation, because

- they do not help close the gap between the integer formulation and its linear programming relaxation;

- it may be easy to find a violated one by an integer VRP solution x^* but very complex when x^* contains fractional numbers.

The first drawback is due to the low number of variables in the left-hand side of the inequality compared to the right-hand value, thus leading to a "weak inequality". The second drawback refers to the "separation problem", which requires an efficient algorithm to deal with the large number of inequalities from equations (7.1). However, in some cases, these drawbacks can be partially eliminated.

One example occurs with the pairing constraint (iii), where an invalid route can be eliminated by the inequalities

$$(7.2) \qquad\qquad \sum_{a \in A(\{0,2n+1\} \cup S)} x_a \le |S|$$

for each set of vertices $S \subset P \cup D$ containing the origin i of a request but not the destination $i + n$ of this request. Inequalities from equations (7.2) are stronger than inequalities from equations (7.1), yet at the same time they have a polynomially solvable separation problem even when x^* contains fractional numbers. To show this result let us consider a subset $S \subset P \cup D$. Using the equations

$$\sum_{a \in \delta^+(j)} x_a = 1 \qquad \forall j \in S,$$

the inequalities (7.2) are linearly equivalent to

$$\sum_{a \in \delta^+(S)} x_a \geq \sum_{j \in S} x_{0,j} + \sum_{j \in S} x_{j,2n+1},$$

which is also equivalent to

$$\sum_{i \in S, j \in S'} x_{i,j} + \sum_{j \in S'} x_{0,j} \geq \sum_{j \in P \cup D} x_{0,j}.$$

Given a customer i, checking whether a given solution x^* satisfies this inequality for each $S \subset P \cup D$ with $i \in S$ and $i + n \in S'$ can be done by solving a min-cut problem. The capacitated directed network G^* for this min-cut problem can be built from G, removing vertex $2n + 1$ and its incident arcs. The capacity of the arc $(0, i)$ is $\sum_{j \in P \cup D} x_{0,j}^*$ units, to force that 0 and i are on the same side of an optimal min-cut solution leading to a violated inequality. The capacity of another arc a in the network G^* is x_a^*. The min-cut problem to be solved has 0 as the source and $n + i$ as the sink. When the capacity of an optimal min-cut solution is smaller than $\sum_{j \in P \cup D} x_{0,j}^*$, a violated inequality from equations (7.2) has been identified. In conclusion, the separation of these inequalities can be done in polynomial time.

Another example occurs with the precedence constraint (iv). To force that the pickup location i of a customer's request is visited before the delivery location $i + n$, we can impose the inequalities

$$(7.3) \qquad \sum_{a \in A(\{0\} \cup S)} x_a \leq |S| - 1$$

for each set of vertices $S \subset P \cup V$ containing $i + n$ and not containing i. A procedure to find inequalities (7.3) violated by a (fractional) solution x^*, if any exists, is similar to the one described for the previous family. To see it, observe that inequalities (7.3) can be rewritten as

$$\sum_{a \in \delta^+(S)} x_a \geq 1 + \sum_{j \in S} x_{0,j},$$

which is also equivalent to

$$\sum_{i \in S, j \in S'} x_{i,j} + \sum_{j \in S'} x_{0,j} + \sum_{j \in S} x_{j,2n+1} \geq 1 + \sum_{j \in P \cup D} x_{0,j}.$$

Again, one min-cut problem must be solved for each customer i on a network G^*. Now the vertex $2n + 1$ and its incident arcs in G should also be considered in G^*. The capacity of the arcs $(0, i)$ and $(n + i, 2n + i)$ is $1 + \sum_{j \in P \cup D} x_{0,j}^*$ units to force that, in an optimal min-cut solution, the vertices 0 and i are on one side and the vertices $2n + 1$ and $n + i$ are on the other side. The capacity of another arc $a \in A$ is x_a^*. The min-cut problem to be solved has 0 as the source and $2n + 1$ as the sink. When the capacity of an optimal min-cut solution is smaller than $1 + \sum_{j \in P \cup D} x_{0,j}^*$, then a violated inequality from equations (7.3) has been identified. Again, the separation problem of these inequalities is polynomially solvable.

Strengthening and separating inequalities (7.1) associated with other constraints may be very complicated. In these situations, it is of interest to have an efficient algorithm to detect an inequality from equations (7.1), violated by a given integer VRP solution, or to prove that one does not exist. This implies checking the DARP validity of a VRP solution and providing a dynamic procedure to approach the DARP. The procedure can be seen as a kind of Benders' decomposition technique, where the MP designs an integer VRP solution and the slave problem (also called subproblem) checks the DARP validity of the (integer) VRP solution. The major difference of this procedure with respect to the classical Benders' decomposition technique is that, depending on the DARP constraints, the slave problem may not be modelled as a linear program. Inequalities (7.1) cannot be determined by dual variables but rather are of a combinatorial type. For that reason, they are named *combinatorial Benders' cuts*, according to Codato and Fischetti [10]. The efficient algorithm mentioned before solves the slave problem, if it is available. The next section addresses this problem, called *the scheduling problem* in the DARP literature.

7.3.2 ▪ The Scheduling Problem

Let us consider a given route starting at 0, ending at $2n + 1$, and visiting q locations from $P \cup D$ in between. It is easy to check whether these locations correspond to the pickups and the deliveries of a subset of customer's requests, with each pickup location visited before its associated delivery location, which is feasible to be performed by a capacitated vehicle. This feasibility can be checked in $O(q)$. Assume that the result of this check on the given route is positive.

The scheduling problem consists of determining the time to start the route from the depot and the time to start serving each customer's request at the pickup location. The time-window constraint (vi), the maximum waiting time constraint (vii), and the maximum ride time constraint (viii) are part of the scheduling problem. The aim is to minimize the route duration. It is a timing problem (see, e.g., Vidal et al. [78]). A mathematical formulation for this optimization problem is as follows.

To simplify the notation, let $i+1$ represent the location in the route after i and $i-1$ the location in the route before i. The route can then be seen as the sequence $0, 1, \ldots, h, 2n+1$. Let T_i be the time at which service begins at location i for each of the h locations in the route. Let T_0 be the time when the vehicle leaves location 0 and T_{h+1} the time when the vehicle enters location $2n + 1$:

$$\text{minimize } T_{h+1}$$

$$
\begin{aligned}
\text{s.t.} \quad & T_i + t_{i,i+1} \leq T_{i+1} && \forall\, i \in P \cup D, \\
& a_i \leq T_i \leq b_i && \forall\, i \in P \cup D, \\
& T_i - a_i \leq W_i && \forall\, i \in P, \\
& (T_{(i+n)-1} + t_{(i+n)-1,i+n}) - T_i \leq R_i && \forall\, i \text{ to } i+n,
\end{aligned}
$$

where, as defined above, W_i is the maximum waiting time acceptable by customer i, and R_i is the maximum ride time.

Firat and Woeginger [28] describe an algorithm with run time $O(q)$ to check the feasibility of this problem and to find a solution when it is feasible. The algorithm is based in expressing the feasibility problem as a shortest path problem in a vertex-weighted interval graph. Starting from the seminar articles by Savelsbergh [67, 68], other articles on this timing problem are Hunsaker and Savelsbergh [36], Haugland and Ho [32], and Tang et al. [75].

7.4 ▪ Solution Methods for Dial-a-Ride Problems

The literature on the DARP is vast, including several surveys. A recent survey was published by Cordeau and Laporte [14]. Another survey with two parts on pickup and delivery problems was published by Parragh, Doerner, and Hartl [53, 54]. This section offers a different classification in heuristic, metaheuristic, and exact methods of the major contributions, with an update of the existing literature on the DARP in chronological order.

7.4.1 ▪ Exact Methods

1980: Psaraftis [59] developed a dynamic programming algorithm to solve the single-vehicle DARP problem without (hard) time constraints. The algorithm was adapted to consider time-window constraints in Psaraftis [60]. Both these articles address the static and online dynamic variants. The objective functions include the customer's dissatisfaction while waiting for service. They are weighted combinations of the time needed to service all customers and the sum of a customer's waiting and riding times. The algorithms run in $O(n^2 3^n)$ time and were able to deal with instances with no more than 10 customer requests.

1986: Desrosiers, Dumas, and Soumis [18] describe a dynamic programming algorithm for the static single-vehicle DARP with time-window, vehicle capacity, and precedence constraints. Note that for a single vehicle, the pairing constraints are automatically satisfied, and the efficiency of a dynamic programming algorithm relies on the use of criteria for the elimination of states which do not satisfy the time-window, vehicle capacity, and precedence constraints. They solved instances with up to 40 customer requests.

1991: An exact approach for the multi-vehicle variant of this problem is described in Dumas, Desrosiers, and Soumis [24]. This approach solves a set-partitioning problem where each row is a customer request and each column is a route for a vehicle. The routes are generated by a subproblem that finds the shortest path on a network with side constraints (pairing, precedence, capacity, and time-window) by a dynamic programming procedure. The algorithm was capable of solving instances to optimality with up to 55 customer requests.

2006: Cordeau [12] presents a mixed integer linear programming formulation for the static multi-vehicle DARP with time constraints. A Branch-and-Cut implementation based on this model solved instances with up to 36 customer requests.

2007: Ropke, Cordeau, and Laporte [66] propose a pure integer linear programming formulation for the same problem. A Branch-and-Cut implementation based on this model solved instances with up to 96 customer requests.

2009: When there are no time windows, the DARP is called a multi-commodity one-to-one pickup-and-delivery routing problem. An exact method based on Benders' decomposition for a single vehicle is described in Hernández-Pérez and Salazar-González [35]. This problem can be solved as a one-commodity pickup-and-delivery routing problem with precedence constraints. A Branch-and-Cut algorithm for this one-commodity problem is described in Hernández-Pérez and Salazar-González [34].

2011: A DARP with a different objective was studied by Garaix et al. [30]. The objective is to maximize the passenger occupancy rate. The problem is motivated from a real-world situation in France where the objective of encouraging people meetings is pursued. They use a column generation approach to solving the problem.

Heilporn, Cordeau, and Laporte [33] developed an integer L-shaped method for the DARP with stochastic customer delays. Here the problem characteristic is that customers may have stochastic delays at their pickup location. When the customer is absent the request is fulfilled by an alternative service. This leads to the effect that the corresponding delivery location has not been serviced. The aim of the problem is to determine an a priori tour minimizing the expected cost of the solution.

7.4.2 ▪ Heuristic Methods

In the past several decades many heuristic and metaheuristic algorithms have been proposed for the static DARP and also for its dynamic and stochastic variants. This section presents an overview of heuristic solution methods, reported in chronological order.

1978: Early heuristic algorithms for the DARP are discussed in Daganzo [17], focusing mainly on a dynamic version of the DARP. This early account analyzed three insertion algorithms: visiting the closest stop next, visiting the closest origin or the closest destination in alternating order, and allowing the insertion of delivery locations only after a fixed number of passengers have been picked up. The second algorithm results in higher waiting to ride time ratios than the first algorithm. However, it also tends to generate lower waiting times for equal ride times when compared with the third algorithm.

1981: One of the first heuristic solution procedures for static multi-vehicle DARP is discussed in Cullen, Jarvis, and Ratliff [16]. They develop an interactive algorithm that follows the "cluster-first route-second" approach. It is based on a set partitioning formulation, solved by means of column generation. The subproblem is an location-allocation problem where clusters must be selected (location) and the customers' requests must be assigned to selected clusters (allocation). This subproblem can only be solved approximately. However, user-related constraints or objectives are not explicitly considered.

1985: Sexton and Bodin [71, 72] developed a heuristic routing and scheduling algorithm for the single-vehicle DARP using Benders' decomposition. The scheduling problem can be solved optimally; the routing problem is solved with a heuristic algorithm.

1986: Jaw et al. [38] propose a sequential insertion procedure. First, customers were ordered by their increasing earliest time for pickup. Second, they could be inserted according to the cheapest feasible insertion criterion. They use the notion of active vehicle periods. In contrast to Sexton and Bodin [71, 72], they consider already multiple vehicles.

1988: Desrosiers, Dumas, and Soumis [19] and Dumas, Desrosiers, and Soumis [23] use the dynamic programming algorithm described in Section 7.4.1 in a column generation approach for heuristically solving the multi-vehicle DARP with time-window constraints. The basic concept is a "mini-cluster", which is a set of geographically and temporally cohesive customer requests that can be served by the same vehicle. A heuristic algorithm groups together nearby customers into a mini-cluster (route segment) that can be served by a single vehicle. A vehicle is empty when it enters and when it leaves a mini-cluster, but it is never empty in between. Then a column generation algorithm constructs vehicle routes by stringing together these mini-clusters. The solved instances with up to 200 customer requests are grouped into 85 mini-clusters.

1995: Ioachim et al. [37] developed an optimization-based mini-clustering algorithm. Here a column generation approach is used to obtain mini-clusters and an enhanced initialization procedure to decrease processing times.

By its very nature, the problem has multiple objectives. In addition to the traditional cost objectives, it is necessary to address also client-centered objectives (such as the minimization of ride times). A heuristic for the multi-objective variant of the DARP is presented by Madsen, Ravn, and Rygaard [42]. They discuss an insertion-based algorithm called REBUS. The objectives considered are the total driving time, the number of vehicles, the total waiting time, the deviation from promised service times, and the cost.

Dial [20] studied the dynamic case of the multi-vehicle DARP. In this paper new transportation requests appear dynamically. Thus, new transportation requests are assigned to clusters according to lowest cost insertion. The routes could be optimized then by using dynamic programming. These authors report the results for a real-life problem instance.

Larger instances are heuristically solved in Ioachim et al. [37]. This approach was based on the "cluster-first route-second" strategy, but where part of the routing problem is used to solve the clustering problem. Mini-clusters are generated by applying dynamic programming to solve shortest path problems with pickups and deliveries and then selected by solving a set-partitioning problem. The heuristic approach was run on instances with up to 250 requests and on real data with 2545 requests.

2004: Diana and Dessouky [22] developed a regret insertion algorithm for the static DARP. In this case, all requests are ranked first according to ascending pickup times. Some swaps in this order are allowed, giving preference to requests that might be difficult to insert later on, because of their spatial location. The first m requests are used as seed customers, with m being the number of vehicles. All the remaining requests are inserted following a regret insertion strategy (see Potvin and Rousseau [58]). The regret insertion-based process is also subject to analysis by Diana [21] in an effort to determine why the performance of this heuristic is superior to that of other insertion rules.

7.4.3 ▪ Metaheuristic Methods

This section presents an overview of metaheuristic solution methods, reported in chronological order.

1996: A first metaheuristic was published by Toth and Vigo [77]. They developed a local search–based metaheuristic, namely a tabu thresholding algorithm, for the static multi-vehicle DARP. Initial solutions were constructed using parallel insertion and that employed the neighborhoods described in Toth and Vigo [76].

2003: Another tabu search algorithm was published in Cordeau and Laporte [13]. The neighborhood used is defined by moving one request to another route. The best possible move serves to generate a new incumbent solution. Reverse moves are declared tabu. However, an aspiration criterion is defined, such that tabu moves that provide a better solution, with respect to all other solutions already constructed by the same move, can constitute a new incumbent solution. This algorithm developed for the static version has been used to solve the dynamic case. The tabu search also has been adapted to the dynamic DARP through parallelization by Attanasio et al. [4].

Starting from now the problems became richer, often focused on the transportation of patients. Melachrinoudis, Ilhan, and Min [46] developed a double request DARP model with soft time windows and its application in health care. The objective of the proposed model was to minimize a weighted sum of total vehicle transportation costs and total clients' inconvenience time. The latter consists of excess riding time, early/late delivery time before service, and late pickup time after service. As a solution technique, this approach used a tabu search technique.

2009: Two years later, another approach is developed and applied to a real-world patient transportation problem. Beaudry et al. [6] considered a patient transportation problem arising in large hospitals. Therefore this study extended the classical DARP by several complicating constraints that were specific to a hospital context and by dynamic requests. The study provided a detailed description of the problem and proposed a two-phase heuristic procedure capable of handling the different features. In the first phase, a simple insertion scheme generated a feasible solution, improved in the second phase with a tabu search algorithm. The same algorithm is used for the decision support system developed by Hanne, Melo, and Nickel [31] as well.

2010: Parragh, Doerner, and Hartl [55] developed solution techniques also for the patient transportation problem in Austria. They started to develop a variable neighborhood search metaheuristic for the classical DARP. The neighborhood concepts are based on swap and ejection chains. As in the real world, besides the classical cost-objective, also the patient related objective is important; this solution concept is adapted to solve the bi-objective case (besides the cost objective, the maximum ride time of the patients will be minimized). For this variant a two-phase method is presented in Parragh et al. [52]: in the first phase a variable neighborhood search is used, and in the second phase path relinking is applied to generate Pareto efficient solutions.

2011: As the real-world patient transportation problem is dynamic and stochastic by nature, Schilde, Doerner, and Hartl [69] developed solution techniques for the dynamic stochastic version of the DARP. The aim is to design vehicle routes to serve partially dynamic transportation requests using a fixed vehicle fleet. In patient transportation each request requires transportation from a patient's home to a hospital (outbound request) or back home from the hospital (inbound request). Some of these requests are known in advance. Some requests are dynamic in the sense that they appear during the day without any prior information. Finally, some inbound requests are stochastic. With a certain probability, each outbound request causes a corresponding inbound request on the same day. Some stochastic information about these return transports is available from historical data. This study investigated whether using this information to design the routes had a significant positive effect on the solution quality. The problem can be modeled as a dynamic stochastic DARP with expected return transports. As solution techniques, a stochastic variable neighborhood search and different variants of the multiple plan and multiple scenario approach appear viable.

A hybrid tabu search with an exact constraint programming developed by Berbeglia, Cordeau, and Laporte [7] is presented to solve the dynamic DARP. An important component of the tabu search consists of three scheduling procedures which are executed sequentially. The constraint programming algorithm is used to accept or reject incoming requests.

2012: In a recent work by Masson, Lehuédé, and Péton [45] the model of the pickup-and-delivery problem is extended by transfer points. The corresponding problem is called the pickup-and-delivery problem with transfers. The solution technique relies on adaptive large neighborhood search. The authors evaluate the method on generated instances and apply it to the transportation of people with disabilities. On these real-life instances they show that the introduction of transfer points can bring significant improvements (up to 9%) to the value of the objective function.

Beyond the multiple objectives and dynamic or stochastic aspects, also driver-related constraints are important for the patient transportation problem, which requires extending the classical DARP by driver-related constraints and different modes of transportation. In Parragh et al. [51] patients may request to be transported either seated, lying in a bed, or in a wheelchair. The driver-related constraints are expressed in terms of maximum route duration limits and mandatory lunch breaks. In this paper a three-index and a set-partitioning formulation of the problem is introduced. The linear programming relaxation of the latter is solved by a column generation algorithm. A variable neighborhood search heuristic and a hybrid method combining column generation are proposed to generate upper bounds. The different transportation modes are introduced by Parragh [50], where also a Branch-and-Cut algorithm is described.

A related problem to the DARP was considered by Kergosien et al. [39]. The problem stems from a real application to the transportation of patients in a large hospital complex. Patients have to be transported between care units and treatment rooms. In large hospital compounds vehicles are required. The problem is dynamic by nature, as not all the transportation requests are known a priori. They consider a heterogeneous fleet, and a vehicle can transport only one patient at a time. Also disinfection operations after the transport of specific patients have to be considered. The problem is solved by using tabu search, and the method is applied to real-world data of a French hospital.

As mentioned before, the DARP without time windows is called a multi-commodity one-to-one pickup-and-delivery routing problem. A metaheuristic algorithm for the single vehicle variant of this problem is described in Rodríguez-Martín and Salazar-González [65].

2013: Parragh and Schmid [56] developed further an innovative hybrid method based on column generation and large neighborhood search for the classical DARP.

In the paper Schmid and Doerner [70] the patient transportation problem is further extended by the scheduling of the patients in the treatment rooms. Hospitalized patients typically undergo several examinations before their actual surgery. Transportation between service units is provided by trained personnel who escort patients. However, valuable resources may be under-utilized if patients arrive too late for scheduled appointments, whereas on the other side, many patients have to wait for a long time before being picked up for, or after, their actual appointment. To improve those deficits, this study adopts on both resource- and client-centered perspectives. In this paper the authors present an integrative combinatorial optimization model combining both scheduling- and routing-related aspects. The problem is solved by using a cooperative hybrid metaheuristic. Traditionally, both underlying subproblems—the scheduling of patients and the transportation of patients—would be solved independently, but the cooperative approach yields substantial advantages over decoupled hierarchical optimization processes.

7.5 ▪ Other Problems Concerning Pickup and Delivery of People

Many situations in practice have motivated the study of several pickup-and-delivery problems related with transporting people. The variety of applications is so extended that it makes it too complex to summarize all of them in a single chapter. To illustrate the current status of the research in the area we have selected two specific topics. One is related to logistic problems arising in school bus routing. The other is related to car pooling, where employees are organized to share some vehicles.

7.5.1 ▪ School Bus Routing

The school bus routing problem is related to the DARP, such that each user (student) must be transported from a given pickup location (home) to a given delivery location (school). For one of the two locations, a time window may be specified. In addition, maximum user ride time limits are usually considered. In contrast with standard DARP situations, the passengers' destinations often coincide (i.e., many children attend the same school), and a hard time-window constraint is always associated with the delivery location (the school). Similar to the classical DARP though, maximum ride time limits can implicitly be considered by artificially constructing a time window at each bus stop.

The *School Bus Routing Problem* (SBRP) consists of routing a given number of buses, such that all children in a certain (rural) area are picked up from the bus stop closest to their homes and delivered to their respective schools. This problem has multiple objectives: on the one hand, the bus company aims to minimize its operating costs, while on the other hand, the children transported must arrive at their schools in time (but not too early with respect to the beginning of the first class). Furthermore, they should spend as little time on the bus as possible. Obviously, these three objectives are conflicting in many cases, and it is difficult to assign a certain weight to any of them a priori. Cost minimization is linked to the total time the drivers spend driving. Combining several transportation requests usually decreases the total route length, but it increases the individual ride times of the children. Arriving exactly at the time school starts instead entails longer ride times for children picked up early and may also increase total route length as well.

This class of problem consists of different subproblems involving bus stop selection, bus route generation, school bell time adjustment, and bus scheduling. A survey by Park and Kim [48] has summarized the various assumptions, constraints, and solution methods used in prior literature on SBRP. The first ideas and concepts of school bus routing solution techniques were published by Bodin and Berman [8] and Swersey and Ballard [74].

Spada, Bierlaire, and Liebling [73] propose a modeling framework where the focus is on optimizing the level of service for a given number of buses, using an automatic procedure to generate a solution to the problem. The procedure first builds a feasible solution, which then can be improved using a heuristic. A simulated annealing technique explores the infeasible solutions.

Fügenschuh [29] presents an integer programming model pursuing the integrated coordination of the school starting times and the public bus services. This approach considered preprocessing techniques, model reformulations, and cutting planes that could be incorporated into a Branch-and-Cut algorithm. Computational results show that fewer buses would be sufficient if the schools started at different times.

Perugia et al. [57] present a model and an algorithm for the design of a home-to-work bus service in a metropolitan area. This type of service must ensure an equilibrium between conflicting criteria such as efficiency, effectiveness, and equity. The authors introduce a multi-objective model in which, among other aspects, equity depends on time windows pertaining to the arrival time of a bus at a stop. Time windows can have other uses too, such as guaranteeing the synchronization of the service with other transportation modes. This "cluster-first route-second" approach models both bus stop location and routing in urban road networks where turn restrictions exist. The resulting multi-objective location-routing model is solved by a tabu search algorithm. A different multi-objective formulation is in Corberán et al. [11].

Martinez and Viegas [44] present an integrated procedure based on traditional formulations of the SBRP. The problem is decomposed: a first step identifies the most suitable concentration points of students, and a second step computes the optimal routes serving those stops. The solution concept is applied to data from Lisbon.

Kim, Kim, and Park [40] introduce a school bus scheduling problem wherein trips for each school are given. Each school has its fixed time window within which trips should be completed. A school bus can serve multiple trips for multiple schools. The school bus scheduling problem seeks to optimize bus schedules to complete all the trips, considering the school time windows. The problem can be modeled as a vehicle routing problem with time windows by treating each trip as a virtual stop. Two exact approaches refer to special cases, and a heuristic algorithm applies to a more general case. This problem also can be extended to mixed loads; see Park, Tae, and Kim [49]. When mixed loads are allowed, students from different schools can get on the same bus at the same time. That study thereby presented an improvement algorithm.

The school bus problem addressing also the issue of the bus stop selection for each child is studied in Riera-Ledesma and Salazar-González [63, 64]. These authors describe a Branch-and-Cut algorithm and a column generation procedure to solve instances with up to 100 children and up to 100 potential stops to optimality.

Fügenschuh [29] examines a pickup-and-delivery problem where the students can be transported by at most two buses. To this end, some locations are considered as transfer points where a passenger can go from one vehicle to another. In this article, the vehicle routes are fixed and the aim is to find the minimum number of buses to cover all the services. Designing the vehicle routes has been addressed by other authors, and the optimization problem is called *Pickup-and-Delivery Problem with Transfers* (PDPT). A mathematical formulation and a Branch-and-Cut algorithm for a general variant have been addressed by Cortés, Matamala, and Contardo in [15], solving instances with up to 6 requests and 2 vehicles to optimality. Mitrović-Minić and Laporte [47] describe a local search approach for an uncapacitated PDPT, solving instances with up to 100 requests. More sophisticated approaches were proposed by Masson, Lehuédé, and Péton [45] and by Qu and Bard [62]. When solving a PDPT, the numbers of requests and vehicles are important, but there are other more crucial features. For example, the problem is simpler when the set of transfer points is much smaller than the set of pickup-and-delivery locations. In some applications the pickup locations may all be different (they are the students' homes), but the delivery locations are very few points (they are the schools). In some cases only the delivery points may offer the service to allow transfers. In other applications there is only one specific location where passengers can be transferred from one vehicle to another. The transfer aspect imposes a synchronization between vehicle routes that is better afforded under the presence of tight time-window constraints.

In the very recent paper of Boegl, Doerner, and Parragh [9] the school bus routing and scheduling problem with transfers is studied. It deals with the planning of the transportation of students from home to their school before it starts under the consideration that they can change buses. Allowing transfers has multiple different consequences. On the one hand, costs can be reduced significantly. On the other hand, transfers clearly have an impact on the service level, i.e., transfers lower the service level, but they may also reduce riding times. The authors develop a heuristic solution framework to solve this problem and compare the results with two standard solution techniques for the DARP and for the open VRP. The main objective is the minimization of costs. The implications on the service level of the pupils, like time loss and number of transfers, are analyzed.

7.5.2 ▪ Car Pooling

A problem class related to the DARP is the car pooling problem, which consists of finding subsets of employees who can share a car, determining the path the driver should follow, and identifying who should be the driver. In contrast with the DARP, either the origin or the destination are the same for all users, depending on whether the trip is from home to the office or back. Two variants can be investigated, either one car pool for both ways or differing to-work and from-work problems.

The client-centered goals (maximum driving/travel time, maximum amount of detours) and the economical goals both have to be considered. In prior literature, several variants of the problem have been distinguished, such as the daily car pooling problem, for which drivers/cars are predefined and the problem consists of assigning passengers to cars; the long-term car-pooling problem, in which each user can act as a driver or a passenger (and drivers/passengers have to be determined); or the problem of defining client pools that share a vehicle (the definition of the driver is left to the passengers) (see Baldacci, Maniezzo, and Mingozzi [5], Wolfler Calvo et al. [79], and Yan and Chen [80]).

Car pooling normally leads to fewer total vehicle kilometers traveled (despite short detours) and is therefore an interesting concept to make transportation activities more environmentally friendly. To increase the number of shared car trips, lanes on many freeway systems in the United States are reserved for vehicles with a defined minimum amount of passengers. Further, car pooling can help guarantee mobility in rural areas, considering that the rise of individual mobility is connected to a weakening demand for public means of transport, which leads to declining offers of public transport activities, especially in rural regions. Despite all these advantages, car-pooling concepts are rarely applied in practice, due to the difficulties of the planning stage (see, e.g., Wolfler Calvo et al. [79]). Literature has mainly focused on car-pooling concepts where either the origin or the destination are the same for all users (many-to-one or one-to-many problems) and only private cars are involved in the planning and solution approaches.

Baldacci, Maniezzo, and Mingozzi [5] propose an exact and a heuristic procedure for this problem. A real-life application is reported in Wolfler Calvo et al. [79]. Maniezzo, Carbonaro, and Hildmann [43] describe an ant colony optimization algorithm which also might resolve the long-term problem.

The problem has only been extended by Yan and Chen [80] to a many-origins-to-many-destinations car pooling problem with multiple vehicles and person types, for which vehicles and persons can be divided into different groups according to gender and smoking status. The idea was to consider the individual characteristics and preferences of passengers and drivers when solving the problem (e.g., non-smokers want to share a car with other non-smokers). The problem can be formulated as an integer multiple commodity network flow problem. An algorithm based on Lagrangian relaxation, a subgradient method, and a heuristic for the upper bound have been developed to solve this problem.

A related problem with the car-pooling problem is the dynamic ride-sharing problem. This problem was recently introduced by Agatz et al. [1]. The difference from the car-pooling problem is the planning horizon. In this problem transportation requests of people with similar itineraries and time schedules to share rides on short notice are considered. This is a new problem which emerged through the increased use of smart-phones. Now it is possible to share rides on short notice. Different strategies are tested by simulations.

7.5.3 ▪ Demand Responsive Transportation and Others

Closely related problems to the DARP are the demand responsive transportation problems. In Errico et al. [25] the different approaches of solving demand responsive transportation problems are surveyed. When demand for people transportation is low, e.g., in rural areas, traditional transit cannot provide an efficient and good-quality level of service, due to the fixed structure. Therefore, public transportation is evolving towards some degree of flexibility. The extension of DARP systems to general public transportation fulfills the requirement of adaptability. New transportation alternatives combining characteristics from both the traditional transit flexible DARP systems start to be introduced. These types of problems are usually called semi-flexible. These new problem characteristics require complex planning activities and a formalization of the decisions. The paper of Errico et al. provides a systematic treatment of the field of semi-flexible systems.

Another related topic is the transportation of people with aircrafts, jets, or helicopters. As an example the following applications will be reported. In Qian et al. [61] employees are transported to and from the offshore installations in the offshore petroleum industry by helicopter. The paper analyzes how to improve transportation safety by solving the helicopter routing problem with a risk objective expressed in terms of expected number of fatalities. A mathematical model is proposed, and a tabu search heuristic is applied to this problem: different objectives, travel time, a passenger risk, and a combined passenger and pilot risk objective. One result show that passenger transportation risk can be reduced by increasing travel time at the expense of pilot risk.

Another application of air transportation is the on-demand air transportation services in which travelers call a few days in advance to schedule a flight. A successful on-demand air transportation service requires an effective scheduling system to construct minimum-cost pilot and jet itineraries for a set of accepted transportation requests. In the paper by Espinoza et al. [26] an integer multicommodity network flow model with side constraints for such dial-a-flight problems is presented. In Espinoza et al. [27] the authors describe how this core optimization technology is embedded in a parallel, large neighborhood, local search scheme to produce high-quality solutions efficiently for large-scale real-life instances.

7.6 ▪ Conclusions and Future Research Directions

This chapter has introduced the basic constraints of a vehicle routing problem dealing with moving people, in which minimizing the customer's dissatisfactions is fundamental. The most studied variant is the Dial-a-Ride Problem (DARP), leading to a wide literature including exact and heuristic approaches. Methodologically speaking, the approaches used to solve problems moving people are not really different from those used to solve problems moving goods. Indeed, most of the formulations are commodity-flow formulations or set-packing models, motivating the implementation of Branch-and-Cut and column generation algorithms as exact methods. However, when dealing with people, there is a deeper multi-objective nature in the problem, as the customer's dissatisfaction does not always fit properly into a numerical expression. Also, these problems are of a high stochastic nature as well: the customer's demand may be known while the vehicles are moving. In some applications there are special features (like, for example, the possibility of transit points) that make the optimization problem more complex to be solved. For that reason, and also for the size of the instances in real cases, there has been an explosion of exact and heuristic approaches trying to generate good solutions. The developed hybrid algorithm of Parragh and Schmid [56] currently provides the best results on the standard DARP instances. In addition, the today telecommunication technology provides lot of online information to both customers and managers, and thus the society is demanding solution approaches for larger and more complex vehicle routing problems. In the next years intermodal aspects will require new solution concepts. Especially the combination of public transportation, demand responsive transportation, and car or bike sharing is of interest for public transportation planners in rural areas.

Bibliography

[1] N. A. H. AGATZ, A. L. ERERA, M. W. P. SAVELSBERGH, AND X. WANG, *Dynamic ride-sharing: A simulation study in metro Atlanta*, Transportation Research Part B: Methodological, 45 (2011), pp. 1450–1464.

[2] N. ASCHEUER, M. FISCHETTI, AND M. GRÖTSCHEL, *A polyhedral study of the asymmetric traveling salesman problem with time windows*, Networks, 36 (2000), pp. 69–79.

[3] ——, *Solving the asymmetric travelling salesman problem with time windows by branch-and-cut*, Mathematical Programming, 90 (2001), pp. 475–506.

[4] A. ATTANASIO, J.-F. CORDEAU, G. GHIANI, AND G. LAPORTE, *Parallel tabu search heuristics for the dynamic multi-vehicle dial-a-ride problem*, Parallel Computing, 30 (2004), pp. 377–387.

[5] R. BALDACCI, V. MANIEZZO, AND A. MINGOZZI, *An exact method for the car pooling problem based on Lagrangean column generation*, Operations Research, 52 (2004), pp. 422–439.

[6] A. BEAUDRY, G. LAPORTE, T. MELO, AND S. NICKEL, *Dynamic transportation of patients to hospitals*, OR Spectrum, 32 (2009), pp. 77–107.

[7] G. BERBEGLIA, J.-F. CORDEAU, AND G. LAPORTE, *A hybrid tabu search and constraint programming algorithm for the dynamic dial-a-ride problem*, INFORMS Journal on Computing, 24 (2012), pp. 343–355.

[8] L. D. BODIN AND L. BERMAN, *Routing and scheduling of school buses by computer*, Transportation Science, 13 (1979), pp. 113–129.

[9] M. BOEGL, K. F. DOERNER, AND S. N. PARRAGH, *The school bus routing and scheduling problem with transfers*, Networks, forthcoming (2014).

[10] G. CODATO AND M. FISCHETTI, *Combinatorial Benders' cuts for mixed-integer linear programming*, Operations Research, 54 (2005), pp. 756–766.

[11] A. CORBERÁN, E. FERNÁNDEZ, M. LAGUNA, AND R. MARTÍ, *Heuristic solutions to the problem of routing buses with multiple objectives*, Journal of the Operational Research Society, 53 (2002), pp. 427–435.

[12] J.-F. CORDEAU, *A branch-and-cut algorithm for the dial-a-ride problem*, Operations Research, 54 (2006), pp. 573–586.

[13] J.-F. CORDEAU AND G. LAPORTE, *A tabu search heuristic for the static multi-vehicle dial-a-ride problem*, Transportation Research Part B: Methodological, 37 (2003), pp. 579–594.

[14] ——, *The dial-a-ride problem: Models and algorithms*, Annals of Operations Research, 152 (2007), pp. 29–46.

[15] C. E. CORTÉS, M. MATAMALA, AND C. CONTARDO, *The pickup and delivery problem with transfers: Formulation and a branch-and-cut solution method*, European Journal of Operational Research, 200 (2010), pp. 711–724.

[16] F. CULLEN, J. JARVIS, AND D. RATLIFF, *Set partitioning based heuristics for interactive routing*, Networks, 11 (1981), pp. 125–143.

[17] C. F. DAGANZO, *An approximate analytic model of many-to-many demand responsive transportation systems*, Transportation Research, 12 (1978), pp. 325–333.

[18] J. DESROSIERS, Y. DUMAS, AND F. SOUMIS, *A dynamic programming solution of the large-scale single-vehicle dial-a-ride problem with time windows*, American Journal of Mathematical and Management Sciences, 6 (1986), pp. 301–325.

[19] ——, *The multiple vehicle dial-a-ride-problem*, in Computer-Aided Transit Scheduling. Lecture Notes in Economics and Mathematical Systems, J. Daduna and A. Wren, eds., vol. 308, Springer, Berlin, 1988, pp. 15–27.

[20] R. DIAL, *Autonomous dial-a-ride transit introductory overview*, Transportation Research Part C: Emerging Technologies, 3 (1995), pp. 261–275.

[21] M. DIANA, *Innovative systems for the transportation disadvantaged: Towards more efficient and operationally usable planning tools*, Transportation Planning and Technology, 27 (2004), pp. 315–331.

[22] M. DIANA AND M. M. DESSOUKY, *A new regret insertion heuristic for solving large-scale dial-a-ride problems with time windows*, Transportation Research Part B: Methodological, 38 (2004), pp. 539–557.

[23] Y. DUMAS, J. DESROSIERS, AND F. SOUMIS, *Large scale multi-vehicle dial-a-ride problems*, Technical Report G-89-30, HEC, Montréal, Canada, 1989.

[24] ——, *The pickup and delivery problem with time windows*, European Journal of Operations Research, 54 (1991), pp. 7–22.

[25] F. ERRICO, T. G. CRAINIC, F. MALUCELLI, AND M. NONATO, *A unifying framework and review of semi-flexible transit systems*, Technical Report CIRRELT-2011-64, Montréal, Canada, 2012.

[26] D. ESPINOZA, R. GARCIA, M. GOYCOOLEA, G. L. NEMHAUSER, AND M. W. P. SAVELSBERGH, *Per-seat, on-demand air transportation part I: Problem description and an integer multicommodity flow model*, Transportation Science, 42 (2008), pp. 263–278.

[27] ——, *Per-seat, on-demand air transportation part II: Parallel local search*, Transportation Science, 42 (2008), pp. 279–291.

[28] M. FIRAT AND G. J. WOEGINGER, *Analysis of the dial-a-ride problem of Hunsaker and Savelsbergh*, Operations Research Letters, 39 (2011), pp. 32–35.

[29] A. FÜGENSCHUH, *Solving a school bus scheduling problem with integer programming*, European Journal of Operational Research, 193 (2009), pp. 867–884.

[30] T. GARAIX, C. ARTIGUES, D. FEILLET, AND D. JOSSELIN, *Optimization of occupancy rate in dial-a-ride problems via linear fractional column generation*, Computers & Operations Research, 38 (2011), pp. 1435–1442.

[31] T. HANNE, T. MELO, AND S. NICKEL, *Bringing robustness to patient flow management through optimized patient transports in hospitals*, Interfaces, 39 (2009), pp. 241–255.

[32] D. HAUGLAND AND S. C. HO, *Feasibility testing for dial-a-ride problems*, in Lecture Notes in Computer Science, vol. 6124/2010, 2010.

[33] G. HEILPORN, J.-F. CORDEAU, AND G. LAPORTE, *An integer L-shaped algorithm for the dial-a-ride problem with stochastic customer delays*, Discrete Applied Mathematics, 159 (2011), pp. 883–895.

[34] H. HERNÁNDEZ-PÉREZ AND J. J. SALAZAR-GONZÁLEZ, *The one-commodity pickup-and-delivery traveling salesman problem: Inequalities and algorithms*, Networks, 50 (2007), pp. 258–272.

[35] ——, *The multi-commodity one-to-one pickup-and-delivery traveling salesman problem*, European Journal of Operational Research, 196 (2009), pp. 987–995.

[36] B. HUNSAKER AND M. W. P. SAVELSBERGH, *Efficient feasibility testing for dial-a-ride problems*, Operations Research Letters, 30 (2002), pp. 169–173.

[37] I. IOACHIM, J. DESROSIERS, Y. DUMAS, M. M. SOLOMON, AND D. VILLENEUVE, *A request clustering algorithm for door-to-door handicapped transportation*, Transportation Science, 29 (1995), pp. 63–78.

[38] J. JAW, A. R. ODONI, H. N. PSARAFTIS, AND N. H. M. WILSON, *A heuristic algorithm for the multi-vehicle advance-request dial-a-ride problem with time windows*, Transportation Research Part B: Methodological, 20 (1986), pp. 243–257.

[39] Y. KERGOSIEN, C. LENTE, D. PITON, AND J.-C. BILLAUT, *A tabu search heuristic for the dynamic transportation of patients between care units*, Discrete Applied Mathematics, 159 (2011), pp. 883–895.

[40] B.-I. KIM, S. KIM, AND J. PARK, *A school bus scheduling problem*, European Journal of the Operational Research, 218 (2012), pp. 577–585.

[41] G. LAPORTE AND Y. NOBERT, *A branch and bound algorithm for the capacitated vehicle routing problem*, OR Spectrum, 5 (1983), pp. 77–85.

[42] O. B. G. MADSEN, H. F. RAVN, AND J. M. RYGAARD, *A heuristic algorithm for a dial-a-ride problem with time windows, multiple capacities, and multiple objectives*, Annals of Operations Research, 60 (1995), pp. 193–208.

[43] V. MANIEZZO, A. CARBONARO, AND H. HILDMANN, *An ANTS heuristic for the long-term car pooling problem*, in New Optimization Techniques in Engineering, G. Onwuboulu and B. Babu, eds., Springer, Heidelberg, Berlin, 2004, pp. 411–430.

[44] L. M. MARTINEZ AND J. M. VIEGAS, *Design and deployment of an innovative school bus service in Lisbon*, Procedia Social and Behavioral Sciences, 20 (2011), pp. 120–130.

[45] R. MASSON, F. LEHUÉDÉ, AND O. PÉTON, *An adaptive large neighborhood search for the pickup and delivery problem with transfers*, Transportation Science, 47 (2013), pp. 344–355.

[46] E. MELACHRINOUDIS, A. B. ILHAN, AND H. MIN, *A dial-a-ride problem for client transportation in a health-care organization*, Computers & Operations Research, 34 (2007), pp. 742–759.

[47] S. MITROVIĆ-MINIĆ AND G. LAPORTE, *The pickup and delivery problem with time windows and transshipments*, INFOR, 44 (2006), pp. 217–228.

[48] J. PARK AND B.-I. KIM, *The school bus routing problem: A review*, European Journal of Operational Research, 202 (2010), pp. 311–319.

[49] J. PARK, H. TAE, AND B.-I. KIM, *A post-improvement procedure for the mixed load school bus routing problem*, European Journal of Operational Research, 217 (2012), pp. 204–213.

[50] S. N. PARRAGH, *Introducing heterogeneous users and vehicles into models and algorithms for the dial-a-ride problem*, Transportation Research Part C: Emerging Technologies, 19 (2011), pp. 912–930.

[51] S. N. PARRAGH, J.-F. CORDEAU, K. F. DOERNER, AND R. F. HARTL, *Models and algorithms for the heterogeneous dial-a-ride problem with driver related constraints*, OR Spectrum, 34 (2012), pp. 593–633.

[52] S. N. PARRAGH, K. F. DOERNER, X. GANDIBLEUX, AND R. F. HARTL, *A heuristic two-phase solution method for the multi-objective dial-a-ride problem*, Networks, 54 (2009), pp. 227–242.

[53] S. N. PARRAGH, K. F. DOERNER, AND R. F. HARTL, *A survey on pickup and deliver problems part I: Transportation between customers and depot*, Journal fuer Betriebswirtschaft, 58 (2008), pp. 21–51.

[54] ———, *A survey on pickup and deliver problems part II: Transportation between pickup and delivery locations*, Journal fuer Betriebswirtschaft, 58 (2008), pp. 81–117.

[55] ———, *Demand responsive transportation*, in Wiley Encyclopedia of Operations Research and Management Science, J. J. Cochran, ed., Wiley, New York, 2010, pp. 1–9.

[56] S. N. PARRAGH AND V. SCHMID, *Hybrid column generation and large neighborhood search for the dial-a-ride problem*, Computers & Operations Research, 40 (2013), pp. 490–497.

[57] A. PERUGIA, L. MOCCIA, J.-F. CORDEAU, AND G. LAPORTE, *Designing a home-to-work bus service in a metropolitan area*, Transportation Research Part B: Methodological, 45 (2011), pp. 1710–1726.

[58] J. Y. POTVIN AND J. M. ROUSSEAU, *A parallel route building algorithm for the vehicle routing and scheduling problem with time windows*, European Journal of Operations Research, 66 (1993), pp. 331–340.

[59] H. N. PSARAFTIS, *A dynamic programming solution to the single vehicle many-to-many immediate request dial-a-ride problem*, Transportation Science, 14 (1980), pp. 130–154.

[60] ———, *An exact algorithm for the single vehicle many to many dial-a-ride problem with time windows*, Transportation Science, 17 (1983), pp. 351–357.

[61] F. QIAN, I. GRIBKOVSKAIA, G. LAPORTE, AND T. HALSKAU SR, *Passenger and pilot risk minimization in offshore helicopter transportation*, Omega, 40 (2012), pp. 584–593.

[62] Y. QU AND J. F. BARD, *A grasp with adaptive large neighborhood search for pickup and delivery problems with transshipment*, Computers & Operations Research, 39 (2012), pp. 2439–2456.

[63] J. RIERA-LEDESMA AND J.-J. SALAZAR-GONZÁLEZ, *Solving school bus routing using the multiple vehicle traveling purchaser problem: A branch-and-cut approach*, Computers & Operations Research, 39 (2012), pp. 391–404.

[64] ———, *A column generation approach for a school bus routing problem with resource constraints*, Computers & Operations Research, 40 (2013), pp. 566–583.

[65] I. RODRÍGUEZ-MARTÍN AND J. J. SALAZAR-GONZÁLEZ, *A hybrid heuristic approach for the multi-commodity one-to-one pickup-and-delivery traveling salesman problem*, Journal of Heuristics, 18 (2012), pp. 849–867.

[66] S. ROPKE, J.-F. CORDEAU, AND G. LAPORTE, *Models and branch-and-cut algorithms for pickup and delivery problems with time windows*, Networks, 49 (2007), pp. 258–272.

[67] M. W. P. SAVELSBERGH, *Local search for routing problems with time windows*, Annals of Operations Research, 4 (1986), pp. 285–305.

[68] ——, *The vehicle routing problem with time windows: Minimizing route duration*, ORSA Journal on Computing, 4 (1992), pp. 146–154.

[69] M. SCHILDE, K. F. DOERNER, AND R. F. HARTL, *Metaheuristics for the dynamic stochastic dial-a-ride problem with expected return transports*, Computers & Operations Research, 38 (2011), pp. 1719–1730.

[70] V. SCHMID AND K. F. DOERNER, *Examination and operating room scheduling including optimization of intrahospital routing*, Transportation Science, 48 (2014), pp. 59–77.

[71] T. SEXTON AND L. D. BODIN, *Optimizing single vehicle many-to-many operations with desired delivery times: I. Scheduling*, Transportation Science, 19 (1985), pp. 378–410.

[72] ——, *Optimizing single vehicle many-to-many operations with desired delivery times: II. Routing*, Transportation Science, 19 (1985), pp. 411–435.

[73] M. SPADA, M. BIERLAIRE, AND T. M. LIEBLING, *Decision-aiding methodology for the school bus routing and scheduling problem*, Transportation Science, 39 (2005), pp. 477–490.

[74] A. J. SWERSEY AND W. BALLARD, *Scheduling school buses*, Management Science, 30 (1984), pp. 844–853.

[75] J. TANG, Y. KONG, H. LAU, AND A. W. H. IP, *A note on a efficient feasibility testing for dial-a-ride problems*, Operations Research Letters, 38 (2010), pp. 405–407.

[76] P. TOTH AND D. VIGO, *Fast local search algorithms for the handicapped persons transportation problem*, in Metaheuristics: Theory and Applications, I. H. Osman and J. P. Kelly, eds., Kluwer, Boston, MA, 1996, pp. 677–690.

[77] ——, *Heuristic algorithms for the handicapped persons transportation problem*, Transportation Science, 31 (1997), pp. 60–71.

[78] T. VIDAL, T. G. CRAINIC, M. GENDREAU, AND C. PRINS, *Timing problems and vehicle routing*, in Informs 2011, Charlotte, NC, 2011.

[79] R. WOLFLER CALVO, F. DE LUIGI, P. HAASTRUP, AND V. MANIEZZO, *A distributed geographic information system for the daily car pooling problem*, Computers & Operations Research, 31 (2004), pp. 2263–2278.

[80] S. YAN AND C.-Y. CHEN, *An optimization model and a solution algorithm for the many-to-many car pooling problem*, Annals of Operations Research, 191 (2011), pp. 37–71.

Chapter 8

Stochastic Vehicle Routing Problems

Michel Gendreau
Ola Jabali
Walter Rei

8.1 ▪ Introduction

Vehicle routing problems (VRPs) have been the subject of numerous research studies since Dantzig and Ramser [16] first presented this general class of optimization problems in a practical setting. Since then, the operations research community has devoted collectively a large effort towards efficiently solving these problems, developing both exact and heuristic methods; see Laporte [40]. The majority of these studies have been conducted under the assumption that all the information necessary to formulate the problems is known and readily available (i.e., one is in a deterministic setting). In practical applications, this assumption is usually not verified given the presence of uncertainty affecting the parameters of the problem. Uncertainty may come from different sources, both from expected variations and unexpected events. Such variations can affect various aspects of the problem under study (e.g., stochastic parameters, which entail additional feasibility requirements and extra costs). This is particularly true in the case of VRP models, which are used both at the tactical level and at the operational level to plan and control logistical operations. In this case, uncertainty is present given the time lag separating the moments where routes are planned and executed, considering the informational flow that defines the problem. For example, if customer demands are uncertain and only revealed when customers are visited, a planned route performed by a vehicle on a given day may turn out to be infeasible if the total observed demand for the customers scheduled on the route exceeds the capacity of the vehicle. When such a situation occurs, additional costs often involving additional decisions must be taken to produce a feasible solution. The need to account for such extra costs when solving VRPs entails developing models that explicitly factor in all relevant uncertainty features of the problem.

Therefore, in recent years, there has been a steadily growing interest in both formulating and solving Stochastic Vehicle Routing Problems (SVRPs); see the surveys of Gendreau, Laporte, and Séguin [25] and Cordeau et al. [15]. This interest has been further spurred by the fact that a SVRP cannot simply be solved using a deterministic approximation model. Louveaux [51] showed that deterministic VRP models may produce arbitrarily bad solutions when used to approximate SVRP. Thus, there exists a need for models that specifically produce solutions that are cost-effective in an uncertain environment. To do so, different modeling paradigms are available.

This chapter focuses on models defined using the stochastic programming paradigm; see Birge and Louveaux [10] for a thorough presentation of this field. Using this paradigm, uncertainty is formulated in optimization problems by introducing stochastic parameters in the models. In the case of the VRP, three sets of parameters are commonly considered to be stochastic:

1. stochastic demands: the product volumes to either be collected or delivered at customers are random;

2. stochastic customers: customers are either present or absent with a given probability;

3. stochastic times: both the service times at customers and the traveling times for the vehicles can be considered stochastic.

As is usually done, such parameters are formulated using suitably defined random variables or through the use of scenarios. The stochastic optimization models are then obtained by first determining the informational process that defines how and when the values of the stochastic parameters are observed. Based on this process, the decision variables are defined in stages, according to when parameters become known. Therefore, the a priori decisions group all decisions taken in the first stage, before any stochastic parameters are observed. The *recourse* decisions refer to the decisions taken in the second stage and onwards that detail how solutions are modified or adjusted as more information becomes available. It should be noted that the recourse decisions, and their associated costs, are directly related to the outcomes of the stochastic parameters.

Once the decision variables are determined, there exist two general modeling approaches to formulate SVRPs using the considered paradigm. The first is based on the use of a *recourse function*, which is defined as the average recourse cost for a given a priori solution. An SVRP model can then be formulated by considering the recourse function in the objective of the problem. The optimal solution to such a model would minimize the expected total cost. The second approach requires the inclusion of *probabilistic constraints* in the model. These constraints impose specified limits on the probabilities associated with particular random events. In the context of SVRPs, such constraints may, for example, take the form of a limit on the probability of observing either that the a priori solution is infeasible or that the value of the recourse function becomes higher than a specified threshold. Such models are used to obtain solutions that guarantee that some risks (defined here as the probabilities of observing specific random events) are limited.

It is important to note the relationship between SVRPs and *dynamic vehicle routing problems*, which cover the broader class of routing problems in which information about problem data becomes available over time. Models and solution approaches for dynamic VRPs are discussed in detail in Chapter 11 of this book.

Given that SVRP models may be quite different from one another, depending on the considered stochastic parameters and on how recourse decisions are defined, providing a unified presentation of all solution techniques proposed for all models appears as a difficult task. Instead, we have opted for a more specialized approach. Therefore, we focus this presentation on the VRP with Stochastic Demands (VRPSD) and simple recourse.

Simple recourse implies that once the vehicle capacity is exceeded a return trip to the depot is performed and the capacity of the vehicle is restored. In the VRPSD, it is usually assumed that the demand of each customer is only revealed when the vehicle arrives at the customer's location, but other schemes for the revelation of demands may be envisioned. Such problems occur in a number of applications, e.g., in the delivery and collection of money to and from banks (Bertsimas [7] and Lambert, Laporte, and Louveaux [39]), in beer distribution and garbage collection (Yang, Mathur, and Ballou [71]), and in the home delivery of oil (Chepuri and Homem de Mello [11]).

In Sections 8.2 to 8.4, we present the main modeling paradigms that have been proposed to formulate the VRPSD and we detail some of the traditional exact solutions methods that have been developed for the problem. Section 8.2 is devoted to the a priori paradigm. In Section 8.3, we describe the *reoptimization* paradigm. Finally, in Section 8.4, a modeling approach based on the use of *chance*, or probabilistic, *constraints*, which control the level of risk accepted by the decision maker, is presented. In doing so, our aim is to provide a general illustration of the differences between the solution strategies for SVRP and their deterministic VRP counterparts.

We later devote a section to each of the previously mentioned stochastic parameters: stochastic customers demands (Section 8.5), stochastic customers (Section 8.6), and stochastic times (Section 8.7). Some problems include two types of stochasticity, e.g., stochastic travel times and demands. In such cases, the paper is included in the section which better matched the paper's main scientific contributions. Finally, we conclude in Section 8.8 by discussing future research directions in this field.

8.2 ▪ A Priori Optimization

One of the most common solution frameworks for stochastic routing problems is a priori optimization, a paradigm initially put forward by Bertsimas [6], Jaillet [35], and Bertsimas, Jaillet, and Odoni [9]. It consists of modeling the problem in two stages. In the first stage, a planned, or a priori, solution is designed. In the second stage, the first stage solution is executed while uncertainties are gradually revealed and recourse actions are taken based on a predetermined policy. As previously mentioned, in the VRPSD, customer demands are revealed upon arrival at customer locations. As a result, a vehicle may reach a customer and not have sufficient capacity to serve its realized demand. Such a situation is referred to as a *failure*. Several recourse policies have been proposed for the VRPSD. A *classical* policy is to return to the depot upon failure, offload, and resume collections by following the planned route starting at the point of failure. Several authors have studied this policy, e.g., Christiansen and Lysgaard [12], Gendreau, Laporte, and Séguin [23, 24], Goodson, Ohlmann, and Thomas [27], Hjorring and Holt [29], Laporte, Louveaux, and Van hamme [42], and Lei, Laporte, and Guo [47]. An important managerial advantage of the classical policy is that it yields stable routes which require little alteration in the event of failure. Furthermore, solving the VRPSD under the classical recourse policy provides a benchmark against which alternative policies can be assessed.

Under the a priori paradigm, the aim of the first-stage problem is to design a set of vehicle routes of least total expected cost. The routes must satisfy the following three conditions: (1) each route must start and end at the depot, (2) each customer is visited exactly once by exactly one vehicle, and (3) the expected demand of each route does not exceed the vehicle capacity. The latter constraint was originally imposed by Laporte, Louveaux, and Van hamme [42] in order to avoid the creation of routes that would systematically fail. The objective function is the sum of the planned routing cost and of the expected recourse cost.

We now present the notation pertaining to this section. The VRPSD is defined on a complete undirected graph $G = (V, E)$, where $V = \{0, 1, \ldots, n\}$ is the vertex set and $E = \{(i, j) : i, j \in V, i < j\}$ is the edge set. Vertex 0 is the depot at which a set K of identical vehicles of capacity Q are based, whereas the remaining vertices represent customers. Each customer $i \in N = V \setminus \{0\}$ has a non-negative stochastic demand ξ_i to be collected. We further assume that these demands are independent random variables with known distributions with expected values μ_i. A travel cost c_{ij} is associated with each edge $(i, j) \in E$. In what follows we present two a priori optimization models for the VRPSD. In Section 8.2.1, we present the network flow formulation, which was put forward by Laporte, Louveaux, and Van hamme [42]. In Section 8.2.2, we present the set partitioning formulation as described by Christiansen and Lysgaard [12]. For each of these formulations, we present the model, an outline of the proposed solution procedure in the corresponding paper, and a summary of the computational results.

8.2.1 ▪ Network Flow Formulation

Let x_{ij} $(i < j)$ be an integer decision variable equal to the number of times edge (i, j) appears in the first-stage solution. The variable x_{ij} must be interpreted as x_{ji} whenever $i > j$. If $i, j > 0$, then x_{ij} can only take the values 0 or 1; if $i = 0$, then x_{ij} can also be equal to 2, representing a situation when a vehicle makes a return trip between the depot and vertex j. Furthermore, let $\mathcal{Q}(x)$ denote the expected recourse cost of solution x.

The model is then

$$(8.1) \quad \text{(VRPSD1)} \quad \text{minimize} \sum_{i < j} c_{ij} x_{ij} + \mathcal{Q}(x)$$

$$(8.2) \quad \text{s.t.} \quad \sum_{j=1}^{n} x_{0j} = 2|K|,$$

$$(8.3) \quad \sum_{i < k} x_{ik} + \sum_{j > k} x_{kj} = 2 \qquad \forall k \in N,$$

$$(8.4) \quad \sum_{i, j \in S} x_{ij} \leq |S| - \left\lceil \sum_{i \in S} \mu_i / Q \right\rceil$$
$$\forall S \subset N, 3 \leq |S| \leq n - 1,$$

$$(8.5) \quad 0 \leq x_{ij} \leq 1 \qquad \forall i, j \in N, i < j,$$

$$(8.6) \quad 0 \leq x_{0j} \leq 2 \qquad \forall j \in N,$$

$$(8.7) \quad x = (x_{ij}) \text{ integer} \qquad \forall i, j \in V, i < j.$$

The objective function (8.1) consists of minimizing the first-stage travel costs and the expected recourse cost. In this model, constraints (8.2) and (8.3) specify the degree of each vertex, whereas constraints (8.4) eliminate subtours and ensure that the expected demand of any route does not exceed the vehicle capacity.

Given a first-stage solution x, the computation of $\mathcal{Q}(x)$ is separable with the routes. The expected cost of route k depends on its orientation, and thus $\mathcal{Q}(x)$ is expressed as

$$(8.8) \quad \mathcal{Q}(x) = \sum_{k=1}^{|K|} \min\{\mathcal{Q}^{k,1}, \mathcal{Q}^{k,2}\},$$

where $\mathcal{Q}^{k,\delta}$ denotes the expected recourse cost of route k for orientation δ. Equation (8.8) implies that, given an undirected first-stage solution, the expected recourse cost of each route is computed for each direction, and the cheapest orientation is selected. The computation of $\mathcal{Q}^{k,1}$ for route k defined by $(i_1 = 0, i_2, \ldots, i_{t+1} = 0)$, assuming that $\xi_i \leq Q$ with probability 1 for all i, is given by

$$(8.9) \qquad \mathcal{Q}^{k,1} = 2 \sum_{j=2}^{t} \sum_{l=1}^{j-1} P\left(\sum_{s=2}^{j-1} \xi_{i_s} \leq lQ < \sum_{s=2}^{j} \xi_{i_s} \right) c_{0i_j}.$$

In the event of a failure, the vehicle performs a return trip to the depot. Therefore, the first factor in the double summation of (8.9) is the probability of incurring the lth failure at customer i_j. This is multiplied by the cost of the return trip to the depot. The value of $\mathcal{Q}^{k,2}$ is computed likewise by reversing the orientation of route k.

If arbitrary distributions are used for demands, the computation of the probabilities appearing in (8.9) becomes rapidly intractable. Therefore, in most applications, the demand probability distributions adhere to the cumulative property; i.e., the sum of two or more independent random variables with a distribution Ψ yields a random variable with distribution Ψ (albeit with a different mean). A number of well-known probability distributions possess the cumulative property, e.g., normal and Poisson distributions. Therefore, this assumption is not overly restrictive.

The described formulation is solved by the integer L-shaped method proposed by Laporte and Louveaux [41]. This is an extension of the L-shaped method of Van Slyke and Wets [68] for continuous stochastic programs, which is an application of Benders' decomposition [4] to stochastic programming. In what follows, we describe the integer L-shaped algorithm, as applied to the VRPSD.

The integer L-Shaped Algorithm. The integer L-shaped algorithm applies Branch-and-Cut to a relaxation for the VRPSD in which the recourse term $\mathcal{Q}(x)$ is bounded below by a variable Θ. In addition, the subtour elimination constraints and the integrality requirements are relaxed. Initially, Θ is set equal to a lower bound L on the expected cost of recourse, and its value Θ^v is computed for the solution of each subproblem solved at iteration v. As is standard in Branch-and-Cut, violated subtour elimination constraints are generated dynamically as they are found to be violated, and integrality is gradually recovered by branching. Optimality cuts are generated at feasible integer solutions. In most applications these cuts are local, and it often pays to also impose lower bounds on the recourse function $\mathcal{Q}(x)$, in the form of linear functionals computed on the basis of infeasible intermediate solutions. In the following summary of the algorithm, CP denotes the current problem.

Step 0 Compute L, and set the iteration counter v equal to 0. Define the CP as a relaxation of VRPSD in which constraints (8.4) and (8.7) are removed, the $\mathcal{Q}(x)$ term of the objective function is replaced with Θ, and the constraint $\Theta \geq L$ is imposed. Set the value of the best known solution to $\bar{z} := \infty$. At this stage, the only pendent node is the initial CP.

Step 1 Select a pendent node from the list. If none exists, stop.

Step 2 Set $v := v + 1$, and solve CP. Let (x^v, Θ) be its optimal solution.

Step 3 Check for any violated subtour elimination constraints, and generate them accordingly. At this stage, valid inequalities or Lower Bounding Functionals (LBFs) may also be generated. If a violated constraint is found, add it to the CP and return to Step 2. Otherwise, if $cx^v + \Theta \geq \bar{z}$, fathom the current node and return to Step 1.

Step 4 If the solution is not integer, then branch on a fractional variable. Append the corresponding subproblems to the list of pendent nodes, and return to Step 1.

Step 5 Compute $Q(x^\nu)$, and set $z^\nu := cx^\nu + Q(x^\nu)$. If $z^\nu < \bar{z}$, then set $\bar{z} := z^\nu$.

Step 6 If $\Theta \geq Q(x^\nu)$, then fathom the current node and return to Step 1. Otherwise, add an optimality cut defined as

$$(8.10) \qquad \sum_{\substack{0 < i < j \\ x^\nu_{ij} = 1}} x_{ij} \leq \sum_{0 < i < j} x^\nu_{ij} - 1$$

and go to Step 2.

A general lower bound L on $\mathcal{Q}(x)$ is described in Proposition 1 of Laporte, Louveaux, and Van hamme [42]. This bound is based on the computation of the probability of failure on each route taken separately. The recourse cost is bounded below by considering the $|K|$ customers closest to the depot and by partitioning the total demand among the $|K|$ vehicles so as to minimize the total cost.

LBFs based on partial routes were first proposed by Hjorring and Holt [29] for the single-vehicle case. A partial route implies a solution which starts and ends at the depot while containing a set of *connected yet not necessarily sequenced customers*. The recourse associated with a partial route is bounded by treating the set of connected yet not necessarily sequenced customers as a single customer, whose demand is equivalent to the sum of the demands of its customers and is distributed accordingly. Furthermore, one associates with this single customer a distance from the depot which is that of the closest customer to the depot, out of the set of connected yet not necessarily sequenced customers. Partial route-based LBFs were then proposed by Laporte, Louveaux, and Van hamme [42] for the multi-vehicle case. Jabali et al. [32] expanded the definition of the LBFs in a way that better exploits the structural information provided by partial routes.

In the context of the VRPSD, Laporte, Louveaux, and Van hamme [42] found that using the normal distribution to model customer demands is more challenging than using the Poisson distribution. We now present a summary of the best results obtained by Jabali et al. [32], where customer demands ξ_i follow a normal distribution $\mathcal{N}(\mu_i, \sigma_i)$ truncated at zero, and all demands are independently distributed. The instances were generated based on the same principles as in Laporte, Louveaux, and Van hamme [42]. Table 8.1 summarizes the results of the 30 instances that were generated for each combination of the number of vertices and $|K|$, for a total of 270 instances. For each combination, several

Table 8.1. *Results for the integer L-shaped algorithm.*

$|V|$	$|K|$	Solved	CPU (min)	%Gap
60	2	24	23	0.3
70	2	17	43	0.5
80	2	13	30	0.5
50	3	16	115	0.7
60	3	6	46	0.7
70	3	9	29	1.5
40	4	9	21	1.5
50	4	5	23	1.9
60	4	3	82	2.0

fill rates are included, but these details are omitted for summary purposes. In Table 8.1, the columns "Solved", "CPU (min)", and "%Gap", respectively, refer to the number of instances solved to optimality, the average CPU times of the algorithm on these instances, and the average gap obtained by the algorithm over all instances of each category. The computation time limit for any given instance was set to 5 hours. The coefficient of variation of the demand distribution was set equal to 30%. Each instance was solved on one out of two Intel(R) Xeon(R) CPU X5675 3.07 GHz processors of a machine with 96 GB of RAM.

The results indicate that the algorithm is able to efficiently solve instances of relatively large sizes, e.g., up to 80 vertices with two vehicles. However, the efficiency of the algorithm deteriorates with the number of vehicles. For instance, the algorithm was able to solve 24 out of 30 instances with 60 vertices and two vehicles, while it was only able to solve three out of 30 instances with 60 vertices and four vehicles.

8.2.2 ▪ The Set Partitioning Formulation

The set partitioning formulation of the VRPSD relies on path-based description of vehicle flows in the graph previously defined. We therefore define a *route* as a path $(0, z_1, \ldots, z_j, 0)$, where $z_1, \ldots, z_j \in N$ and $z_h \neq z_{h+1}$ for $h = 1, \ldots, j-1$. We say that a route is elementary if $z_1, \ldots, z_j$ are different; otherwise we say that the route is *non-elementary*. A route is said to be feasible if $\sum_{h=1}^{j} \mu_{z_h} \leq Q$. For each feasible and elementary route r, let c_r denote its expected cost. Let Ω_e denote the set of all feasible and elementary routes. Furthermore, let a_{ir} be a parameter equal to 1 if route r visits customer i and to 0 otherwise. Finally, let λ_r be a variable which takes the value 1 if route r is chosen and 0 otherwise. The set partitioning formulation for the VRPSD can then be written as

$$(8.11) \quad \text{(VRPSD2)} \quad \text{minimize} \sum_{r \in \Omega_e} c_r \lambda_r$$

$$(8.12) \quad \text{s.t.} \quad \sum_{r \in \Omega_e} a_{ir} \lambda_r = 1 \qquad \forall i \in N,$$

$$(8.13) \quad \lambda_r \in \{0, 1\} \qquad \forall r \in \Omega_e.$$

The objective function (8.11) minimizes the total expected distribution cost. Constraints (8.12) guarantee that each customer is visited exactly once by one vehicle. Constraints (8.13) ensure that the decision variables are binary.

Assuming each customer's demand is only revealed upon reaching it, the total expected cost c_r of a route can be decomposed into two elements. The first element is the deterministic travel cost of the route, while the second is the recourse cost. Since the former is straightforward to compute, we focus on the computation of the latter. Given that each ξ_i follows an independent probability distribution, Christiansen and Lysgaard [12] showed that the probability that the total demand along a path $\{0, z_1, \ldots, z_j\}$ is less than lQ is only dependent on the total demand along this path. This implies that this probability is independent of the demand allocation along the path. This observation is used to compute the recourse cost.

Given an elementary path $\{0, z_1, \ldots, z_j\}$, let μ and σ denote for this path the total demand expected value and standard deviation, respectively. Let $F(\mu, \sigma^2, lQ) = P(\sum_{h=1}^{j} \xi_{z_h} \leq lQ)$ define the probability that the cumulative demand at customer z_j is less than or equal to lQ. Furthermore, let σ_{z_j} denote the standard deviation of the demand of customer z_j.

The expected recourse cost for customer z_j is then computed as

$$(8.14)\qquad \text{ERC}(\mu,\sigma^2,z_j)=2c_{0z_j}\sum_{l=1}^{\infty}F(\mu-\mu_{z_j},\sigma^2-\sigma^2_{z_j},lQ)-F(\mu,\sigma^2,lQ).$$

Equation (8.14) is approximated by replacing ∞ with a sufficiently large number. The summation provides the expected number of failures at customer z_j. Therefore, the above computation expresses the total expected failure cost caused by customers along the path. This is used to establish a dominance criterion among paths characterized by (μ,σ^2,z_j) and the set of customers visited by each path.

The Branch-and-Price Algorithm.   The solution procedure is based on the Dantzig–Wolfe decomposition and column generation. The master problem considered at any given iteration is established by (i) considering only a subset, denoted by Ω, of the set of all feasible routes, which is enlarged by allowing non-elementary routes; (ii) relaxing the integrality constraints (8.13); (iii) changing the set partitioning constraints to set covering constraints (8.12); and (iv) introducing the coefficient α_{ir}, which denotes the number of times that customer i is visited by route r. The master problem is then

$$(8.15)\qquad (\text{VRPSD2 } M_p)\quad \text{minimize}\sum_{r\in\Omega}c_r\lambda_r$$

$$(8.16)\qquad\qquad\qquad \text{s.t.}\ \sum_{r\in\Omega}a_{ir}\lambda_r\geq 1 \qquad\qquad \forall i\in N,$$

$$(8.17)\qquad\qquad\qquad\qquad \lambda_r\geq 0 \qquad\qquad\qquad \forall r\in\Omega.$$

The set Ω contains feasible non-elementary routes without 2-cycles (i,j,i). Note that the computation of the expected failure cost is not affected by allowing non-elementary routes. Following the usual principles of column generation, the dual prices $(\pi_1,\ldots,\pi_n)$ obtained from solving M_p are used in a subproblem in the search for one or more routes (columns) with negative reduced costs. If such columns are found, they are added to the master problem, which is then reoptimized. The steps of column generation are repeated until no columns with negative reduced costs can be identified. At this point the current solution is the optimal solution for the master problem.

If the master problem solution is integer and constraints (8.16) are satisfied with equality, then the current solution is optimal for the original problem. If the current solution is fractional, branching is then performed. In the following subsection, we describe the column generation procedure proposed by Christiansen and Lysgaard [12].

The column generation subproblem is solved by dynamic programming. The authors assume that the expected demand and variance of each customer have integer values. Furthermore, an upper bound $V_{\max}$ on the total variance on any feasible elementary route is established by solving a 0-1 knapsack problem.

In order to solve the column generation subproblem, the authors use graph $G_S=(V_S,A_S)$, where V_S has $(n+1)QV_{\max}+1$ vertices. Except for source vertex $v(0,0,0)$, each vertex is denoted by $v(\mu,\sigma^2,i)$ for $\mu=1,\ldots,Q$, $\sigma^2=1,\ldots,V_{\max}$, and $i=0,\ldots,n$. Each vertex represents a particular set of paths from 0 to i, with a total expected demand of μ and a total variance of σ^2. The arc set A_S is constructed as follows:

Step 1 $A_S := \emptyset$.

Step 2 For $i = 1,\ldots,n$, add an arc from $v(0,0,0)$ to $v(\mu_i,\sigma_i^2,i)$ and set its cost to $c_{0i} + \text{ERC}(\mu_i,\sigma_i^2,i) - \pi_i$.

Step 3 For each ordered pair $i,j \in N$, $i \neq j$, each $\mu = 1,\ldots,Q-1$, and $\sigma^2 = 1,\ldots,V_{\max}$, add an arc from $v(\mu,\sigma^2,j)$ to $v(\mu+\mu_i,\sigma^2+\sigma_i^2,i)$ (provided that $\mu+\mu_i \leq Q$ and $\sigma^2+\sigma_i^2 \leq V_{\max}$) and set its cost to $c_{ij} + \text{ERC}(\mu+\mu_i,\sigma^2+\sigma_i^2,i) - \pi_i$.

Step 4 For each $\mu = 1,\ldots,Q$, each $\sigma^2 = 1,\ldots,V_{\max}$, and each $j = 1,\ldots,n$, add an arc from $v(\mu,\sigma^2,j)$ to $v(\mu,\sigma^2,0)$ and set its cost to c_{j0}.

The shortest path in G_S from $v(0,0,0)$ to $v(\mu,\sigma^2,0)$, for all $\mu = 1,\ldots,Q$ and $\sigma^2 = 1,\ldots,V_{\max}$, is the route of least reduced cost among routes with total demand μ and total variance equaling σ^2. Therefore, this shortest path solves the column generation subproblem. This can be done in $O(n^2QV_{max})$ time and also in the case where 2-cycles are prohibited.

The algorithm was tested under the assumption that demands follow Poisson distributions. Since this assumption implies that $\mu = \sigma^2$, the number of vertices in G_S reduces to $(n+1)Q+1$. The considered test instances were derived from the instances of Augerat et al. [2] and of Christofides and Eilon [13] having at most 60 customers. The expected demand of each customer was set to the deterministic value of the demand in the original instance. Computational results showed that the proposed approach could solve 18 of the 40 benchmark instances in running times up to 20 minutes on a Pentium Centrino 1500MHz computer with 480MB of RAM. In general, the algorithm yielded better results on instances with tight capacity constraints.

Recently, Gauvin, Desaulniers, and Gendreau [22] reimplemented the Christiansen and Lysgaard method using the most recent techniques for solving the column generation subproblem, including bidirectional labeling, a combination of 2-cycle elimination with ng-routes, and the application of a tabu search heuristic. They also introduced a new aggregate dominance rule and added capacity and subset-row inequalities dynamically in order to strengthen the linear relaxation of the master problem. Their computational results show that their algorithm is much more effective than the Christiansen and Lysgaard original code: it solves 38 out of the 40 instances of the benchmark set, compared to 18, and CPU times for previously closed instances are in general much smaller. For example, an instance with 60 customers and 15 vehicles was solved in less than 10 seconds. These computational results are summarized in Table 8.2, where the results are aggregated according to the number of customers in the instances: instances ranging from 16 to 50 customers (Category 16-50) and from 51 to 60 customers (Category 51-60). In Table 8.2, the columns "Max Veh", "Number", and "Solved", respectively, refer to the maximum number of vehicles used in the obtained solution, the number of instances in the category, and the number of these instances that were solved, while the columns "B&B nodes", "cuts", and "CPU(s)" are, respectively, the average number of nodes in the search tress, cuts added

Table 8.2. *Results for the Branch-and-Price algorithm.*

Category	Max Veh	Number	Solved	B&B nodes	Cuts	CPU(s)
16-50	10	29	29	2.0	77.0	70.4
51-60	15	11	9	10.2	165.5	160.1
16-60	15	40	38	4.3	101.6	91.7

by the algorithm, and average computational times over all instances. Experiments were performed on a Intel i7-2600 processor with 3.4 GHz and 16 GB of RAM.

It is interesting to recall that the integer L-shaped algorithm could solve within reasonable CPU time instances with 80 vertices and two vehicles, as well as some instances with 60 vertices and four vehicles. These results together with those just presented indicate that the Branch-and-Price algorithm works well on instances with a large number of vehicles, while the integer L-shaped algorithm performs effectively on instances with a relatively small number of vehicles.

8.3 ▪ The Reoptimization Model

Considering the informational process traditionally assumed when formulating the VRPSD (i.e., each demand becoming known when a vehicle arrives at the associated customer location), a company may decide to sequence the customers in each route in a dynamic fashion (i.e., as the demands are revealed during operations). As such, when considering the overall decisions made when solving the VRPSD, i.e., the assignment of customers to vehicles and both the routing and the replenishment decisions that determine the sequence of vertices visited by the vehicles to produce feasible routes, a company looking to minimize the average cost of each route (i.e., the average distance traveled) would benefit from making the latter decisions dynamically; see Secomandi [58]. As more and more efficient information and communication technologies become accessible to companies, the prospect of applying such dynamic models to perform routing activities becomes increasingly attractive; see Psaraftis [56] and Chapter 11 of this book.

In this section, we first present the general stochastic shortest path problem (SSPP), originally developed by Secomandi in [58, 59], to formulate the case where routes are constructed for the VRPSD using a reoptimization approach. This model is obtained by making no particular assumptions concerning either the routing or replenishment decisions to be made. For example, a vehicle may service the observed demand of a customer in more than one visit, scheduled consecutively or not. Therefore, as opposed to the a priori paradigm, where fixed routes are used during operations by applying reactive recourse actions when replenishments at the depot are necessary (i.e., whenever a vehicle cannot service the demand observed at a particular customer), the reoptimization approach sequences routes dynamically by deciding at each step (occurring when the vehicle leaves a particular location) which customer to visit next and whether a replenishment action should be performed (i.e., a visit to the depot). Therefore, this approach defines a more flexible and proactive recourse strategy when compared to the classical reactive recourse strategy detailed in Section 8.2. We conclude by presenting the general solution approaches developed to solve such dynamic optimization problems.

Before describing the stochastic shortest path formulation, there is an important point to be made. Since the reoptimization approach was first introduced in Dror, Laporte, and Trudeau [17] and formulated as a Markov Decision Process (MDP), the majority of solution methods that have been proposed to solve the VRPSD under this recourse strategy consider the particular case where a single vehicle is located at the depot; see the studies of Secomandi [58, 59], Secomandi and Margot [60], and Novoa and Storer [54]. This fact can be explained by the complexity associated with the problem of solving the multiple vehicle case (i.e., the extremely large number of possible states defined in the dynamic formulation). In order to present all modeling paradigms in a unified way, a general assumption is therefore made here with regard to the assignment decisions. As such, it is assumed that for each vehicle considered in the problem, the subset of customers to be serviced by the vehicle has been determined beforehand (i.e., all assignment decisions have been made prior to conducting the routing operations). These decisions may reflect specific choices made by the company or may be the result of a clustering model. With regard to the a priori paradigm, the assignment decisions can be viewed in the present context as defining the a priori plan to be used. In all cases, by fixing the assignment decisions, the problem decomposes by vehicle.

We now present the model as it is defined for a single vehicle in Secomandi [58]. A vehicle of capacity Q is located at the depot and is available to service the customers in the set N. Let ξ_i for all $i \in N$, define again the random variables that are used to represent the demands of the customers. It is assumed here that these variables are all discrete and are defined such that for all $i \in N$, $p_i(k) = Pr\{\xi_i = k\}$, where the support is $[0, \mathcal{K}]$ with $\mathcal{K} \leq Q$. Furthermore, all random variables are considered to be independent from each other and their realizations are only observed on the first visit of the vehicle at each customer location. The vehicle must start and end its route at the depot. While serving a particular customer, if the available capacity is either reached or exceeded, a return trip to the depot is performed to replenish the capacity up to level Q. It is important to stress that in the present problem, if a failure in service occurs, once the return trip to the depot is performed, the vehicle does not necessarily resume service at the location of the failure. Therefore, all observed demands can be served through multiple visits to the associated customers that are not necessarily scheduled consecutively. The problem consists of finding a dynamic routing policy $\mathbf{p}$ for the vehicle such that all customer demands are serviced and the expected distance traveled is minimized.

The SSPP model proposed by Secomandi [58, 59] represents a special case of the MDP. It is defined as a discrete-time dynamic system where the transitions between two states depend on a given control. Let the state of the system be defined as vector $\mathbf{x} = (l, d_l, j_1, \ldots, j_n)$, where $l \in \{0, 1, \ldots, n\}$ represents the current location of the vehicle, $d_l \in \{0, 1, \ldots, Q\}$ defines the residual capacity of the vehicle before performing the service at l, and values $j_i \in \{?, 0, 1, \ldots, \mathcal{K}\}$, for all $i \in N$, represent the amount of unserviced demand for each customer considered (where ? represents the demand of a customer yet to be visited for which the only available information is the associated distribution of its demand). As such, the starting and finishing states for the system are respectively defined as $(0, Q, ?, ?, \ldots, ?)$ and $(0, Q, 0, 0, \ldots, 0)$. Given a state $\mathbf{x}$, an associated control set is defined as follows: $U(\mathbf{x}) = \{\{m \in \{1, 2, \ldots, n\} \mid j_m \neq 0\} \cup \{0\}\} \times \{a : a \in \{0, 1\}\}$. Therefore, a control $u \in U(\mathbf{x})$ takes the form of a pair of values (m, a). The value m represents the next vertex to be visited along the route. This vertex can either be a customer that has yet to be visited or one that has been visited but that has part of its demand still unserved. Vertex m can be visited either directly (i.e., $a = 0$) or by first performing a replenishment visit to the depot (i.e., $a = 1$). It should be noted that m can also be set to the depot to account for the case where all demands have been serviced and the system enters the terminal state.

When transiting from one state to another, if the system moves from $\mathbf{x} = (l, d_l, j_1, \ldots, j_l, \ldots, j_m, \ldots, j_n)$ to $\mathbf{x}' = (m, d_m, j_1, \ldots, j_l', \ldots, j_m', \ldots, j_n)$, assuming with no loss of generality that $l \leq m$, then the transition cost under control $u \in U(\mathbf{x})$ is defined as

$$(8.18) \qquad g(\mathbf{x}, u, \mathbf{x}') = \begin{cases} c_{lm} & \text{if } u = (m, 0), \\ c_{l0} + c_{0m} & \text{if } u = (m, 1). \end{cases}$$

In such a transition, given control u, the capacity in state $\mathbf{x}'$ becomes

$$(8.19) \qquad d_m = \begin{cases} \max(0, d_l - j_l) & \text{if } u = (m, 0), \\ \max(d_l + Q - j_l, Q) & \text{if } u = (m, 1). \end{cases}$$

If customer m has not yet been visited in state $\mathbf{x}$ (i.e., $j_m = ?$), then the value of its unserviced demand in the following state (i.e., j'_m) is set to be the realization of the random variable ξ_m; otherwise this value is already known exactly (i.e., $j'_m = j_m$). In the case of j'_l, the following update is applied: $j'_l = \max(0, j_l - d_l)$. The transition probabilities are then defined as

$$(8.20) \qquad p_{\mathbf{x}\mathbf{x}'} = \begin{cases} 1 & \text{if } j_m \text{ is known,} \\ Pr\{\xi_m = j'_m\} & \text{otherwise.} \end{cases}$$

The SSPP consists of finding an optimal sequence of controls that will bring the system to the termination state. The number of transitions that the system will take to reach the termination state is a random variable, denoted by R, which is dependent on both the random demands of the VRPSD and the controls that are used. Let $\mathbf{x}_k$ define the state of the system at stage k and $\mathbf{x}_R$ define the termination state; then a policy is defined as follows: $\mathbf{p} = \{\mathbf{u}_0, \mathbf{u}_1, \ldots, \mathbf{u}_{R-1}\}$. For each stage k, $\mathbf{u}_k$ defines a function that associates a control $\mathbf{u}_k(\mathbf{x}_k) \in U_k(\mathbf{x}_k)$ with all possible states $\mathbf{x}_k$. Using the transition cost defined in (8.18), the objective function can now be defined as

$$(8.21) \qquad J_R^{\mathrm{p}}(\mathbf{x}) = E\left[\sum_{k=0}^{R-1} g(\mathbf{x}_k, \mathbf{u}_k(\mathbf{x}_k), \mathbf{x}_{k+1}) \mid \mathbf{x}_0 = \mathbf{x}\right]$$

for all possible states $\mathbf{x} \in S$, where set S defines the state space. The expectation is formulated according to the probability distribution of the underlying Markov chain $\{\mathbf{x}_0, \mathbf{x}_1, \ldots, \mathbf{x}_R\}$, which again depends on both the initial state and on the considered policy $\mathbf{p}$. The optimal R-stage cost-to-go function from $\mathbf{x}$ is defined as

$$(8.22) \qquad J_R^{\star}(\mathbf{x}) = \min_{\mathbf{p}} J_R^{\mathrm{p}}(\mathbf{x}).$$

It can be shown that (8.22) satisfies Bellman's equation, and we can therefore write

$$(8.23) \qquad J_R^{\star}(\mathbf{x}) = \min_{u_k \in U_k(\mathbf{x}_k)} \sum_{\mathbf{x}_{k+1} \in S} p_{\mathbf{x}_k \mathbf{x}_{k+1}}(u_k)\{g(\mathbf{x}_k, u_k, \mathbf{x}_{k+1}) + J_R^{\star}(\mathbf{x}_{k+1}) \mid \mathbf{x}_k = \mathbf{x}\}$$

for all $\mathbf{x} \in S$. Provided (8.23) is known for all possible states $\mathbf{x} \in S$, then an optimal control associated with each state and at each stage k (i.e., $\mathbf{u}_k^{\star}(\mathbf{x})$) can be obtained by solving the following problem:

$$\mathbf{u}_k^{\star}(\mathbf{x}) = \arg \min_{u_k \in U_k(\mathbf{x}_k)} \sum_{\mathbf{x}_{k+1} \in S} p_{\mathbf{x}_k \mathbf{x}_{k+1}}(u_k)\{g(\mathbf{x}_k, u_k, \mathbf{x}_{k+1}) + J_R^{\star}(\mathbf{x}_{k+1}) \mid \mathbf{x}_k = \mathbf{x}\}.$$

Thus, one obtains an optimal policy for the problem $\mathbf{p}^{\star} = \{\mathbf{u}_0^{\star}, \mathbf{u}_1^{\star}, \ldots, \mathbf{u}_{R-1}^{\star}\}$.

As presented in Secomandi [58], neuro-dynamic programming has successfully been applied to solve the SSPP. Given the size of the state space S, exact computation of the cost-to-go function (8.22) is prohibitive from a computational perspective. Neuro-dynamic programming methods rely on the use of approximation functions, obtained using simulation techniques and parametric approximations, to compute (8.22). Such approximation functions are then used in specific solution strategies, such as the Policy Iteration Algorithm (PIA), to heuristically produce policies for the considered problem.

Assuming that an initial policy is available, the PIA searches to improve it through an iterative process. Therefore, at iteration ν of the PIA, let the considered policy be

$\mathbf{p}^{\nu} = \{\mathbf{u}_0^{\nu}, \mathbf{u}_1^{\nu}, \ldots, \mathbf{u}_{R-1}^{\nu}\}$. Policy $\mathbf{p}^{\nu}$ is first evaluated by solving the following equation for all states $\mathbf{x} \in S$:

$$J^{\mathbf{p}^{\nu}}(\mathbf{x}) = \sum_{\mathbf{x}_{k+1} \in S} p_{\mathbf{x}_k \mathbf{x}_{k+1}}(\mathbf{u}_k^{\nu}(\mathbf{x}_k))\{g(\mathbf{x}_k, \mathbf{u}_k^{\nu}(\mathbf{x}_k), \mathbf{x}_{k+1}) + J^{\mathbf{p}^{\nu}}(\mathbf{x}_{k+1}) \mid \mathbf{x}_k = \mathbf{x}\}.$$

An improvement step is then performed to produce the next policy to consider $\mathbf{p}^{\nu+1} = \{\mathbf{u}_0^{\nu+1}, \mathbf{u}_1^{\nu+1}, \ldots, \mathbf{u}_{R-1}^{\nu+1}\}$. This step is defined by solving the following problem again for all states $\mathbf{x} \in S$:

$$\mathbf{u}_k^{\nu+1}(\mathbf{x}) = \arg \min_{u_k \in U_k(\mathbf{x}_k)} \sum_{\mathbf{x}_{k+1} \in S} p_{\mathbf{x}_k \mathbf{x}_{k+1}}(u_k)\{g(\mathbf{x}_k, u_k, \mathbf{x}_{k+1}) + J^{\mathbf{p}^{\nu}}(\mathbf{x}_{k+1}) \mid \mathbf{x}_k = \mathbf{x}\}.$$

When both the evaluation and improvement steps are performed exactly, the PIA produces a sequence of policies that converges to the optimal one $(\mathbf{p}^0, \mathbf{p}^1, \ldots, \mathbf{p}^\star)$. However, as previously mentioned, in an effort to speed up the solution process, neuro-dynamic programming methods perform these steps using approximations.

Several solution approaches, based on the policy improvement strategy, have been successfully applied to solve the SSPP. Rollout algorithms have been proposed by Secomandi [58, 59] in this context. Starting from an initial policy obtained through an a priori route for the VRPSD, a rollout algorithm performs a single policy improvement step using an approximation of the cost-to-go function for each considered state. The particularity of these algorithms is that the improvement step is performed using the states that are generated while the procedures are executed. By proceeding in this way, the solution times are considerably reduced. However, this is done at the expense of the precision of the improvement step. Enhancements to this method were proposed by Novoa and Storer [54] by performing a two-improvement step approach which applies pruning rules on the set of controls considered, but also uses multiple initial policies derived from different a priori routes to produce a better upper bound on the routing cost.

Finally, Secomandi and Margot [60] proposed applying a partial reoptimization strategy based on the idea of obtaining an optimal policy for a restricted set of states for the MDP. By doing so, the VRPSD is heuristically solved. In the model proposed in [60], it was assumed that when a failure happens at a particular customer the vehicle, after performing a partial delivery, returns to the depot to replenish its capacity and resumes service at the customer where the failure occurred. Considering this special case of the problem, the authors proposed two strategies aimed at reducing the state space. Given an a priori route, these strategies are obtained by using either disjoint or overlapping blocks of customers along the considered route. When disjoint blocks are used, states are generated such that the sequence of customer blocks in the a priori route is followed but the order in which the customers in each block are visited may be dynamic. As for the strategy using overlapping blocks, states are produced such that each customer along the a priori route may be visited in a dynamic fashion within a sliding window defined according to the considered blocks. Based on these two strategies, heuristic algorithms are developed for the problem.

Given that the dynamic formulations for the VRPSD have been applied to the single-vehicle case, it is hard to compare the results obtained using these models with those reported for the a priori models presented earlier. However, a particular route can always be evaluated using either an a priori or a dynamic recourse formulation. Such experimentations were performed by Secomandi [58] to assess the potential improvements, in terms of the average cost of routes, when using the SSPP model proposed. Using the generator originally developed by Gendreau, Laporte, and Séguin [23], instances of different sizes

Table 8.3. *Relative improvements of RP compared to HP.*

$\overline{F}$	n		
	5	10	15
1.0	8.0401	8.0968	8.8898
1.5	5.9113	6.2811	8.1915
2.0	5.4249	5.2688	7.1950

were produced, $n \in \{5, 10, 15\}$, where the expected number of route failures $\overline{F}$ was set to $\{1.0, 1.5, 2.0\}$. For each considered combination, a total of five instances were randomly generated. All instances were first solved to produce a Rollout Policy (RP), obtained by applying the rollout algorithm proposed in [58]. A Heuristic Policy (HP) was then constructed by running a traveling salesman heuristic that produced a route while ignoring the stochastic demands. The route was then evaluated using the classical a priori recourse strategy. Table 8.3 reports the results presented in [58], which correspond to the percentage improvements in the average distances traveled when applying RP compared to HP. Over all instances tested, the average percentage improvement reported was 7.0332. These results show that as the number of customers to be served increases, so does the average improvement observed. Also, when varying the expected number of route failures, one observes that as the value of $\overline{F}$ increases the average improvement decreases.

Finally, considering the size of the problems solved, given the complexity of the SSPP, the instances initially considered by Secomandi [58, 59] were relatively small (i.e., $5 \le n \le 15$). The enhancements proposed by Novoa and Storer [54] were able to improve the algorithms to solve larger instances (i.e., $n \in \{5, 8, 20, 30, 40, 60\}$). However, the current state-of-the-art strategies available are the heuristic algorithms proposed by Secomandi and Margot [60], which were shown to efficiently solve instances with up to 100 customers in a few seconds, thus clearly illustrating the advantages associated with the partial reoptimization strategy.

8.4 ▪ Probabilistic Formulation

Accounting for stochastic demands in routing problems is often driven by the paradigm of guaranteeing that a given route will not fail with a given probability. This paradigm is useful when the violation of the capacity constraint is not well defined (Stewart and Golden [62]). Therefore, the Vehicle Routing Problem with Stochastic Demands and Probabilistic Constraints (VRPSDPC) does not explicitly account for the worst-case scenarios, but rather ensures that the feasibility of routes is achieved within a predetermined threshold. Laporte, Louveaux, and Mercure [43] proposed an exact algorithm for the stochastic location-routing problem, which consists of simultaneously locating a depot among a set of potential sites, as well as routing vehicles to service customers with stochastic demands, thus modeling and solving a broader problem than the VRPSD. For consistency reasons, we will omit the location decisions. In addition to the previously defined notation, we impose that the probability that a planned route fails at least once does not exceed β. Finally, we define $V_\beta(S)$ as the minimum number of vehicles required to serve $S \subseteq N$, so that the probability of failure in S does not exceed β. This definition implies that $V_\beta(S)$ is the smallest integer satisfying

$$P\left(\sum_{i \in S} \xi_i > Q V_\beta(S)\right) \le \beta.$$

In what follows we present the single location adaptation of the model presented in Laporte, Louveaux, and Mercure [43]:

(8.24)   (VRPSDPC)   minimize $\sum_{i<j} c_{ij} x_{ij}$

(8.25)   s.t. $\sum_{j=1}^{n} x_{0j} = 2|K|,$

(8.26) $\sum_{i<k} x_{ik} + \sum_{j>k} x_{kj} = 2$ $\forall k \in N,$

(8.27) $\sum_{\substack{i \in S, j \in \bar{S} \\ \text{or} \\ i \in \bar{S}, j \in S}} x_{ij} \geq 2V_{\beta}(S)$

$$\forall S \subset N, 3 \leq |S| \leq n-1,$$

(8.28) $0 \leq x_{ij} \leq 1$ $\forall i,j \in N, i < j,$

(8.29) $0 \leq x_{0j} \leq 2$ $\forall j \in N,$

(8.30) $x = (x_{ij})$ integer $\forall i,j \in V, i < j.$

In this model, constraints (8.25) and (8.26) specify the degree of each vertex. Constraints (8.27) are probabilistic connectivity constraints; their interpretation is that if it is known that at least $V_{\beta}(S)$ vehicles will visit the customer set S, then there must be at least $2V_{\beta}(S)$ crossings between S and $\bar{S}$. Therefore, constraints (8.27) are an extension of the classical deterministic connectivity constraints used in the VRP; see Laporte, Nobert, and Desrochers [46].

The customer demands are assumed to be independently and identically distributed random variables of finite mean μ_i and finite variance σ_i^2. If $|S|$ is sufficiently large or if customer demands follow a normal distribution, then $\sum_{i \in S} d_i$ is normally distributed. In both cases, $V_{\beta}(S)$ can be expressed as

$$V_{\beta}(S) = \left\lceil \frac{z_{\beta}(\sum_{i \in S} \sigma_i^2)^{1/2} + \sum_{i \in S} \mu_i}{Q} \right\rceil,$$

where z_{β} is the order β fractile of the standard normal distribution. Laporte, Louveaux, and Mercure [43] proposed a Branch-and-Bound algorithm to solve the VRPSDPC. As previously mentioned, the VRPSPC reduces to a deterministic VRP. Therefore, we will not present the results of Laporte, Louveaux, and Mercure [43], because state-of-art algorithms for the VRP are likely to outperform the reported results. Furthermore, given that this model can be solved as a deterministic problem, the VRPSDPC is thus much easier to tackle than the previously presented VRPSD models.

It should be noted that Laporte, Louveaux, and Mercure [43] present another model that minimizes the sum of the depot operating costs, vehicle fixed costs, planned routing costs, and expected failure costs. Furthermore, they consider two operational modes with respect to demand revelation and failures. The first resembles the one described in Section 8.2; i.e., customer demand is only revealed upon arriving at the customer location and failures occur whenever vehicle capacity is exceeded. The second operating mode considers that a priori information with respect to customer demand is available. Since it is very often desired that drivers take the same route every day on a routine basis, the authors assume that the routes determined in the first stage will remain unchanged. However, this operating mode differs from the previous one, in that a vehicle never proceeds directly from one customer to the next if it is known in advance that its capacity will be exceeded.

The VRPSDPC was also studied by Golden and Yee [26], who introduced probabilistic constraints considering several demand distributions. The authors also study the case of correlated demands. Stewart and Golden [62] present a three-index formulation to the VRPSDPC, where the probabilistic constraints are non-linear. The authors then show that the probabilistic constraints may be linearized for several distributions, e.g., Poisson, binomial, negative binomial, and gamma. The authors propose two heuristic algorithms to solve the problem. The first is based on the Clarke and Wright algorithm, while the second makes use of Lagrange multipliers.

Bastian and Rinnooy Kan [3] further studied the VRPSDPC for the single-vehicle case and showed that the problem is equivalent to the time-dependent TSP.

8.5 ▪ Stochastic Demands

The VRPSD is by far the most studied variant of SVRP. The first study on this subject is attributed to Tillman [67], who addressed the problem in the multi-depot context. The author proposed a solution procedure based on an extension of the Clarke and Wright savings heuristic [14]. Dror and Trudeau [18] furthered the development of heuristics based on the same algorithm and studied the impact of route failures on the average routing cost. The authors were the first to show that the average cost of a route is dependent on its orientation. One of the first dissertations dealing with the VRPSD is that of Bertsimas [6], which derives a series of bounds, asymptotic results, and theoretical properties for problems where demands are unit random variables (i.e., for customer i, $\xi_i = 1$ with probability p_i and $\xi_i = 0$ with probability $1 - p_i$). This was followed by Dror, Laporte, and Trudeau [17], who derived a general discussion on possible formulations for the VRPSD and studied the properties of these models. Noteworthy theoretical results included the extension of properties that were originally established in the context of the probabilistic TSP by Jaillet [34, 35] and Jaillet and Odoni [36]. Specifically, in an optimal solution of the VRPSD,

1. an optimal route may intersect itself;

2. considering a Euclidean problem, the visits defined by the optimal routes do not necessarily reflect the order in which the customers appear on the convex hull of vertices;

3. if considered separately, the segments of an optimal route are not necessarily optimal.

The first two results provide insights in the structure of optimal solutions of the VRPSD. As for the third result, its impact is mainly on the design of solution procedures.

Following these initial studies, as presented in Sections 8.2–8.4, several papers, providing various optimization models and solution approaches, have been proposed for the VRPSD. In this section, we complete the presentation of these methods by focusing on the studies that consider either alternative definitions of the recourse decisions or problem characteristics other than the ones previously presented (i.e., added problem dimensions or extra constraints).

Regarding the recourse definition, we have already seen that the manner in which both the routing and replenishment decisions are made leads to different formulations for the VRPSD. Recall that the a priori paradigm assumes that both decisions are determined before any information regarding demands is available (i.e., routes are fixed and the replenishment decisions follow the simple recourse strategy), while the reoptimization paradigm allows both decisions to be made dynamically as demands are observed.

Yang, Mathur, and Ballou [71] defined an in-between approach where routes are fixed beforehand and replenishment decisions are made dynamically. As a result, preventive restocking can be performed in order to reduce the average routing cost. Under these principles, the authors showed that, for a fixed sequence of visits to customers, the optimal replenishment decisions can be formulated as a threshold-based policy. Following such a policy, a vehicle, performing the considered sequence, makes a preventive return to the depot if its residual capacity when leaving a particular customer is below the associated customer's threshold. Two heuristics were presented for the single- and multi-vehicle cases. These procedures were tested on instances that included from 10 to 60 customers with different route length constraints.

An extension to the simple recourse strategy was proposed by Ak and Erera [1], where fixed routes are paired a priori to serve their assigned customers. Following this strategy, which the authors refer to as the *paired locally coordinated operating scheme*, routes are fixed and then paired before observing the informational flow (i.e., demands become known). In each pair, routes are assigned a tour type (i.e., one to Type I and the other to Type II). The vehicle performing the route assigned to Type I visits the customers according to the fixed sequence and simply returns to the depot if a failure occurs (thus finishing the tour). All unvisited customers are appended to the end of the Type II route. The other vehicle, performing the route assigned to Type II, serves all customers assigned to it in the fixed sequence plus the unvisited customers of the Type I route. In this case, failures are handled by applying the simple recourse strategy. The rationale behind this approach is that, at the expense of added coordination between vehicles, the total average cost of a solution can be reduced by pooling the capacity of vehicles in pairs. A tabu search procedure was proposed to solve the VRPSD under this recourse strategy. Instances sized from 20 to 150 customers were solved, and the results showed improvements in terms of the expected cost of routes operated using the proposed recourse strategy compared to the simple one. These improvements ranged from approximately 1% to 17% according to the size of the instances and the probability distributions chosen for the demands.

This type of paired-recourse strategy was applied to the case of the VRPSD and split deliveries by Lei, Laporte, and Guo [48]. Using the a priori paradigm, the authors proposed a two-stage model where routes are constructed and paired in the first stage and then used in the second stage to serve customers while allowing demands to be split among paired routes. A large neighborhood search heuristic was developed to solve the problem.

Finally, multi-stage formulations have been proposed for the VRPSD by Hvattum, Løkketangen, and Laporte [30, 31]. The authors considered a problem where customer demands (requests) appear over a given time period. Routes are therefore constructed dynamically by integrating the requests in the tours at predefined stages. In this case, the recourse decisions are defined in multiple stages and correspond to the adjustments made to the routes as additional demands are observed. The authors proposed two heuristic strategies: a sample-based algorithm that applies scenario decomposition to the model [30] and a Branch-and-Regret procedure [31].

In addition to providing alternative recourse formulations, recent studies have also focused on extending VRPSD formulations to include additional problem characteristics. Tatarakis and Minis [66] proposed two dynamic programming models to compute the expected cost of a fixed route for the multiple product VRPSD. In this case, customers have a distinct stochastic demand associated with each product. Therefore, when operating a fixed route, a failure occurs when at least one of the product demands for a given customer cannot be serviced by the vehicle. Two variants of this problem were solved: the compartmentalized load case (i.e., each product is loaded into a distinct compartment on a vehicle) and the unified load case (i.e., all products are loaded in a single compartment on a vehicle). The authors showed that, for one vehicle and two products, the optimal replenishment decisions for a fixed route take the form of a threshold-based policy similar to the one proposed by Yang, Mathur, and Ballou [71]. This result was then generalized by Pandelis, Kyriakidis, and Dimitrakos [55] to the case of an arbitrary number of products. Considering this problem, Mendoza et al. [53] proposed a formulation based on the a priori paradigm for the compartmentalized load case. The authors developed a memetic algorithm and solved instances with up to 484 customers.

Given the presence of route failures in the context of the VRPSD, when a route is operated, its actual duration tends to be longer than planned. Therefore, time requirements, related either to the service of customers or to the workload performed by the vehicles, can become important issues. Several studies have consequently focused on introducing time constraints in VRPSD formulations.

Erera, Savelsbergh, and Uyar [20] addressed the case where both route duration requirements, imposed by driver work rules, and hard time-window constraints are considered when solving the VRPSD. The authors proposed a two-stage model based on the a priori paradigm. In the first stage, a set of fixed routes are created for a subset of customers deemed regular (i.e., ruling out customers that have a low probability of placing a request). In the second stage, given the set of planned routes and the observed demands, a set of operational routes are created to enforce the constraints. As described in [20], routes are constructed in the second stage so as to retain, as much as possible, the characteristics defined by the planned routes (i.e., the assignment and the sequence of customers in the routes).

Erera, Morales, and Savelsbergh [19] showed that, when constructing a priori routes for the VRPSD, imposing tour duration constraints that ensure feasibility for all demand realizations can be done efficiently by solving an adversarial optimization problem. The authors also investigated the impact that such constraints have on the size of the fleet necessary to obtain feasible solutions for the problem.

Lei, Laporte, and Guo [47] proposed an a priori formulation for the VRPSD in the presence of soft time-window constraints. Whenever a vehicle fails to meet a customer's time window, an additional cost equivalent to a route failure is assumed (thus yielding the case where the customer is served by a separate direct trip from the depot). Under this assumption, an efficient method for computing the total expected cost of a fixed route was developed and embedded within an adaptive large neighborhood heuristic. The proposed algorithm was shown to be efficient on a set of modified Solomon instances.

Finally, Goodson, Ohlmann, and Thomas [28] considered route duration limits in the context of the VRPSD. The problem was formulated using the reoptimization strategy. The authors developed a series of rollout heuristics to solve the model and presented numerical results showing the relative superiority of the proposed methods over both a rolling horizon solution strategy and fixed-route approaches.

8.6 ▪ Stochastic Customers

The Vehicle Routing Problem with Stochastic Customers (VRPSC) considers that customers are present with a given probability. Traditionally, the VRPSC is modeled as a two-stage stochastic programming problem. The first stage consists of determining the routes that adhere to the VRP constraints. Given the realization of present and absent customers, the second stage solution is to follow up the routes set by the first stage, while skipping the absent customers. Unless mentioned otherwise the problem assumes deterministic demand.

We will first present some of the relevant literature on the Traveling Salesman Problem with Stochastic Customers (TSPSC), a special case of the VRPSC. This problem was introduced by Jaillet [34], who derived several models, bounds, and properties. Rossi and Gavioli [57] and Jézéquel [37] adapted the Clarke and Wright heuristic [14] to the TSPSC. Further heuristic algorithms were proposed by Bertsimas [6] and Bertsimas and Howell [8]. An exact Branch-and-Cut algorithm for the problem was developed by Laporte, Louveaux, and Mercure [45].

In the VRPSC, customer demand is often considered to be of unit size. This setting was studied by Jézéquel [37] and by Jaillet and Odoni [36]. The latter found that, even for symmetric distances, the cost of a solution depends on the orientation of travel. Jaillet and Odoni [36] also stated that large vehicle capacities may yield large solution costs. Bertsimas [6] describes several properties, bounds, and heuristics for the problem. Waters [70] studied the problem with general integer demands and considered three operating policies. The first follows the planned routes, regardless of the presence or absence of customers. The second follows the planned route while skipping absent customers, and the third reoptimizes the route whenever the absence of a customer is observed.

The Vehicle Routing Problem with Stochastic Customers and Demands (VRPSCD) combines stochastic customers and stochastic demands. The VRPSCD was mentioned by Jézéquel [37] and Jaillet and Odoni [36], but it was formalized by Bertsimas [7]. As for the VRPSC, the VRPSCD is formulated as a two-stage stochastic programming model. The first stage designs routes that visit all customers, the set of present customers is revealed before routes are executed, and customer demands are revealed upon the arrival of the vehicle. The routes are followed while skipping the absent customers, and the vehicle returns to the depot to replenish when its capacity is exhausted. The computation of the objective function has been shown to be difficult. Séguin [61] and Gendreau, Laporte, and Séguin [23] proposed the first exact algorithm for the problem. They computed efficient bounds on the first-stage travel times and solved the problem with the integer L-shaped algorithm. Gendreau, Laporte, and Séguin [24] later developed a tabu search algorithm for the problem, exploiting an approximation that facilitated the computation of the objective function. Finally, Benton and Rossetti [5] considered a version of the VRPSCD in which route reoptimization is allowed when demands are revealed.

Sungur et al. [63] considered the Courier Delivery Problem with uncertainty on the presence of customers and service times. Customers have soft time windows while a hard constraint is considered on the route duration. Uncertainty is represented by scenarios. The objective is to produce a master plan that maximizes the number of served customers, minimizes the total time spent by couriers and penalties for early or late arrival at customers, and maximizes a measure of consistency of the routes for each scenario. The authors proposed a two-phase heuristic where in the first phase an initial master plan is obtained by an insertion heuristic. Recourse subproblems are then solved by an insertion heuristic for each scenario where absent customers are skipped. In the second phase, an iterative procedure provides feedback from the scenarios insertion algorithm to the master insertion algorithm regarding unserved customers. The master plan heuristic gives priority to unserved customers, and new subproblems for each scenario are solved. Tabu search is used to improve the solutions of the insertion heuristics of the scenario subproblems.

8.7 • Stochastic Travel Times

In the classical VRP, travel times are assumed to be deterministic and are usually assumed to be linearly correlated with the distance. Therefore, travel time cost is often referred to as distance. However, in many real-life applications a sizeable degree of variability in travel times is observed, and thus static, deterministic travel times do not represent an accurate approximation of actual travel time. Malandraki and Daskin [52] classified potential causes of variability in travel times into two components. The first component covers temporal variations resulting from hourly, daily, weekly, or seasonal cycles in the average traffic volumes. Because these variations are of a repetitive nature, they can be modeled deterministically by incorporating time-dependent travel times; see, e.g., Malandraki and Daskin [52]. The second component of variations is due to accidents, weather conditions, or other random events; these are the focus of this section. Furthermore, variability may be observed in service times. This variability is fundamental since it influences vertex departures and consequently the resulting travel times. In the remainder of this section, we present an overview of the literature on VRPs with stochastic travel or service times.

The VRP with stochastic travel and service times was first introduced by Laporte, Louveaux, and Mercure [44]. These authors considered the a priori framework, whereby routes are designed before random travel and service times are revealed. After the realization of the random travel and service times, the vehicles follow their a priori routes. No capacity constraints were considered, but vehicles incur a penalty if the route duration exceeds a given deadline. The penalty is proportional to the elapsed route duration in excess of the deadline. The authors presented a chance constraint model and two recourse models. The chance constraint model follows principles similar to those presented in Section 8.4. The objective is to minimize the routing costs, while ensuring that the probability of exceeding the route duration deadline does not exceed β. This feasibility requirement is incorporated in the subtour elimination constraints. Furthermore, the authors propose a three-index simple recourse model and a two-index recourse model. The first-stage decisions of the recourse models consist of determining the number of vehicles and their routes. In the second stage, random variables corresponding to stochastic travel and service times are realized and penalties are incurred for excess duration. A general Branch-and-Cut algorithm was proposed for all three models. Test instances with up to 20 customers and two to five travel time scenarios were solved.

Kenyon and Morton [38] later studied the VRP with uncertainty in travel and service times. The authors proposed two a priori models and, as in Laporte, Louveaux, and Mercure [44], they considered uncapacitated vehicles. The two models correspond to two different objectives. The first minimizes the maximum expected completion time of all routes, while the second maximizes the probability that the operation is completed no later than a pre-specified target time. A three-index formulation is proposed for both models. The authors showed that the solution of the first model with a single vehicle (i.e., the TSP with stochastic travel time) is equivalent to solving the TSP in which the travel and service times are replaced by their means. Considering the second model with one vehicle, the authors developed a methodology based on solving a deterministic non-linear integer program whose continuous relaxation is a convex program. The solution method embeds a Branch-and-Cut scheme within a Monte Carlo sampling-based procedure for the multi-vehicle case. The performance of the algorithm was validated on a 28-vertex instance.

Lambert, Laporte, and Louveaux [39] studied the problem of designing vehicle routes to collect deposits from bank branches and deliver them to a central office. Considering route duration constraints and uncapacitated vehicles, the objective of the problem is to minimize vehicle fixed costs, vehicle costs, and penalty cost for lost interest. The authors considered that congestion may occur with some probability on some arcs and worked with two travel time scenarios for subsets of the arcs. The problem was cast into an a priori two-index model and solved by adapting the Clarke and Wright heuristic. The algorithm was tested on two data sets of instances with 29 and 44 vertices. These data were derived from a practical application.

Several heuristics were developed for the VRP with stochastic travel time or service time. Lei, Laporte, and Guo [49] considered capacity constraints, stochastic service times, and a maximum route duration limit. The objective is to minimize the sum of travel costs, expected service cost, and expected recourse cost. The authors provide a closed form expression for the expected cost of a given route in case of stochastic service times and propose a generalized variable neighborhood search heuristic to solve the problem. The latter heuristic was shown to outperform both a variable neighborhood descent and a variable neighborhood search heuristic. Jabali et al. [33] studied the capacitated VRP with time-dependent travel times, route duration limits, and stochastic service times. In this case, the stochastic service times were modeled as unexpected delays at customer locations by considering a single disruption per route. A tabu search algorithm was adapted to the problem and tested on instances from Augerat et al. [2].

The VRP with soft time windows and stochastic travel times was studied by a number of authors. The uncapacitated version of the problem was considered by Wang and Regan [69]. Li, Tian, and Leung [50] addressed a stochastic capacitated VRP with soft time windows and stochastic service and travel times. All random variables were assumed to be independent and normally distributed. They considered a chance constraint for each time window and another for the route duration. The authors also proposed a two-stage program with recourse, where a penalty is incurred in the case of time-window or route duration violations. They adapted a tabu search–based heuristic for the solution of both models, while using Monte Carlo simulation to evaluate the random events. The VRP with soft time windows and stochastic travel times was also studied by Taş et al. [64]. These authors considered the objective of minimizing customer inconvenience, expressed as expected earliness and lateness at customers locations, and operational costs, expressed as expected driver overtime and vehicle and travel costs. Travel times were assumed to follow a gamma distribution. The authors proposed a three-phase tabu search–based algorithm. In the first phase, a solution is constructed and then improved with respect to the total operating cost, thus yielding an initial feasible solution. This solution is improved by tabu search in the second phase. In the third phase, a post-optimization procedure, accounting for customer inconvenience, is applied to further improve the solution obtained by the tabu search algorithm. In a more recent paper, Taş et al. [65] applied a column generation procedure to optimally solve the problems proposed in Taş et al. [64]. The master problem is modeled as a classical set partitioning problem. The pricing subproblem, for each vehicle, corresponds to an elementary shortest path problem with resource constraints. The column generation procedure is embedded within a Branch-and-Price algorithm.

Errico et al. [21] were the first to propose a formulation for the VRP with stochastic service times and hard time windows, which is considerably more difficult than the case with soft time windows. The difficulty arises from the fact that, given a route, the probability distributions of arrival times at customers locations have to be truncated because of the hard time windows, thus prohibiting the use of convolution properties when summing the random variables. The problem was cast into a chance constraint framework that considers a minimum success probability of the set of vehicle routes. The model is based on the set partitioning approach for the VRP and relies on a transformation of the minimum success chance constraint into a constraint that allocates a probabilistic resource among routes. A Branch-and-Cut-and-Price algorithm was developed. In the column (route) generation subproblem, one accounts for the consumption of probabilistic resource by extending the label dimension and by providing specialized dominance rules.

8.8 ▪ Conclusions and Future Research Directions

In the VRP, stochasticity is observed in customer demands, customer presence, and travel or service times. Formulating this stochasticity may follow several paradigms. Three main modeling paradigms emerge from the literature: a priori optimization, reoptimization, and probabilistic modeling. The choice of a paradigm depends on the informational flow and on the operational policies that govern the application.

The a priori optimization paradigm follows a two-stage model, in which an a priori solution is determined in the first stage. This solution is executed in the second stage while uncertainties are revealed and recourse actions are taken according to a chosen policy. The choice of the recourse policy is paramount in determining the expected total cost of the solutions. These models constitute fundamental building blocks in the SVRP literature. Their results usually serve as benchmarks to more elaborate approaches.

The reoptimization paradigm assumes that routing decisions are reoptimized as stochastic events unfold. In this context, dynamic models are likely to outperform a priori optimization. This superiority stems from the fact that the solution space considered by a priori models is contained in the solution space considered by dynamic models. The sizeable solution space of the latter is precisely the reason why studying dynamic models in absolute terms is a challenging task.

The probabilistic paradigm is steered by the notion that solutions should guarantee a certain level of service. In the context of SVRP, quality of service is measured through the probability of route failure. Therefore, probabilistic models guarantee a certain level of protection against undesired outcomes.

As detailed in Sections 8.5 to 8.7, most of the research on SVRPs has focused on the existence of a single stochastic parameter. Realistic extensions could focus on simultaneously considering a number of stochastic parameters. The interactions between these parameters will yield challenging models. Furthermore, most models consider independent random variables, yet in applications random events are often correlated; e.g., bad weather or road accidents may impact the travel times on a number of arcs at the same time. Finally, many important variants of the VRP have not yet been addressed in a stochastic context.

Bibliography

[1] A. AK AND A. L. ERERA, *A paired-vehicle recourse strategy for the vehicle-routing problem with stochastic demands*, Transportation Science, 41 (2007), pp. 222–237.

[2] P. AUGERAT, J. M. BELENGUER, E. BENAVENT, Á. CORBÉRAN, AND D. NADDEF, *Separating capacity constraints in the cvrp using tabu search*, European Journal of Operational Research, 106 (1998), pp. 546–557.

[3] C. BASTIAN AND A. H. G. RINNOOY KAN, *The stochastic vehicle routing problem revisited*, European Journal of Operational Research, 56 (1992), pp. 407–412.

[4] J. F. BENDERS, *Partitioning procedures for solving mixed-variables programming problems*, Numerische Mathematik, 4 (1962), pp. 238–252.

[5] W. C. BENTON AND M. D. ROSSETTI, *The vehicle scheduling problem with intermittent customer demands*, Computers & Operations Research, 19 (1992), pp. 521–531.

[6] D. J. BERTSIMAS, *Probabilistic combinatorial optimization problems*, PhD thesis, Operations Research Center, Massachusetts Institute of Technology, Cambridge, MA, 1988.

[7] ——, *A vehicle routing problem with stochastic demand*, Operations Research, 40 (1992), pp. 574–585.

[8] D. J. BERTSIMAS AND L. H. HOWELL, *Further results on tile probabilistic traveling salesman problem*, European Journal of Operational Research, 65 (1993), pp. 68–95.

[9] D. J. BERTSIMAS, P. JAILLET, AND A. R. ODONI, *A priori optimization*, Operations Research, 38 (1999), pp. 1019–1033.

[10] J. R. BIRGE AND F. V. LOUVEAUX, *Introduction to Stochastic Programming*, Springer Series in Operations Research and Financial Engineering, Springer, New-York, second ed., 2011.

[11] K. CHEPURI AND T. HOMEM DE MELLO, *Solving the vehicle routing problem with stochastic demands using the cross entropy method*, Annals of Operations Research, 134 (2005), pp. 153–181.

[12] C. H. CHRISTIANSEN AND J. LYSGAARD, *A branch-and-price algorithm for the capacitated vehicle routing problem with stochastic demands*, Operations Research Letters, 35 (2007), pp. 773–781.

[13] N. CHRISTOFIDES AND S. EILON, *An algorithm for the vehicle-dispatching problem*, Journal of the Operational Research Society, 20 (1969), pp. 309–318.

[14] G. CLARKE AND J. W. WRIGHT, *Scheduling of vehicles from a central depot to a number of delivery points*, Operations Research, 12 (1964), pp. 568–581.

[15] J.-F. CORDEAU, G. LAPORTE, M. SAVELSBERGH, AND D. VIGO, *Vehicle routing*, in Transportation, C. Barnhart and G. Laporte, eds., vol. 14 of Handbooks in Operations Research & Management Science, North-Holland, Amsterdam, 2007, ch. 6, pp. 367–428.

[16] G. B. DANTZIG AND R. RAMSER, *The truck dispatching problem*, Management Science, 6 (1959), pp. 80–91.

[17] M. DROR, G. LAPORTE, AND P. TRUDEAU, *Vehicle routing with stochastic demands: Properties and solution frameworks*, Transportation Science, 23 (1989), pp. 166–176.

[18] M. DROR AND P. TRUDEAU, *Stochastic vehicle routing with modified savings algorithm*, European Journal of Operational Research, 23 (1986), pp. 228–235.

[19] A. L. ERERA, J. C. MORALES, AND M. W. P. SAVELSBERGH, *The vehicle routing problem with stochastic demand and duration constraints*, Transportation Science, 44 (2010), pp. 474–492.

[20] A. L. ERERA, M. W. P. SAVELSBERGH, AND E. UYAR, *Fixed routes with backup vehicles for stochastic vehicle routing problems with time constraints*, Networks, 54 (2009), pp. 270–282.

[21] F. ERRICO, G. DESAULNIERS, M. GENDREAU, W. REI, AND L.-M. ROUSSEAU, *The vehicle routing problem with hard time windows and stochastic service times*, Cahier du GERAD, G-2013-45, 2013.

[22] C. GAUVIN, G. DESAULNIERS, AND M. GENDREAU, *A branch-cut-and-price algorithm for the vehicle routing problem with stochastic demands*, Computers & Operations Reserarch, 50 (2014), pp. 141–153.

[23] M. GENDREAU, G. LAPORTE, AND R. SÉGUIN, *An exact algorithm for the vehicle routing problem with stochastic demands and customers*, Transportation Science, 29 (1995), pp. 143–155.

[24] ——, *A tabu search heuristic for the vehicle routing problem with stochastic demands and customers*, Operations Research, 44 (1996), pp. 469–477.

[25] ——, *Stochastic vehicle routing*, European Journal of Operational Research, 88 (1996), pp. 3–12.

[26] B. L. GOLDEN AND J. R. YEE, *A framework for probabilistic vehicle routing*, AIIE Transactions, 11 (1979), pp. 109–112.

[27] J. C. GOODSON, J. W. OHLMANN, AND B. W. THOMAS, *Cyclic-order neighborhoods with application to the vehicle routing problem with stochastic demand*, European Journal of Operational Research, 217 (2012), pp. 312–323.

[28] ——, *Rollout policies for the dynamic solutions to the multivehicle routing problem with stochastic demand and duration limits*, Operations Research, 61 (2013), pp. 138–154.

[29] C. HJORRING AND J. HOLT, *New optimality cuts for a single-vehicle stochastic routing problem*, Annals of Operations Research, 86 (1999), pp. 569–584.

[30] L. M. HVATTUM, A. LØKKETANGEN, AND G. LAPORTE, *Solving a dynamic and stochastic vehicle routing problem with a sample scenario hedging heuristic*, Transportation Science, 40 (2006), pp. 421–438.

[31] ——, *A branch-and-regret heuristic for stochastic and dynamic vehicle routing problems*, Networks, 49 (2007), pp. 330–340.

[32] O. JABALI, W. REI, M. GENDREAU, AND G. LAPORTE, *Partial-route inequalities for the multi-vehicle routing problem with stochastic demands*, Discrete Applied Mathematics, 177 (2014), pp. 121–136.

[33] O. JABALI, T. VAN WOENSEL, A. G. DE KOK, C. LECLUYSE, AND H. PEREMANS, *Time-dependent vehicle routing subject to time delay perturbations*, IIE Transactions, 41 (2009), pp. 1049–1066.

[34] P. JAILLET, *Probabilistic traveling salesman problem*, PhD thesis, Operations Research Center, Massachusetts Institute of Technology, Cambridge, MA, 1985.

[35] ——, *A priori solution of a traveling salesman problem in which a random subset of the customers are visited*, Operations Research, 36 (1988), pp. 929–936.

[36] P. JAILLET AND A. R. ODONI, *The probabilistic vehicle routing problem*, in Vehicle Routing: Methods and Studies, B. L. Golden and A. A. Assad, eds., North–Holland, Amsterdam, 1988, pp. 293–318.

[37] A. JÉZÉQUEL, *Probabilistic vehicle routing problems*, M.Sc. Dissertation, Department of Civil Engineering, Massachusetts Institute of Technology, Cambridge, MA, 1985.

[38] A. S. KENYON AND D. P. MORTON, *Stochastic vehicle routing with random travel times*, Transportation Science, 37 (2003), pp. 69–82.

[39] V. LAMBERT, G. LAPORTE, AND F. V. LOUVEAUX, *Designing collection routes through bank branches*, Computers & Operations Research, 20 (1993), pp. 783–791.

[40] G. LAPORTE, *Fifty years of vehicle routing*, Transportation Science, 43 (2009), pp. 408–416.

[41] G. LAPORTE AND F. V. LOUVEAUX, *The integer L-shaped method for stochastic integer programs with complete recourse*, Operations Research Letters, 13 (1993), pp. 133–142.

[42] G. LAPORTE, F. V. LOUVEAUX, AND L. VAN HAMME, *An integer L-shaped algorithm for the capacitated vehicle routing problem with stochastic demands*, Operations Research, 50 (2002), pp. 415–423.

[43] G. LAPORTE, F. V. LOUVEAUX, AND H. MERCURE, *Models and exact solutions for a class of stochastic location-routing problems*, European Journal of Operational Research, 39 (1989), pp. 71–78.

[44] ———, *The vehicle routing problem with stochastic travel times*, Transportation Science, 26 (1992), pp. 161–170.

[45] ———, *A priori optimization of the probabilistic traveling salesman problem*, Operations Research, 42 (1994), pp. 543–549.

[46] G. LAPORTE, Y. NOBERT, AND M. DESROCHERS, *Optimal routing under capacity and distance restrictions*, Operations Research, 33 (1985), pp. 1050–1073.

[47] H. LEI, G. LAPORTE, AND B. GUO, *The capacitated vehicle routing problem with stochastic demands and time windows*, Computers & Operations Research, 38 (2011), pp. 1775–1783.

[48] ———, *The vehicle routing problem with stochastic demands and split deliveries*, INFOR, 50 (2012), pp. 59–71.

[49] ———, *A generalized variable neighborhood search heuristic for the capacitated vehicle routing problem with stochastic service times*, TOP, 20 (2012), pp. 99–118.

[50] X. LI, P. TIAN, AND S. C. H. LEUNG, *Vehicle routing problems with time windows and stochastic travel and service times: Models and algorithm*, International Journal of Production Economics, 125 (2010), pp. 137–145.

[51] F. V. LOUVEAUX, *An introduction to stochastic transportation models*, in Operations Research and Decision Aid Methodologies in Traffic and Transportation Management, M. Labbé, G. Laporte, K. Tanczos, and P. Toint, eds., vol. 166 of Computer and Systems Sciences, Springer-Verlag, Berlin, 1998, pp. 244–263.

[52] C. MALANDRAKI AND M. S. DASKIN, *Time dependent vehicle routing problems: Formulations, properties and heuristic algorithms*, Transportation Science, 26 (1992), pp. 185–200.

[53] J. E. MENDOZA, B. CASTANIER, C. GUÉRET, A. L. MEDAGLIA, AND N. VELASCO, *A memetic algorithm for the multi-compartment vehicle routing problem with stochastic demands*, Computers & Operations Research, 37 (2010), pp. 1886–1898.

[54] C. NOVOA AND R. STORER, *An approximate dynamic programming approach for the vehicle routing problem with stochastic demands*, European Journal of Operational Research, 196 (2009), pp. 509–515.

[55] D. G. PANDELIS, E. G. KYRIAKIDIS, AND T. D. DIMITRAKOS, *Single vehicle routing problems with a predefined customer sequence, compartmentalized load and stochastic demands*, European Journal of Operational Research, 217 (2012), pp. 324–332.

[56] H. N. PSARAFTIS, *Dynamic vehicle routing: Status and prospects*, Annals of Operations Research, 61 (1995), pp. 143–164.

[57] F. ROSSI AND I. GAVIOLI, *Aspects of heuristic methods in the "probabilistic traveling salesman problem" (PTSP)*, in Stochastics in Combinatorial Optimization, World Scientific, Singapore, 1987, pp. 214–227.

[58] N. SECOMANDI, *Comparing neuro-dynamic programming algorithms for the vehicle routing problem with stochastic demands*, Computers & Operations Research, 27 (2000), pp. 1201–1225.

[59] ———, *A rollout policy for the vehicle routing problem with stochastic demands*, Operations Research, 49 (2001), pp. 796–802.

[60] N. SECOMANDI AND F. MARGOT, *Reoptimization approaches for the vehicle-routing problem with stochastic demands*, Operations Research, 57 (2009), pp. 214–230.

[61] R. SÉGUIN, *Problèmes stochastiques de tournées de véhicules*, PhD thesis, Département d'informatique et de recherche operationnelle, Université de Montréal, Canada, 1994.

[62] W. R. STEWART AND B. L. GOLDEN, *Stochastic vehicle routing: A comprehensive approach*, European Journal of Operational Research, 14 (1983), pp. 371–385.

[63] I. SUNGUR, Y. REN, F. ORDÓÑEZ, M. DESSOUKY, AND H. ZHONG, *A model and algorithm for the courier delivery problem with uncertainty*, Transportation Science, 44 (2010), pp. 193–205.

[64] D. TAŞ, N. DELLAERT, T. VAN WOENSEL, AND A. G. DE KOK, *Vehicle routing problem with stochastic travel times including soft time windows and service costs*, Computers & Operations Research, 40 (2012), pp. 214–224.

[65] D. TAŞ, M. GENDREAU, N. DELLAERT, T. VAN WOENSEL, AND A. G. DE KOK, *Vehicle routing with soft time windows and stochastic travel times: A column generation and branch-and-price solution approach*, European Journal of Operational Research, (2013).

[66] A. TATARAKIS AND I. MINIS, *Stochastic single vehicle routing with a predefined customer sequence an multiple depot returns*, European Journal of Operational Research, 197 (2009), pp. 557–571.

[67] F. A. TILLMAN, *The multiple terminal delivery problem with probabilistic demands*, Transportation Science, 3 (1969), pp. 192–204.

[68] R. M. VAN SLYKE AND R. WETS, *L-shaped linear programs with applications to optimal control and stochastic programming*, SIAM Journal on Applied Mathematics, 17 (1969), pp. 638–663.

[69] X. WANG AND A. C. REGAN, *Assignment models for local truckload trucking problems with stochastic service times and time window constraints*, Transportation Research Record: Journal of the Transportation Research Board, 1771 (2001), pp. 61–68.

[70] C. D. J. WATERS, *Vehicle-scheduling problems with uncertainty and omitted customers*, Journal of the Operational Research Society, 40 (1989), pp. 1099–1108.

[71] W. H. YANG, K. MATHUR, AND R. H. BALLOU, *Stochastic vehicle routing with restocking*, Transportation Science, 34 (2000), pp. 99–112.

Chapter 9

Four Variants of the Vehicle Routing Problem

Stefan Irnich
Michael Schneider
Daniele Vigo

9.1 ▪ Introduction

In this chapter, we review four important variants of the *Vehicle Routing Problem* (VRP):

1. *VRP with Backhauls* (VRPB);

2. *Heterogeneous or mixed Fleet VRP* (HFVRP);

3. *Periodic VRP* (PVRP);

4. *Split Delivery VRP* (SDVRP).

Each family plays an important role in the literature and considers a characteristic feature often encountered, both alone and combined, in real-world applications. The resulting VRP variants are non-trivial extensions of the basic VRP, and they all deserve specifically tailored algorithmic approaches for their effective solution.

In the first problem class, the *VRP with Backhauls* (VRPB), the customers are partitioned into two sets, the linehaul and backhaul customers. The linehaul customers require the delivery of goods from the depot, and the backhaul customers have goods to be picked up and transported to the depot. Since a rearrangement of the load inside a vehicle is often impossible or undesirable, e.g., because the vehicle's loading space is only accessible from the rear, a precedence relation on the service of linehaul and backhaul customers is imposed: In a feasible VRPB solution the combined service of linehauls and backhauls in the same route is possible, but all linehaul customers receive their demand before load at backhaul customers is picked up. The VRPB belongs to the more general class of delivery and collection VRP, which clearly introduces considerable improvements over separate distribution and collection services in different routes.

In the second variant, the family of *Heterogeneous or mixed Fleet VRP* (HFVRP), the fleet consists of different vehicles or vehicle types, each with a specific capacity and costs.

The choice of the appropriate vehicle to be assigned to each route can have a considerable impact on the overall cost of the solution both at the operational level, at which the heterogeneous fleet is given, and at the strategic level, at which the optimal fleet composition has to be determined.

The third family we examine is the *Periodic VRP* (PVRP). Here deliveries occur in a multi-day service horizon: each customer has to be assigned to one or several specific days for deliveries, and routing must be performed for each day. Hence, the PVRP comprises three types of decisions: the choice of visiting days for each customer, the clustering of customers into tours on each day, and the routing of the vehicle for each cluster.

The last variant treated in this chapter is the *Split Delivery VRP* (SDVRP), in which customers may be served more than once by dividing the delivery quantity among the different visits. Hence, the solution of SDVRP requires the determination of the number of visits to each customer, the portion of the demand to deliver to the customer, and the routing of the vehicles. Of course, the three above types of decisions are intertwined, and the practical hardness of the SDVRP can be attributed to this interdependency.

Both heterogeneous and backhaul variants were introduced early in the VRP literature. However, despite their practical relevance, they received relatively little attention from the research community in the last decade. Thus, besides from surveying the relatively few recent papers on these two families, our goal is to revive the attention and possibly stimulate additional research on them. On the contrary, the PVRP and SDVRP variants were the subject of intense research efforts in recent years.

Therefore, the material presented in this chapter is structured as follows: Section 9.2 provides a survey on the VRPB referring to a longer time period starting approximately in 2002. Due to good and very recent surveys available for HFVRP, PVRP, and SDVRP, Sections 9.3, 9.4, and 9.5 only provide an update on these variants by discussing the new contributions that were presented afterwards.

In order to make the four sections self-contained, we decided to re-introduce names and acronyms in each section.

9.2 ▪ VRP with Backhauls

The *VRP with Backhauls* (VRPB), also known as the linehaul-backhaul problem, is an extension of the *Capacitated VRP* (CVRP) in which the customer set is partitioned into two subsets. The first one contains the *linehaul customers*, each requiring a given quantity of product to be delivered. The second subset contains the *backhaul customers*, where a given quantity must be picked up. This customer partition is very frequent in practical situations, such as in the grocery industry, where supermarkets and shops are the linehaul customers, and grocery suppliers are the backhaul customers. In this mixed distribution and collection context a significant saving in terms of transportation costs can be achieved by visiting backhaul customers in distribution routes.

More precisely, the VRPB can be stated as the problem of determining a set of vehicle routes visiting all customers such that (i) each vehicle performs one route; (ii) each route starts and ends at the depot; (iii) for each route the total load associated with linehaul and backhaul customers does not exceed, separately, the vehicle capacity; (iv) on each route the backhaul customers, if any, are visited after all linehaul customers; (v) the total distance traveled by the vehicles is minimized. The *precedence constraint* (iv) is practically motivated by the fact that often vehicles are rear-loaded. Hence, the on-board load rearrangement required by a mix of collection and distribution is difficult, or impossible, to carry out at customer locations. Another practical reason is that, in many applications, linehaul customers have a higher service priority than backhaul customers. Note that the mixed vehicle routes, which visit both linehaul and backhaul customers, are implicitly oriented due to precedence constraint (iv).

In general, VRPB denotes the symmetric version of the problem, in which traveling costs between each pair of customers are equal in both directions, while the case in which such costs may be different is indicated as the *Asymmetric VRPB* (AVRPB). Both VRPB and AVRPB are NP-hard in the strong sense, as they generalize the symmetric and asymmetric CVRP, respectively, arising when no backhauls are present. Toth and Vigo [141] present an extensive survey of the literature on VRPB until the end of the last century. Thus, we mainly discuss the more recent contributions. Computational testing for these problems is generally performed on a set of randomly generated, symmetric instances proposed by Goetschalckx and Jacobs-Blecha [69] (called the GJ set) and on symmetric and asymmetric instances derived from CVRP and ACVRP benchmark sets by Toth and Vigo [138, 139, 140], called the TVS and TVA sets.

Exact Methods   Few exact approaches have been proposed for VRPB and, to the best of our knowledge, none during the last decade. These are all surveyed in [141], but since they still represent the state of the art for the optimal solution of this problem variant we briefly recall their main characteristics. The first exact method for both VRPB and AVRPB is a Branch-and-Bound introduced in [139] based on an effective Lagrangian bound for the problem. A valid model for VRPB is obtained by transforming it into a directed CVRP in which all the arcs connecting backhaul to linehaul customers are removed as a consequence of precedence constraint (iv) above: to this end any formulation for directed CVRP, such as model VRP1 presented in Chapter 1, can be used. The Lagrangian bound in [139] extends methods previously proposed for the CVRP (see Section 2.2.3 and [141]) and is derived from combinatorial relaxations based on projection and leading to the solution of directed trees over the linehaul and the backhaul customers set separately. The resulting Branch-and-Bound algorithm is able to solve VRPB test instances from the GJ and TVS sets and AVRPB instances from set TVA with up to 70 customers in total. It is worth noting that the Lagrangian bound is also used in [140] as a base for an effective heuristic algorithm for both variants of the problem.

The other existing exact method for VRPB is due to Mingozzi, Giorgi, and Baldacci [96], who developed a set-partitioning-based approach starting in which routes are obtained by connecting paths visiting only either linehaul or backhaul customers. The resulting integer program is possibly solved through a complex procedure making use of dynamic programming and dual approximation of relaxations of the problem. The results of the computational testing show that the approach is capable of solving undirected problems from the GJ and TVS sets with up to 70 vertices.

Heuristics and Metaheuristics   The first heuristic for VRPB is an extension of the savings algorithm proposed by Deif and Bodin [49], followed by a few other early approaches reviewed in Toth and Vigo [141]. More recent approximate algorithms for VRPB adopt metaheuristic paradigms. One of the first is by Osman and Wassan [106], who introduce a *Reactive Tabu Search* (RTS) initialized by two savings-based heuristics. The algorithm uses several data structures to speed up the search and is able to obtain new best solutions for the larger instances of the GJ and TVS sets. The RTS algorithm is later combined with an *Adaptive Memory Programming* (AMP) scheme by Wassan [149], further improving its performance both in terms of solution quality and required time. Another *Tabu Search* (TS) approach is proposed by Brandão [31], who initializes it with two interesting construction algorithms: one is based on the solution of open VRP on the two set of customers, and the other one is based on a K-tree relaxation. The overall TS proves competitive with previous methods finding some new best solutions on the GJ and TVS test instances. Ropke and Pisinger [121] extend their *Adaptive Large Neighborhood Search* (ALNS) framework, which they previously used for several other VRP variants, to the VRPB. The unified heuristic is able to solve six different variants of the VRPB obtaining state-of-the-art results for all of them. A *multi-ant colony system* was proposed by Gajpal and Abad [66] in which a combination of route-ants and vehicle-ants are used to construct solutions respecting the maximum number of vehicles: the resulting algorithm achieves good average performance on the GJ and TVS instances. Another effective method for VRPB is the *route promise algorithm* by Zachariadis and Kiranoudis [154], which is based on an efficient implementation of the *variable length bone exchange* local search.

Vidal et al. [145] present a sophisticated *Hybrid Genetic Search with Adaptive Diversity Control* (HGSADC), which is the current best in class not only for the VRPB but also for a wide class of other VRPs. It yields the best solutions on the GJ test set, but is not tested on TVS and TVA instances. HGSADC uses two subpopulations of feasible and infeasible individuals which are encoded by means of a giant tour representation, to allow for the application of simple crossover operators. A key feature of their method lies in the fitness evaluation, which incorporates the contribution of an individual to the diversity of the population (based on the distance to other individuals in the subpopulation). Tests show that this approach helps to avoid premature convergence and to achieve higher solution quality in shorter time compared to traditional approaches for diversity management. In the so-called education phase, route improvement methods are applied and resulting infeasible solutions are repaired with a given probability. Further details on this heuristic approach can be found in Chapter 4.

A few variants of the VRPB are studied in the literature: The first type of modification is related to the relaxation of the precedence constraint (iv) resulting in problems in which backhaul customers may be served either when sufficient space is available in the vehicle, or even in any position of the tour. After some initial interest on these variants, culminated in the paper by Wade and Salhi [147], who present a specialized insertion algorithm, they have not attracted further research efforts. The only notable exception is the paper by Ropke and Pisinger [121], who call these variants the *Mixed VRPB* (MVRPB). Similarly, it seems that the variant with multiple depots has only been considered in [121] and by Salhi and Nagy [125]. Computational experiments for these variants are generally performed by using the GJ instances disregarding precedences and the new ones proposed by Nagy and Salhi [100].

The *VRPB with time windows* has been extensively studied in the late nineties. Thangiah, Potvin, and Sun [137] and Duhamel, Potvin, and Rousseau [58] consider problems both with and without the strict precedence constraint between linehauls and backhauls, for which test instances are also proposed. More recently, this problem was considered by Zhong and Cole [155], whose *guided local search* approach slightly improves some results obtained with previous metaheuristics. Finally, Ropke and Pisinger [121] use their ALNS algorithm to solve the time-window variant of both VRPB and MVRPB.

Another interesting variant of VRPB is that with heterogeneous fleet, for which Tavakkoli-Moghaddam, Saremi, and Ziaee [136] derive a *memetic algorithm* which is also applied to the problem with homogeneous fleet. Wang and Wang [148] consider the variant of VRPB in which the travel speed of the vehicle is time-dependent. To solve the problem they develop a two-phase approach based on the RTS of Wassan [149]. Finally, Anbuudayasankar et al. [6] describe an interesting real-world problem arising in the replenishment of automated teller machines which turn out to be a bi-objective VRPB, in which both cost and duration of the routes are minimized. For this problem the authors develop a simple saving-based heuristic and a *genetic algorithm* which are tested on real-world and randomly generated instances.

9.3 ▪ Heterogeneous or Mixed Fleet VRP

This section considers the variant of the VRP in which a *heterogeneous fleet* of vehicles, each having a possibly different capacity and cost, is available for the distribution activities. The problem has been known since the early VRP literature, and it was first studied in detail by Golden et al. [70]. Several specific variants have been developed over time to consider the two main "features" of this problem. On the one hand, there is the strategic issue of finding the best assortment of vehicles to be used for the long-term sizing of a fleet. This results in *Fleet Size and Mix* (FSM) problems, in which the vehicle fleet is assumed to be unlimited. On the other hand, one can consider the tactical issue of using the most appropriate vehicles from a limited fleet, which results in the *Heterogeneous VRP* (HVRP). We also use the term *Heterogeneous or mixed Fleet VRP* (HFVRP) to identify the whole problem family.

An extensive survey on HFVRP, covering the main results until 2007, is given by Baldacci, Battarra, and Vigo [19]. Therefore, this section is aimed at providing an update of the literature on the subject. In the following, we mainly concentrate on the problems including only capacity constraints but no additional features, which received more attention in the literature. We also briefly review the approaches proposed to include additional constraints such as time windows and multiple depots. Several case studies and applications which included the presence of a heterogeneous fleet, often in combination with other features, are listed in [19], whereas recent ones are discussed in Tarantilis and Kiranoudis [135], Belfiore and Yoshizaki [25], Oppen, Løkketangen, and Desrosiers [104], and Hertz, Uldry, and Widmer [78]. A decision support system that is capable of solving variants of HVRP with and without backhauls is presented by Tütüncü [143]. We finally mention an interesting application of continuous approximation to fleet sizing described in Jabali, Gendreau, and Laporte [82].

In the HFVRP, we are given a set of n customers, each with a demand q_i, and a fleet of vehicles made up of $|P|$ different vehicle types; i.e., the fleet K is partitioned into subsets of homogeneous vehicles $K = K^1 \cup K^2 \cup \cdots \cup K^{|P|}$. Each vehicle type $p = 1, \ldots, |P|$ has capacity Q^p, and may also have a fixed cost FC^p and a specific traveling cost c_{ij}^p along each arc of the graph modeling the road network.

As already mentioned, many specific variants of the problem were studied in the literature, often known with specific acronyms, depending on the available fleet and the type of considered costs. In particular, the following characteristics vary across different problems:

(i) the vehicle fleet may be either *limited*, i.e., at most $|K^p|$ vehicles of type p may be used, or *unlimited*, i.e., $|K^p| \geq n$ for all $p = 1, \ldots, |P|$;

(ii) the *fixed costs* FC^p of the vehicles may be either considered or ignored, i.e., $FC^p = 0$ for all $p = 1, \ldots, |P|$;

(iii) the routing costs on arcs may be *vehicle-dependent*, i.e., possibly different for each vehicle type, or *vehicle-independent*, i.e., $c_{ij}^{p_1} = c_{ij}^{p_2} = c_{ij}$ for all $p_1 \neq p_2$ and for all $(i, j) \in A$, or *site-dependent* when for each customer i, only a limited subset of vehicle types may be used, i.e., c_{ij}^p are equal to $+\infty$ for some $p = 1, \ldots, |P|$.

Table 9.1. *Heterogeneous VRP variants presented in the literature.*

Acronym	Problem Name	Fleet Size	Fixed Costs	Routing Costs
HVRPFD	Heterogeneous VRP with Fixed Costs and Vehicle-Dependent Routing Costs	*Limited*	*Considered*	*Dependent*
HVRPD	Heterogeneous VRP with Vehicle-Dependent Routing Costs	*Limited*	*Ignored*	*Dependent*
FSMFD	Fleet Size and Mix VRP with Fixed Costs and Vehicle-Dependent Routing Costs	*Unlimited*	*Considered*	*Dependent*
FSMD	Fleet Size and Mix VRP with Vehicle-Dependent Routing Costs	*Unlimited*	*Ignored*	*Dependent*
FSMF	Fleet Size and Mix VRP with Fixed Costs	*Unlimited*	*Considered*	*Independent*

Following the naming convention proposed in Baldacci, Battarra, and Vigo [19], we summarize in Table 9.1 the different problem variants that were actually considered in the literature. The variants with time windows are denoted by adding TW to the end of the acronym of the specific problem, those with multiple depot by adding MD at the beginning of the acronym, and those with backhauls by adding B at the end.

All the problems described above are NP-hard, as they are generalizations of the *Capacitated VRP* (CVRP), which arises when only one vehicle type is present.

Computational testing for HFVRP problems is generally conducted by using a benchmark proposed by Golden et al. [70] that includes a set of 20 test instances for FSM, called the GF set. Instances GF1, GF2, and GF7-12 are adapted from VRP ones with explicit distance matrix and 12 to 30 customers, whereas instances GF3-6 and GF13-20 have a Euclidean distance matrix with 20 to 100 customers. The adaptation of these Euclidean test instances to HVRPD, resulting in set GH, is described in Taillard [132], whereas Li, Golden, and Wasil [88] extend them to HVRPFD and propose five new large instances with 200 to 360 customers, called the LH set. Other authors propose new instances for these problems, and we report them in the following text.

Exact Methods Research on exact approaches for HFVRP has been rather limited in the literature and mostly devoted to FSM problems. The first lower bounds for FSMF were proposed by Golden et al. [70], who describe a relaxation whose optimal solution is found by solving a shortest path on a suitably defined auxiliary graph. Other early approaches are described by Baldacci, Battarra, and Vigo [19]. More recently, the same authors in [20] propose bounds for FSM based on a new two-commodity formulation strengthened with specifically designed cut families. The resulting lower bounds are equivalent in quality to previous ones from the literature, but require considerably less computing time when applied to the twelve Euclidean GF instances.

As in other VRP variants, the most successful exact approaches for the HFVRP are currently those based on set partitioning formulations. Pessoa, Poggi, and Uchoa [110] present a Branch-and-Cut-and-Price algorithm for the FSMF in which the columns are associated with the so-called q-routes (see Christofides, Mingozzi, and Toth [41]), as previously done by Choi and Tcha [39]. The q-routes are (not necessarily simple) circuits covering the depot and a subset of customers, whose total demand is equal to q. The formulation is strengthened by using new families of cuts producing an overall improvement of the bound that allows for the optimal solution of instances from the GF set with up to 75 customers.

Currently the most effective exact approach for the solution of the HFVRP is the Branch-and-Cut-and-Price algorithm by Baldacci and Mingozzi [22]. The algorithm extends approaches proposed by the same authors for the CVRP (see, e.g., Baldacci,

Christofides, and Mingozzi [21]) and uses both new relaxations of the set partitioning formulation and reduction rules to resize the vehicle fleet that are particularly effective when, as happens in most HFVRP variants, the vehicle fixed cost contribution to the total cost is relevant. The resulting exact method permits the solution of most Euclidean test instances from the literature for FSMF, FSMD, HVRPFD, and HVRPD.

Few other variants of heterogeneous VRP are studied with the purpose of developing exact approaches. A notable exception is the MDHVRPFDTW, i.e., the variant of HVRPFD with multiple depots and time windows, for which Bettinelli, Ceselli, and Righini [28] present a Branch-and-Cut-and-Price which is used to solve test instances with up to 100 customers derived from an existing benchmark proposed for the *VRP with Time Windows* (VRPTW).

Finally, Dondo and Cerdá [52] examine the MDHVRPFDTW and develop a three-phase matheuristic. In the first phase, customers are heuristically grouped into a limited number of clusters. In the next phase, these clusters are assigned to vehicles and depots through the exact solution of a small-sized ILP model. Finally, the detailed routing and scheduling of each tour is determined heuristically in the last phase. The method is applied to some small-sized test instances derived from VRPTW instances.

Heuristics and Metaheuristics   Most solution approaches presented so far in the literature are heuristic algorithms. These are often adaptations or extensions of the methods proposed for the main VRP variants, like the CVRP and the VRPTW.

As previously mentioned, the first study of HFVRP is due to Golden et al. [70], who present a construction heuristic based on an adaptation of the *savings algorithm* by Clarke and Wright [43] and the *giant-tour*-based approach of Beasley [23] to FSMF. In this section, we review the most recent contributions to the approximate solution of HFVRP; for other early approaches see Baldacci, Battarra, and Vigo [19].

Li, Tian, and Aneja [90] address the HVRPFD by means of a *Multi-start Adaptive Memory Programming* (MAMP) metaheuristic, which may be seen as an evolution of the heuristic column generation technique of Taillard [132] and Choi and Tcha [39]. A wide set of routes is generated with construction algorithms, followed by *Tabu Search* (TS) and *path relinking*. Then complete solutions are selected by heuristically solving a set covering problem. The MAMP improved the results on seven out of eight GH instances and is also applied to 19 new random instances. The most recent example of such an approach is represented by the method of Subramanian et al. [131]. The algorithm is based on an *iterated local search* with random neighborhood ordering used to generate the large set of routes which are later selected through the set covering. The algorithm is successfully applied to all variants listed in Table 9.1.

Another successful approach are *Memetic Algorithms* (MAs), which combine population-based evolutionary methods and local search and are developed by Prins [117] for FSMF, FSMD, FSMFD, and HVRPD. Each solution in the population is represented as a giant tour visiting all customers that is later optimally split into a feasible solution through a specialized shortest path procedure. Two different mechanisms to favor population diversification are implemented: one based on solution cost difference and the second based on their distance in the solution space. This latter diversification mechanism yields the best results which, for FSMFD, are competitive with the previous one from the literature. As to the remaining variants, the MA results are close to the best ones and generally require less computing effort. Another algorithm of this type is described in Liu, Huang, and Ma [94], who apply it to FSMF and FSMD. The initial population is built by using simple construction algorithms, and specialized crossover and local search operators are implemented. The resulting algorithm generates solutions whose quality is comparable with the best previous methods.

TS approaches, which were initially quite popular for this problem family, are revived by Brandão in [32, 33]. The first paper reports the implementation of a TS for FSMF and FSMD based on standard insertion and swap neighborhoods. To speed up the search, the neighborhoods are appropriately restricted, as they consider only moves in which some of the involved customers are close to each other. The computational results show that the algorithm compares favorably with previous ones from the literature. The TS algorithm is applied in [33] to HVRPD and obtains good results on the GH set, the LH set, and five new random instances.

A *Variable Neighborhood Search* (VNS) algorithm for the solution of all variants with fixed and variable fleet is proposed by Imran, Salhi, and Wassan [80]. They use the giant-tour representation, whose optimal split is again determined through the solution of a shortest path problem on a suitably defined network. The VNS employs six different interchange and shift neighborhoods in the shaking step and six classical customer insertion and arc exchange neighborhoods in the local search step. The computational testing is performed on all test instances from the literature and shows that the algorithm is competitive with previous methods.

The most recent research on heuristics for heterogeneous routing follows the current trend towards hybridization of various frameworks which achieve both a better overall performance and the possibility of handling several variants with the same algorithm. A very good example of such approaches is the method by Duhamel, Lacomme, and Prodhon [57], who propose a combination of a *greedy randomized adaptive search procedure* and an *evolutionary local search*. The resulting algorithm, which is successfully applied to FSMFD, FSMF, FSMD, and HVRPFD, works on a giant-tour representation of a solution. Two implementations of the split procedure are discussed. The testing is conducted on all available instance classes for the considered variants and demonstrates that the method obtains a solution quality comparable to that of existing approaches. The authors also propose a new set of 96 instances based on actual town locations within French counties and real street distances.

The currently most powerful approach for FSM variants is the *Hybrid Genetic Search with Adaptive Diversity Control* (HGSADC) algorithm by Vidal et al. [145], already described in Section 9.2.

Finally, Naji-Azimi and Salari [101] describe an *Integer Linear Programming* (ILP) based procedure which can be used to improve the heuristic solutions found by other methods within relatively short additional computing time.

We conclude this section by examining the heuristics that were developed for other variants of heterogeneous VRP. Most of the early work in this area refers to the inclusion of time windows to FSMF, denoted as FSMFTW, which is first considered in Liu and Shen [92]. In addition to a first algorithm for this problem, they also propose a new large set of 168 instances with TWs, called LS168. Other early approaches for this problem variant are discussed in Baldacci, Battarra, and Vigo [19].

A heuristic column generation algorithm, derived from the exact approach for the multiple depot variant, is applied by Bettinelli, Ceselli, and Righini [28] to FSMFTW. The method is able to improve previous approaches on the LS168 test set. Bräysy et al. [35] introduce a metaheuristic designed to solve large-scale FSMFTW instances. The algorithm constructs an initial solution through a savings approach followed by a local search step. Both steps take into account proximity measures of the involved customers, which avoid considering merging or moving far away customers. The metaheuristic framework is a simplified version of a combined *threshold accepting* and *guided local search* approach previously proposed by Bräysy et al. [34]. Computational testing on the LS168 set, and on a new set of 600 large-scale instances with up to 1000 customers, shows that this fast metaheuristic is capable of obtaining state-of-the-art results in few seconds of CPU time even for the larger instances. Paraskevopoulos et al. [108] discuss a *Reactive Variable Neighborhood TS* (RVNTS) which is tested on both FSMFTW and HVRPFDTW. They use a two-phase approach in which an initial set of high-quality solutions is determined through a parallel construction method, followed by a route elimination procedure to possibly improve the fleet's utilization. In the second phase, the RVNTS is used to further improve routing costs of the selected solutions based on eight classical neighborhoods. Testing for FSMFTW is performed on the LS168 set and shows that the proposed method considerably improves existing solutions. As to HVRPFDTW, the authors again use a benchmark based on LS168, where the best fleet found by Liu and Shen [92] for FSMFTW is imposed as fixed. In this case, the RVNTS also obtains very competitive solutions. The most recent approach for FSMFTW is an *adaptive memory programming* algorithm described by Repoussis and Tarantilis [119]. The algorithm generates a large set of solutions through a semi-parallel construction algorithm followed by a short-term memory TS improvement. Then, the routes are heuristically selected from the adaptive memory and the obtained solutions are improved through an *iterated TS*. Testing on LS168 show the excellent performance of the proposed method.

The MDFSMF, to the best of our knowledge, is only studied by Salhi and Sari [126]. They propose a multi-level heuristic in which successive refinement steps involving increasingly complex neighborhoods are applied sequentially. The testing is performed on instances derived from a benchmark of 23 MDVRP instances and shows a moderate benefit through the introduction of the heterogeneous fleet compared to the homogeneous vehicles case.

Recently, the heterogeneous version of other important variants of VRP have been studied. For example, Li, Leung, and Tian [89] examine the *open* variant of HVRPFD for which they develop a MAMP algorithm and test it on randomly generated instances. In addition, the FSMFD incorporating two-dimensional loading constraints (2L-HVRP) is introduced by Leung et al. [87]. They use a *simulated annealing* algorithm for the routing component, in conjunction with six heuristics for item loading and a local search refining step, to solve test instances derived from 2L-CVRP benchmarks.

9.4 ▪ Periodic Routing Problems

In periodic routing problems customers require repeated visits during the planning horizon. The days of visit for each customer can be chosen from a given set of feasible *visiting patterns*. For example, assuming a weekly planning horizon, a customer that requires two visits and whose feasible patterns are given as (2,4) and (3,5) must either be visited Tuesday and Thursday, or Wednesday and Friday. Thus, periodic routing problems require three types of decisions: (i) the selection of visiting patterns for each customer, (ii) the assignment of the chosen day-customer combinations to tours, and (iii) the routing of vehicles for each day of the planning horizon. The latter includes that the classical VRP constraints are respected, i.e., each route has to start and end at the depot and has to observe capacity and route duration restrictions. The objective is to minimize the total traveled distance, while the number of employed vehicles on each day must not exceed the fleet size.

These problems are denoted as *Period or periodic VRP* (PVRP). For a detailed discussion of modeling techniques of the PVRP, we refer the reader to Francis, Smilowitz, and Tzur [64]. Applications of practical importance arise in waste and recyclables collection (see Angelelli and Speranza [7], Nuortio et al. [103], and Coene, Arnout, and Spieksma [44]), product distribution and collection (see Alegre, Laguna, and Pacheco [1], Claassen and Hendriks [42], and Ronen and Goodhart [120]), maintenance operations (see Hadjiconstantinou and Baldacci [75] and Blakeley et al. [29]), and health care (see Hemmelmayr et al. [77], Pacheco et al. [107], Shao, Bard, and Jarrah [128], and Maya, Sörensen, and Goos [95]).

The PVRP was first mentioned in Beltrami and Bodin [26] and Foster and Ryan [60] and later formalized in Russell and Igo [123] and Christofides and Beasley [40]. Early solution methods for PVRP were mainly simply construction and improvement heuristics (see Russell and Gribbin [122], Tan and Beasley [134], and Gaudioso and Paletta [67]), which were later replaced by metaheuristic approaches (see Chao, Golden, and Wasil [37], Cordeau, Gendreau, and Laporte [45], and Drummond, Ochi, and Vianna Dalessandro [56]). More recently, mathematical programming–based solution methods have been proposed (see Francis and Smilowitz [61], Francis, Smilowitz, and Tzur [62], and Mourgaya and Vanderbeck [99]). For the assessment of the quality of the proposed solution methods, the benchmark instance set presented in Cordeau, Gendreau, and Laporte [45] is well established. It consists of a set of so-called "old" instances proposed in Chao, Golden, and Wasil [37], Christofides and Beasley [40], Russell and Igo [123], and Russell and Gribbin [122], containing between 20 and 417 customers, and a set of ten "new" instances introduced by Cordeau, Gendreau, and Laporte [45] and containing between 48 and 288 customers.

The literature until roughly 2008 was surveyed in great detail by Francis, Smilowitz, and Tzur [64]. The goal of the section at hand is to review the most important advances concerning the PVRP and its variants published afterwards. In Section 9.4.1, we review work on the classical PVRP. Section 9.4.2 is concerned with the *PVRP with Time Windows* (PVRPTW), which constitutes the most studied variant of the PVRP. Further important PVRP variants are addressed in Section 9.4.3, and, finally, an overview of recent PVRP-related case studies is provided in Section 9.4.4.

9.4.1 ▪ The Standard PVRP

Solution methods for the basic PVRP are reviewed in this section.

Exact Methods and Matheuristics   Baldacci et al. [18] are, to the best of our knowledge, the first and only authors to propose bounding techniques and an exact algorithm for the PVRP. They use different relaxations of an extended set-partitioning formulation to derive five bounding procedures, which are used to generate a reduced problem that can be solved by an integer programming solver (see Baldacci, Christofides, and Mingozzi [21]). Tests on the old PVRP instances with up to 153 customers and a set of 20 practice-inspired instances with up to 199 customers for a special case of the problem, called the *tactical planning VRP*, show the effectiveness of their method. The obtained lower bounds are tight (on average within 1% of optimality), several test instances were solved, and some new best known solutions were found.

Gulczynski, Golden, and Wasil [74] present a hybrid method for the PVRP and two practice-inspired problem variants. The first variant addresses the reassignment of customers to new routes while restricting the changes to potentially long established delivery patterns. The second aims at maintaining balanced workloads of the different drivers. Their approach combines (i) a *large neighborhood search* that is based on an integer programming model for reassigning multiple customers to new visiting patterns and moving customers to new routes in order to decrease the total distance traveled and (ii) a *record-to-record travel algorithm* for improving the daily routes as introduced in Groër, Golden, and Wasil [71]. The presented algorithm shows a convincing performance on the old PVRP instances.

Cacchiani, Hemmelmayr, and Tricoire [36] present another matheuristic based on fixing and releasing variables in the LP-relaxation of a set-covering formulation of the PVRP. Columns are generated in a heuristic fashion using an *Iterated Local Search* (ILS) algorithm. On the old PVRP test set, the authors achieve a performance that comes close to that of the hybrid *Genetic Algorithm* (GA) of Vidal et al. [144]; however, their algorithm requires clearly higher run times. On the realistic instances of Pacheco et al. [107] (see below), their matheuristic is able to obtain better solution quality than the approaches presented in Alegre, Laguna, and Pacheco [1], Cordeau, Gendreau, and Laporte [45], Hemmelmayr, Doerner, and Hartl [76], and Pacheco et al. [107], again requiring significantly higher computational effort.

Metaheuristics Hemmelmayr, Doerner, and Hartl [76] use a *Variable Neighborhood Search* (VNS) approach with an acceptance criterion based on *Simulated Annealing* (SA) to tackle the PVRP. Their shaking phase is based on random changes of customers' visit patterns and string relocate and cross-exchange operators for perturbation of the routes (see Savelsbergh [127] and Taillard et al. [133]). The subsequent local search applies the 3-opt operator (see Lin [91]). The authors conduct tests on the old and new PVRP instances, obtaining competitive results on all instances. Moreover, they provide new best solutions for almost all instances in extensive tests. Their method shows a good scaling behavior and thus achieves very good results in short run time for the larger instances. Pirkwieser and Raidl [115] present another VNS approach that uses a multi-level refinement strategy to improve scalability and thus be capable of tackling large-scale instances. The proposed method achieves a solution quality comparable to that of Hemmelmayr, Doerner, and Hartl [76] on standard PVRP instances if both methods are run for the same number of iterations. Moreover, the authors report results on a set of larger PVRP instances with up to 576 customers, which are generated following the procedure of Cordeau, Gendreau, and Laporte [45].

Vidal et al. [144] use their *Hybrid Genetic Search with Adaptive Diversity Control* (HGSADC) to address the PVRP, the *Multi-Depot PVRP* (MDPVRP), and many other VRP variants as mentioned in Sections 9.3 and 9.2. For all available PVRP benchmark instances, HGSADC outperforms previous methods producing best average results for 41 of the 42 instances in reasonable run times and providing 20 new best solutions.

A parallel, iterated version of the unified *Tabu Search* (TS) of Cordeau, Laporte, and Mercier [46] for solving VRP, PVRP, MDVRP, the *site-dependent VRP* and the time-window versions of the problems is presented by Cordeau and Maischberger [48]. The TS is embedded in an ILS and exploits the concurrent computation capabilities of modern computers by means of a simple parallel computation framework. It achieves reasonable results on the standard VRP, PVRP, MDVRP, and *VRP with Time Windows* (VRPTW) instances and is quite competitive for the other problems addressed. The computational experiments show that the integration into the ILS framework and the parallelization clearly improve the solution quality compared to the original TS.

Several variants of the PVRP exist, which differ either in the considered objective or the side constraints and the types of solutions which are searched (see Mourgaya and Vanderbeck [98]). The following section reviews the most studied variant, namely the PVRPTW, while other variants are discussed in Section 9.4.3.

9.4.2 ▪ The PVRP with Time Windows

In the PVRPTW, service at each customer must begin within an associated time interval and every vehicle has to leave and return to the depot within a given scheduling horizon.

The PVRPTW was first studied in Cordeau, Laporte, and Mercier [46], who propose a unified TS heuristic for several routing problems with time windows. In the same work the standard PVRPTW benchmark based on the PVRP instances of Cordeau, Gendreau, and Laporte [45] was introduced. The benchmark consists of 20 instances, 10 of them featuring wide time windows and the remainder having tight ones.

Exact Methods and Matheuristics   We are not aware of exact methods for PVRPTW; however, a bounding procedure and several matheuristics have been proposed. Pirkwieser and Raidl [114] present a set covering formulation of PVRPTW and use column genera- tion to solve the LP-relaxation of the master problem. The pricing subproblem resembles an *Elementary Shortest Path Problem with Resource Constraints* (ESPPRC) (see Feillet et al. [59] and Irnich and Desaulniers [81]). Columns are generated by solving the ESPPRC both in an exact way using *dynamic programming* and heuristically with a *Greedy Ran- domized Adaptive Search Procedure* (GRASP) approach. In numerical tests on a partially reduced version of the standard PVRPTW benchmark of Cordeau, Laporte, and Mercier [46] (considering only a subset of customers and a reduced vehicle number), the authors are able to provide strong lower bounds for some instances.

Pirkwieser and Raidl [113] develop a hybrid of their VNS heuristic [111] and a generic integer programming solver using the set covering formulation presented in [114]. The VNS transfers feasible solutions and the currently best one to the solver, which tries to improve the route combinations. If the solver is successful, the improved solution is trans- ferred back to the VNS for further optimization. To test the approach, the authors pro- pose a new set of 30 PVRPTW benchmark instances with a four or six day planning hori- zon based on the well-known Solomon instances for VRPTW (see Solomon [130]). In Pirkwieser and Raidl [112], this approach is enhanced by using a *Multiple VNS* (MVNS) approach, in which several cooperative VNS threads are running in the framework of a se- quential cooperative multi-start search. Both the standalone MVNS and an MVNS solver hybrid are able to significantly improve the results of the pure VNS on the benchmark instances introduced in [113], while the hybrid is able to yield better results for most in- stances without dominating the pure MVNS. In [116], the authors additionally propose a hybrid between an evolutionary algorithm and the column generation approach of Pirk- wieser and Raidl [114], as well as 15 additional test instances with a planning horizon of eight days.

Metaheuristics   The unified TS method of Cordeau, Laporte, and Mercier [46] is en- hanced in Cordeau, Laporte, and Mercier [47] and proves able to clearly improve upon the results obtained with the original heuristic. As described above, Cordeau and Mais- chberger [48] present a parallel iterated version of the TS, which yields very competitive results. Pirkwieser and Raidl [111] propose a VNS for the PVRPTW that uses an SA-based acceptance criterion and a random ordering of the shaking neighborhoods. The method can clearly improve the solution quality of state-of-the-art approaches and shows that the random variant in most cases outperforms a fixed ordering of the shaking neighborhoods. Yu and Yang [153] propose an *ant colony optimization* algorithm that improves upon the old solutions presented in Cordeau, Laporte, and Mercier [46], but is inferior compared to the methods in Cordeau, Laporte, and Mercier [47] and Pirkwieser and Raidl [111].

Nguyen, Crainic, and Toulouse [102] propose a hybrid GA with two new crossover operators, one aiming at exploration and the other at exploiting the high-quality features of the parents. To educate the offspring towards feasibility and higher fitness, the TS of Cordeau, Laporte, and Mercier [46] and the VNS of Pirkwieser and Raidl [111] are integrated into the algorithm. Tests on the standard instances described in [46] and [116] show a superior solution quality of the hybrid GA compared to previously published methods, with the GA requiring very high run times. However, the authors show that neither the TS in [46] nor the VNS in [111] are able to achieve the same quality when given the same time limit.

Vidal et al. [146] extend the HGSADC proposed in Vidal et al. [144] to be capable of solving VRPTW, PVRPTW, and the time-window versions of MDVRP and the site-dependent VRP. The education phase of the GA, which makes up for most of the computational effort, is specifically tailored to efficiently handle problems with time windows. The authors apply methods for restricting the neighborhood to be searched, use memories to prevent unnecessary re-evaluations of insertion costs, and use a new procedure for determining the change in time-window and route duration violations in amortized constant time. The latter works for all neighborhood operators that exchange a bounded number of arcs or relocate a bounded number of nodes. For solving larger instances, structural and geographical decomposition phases are applied. In numerical tests on the standard benchmarks of PVRPTW, MDVRPTW, and the site-dependent VRPTW, the method outperforms all existing approaches concerning solution quality while using similar computation time. Finally, the authors propose new larger instances for PVRPTW, MDVRPTW, and the site-dependent VRPTW with 360 to 960 customers using the generation procedure described in Cordeau, Gendreau, and Laporte [45] and Cordeau, Laporte, and Mercier [46].

9.4.3 ▪ Other PVRP Variants

The MDPVRP has gained increasing attention in recent years. It extends the PVRP by a given set of depots with homogeneous vehicles which are fixedly allocated to the depots over the planning horizon. The task is to assign to each customer a visiting pattern and a depot, which serves the customer in all periods. The objective is the minimization of total traveled distance. Hadjiconstantinou and Baldacci [75] introduced the MDPVRP in the context of preventive maintenance of a utility company and presented a four-phase algorithm involving a TS for the routing decision. Parthanadee and Logendran [109] consider an MDPVRP with multiple trips, time windows, limited product supplies, and interdependent depot operations; i.e., customers are not fixedly assigned to depots. The problem is addressed by multiple TS variants, which are tested on sets of practice-inspired instances. The authors find that operating depots interdependently can be beneficial, in particular if product shortages occur.

Vidal et al. [144] show that any MDPVRP can be converted into an equivalent PVRP that involves one period for each period-depot pair of the MDPVRP. They use this insight to solve the MDPVRP with their HGSADC using a new set of MDPVRP instances generated by combining the PVRP and MDVRP instances of Cordeau, Gendreau, and Laporte [45]. Lahrichi et al. [85] present a parallel cooperative search framework for solving rich combinatorial optimization problems, which involves several heuristics and exact methods that address subproblems generated by attribute-based decomposition of the original problem and recombine and improve the partial solutions. The authors test their framework on the MDPVRP instances proposed in [144] using a generalized version of the TS of [46] and HGSADC as solvers and clearly improve upon the solution quality of the stand-alone methods. Rahimi-Vahed et al. [118] propose a Path Relinking (PR) method that can either be used as a stand-alone algorithm or in the cooperative search framework of Lahrichi et al. [85], achieving competitive solution quality on the MDPVRP instances of [144].

Alonso, Alvarez, and Beasley [5] extend the TS for PVRP proposed in Cordeau, Gendreau, and Laporte [45] to be able to solve the *site-dependent multi-trip PVRP*, or more generally, all combinations of VRP with accessibility restrictions, multiple use of vehicles, and multiple periods. To assess the quality of their solution method, they conduct computational tests on available benchmark instances of PVRP, the site-dependent VRP, and the *multi-trip VRP* and CVRP, showing very competitive performance for all problems except for the CVRP (for which they still achieve a reasonable average deviation of 1% from the best known solution).

Smilowitz, Nowak, and Jiang [129] present a method for quantifying the effect of workforce management in route construction, i.e., to place a value on the benefit of visiting a customer repeatedly with the same driver. Thus, they are able to balance the resulting consistency benefits against additional routing costs like, e.g., increased distance. They propose three different PVRP variants that include consistency metrics as a part of the objective function to be compared against a base PVRP minimizing traveled distance. Similar metrics were considered in Francis, Smilowitz, and Tzur [63]; however, they were calculated a posteriori to evaluate routing solutions and the presented models did not attempt to optimize with respect to these measures.

9.4.4 ▪ Case Studies

In recent years, several practical problem settings closely related to the PVRP have been studied in the context of waste management, product distribution, health care, and similar operations. Nuortio et al. [103] study a waste-collection problem in Eastern Finland, which is modeled as a stochastic PVRPTW with a fixed number of vehicles. For solving the problem, they use a guided variable neighborhood thresholding method specifically developed for large-scale VRP. Their method is tested on real-life data of a Finnish company and achieves drastic savings in traveled distance compared to the manual optimization previously employed at the company.

Alegre, Laguna, and Pacheco [1] consider a PVRP in the context of a car manufacturer's raw material collection operations. Contrary to the problems and associated benchmark instances addressed in the literature, the focus here lies on long planning horizons of up to 90 days, resulting in a complex visiting-pattern assignment problem, which is followed by a relatively simple routing decision. As a solution approach, the authors propose a scatter search for determining the collection schedules and use a simple local search method using the cross exchanges of Taillard et al. [133] and Or exchanges of Or [105] to improve the vehicle routes. The proposed approach is competitive with the state-of-the-art methods on standard PVRP instances at the time of publication and outperforms the best performing method of Cordeau, Gendreau, and Laporte [45] on real-world problem instances with long planning horizons.

Pacheco et al. [107] address a variant of the PVRP faced by a Spanish bakery company planning its deliveries to distribution centers, in which delivery to a customer has to take place within a certain time before a deadline associated with the customer. To decide the delivery day for each customer and construct the daily routes, the authors propose a two-phase algorithm, consisting of a GRASP phase with local improvement to generate a set of high-quality and diverse solutions and a PR phase to intensify the search. Their approach is able to outperform the methods in [1, 45] and in Hemmelmayr, Doerner, and Hartl [76] on a set of reality-based instances. Moreover, they show that a reduction of travel distance of about 20% is possible by relaxing the customer delivery deadlines previously used by the company and additionally allowing service one day prior to the deadline.

Finally, Maya, Sörensen, and Goos [95] address a variant of the MDPVRP in the field of providing mobile teaching assistance. The authors develop an auction heuristic enhanced by VNS and GRASP concepts that yields a reduction in traveled distance of more than 20% in comparison to the currently implemented solution. A similar problem is studied in Shao, Bard, and Jarrah [128], where the authors investigate the assignment and routing of therapists providing home health care.

9.5 ▪ VRP with Split Deliveries

In the classical VRP, customers are visited exactly once and each vehicle delivers the entire demand to the respective customer. The *Split Delivery Vehicle Routing Problem* (SDVRP) is a relaxation of the *Capacitated VRP* (CVRP) in the sense that a customer may be visited more than once. In this case the demand is covered by two or more *split deliveries*.

The SDVRP has been introduced by Dror and Trudeau [54, 55]. The interest in the SDVRP, and more generally in VRP variants which allow split deliveries, can be explained by the fact that significant savings in costs and the number of routes are possible compared to the problems without splitting possibility. Therefore, several papers have been devoted to worst-case analyses in order to quantify how much worse a solution without split deliveries can be compared to an SDVRP solution. Moreover, some exact approaches and several heuristics and metaheuristics addressing SDVRP and its variants can be found in the literature.

The SDVRP has been surveyed by Archetti and Speranza [14, 15]. We briefly summarize their results but put the main focus on recent results not included in the surveys. A first step is the classification of SDVRP and its basic variants provided in Section 9.5.1. Hereafter, we present theoretical results in Section 9.5.2 including properties of optimal solutions, computational complexity, and worst-case results. A description of the most recent exact and heuristic approaches is given in Section 9.5.3.

9.5.1 ▪ The SDVRP and Its Basic Variants

For a more formal definition of SDVRP, we follow the notation of Chapter 1: Let 0 be the depot and $N = \{1, 2, \ldots, n\}$ be the set of customers together forming the node set $V = \{0, 1, \ldots, n\}$. The customer *demands* q_i for all $i \in N$ are assumed to be positive integers. Moreover, the demand q_i for $i \in N$ must not exceed the *vehicle capacity* Q (also an integer number), where an unlimited homogeneous fleet is assumed. Then, for a given cost matrix $(c_{ij})_{i,j \in V}$, the SDVRP is the following problem: determine a set of minimum-cost routes and associated *delivery quantities* so that (i) each route starts at the depot 0, visits a subset of the customers, and ends at the depot 0 again, (ii) the delivery quantities of each route do not exceed the vehicle capacity Q, and (iii) the delivery amounts of all routes serving a customer $i \in N$ sum up to the demand q_i.

Since the vehicle fleet is assumed unlimited (*Unlimited Fleet*, UF) in the SDVRP, it always has a feasible and herewith an optimal solution. The version with *Limited Fleet* (LF) is denoted by SDVRP-LF. A feasible solution to the SDVRP-LF with $|K|$ vehicles exists if and only if $\sum_{i \in N} q_i \leq |K| Q$ holds.

For a given solution, let n_i denote the *number of visits* to the customer $i \in N$. In the classical VRP, we have $n_i = 1$ for all $i \in N$, while n_i may be any positive number in the SDVRP.

Split deliveries allow one to handle customers with large demands q_i exceeding the capacity Q. Following Archetti, Savelsbergh, and Speranza [12], SDVRP$^+$ is the problem variant in which at least one customer $j \in N$ has a demand $q_j > Q$. In the SDVRP$^+$, the number n_i of visits to customers $i \in N$ can be any positive integer. In contrast, VRP$^+$ denotes the variant in which the number of visits to every customer $i \in N$ is limited to the minimum number of necessary visits, which is equal to $\lceil q_i/Q \rceil$.

A possible heuristic solution to both the SDVRP$^+$ and VRP$^+$ results from first splitting demands $q_i > Q$ into $\lfloor q_i/Q \rfloor \cdot Q + (q_i \bmod Q)$. $\lfloor q_i/Q \rfloor$ routes serve customer i with full loads Q, and only routes for the remaining demands $(q_i \bmod Q) \neq 0$, $i \in N$ (if any), have to be planned because the full loads are served by *direct trips* $(0, i, 0)$, a.k.a. *out-and-back tours*. The remaining demands (all not greater than Q) are then served with an optimal SDVRP or VRP solution. We denote by H$^{SDVRP^+}$ the respective heuristics and problem variants resulting from the initial splitting. Similarly, H$^{VRP^+}$ is a restricted H$^{SDVRP^+}$ in which the number of visits to customer $i \in N$ is limited to $\lceil q_i/Q \rceil$.

Finally, it may be desirable to restrict the granularity of demand splits. Gulczynski, Golden, and Wasil [73] introduce a variant in which, for any real number $0 \leq p \leq 1$, a *Minimum Delivery Amount* (MDA) is defined by pq_i. A feasible solution to the SDVRP-MDA$_p$ requires that every route visiting customer i must deliver at least pq_i units. For example, if demand is $q_i = 10$ and $p = 0.2$, not more than five routes must deliver at least two units each. Obviously, $p = 0$ provides full delivery flexibility as modeled by the SDVRP, while for $p > 1/2$ the SDVRP-MDA$_p$ and the VRP coincide. Interesting new variants result for $0 < p \leq 1/2$.

Table 9.2 summarizes the VRP variants defined above.

Table 9.2. *SDVRP and VRP variants used in worst-case analyses.*

VRP Variant	Demand	Num. Visits	MDA	Fleet
SDVRP	$\forall i \in N : 1 \leq q_i \leq Q$	any number	0	unlimited
VRP	$\forall i \in N : 1 \leq q_i \leq Q$	1	d_i	unlimited
SDVRP-LF	$\forall i \in N : 1 \leq q_i \leq Q$	any number	0	limited
SDVRP-MDA$_p$	$\forall i \in N : 1 \leq q_i \leq Q$	$1, \ldots, \lfloor 1/p \rfloor$	pq_i	unlimited
SDVRP$^+$	$\exists j \in N : q_j > Q$	any number	0	unlimited
VRP$^+$	$\exists j \in N : q_j > Q$	$\lceil q_i/Q \rceil$	$(q_i \bmod Q)$	unlimited
H$^{SDVRP^+}$	initial splitting $q_i =$ $Q + \cdots + Q + (q_i \bmod Q)$	any number	0	unlimited
H$^{VRP^+}$	initial splitting $q_i =$ $Q + \cdots + Q + (q_i \bmod Q)$	$\lceil q_i/Q \rceil$	$(q_i \bmod Q)$	unlimited

9.5.2 ▪ Theoretical Results

Several heuristic and exact algorithms benefit from the following properties allowing one to restrict the search for optimal solutions.

Property 1 There exists an optimal solution in which the delivery quantities are integer numbers (see Archetti, Savelsbergh, and Speranza [12] and Archetti, Bouchard, and Desaulniers [9]) (demands and capacity are assumed to be integer).

For a given solution to the SDVRP, let $i_1, i_2, \ldots, i_k \in N$ be $k \geq 2$ different *split customers*, i.e., customers served by at least two different routes. If there exist k routes $r_1, \ldots, r_k$ such that route r_j contains the customers i_j and i_{j+1} for all $j = 1, 2, \ldots, k$ (defining $i_{k+1} := i_1$), then $(i_1, i_2, \ldots, i_k, i_1)$ is a *k-split cycle*.

All the following properties are based on the assumption that the routing costs c_{ij} satisfy the *triangle inequality*, i.e., $c_{ij} + c_{jk} \geq c_{ik}$ for all $i, j, k \in V$. Then there exists an optimal solution to the SDVRP ...

Property 2 ...which does not contain k-split cycles, $k \geq 2$ (see Dror and Trudeau [54]).

Property 3 ...where no two routes share more than one split customer (see Dror and Trudeau [55]).

Property 4 ...where each arc (i, j) between two customers $i, j \in N$ is traversed at most once (see Gendreau et al. [68]).

Property 5 ...where for each pair of reverse arcs (i, j) and (j, i) at most one of them is traversed (see Desaulniers [51]).

Property 6 ...where the number of vehicles used is not greater than $2\lceil \sum_{i \in N} q_i / Q \rceil$ (see Archetti, Bianchessi, and Speranza [8]).

Property 7 ...where the number of splits is less than the number of routes (see Archetti, Savelsbergh, and Speranza [12]).

Maximum Savings by Split Deliveries   Table 9.3 summarizes the maximum savings results presented in the literature. All results are based on the assumption that the triangle inequality holds for the routing costs. Most of them, except for some special cases, say that savings of up to 50% for the overall routing distance are possible by splitting deliveries compared to optimal solutions of the respective VRP variant with identical input data.

Table 9.3. *Summary of worst-case analyses.*

Objective and Reference	Precondition	Result
Routing Costs $z(\cdot)$		
[12]		$z(\text{VRP}) \leq 2z(\text{SDVRP})$
[12]		$z(\text{VRP}^+) \leq 2z(\text{SDVRP}^+)$
[12]		$z(\text{H}^{VRP^+}) \leq 2z(\text{VRP}^+)$
[12]		$z(\text{H}^{SDVRP^+}) \leq 2z(\text{SDVRP}^+)$
[12]	$Q = 3$	$z(VRP) \leq \frac{3}{2}z(\text{SDVRP})$
[12]	$Q = 3$	$z(VRP^+) \leq \frac{3}{2}z(\text{SDVRP}^+)$
[152]	$0 < p < 1/2$	$z(\text{VRP}) \leq 2z(\text{SDVRP-MDA}_p)$
[152]	$p = 1/2$	$z(\text{VRP}) \leq \frac{3}{2}z(\text{SDVRP-MDA}_p)$
[72]		$z(\text{VRP})/z(\text{SDVRP-LF})$ can be arbitrarily large
Number of Routes $r(\cdot)$		
[13]		$r(\text{SDVRP}) = \lceil \sum_{i \in N} q_i / Q \rceil$
[13]		$r(\text{VRP}) \leq 2r(\text{SDVRP-LF})$

The SDVRP allows, in principle (disregarding Property 1), that each demand is split into any number of smaller deliveries performed by different vehicles. Xiong et al. [152] show that if each delivery amount d can only be split into two equal deliveries of size $\frac{1}{2}d$, maximum savings of $33\frac{1}{3}$% are possible. Interestingly, for every minimum fraction $p \in (0, \frac{1}{2})$ into which the demand is split (i.e., for the SDVRP-MDA$_p$ as defined above) the same maximal saving of 50% is possible as in the SDVRP case. Note finally that all presented bounds in Table 9.3 have been proven to be tight.

Computational Complexity   The VRP and SDVRP are NP-hard because they generalize the *Traveling Salesman Problem* (TSP). However, both are polynomially solvable for the case with vehicle capacity $Q = 2$ (see Archetti, Mansini, and Speranza [11]). Note that for the SDVRP with $Q = 2$, H^{SDVRP} provides an optimal solution (Archetti, Mansini, and Speranza [11] name this property *reducibility*). Both VRP and SDVRP remain NP-hard for $Q \geq 3$ even in case of unit demands [11]. The computational complexity of the SDVRP on classes of instances, where the underlying graph is a circle ($\mathcal{O}(n^2)$), a half-line, and a line ($\mathcal{O}(n)$), a star (NP-hard for LF and $\mathcal{O}(n)$ for UF), and a tree (NP-hard), are shown by Archetti et al. [10], where corresponding results for CVRP (with LF and UF) are also presented.

9.5.3 ▪ Solution Methods for the SDVRP

Exact Methods and Matheuristics   Early exact algorithms for the SDVRP were already surveyed by Archetti and Speranza [14, 15]: Based on an arc-flow formulation for the SDVRP, Dror, Laporte, and Trudeau [53] provide a first exact solution approach. Valid inequalities and a cutting-plane algorithm to provide lower bounds on the SDVRP are presented by Belenguer, Martinez, and Mota [24]. Lee et al. [86] develop a dynamic programming–based solution approach, Jin, Liu, and Bowden [83] and Liu [93] a two-stage approach, and Jin, Liu, and Eksioglu [84] a *Column Generation* (CG) approach. The most powerful approach is, at the moment, the Branch-and-Cut-and-Price by Archetti, Bianchessi, and Speranza [8]. The authors are able to consistently solve instances with approximately 20–30 customers, but also some larger instances, where the largest has 144 customers. Hybrid algorithms using *mixed integer programming* to partially optimize solutions are presented by Archetti, Speranza, and Savelsbergh [17], Chen, Golden, and Wasil [38] and Jin, Liu, and Eksioglu [84].

The *SDVRP with Time Windows* (SDVRPTW) is the extension where time-window constraints are added. The extended set covering formulation of Gendreau et al. [68] exploits Property 4 and uses two types of variables, one for all feasible vehicle routes and another one for the delivery quantities provided by each chosen route. The alternative formulation by Desaulniers [51] is an extended set covering problem with variables that represent both routes and associated delivery quantities. Herein, only *extreme delivery patterns* need to be considered; i.e., only one customer (if any) on the route receives a positive quantity below its entire demand, while all other customers are either served with their entire demand or visited only so that the delivery quantity is zero. Combinations of extreme delivery patterns of a route are then convex-combined into regular patterns in the CG master program, which also exploits Property 5. Later Archetti, Bouchard, and Desaulniers [9] improve the approach [51] by introducing a powerful *Tabu Search* (TS) algorithm to solve the CG subproblem and devising several classes of valid inequalities for the SDVRPTW and corresponding separation procedures.

Salani and Vacca [124] introduce the *VRPTW with Discrete Splits* (DSDVRPTW), in which the demand of a customer consists of several items that may be delivered by several vehicles, but items cannot be split further. If all items have weight one (unit demand case), then the resulting problem is an SDVRPTW, while in general the SDVRPTW is a relaxation of the DSDVRPTW. Furthermore, there may be the restriction to deliver only specific *combinations of items*, called *orders*, to a customer at the same time. The set partitioning formulation of the DSDVRPTW is solved with Branch-and-Price, where the master program has partitioning constraints for each item. The subproblem is an elementary shortest path problem with resource constraints (see Irnich and Desaulniers [81]) with one node per order and elementary requirements with respect to all orders of the same customer. Order-specific service times can easily be integrated. The approach seems to work well for not too small orders but is not suited for solving the SDVRPTW.

Heuristics and Metaheuristics   Heuristics developed prior to 2011 are thoroughly surveyed in Archetti and Speranza [15], including local search (see Dror and Trudeau [54] and Dror and Trudeau [55]), TS (see Archetti, Speranza, and Hertz [16]), *memetic algorithm* (see Boudia, Prins, and Reghioui [30]), and *scatter search* (see Mota, Campos, and Corberán [97]). Derigs, Li, and Vogel [50] compare implementations of *simulated annealing*, *threshold accepting*, *record-to-record travel*, attribute-based local *beam search*, and attribute-based *hill climber*, where the last performs best. In a series of works [2, 3, 4], Aleman and coauthors propose several metaheuristics including *adaptive memory programming*, *variable neighborhood descent*, and a population-based algorithm. In the following, we focus on three approaches not surveyed before.

Berbotto, García, and Nogales [27] present a *randomized granular TS*. The approach follows the granular-TS idea (see Toth and Vigo [142]) of restricting the set of edges available for insertion in a local search move to promising edges depending on whether intensification and diversification is desirable. A novelty is that the authors control the set of short edges using a threshold parameter that also depends on the respective route and its residual capacity, the cost, and the number of edges of the current solution. Instead of only one neighborhood traditionally used in TS, the authors consider seven different neighborhood operators together: Three are classical VRP operators (node relocation, node exchange or swap, and 2-opt*; see, e.g., Funke, Grünert, and Irnich [65]), two others are known SDVRP-specific operators (exchange split and delete worst split; see Ho and Haugland [79] and Archetti, Speranza, and Hertz [16]), and the operators *delete split new* and *delete split* are newly introduced. The moves involve three different routes and simultaneously consider the relocation of a node and the modification of delivery quantities. All seven operators are used at the same time, and a randomization mechanism selects the next neighbor solution. Infeasible solutions regarding vehicle capacity are allowed, a capacity correction phase tries to eliminate such infeasibilities, a further improvement phase eliminates 2-split cycles (see Property 2), and an exact TSP solver is applied for optimizing the individual routes. Overall, the approach achieves very remarkable results because it improves many best known solutions on several benchmark sets from the literature.

Wilck and Cavalier [150] develop a two-phase construction heuristic, where in the first phase customer clusters and delivery quantities are determined, and in the second phase routes are formed for each cluster by solving a TSP. The first phase for the demand partitioning is solved by a rule-based procedure that iteratively adds customers to the current partial solution (clusters and delivery quantities). The three main decisions taken are (i) selection of customers with unsatisfied demand for initializing a route, (ii) sharing residual capacities between the routes, and (iii) insertion of one, two, or three customers with unsatisfied demand into existing routes. The three decisions are made using different rules leading to 36 possible construction algorithms. The authors report nine of them being beneficial. The main advantage of the proposed algorithms is that they are fast and compute relatively good solution.

In Wilck and Cavalier [151], the same authors present *genetic algorithms* for the SD-VRP. Two variants differing in the fitness computation perform well compared to the older approaches Chen, Golden, and Wasil [38] and Jin, Liu, and Eksioglu [84] but are not fully convincing compared to the newer approach by Berbotto, García, and Nogales [27] because only few and small improvements are achieved often using longer computation times.

9.6 ▪ Conclusions and Future Research Directions

Four important variants of the VRP are reviewed in this chapter. The VRPB and HFVRP received relatively little attention in the recent literature. For both of them further research on exact methods may highlight structural properties and problem-specific classes of valid inequalities that can improve the quality of currently available methods. In fact, in the heterogeneous case the existing approaches are based on general frameworks that may benefit from further insight on the HFVRP properties, whereas the exact methods for VRPB may be updated to current state of the art. Several aspects of these problems merit further attention also from a heuristic point of view. The specific impact of vehicle-dependent travel times in HVRP, or the multi-depot and mixed service variants of VRPB, are relevant examples in this direction.

The PVRP has continuously received attention by the scientific community, and many interesting solution approaches covering exact methods, matheuristics, and metaheuristics have been published in recent years. The practical importance of the problem is further underlined by a high number of case studies that deal with real-life PVRP. An interesting topic for future research could be the development of exact solution approaches for the most popular PVRP variants, namely the *PVRP with time windows* and the *multi-depot PVRP*. Also, new practically inspired variants that, e.g., incorporate consistency constraints (see Smilowitz, Nowak, and Jiang [129]) or the deviation from existing delivery patterns (see Gulczynski, Golden, and Wasil [74]) should be further investigated.

The SDVRP and its variants are well studied in the literature: Many worst-case analyses and also several empirical studies indicate that substantial savings can result when customers are allowed to be visited more than once. On the downside, multiple visits may be undesirable from a customer point of view. Moreover, the algorithmic handling of possible split deliveries requires specifically tailored programming components in both exact and (meta)heuristic approaches. In particular for the latter, it is not yet clear which algorithmic concepts are best suited for representing (partial) tours and solutions so that fundamental operations in local search and population-based heuristics can be performed in an efficient and effective way. We expect further research in this direction.

Bibliography

[1] J. ALEGRE, M. LAGUNA, AND J. PACHECO, *Optimizing the periodic pick-up of raw materials for a manufacturer of auto parts*, European Journal of Operational Research, 179 (2007), pp. 736–746.

[2] R. E. ALEMAN, *A guided neighborhood search applied to the split delivery vehicle routing problem*, PhD thesis, Wright State University, Dayton, OH, 2009.

[3] R. E. ALEMAN AND R. R. HILL, *A tabu search with vocabulary building approach for the vehicle routing problem with split demands*, International Journal of Metaheuristics, 1 (2010), pp. 55–80.

[4] R. E. ALEMAN, X. ZHANG, AND R. R. HILL, *An adaptive memory algorithm for the split delivery routing problem*, Journal of Heuristics, 16 (2010), pp. 441–473.

[5] F. ALONSO, M. J. ALVAREZ, AND J. E. BEASLEY, *A tabu search algorithm for the periodic vehicle routing problem with multiple vehicle trips and accessibility restrictions*, Journal of the Operational Research Society, 59 (2008), pp. 963–976.

[6] S. P. ANBUUDAYASANKAR, K. GANESH, S. C. LENNY KOH, AND Y. DUCQ, *Modified savings heuristics and genetic algorithm for bi-objective vehicle routing problem with forced backhauls*, Expert Systems with Applications, 39 (2012), pp. 2296–2305.

[7] E. ANGELELLI AND M. G. SPERANZA, *The application of a vehicle routing model to a waste-collection problem: Two case studies*, Journal of the Operational Research Society, 53 (2002), pp. 944–952.

[8] C. ARCHETTI, N. BIANCHESSI, AND M. G. SPERANZA, *A column generation approach for the split delivery vehicle routing problem*, Networks, 58 (2011), pp. 241–254.

[9] C. ARCHETTI, M. BOUCHARD, AND G. DESAULNIERS, *Enhanced branch and price and cut for vehicle routing with split deliveries and time windows*, Transportation Science, 45 (2011), pp. 285–298.

[10] C. ARCHETTI, D. FEILLET, M. GENDREAU, AND M. G. SPERANZA, *Complexity of the VRP and SDVRP*, Transportation Research Part C: Emerging Technologies, 19 (2011), pp. 741–750.

[11] C. ARCHETTI, R. MANSINI, AND M. G. SPERANZA, *Complexity and reducibility of the skip delivery problem*, Transportation Science, 39 (2005), pp. 182–187.

[12] C. ARCHETTI, M. W. P. SAVELSBERGH, AND M. G. SPERANZA, *Worst-case analysis for split delivery vehicle routing problems*, Transportation Science, 40 (2006), pp. 226–234.

[13] ——, *To split or not to split: That is the question*, Transportation Research Part E: Logistics and Transportation Review, 44 (2008), pp. 114–123.

[14] C. ARCHETTI AND M. G. SPERANZA, *The split delivery vehicle routing problem: A survey*, in The Vehicle Routing Problem: Latest Advances and New Challenges, B. L. Golden, S. Raghavan, and E. A. Wasil, eds., vol. 43 of Operations Research/Computer Science Interfaces Series, Springer, New York, 2008, pp. 103–122.

[15] ——, *Vehicle routing problems with split deliveries*, International Transactions in Operational Research, 19 (2012), pp. 3–22.

[16] C. ARCHETTI, M. G. SPERANZA, AND A. HERTZ, *A tabu search algorithm for the split delivery vehicle routing problem*, Transportation Science, 40 (2006), pp. 64–73.

[17] C. ARCHETTI, M. G. SPERANZA, AND M. W. P. SAVELSBERGH, *An optimization-based heuristic for the split delivery vehicle routing problem*, Transportation Science, 42 (2008), pp. 22–31.

[18] R. BALDACCI, E. BARTOLINI, A. MINGOZZI, AND A. VALLETTA, *An exact algorithm for the period routing problem*, Operations Research, 59 (2011), pp. 228–241.

[19] R. BALDACCI, M. BATTARRA, AND D. VIGO, *Routing a heterogeneous fleet of vehicles*, in The Vehicle Routing Problem: Latest Advances and New Challenges, B. L. Golden, S. Raghavan and E. A. Wasil, eds., vol. 43 of Operations Research/Computer Science Interfaces Series, Springer, New York, 2008, pp. 3–27.

[20] ——, *Valid inequalities for the fleet size and mix vehicle routing problem with fixed costs*, Networks, 54 (2009), pp. 178–189.

[21] R. BALDACCI, N. CHRISTOFIDES, AND A. MINGOZZI, *An exact algorithm for the vehicle routing problem based on the set partitioning formulation with additional cuts*, Mathematical Programming, 115 (2008), pp. 351–385.

[22] R. BALDACCI AND A. MINGOZZI, *A unified exact method for solving different classes of vehicle routing problems*, Mathematical Programming A, 120 (2009), pp. 347–380.

[23] J. E. BEASLEY, *Route-first cluster-second methods for vehicle routing*, Omega, 11 (1983), pp. 403–408.

[24] J.-M. BELENGUER, M. C. MARTINEZ, AND E. MOTA, *A lower bound for the split delivery vehicle routing problem*, Operations Research, 48 (2000), pp. 801–810.

[25] P. BELFIORE AND H. T. Y. YOSHIZAKI, *Scatter search for a real-life heterogeneous fleet vehicle routing problem with time windows and split deliveries in Brazil*, European Journal of Operational Research, 199 (2009), pp. 750–758.

[26] E. J. BELTRAMI AND L. D. BODIN, *Networks and vehicle routing for municipal waste collection*, Networks, 4 (1974), pp. 65–94.

[27] L. BERBOTTO, S. GARCÍA, AND F. J. NOGALES, *A randomized granular tabu search heuristic for the split delivery vehicle routing problem*, Annals of Operations Research, (2013), pp. 1–21.

[28] A. BETTINELLI, A. CESELLI, AND G. RIGHINI, *A branch-and-cut-and-price algorithm for the multi-depot heterogeneous vehicle routing problem with time windows*, Transportation Research Part C: Emerging Technologies, 19 (2011), pp. 723–740.

[29] F. BLAKELEY, B. ARGÜELLO, B. CAO, W. HALL, AND J. KNOLMAJER, *Optimizing periodic maintenance operations for Schindler elevator corporation*, Interfaces, 33 (2003), pp. 67–79.

[30] M. BOUDIA, C. PRINS, AND M. REGHIOUI, *An effective memetic algorithm with population management for the split delivery vehicle routing problem*, in Hybrid Metaheuristics, T. Bartz-Beielstein, M. Blesa Aguilera, C. Blum, B. Naujoks, A. Roli, G. Rudolph, and M. Sampels, eds., vol. 4771 of Lecture Notes in Computer Science, Springer, Berlin, Heidelberg, 2007, pp. 16–30.

[31] J. BRANDÃO, *A new tabu search algorithm for the vehicle routing problem with backhauls*, European Journal of Operational Research, 173 (2006), pp. 540–555.

[32] ——, *A deterministic tabu search algorithm for the fleet size and mix vehicle routing problem*, European Journal of Operational Research, 195 (2009), pp. 716–728.

[33] ——, *A tabu search algorithm for the heterogeneous fixed fleet vehicle routing problem*, Computers & Operations Research, 38 (2011), pp. 140–151.

[34] O. BRÄYSY, W. DULLAERT, G. HASLE, D. MESTER, AND M. GENDREAU, *An effective multi-restart deterministic annealing metaheuristic for the fleet size and mix vehicle routing problem with time windows*, Transportation Science, 42 (2008), pp. 371–386.

[35] O. BRÄYSY, P. P. PORKKA, W. DULLAERT, P. P. REPOUSSIS, AND C. D. TARAN-
TILIS, *A well-scalable metaheuristic for the fleet size and mix vehicle routing problem
with time windows*, Expert Systems with Applications, 36 (2009), pp. 8460–8475.

[36] V. CACCHIANI, V. C. HEMMELMAYR, AND F. TRICOIRE, *A set-covering based
heuristic algorithm for the periodic vehicle routing problem*, Discrete Applied Mathematics, 163 (2014), pp. 53–64.

[37] I.-M. CHAO, B. L. GOLDEN, AND E. A. WASIL, *An improved heuristic for the
period vehicle routing problem*, Networks, 26 (1995), pp. 25–44.

[38] S. CHEN, B. L. GOLDEN, AND E. A. WASIL, *The split delivery vehicle routing problem: Applications, algorithms, test problems, and computational results*, Networks, 49
(2007), pp. 318–329.

[39] E. CHOI AND D. TCHA, *A column generation approach to the heterogeneous fleet
vehicle routing problem*, Computers & Operations Research, 34 (2007), pp. 2080–
2095.

[40] N. CHRISTOFIDES AND J. E. BEASLEY, *The period routing problem*, Networks, 14
(1984), pp. 237–256.

[41] N. CHRISTOFIDES, A. MINGOZZI, AND P. TOTH, *Exact algorithms for the vehicle
routing problem based on spanning tree and shortest path relaxation*, Mathematical
Programming, 10 (1981), pp. 255–280.

[42] G. D. H. CLAASSEN AND T. H. B. HENDRIKS, *An application of special ordered
sets to a periodic milk collection problem*, European Journal of Operational Research,
180 (2007), pp. 754–769.

[43] G. CLARKE AND J. WRIGHT, *Scheduling of vehicles from a central depot to a number
of delivery points*, Operations Research, 12 (1964), pp. 568–581.

[44] S. COENE, A. ARNOUT, AND F. C. R. SPIEKSMA, *On a periodic vehicle routing
problem*, Journal of the Operational Research Society, 61 (2010), pp. 1719–1728.

[45] J.-F. CORDEAU, M. GENDREAU, AND G. LAPORTE, *A tabu search heuristic for
periodic and multi-depot vehicle routing problems*, Networks, 30 (1997), pp. 105–
119.

[46] J.-F. CORDEAU, G. LAPORTE, AND A. MERCIER, *A unified tabu search heuristic
for vehicle routing problems with time windows*, Journal of the Operational Research
Society, 52 (2001), pp. 928–936.

[47] ———, *An improved tabu search algorithm for the handling of route duration constraints in vehicle routing problems with time windows*, Journal of the Operational
Research Society, 55 (2004), pp. 542–546.

[48] J.-F. CORDEAU AND M. MAISCHBERGER, *A parallel iterated tabu search heuristic for vehicle routing problems*, Computers & Operations Research, 39 (2012),
pp. 2033–2050.

[49] I. DEIF AND L. D. BODIN, *Extension of the Clarke and Wright algorithm for solving
the vehicle routing problem with backhauling*, in Proceedings of the Babson Conference on Software Uses in Transportation and Logistic Management, A. Kidder, ed.,
Babson Park, FL, 1984, pp. 75–96.

[50] U. DERIGS, B. LI, AND U. VOGEL, *Local search-based metaheuristics for the split delivery vehicle routing problem*, Journal of the Operational Research Society, 61 (2010), pp. 1356–1364.

[51] G. DESAULNIERS, *Branch-and-price-and-cut for the split-delivery vehicle routing problem with time windows*, Operations Research, 58 (2010), pp. 179–192.

[52] R. DONDO AND J. CERDÁ, *A cluster-based optimization approach for the multi-depot heterogeneous fleet vehicle routing problem with time windows*, European Journal of Operational Research, 176 (2007), pp. 1478–1507.

[53] M. DROR, G. LAPORTE, AND P. TRUDEAU, *Vehicle routing with split deliveries*, Discrete Applied Mathematics, 50 (1994), pp. 239–254.

[54] M. DROR AND P. TRUDEAU, *Savings by split delivery routing*, Transportation Science, 23 (1989), pp. 141–145.

[55] ———, *Split delivery routing*, Naval Research Logistics, 37 (1990), pp. 383–402.

[56] L. M. A. DRUMMOND, L. S. OCHI, AND S. VIANNA DALESSANDRO, *An asynchronous parallel metaheuristic for the period vehicle routing problem*, Future Generation Computer Systems, 17 (2001), pp. 379–386.

[57] C. DUHAMEL, P. LACOMME, AND C. PRODHON, *A hybrid evolutionary local search with depth first search split procedure for the heterogeneous vehicle routing problems*, Engineering Applications of Artificial Intelligence, 25 (2012), pp. 345–358.

[58] C. DUHAMEL, J.-Y. POTVIN, AND J.-M. ROUSSEAU, *A tabu search heuristic for the vehicle routing problem with backhauls and time windows*, Transportation Science, 31 (1997), pp. 49–59.

[59] D. FEILLET, P. DEJAX, M. GENDREAU, AND C. GUEGUEN, *An exact algorithm for the elementary shortest path problem with resource constraints: Application to some vehicle routing problems*, Networks, 44 (2004), pp. 216–229.

[60] B. A. FOSTER AND D. M. RYAN, *An integer programming approach to the vehicle scheduling problem*, Operational Research Quarterly, 27 (1976), pp. 367–384.

[61] P. M. FRANCIS AND K. R. SMILOWITZ, *Modeling techniques for periodic vehicle routing problems*, Transportation Research Part B: Methodological, 40 (2006), pp. 872–884.

[62] P. M. FRANCIS, K. R. SMILOWITZ, AND M. TZUR, *The period vehicle routing problem with service choice*, Transportation Science, 40 (2006), pp. 439–454.

[63] ———, *Flexibility and complexity in periodic distribution problems*, Naval Research Logistics, 54 (2007), pp. 136–150.

[64] ———, *The period vehicle routing problem and its extensions*, in The Vehicle Routing Problem: Latest Advances and New Challenges, B. L. Golden, S. Raghavan, and E. A. Wasil, eds., vol. 43 of Operations Research/Computer Science Interfaces Series, Springer, New York, 2008, pp. 73–102.

[65] B. FUNKE, T. GRÜNERT, AND S. IRNICH, *Local search for vehicle routing and scheduling problems: Review and conceptual integration*, Journal of Heuristics, 11 (2005), pp. 267–306.

[66] Y. GAJPAL AND P. L. ABAD, *Multi-ant colony system (macs) for a vehicle routing problem with backhauls*, European Journal of Operational Research, 196 (2009), pp. 102–117.

[67] M. GAUDIOSO AND G. PALETTA, *A heuristic for the periodic vehicle routing problem*, Transportation Science, 26 (1992), pp. 86–92.

[68] M. GENDREAU, P. DEJAX, D. FEILLET, AND G. GUEGUEN, *Vehicle routing with time windows and split deliveries*, Technical Report 2006-851, Laboratoire d'Informatique d'Avignon, Avignon, France, 2006.

[69] M. GOETSCHALCKX AND C. JACOBS-BLECHA, *The vehicle routing problem with backhauls*, European Journal of Operational Research, 42 (1989), pp. 39–51.

[70] B. L. GOLDEN, A. A. ASSAD, L. LEVY, AND F. G. GHEYSENS, *The fleet size and mix vehicle routing problem*, Computers & Operations Research, 11 (1984), pp. 49–66.

[71] C. GROËR, B. L. GOLDEN, AND E. A. WASIL, *A library of local search heuristics for the vehicle routing problem*, Mathematical Programming Computation, 2 (2010), pp. 79–101.

[72] C. GUEGUEN, *Méthodes de résolution exacte pour les problèmes de tournées de véhicules*, PhD thesis, École Centrale Paris, France, 1999.

[73] D. GULCZYNSKI, B. L. GOLDEN, AND E. A. WASIL, *The split delivery vehicle routing problem with minimum delivery amounts*, Transportation Research Part E: Logistics and Transportation Review, 46 (2010), pp. 612–626.

[74] ———, *The period vehicle routing problem: New heuristics and real-world variants*, Transportation Research Part E: Logistics and Transportation Review, 47 (2011), pp. 648–668.

[75] E. HADJICONSTANTINOU AND R. BALDACCI, *A multi-depot period vehicle routing problem arising in the utilities sector*, Journal of the Operational Research Society, 49 (1998), pp. 1239–1248.

[76] V. C. HEMMELMAYR, K. DOERNER, AND R. F. HARTL, *A variable neighborhood search heuristic for periodic routing problems*, European Journal of Operational Research, 195 (2009), pp. 791–802.

[77] V. C. HEMMELMAYR, K. DOERNER, R. F. HARTL, AND M. W. P. SAVELSBERGH, *Delivery strategies for blood products supplies*, OR Spectrum, 31 (2009), pp. 707–725.

[78] A. HERTZ, M. ULDRY, AND M. WIDMER, *Integer linear programming models for a cement delivery problem*, European Journal of Operational Research, 222 (2012), pp. 623–631.

[79] S. C. HO AND D. HAUGLAND, *A tabu search heuristic for the vehicle routing problem with time windows and split deliveries*, Computers & Operations Research, 31 (2004), pp. 1947–1964.

[80] A. IMRAN, S. SALHI, AND N. A. WASSAN, *A variable neighborhood-based heuristic for the heterogeneous fleet vehicle routing problem*, European Journal of Operational Research, 197 (2009), pp. 509–518.

[81] S. IRNICH AND G. DESAULNIERS, *Shortest path problems with resource constraints*, in Column Generation, G. Desaulniers, J. Desrosiers, and M. M. Solomon, eds., Springer, New York, 2005, ch. 2, pp. 33–65.

[82] O. JABALI, M. GENDREAU, AND G. LAPORTE, *A continuous approximation model for the fleet composition problem*, Transportation Research Part B: Methodological, 46 (2012), pp. 1591–1606.

[83] M. JIN, K. LIU, AND R. O. BOWDEN, *A two-stage algorithm with valid inequalities for the split delivery vehicle routing problem*, International Journal of Production Economics, 105 (2007), pp. 228–242.

[84] M. JIN, K. LIU, AND B. EKSIOGLU, *A column generation approach for the split delivery vehicle routing problem*, Operations Research Letters, 36 (2008), pp. 265–270.

[85] N. LAHRICHI, T. G. CRAINIC, M. GENDREAU, W. REI, G. CERASELA CRIŞAN, AND T. VIDAL, *An integrative cooperative search framework for multi-decision- attribute combinatorial optimization*, Technical Report 2012-42, CIRRELT, Montréal, Canada, 2012.

[86] C. G. LEE, M. A. EPELMAN, C. C. WHITE III, AND Y. A. BOZER, *A shortest path approach to the multiple-vehicle routing problem with split pick-ups*, Transportation Research Part B: Methodological, 40 (2006), pp. 265–284.

[87] S. C. H. LEUNG, Z. ZHANG, D. ZHANG, X. HUA, AND M. K. LIM, *A meta-heuristic algorithm for heterogeneous fleet vehicle routing problems with two-dimensional loading constraints*, European Journal of Operational Research, 225 (2013), pp. 199–210.

[88] F. LI, B. L. GOLDEN, AND E. A. WASIL, *A record-to-record travel algorithm for solving the heterogeneous fleet vehicle routing problem*, Computers & Operations Research, 34 (2007), pp. 2734–2742.

[89] X. LI, S. C. H. LEUNG, AND P. TIAN, *A multistart adaptive memory-based tabu search algorithm for the heterogeneous fixed fleet open vehicle routing problem*, Expert Systems with Applications, 39 (2012), pp. 365–374.

[90] X. LI, P. TIAN, AND Y. P. ANEJA, *An adaptive memory programming metaheuristic for the heterogeneous fixed fleet vehicle routing problem*, Transportation Research Part E: Logistics and Transportation Review, 46 (2010), pp. 1111–1127.

[91] S. LIN, *Computer solutions of the traveling salesman problem*, Bell System Technical Journal, 44 (1965), pp. 2245–2269.

[92] F. H. LIU AND S. Y. SHEN, *The fleet size and mix vehicle routing problem with time windows*, Journal of the Operational Research Society, 50 (1999), pp. 721–732.

[93] K. LIU, *A study on the split delivery vehicle routing problem*, PhD thesis, Mississippi State University, Starkville, MS, 2005.

[94] S. LIU, W. HUANG, AND H. MA, *An effective genetic algorithm for the fleet size and mix vehicle routing problems*, Transportation Research Part E: Logistics and Transportation Review, 45 (2009), pp. 434–445.

[95] P. MAYA, K. SÖRENSEN, AND P. GOOS, *A metaheuristic for a teaching assistant assignment-routing problem*, Computers & Operations Research, 39 (2012), pp. 249–258.

[96] A. MINGOZZI, S. GIORGI, AND R. BALDACCI, *An exact method for the vehicle routing problem with backhauls*, Transportation Science, 33 (1999), pp. 315–329.

[97] E. MOTA, V. CAMPOS, AND A. CORBERÁN, *A new metaheuristic for the vehicle routing problem with split demands*, in Evolutionary Computation in Combinatorial Optimization, C. Cotta and J. van Hemert, eds., vol. 4446 of Lecture Notes in Computer Science, Springer, Berlin, Heidelberg, 2007, pp. 121–129.

[98] M. MOURGAYA AND F. VANDERBECK, *The periodic vehicle routing problem: Classification and heuristic*, RAIRO - Operations Research, 40 (2006), pp. 169–194.

[99] ———, *Column generation based heuristic for tactical planning in multi-period vehicle routing*, European Journal of Operational Research, 183 (2007), pp. 1028–1041.

[100] G. NAGY AND S. SALHI, *Heuristic algorithms for single and multiple depot vehicle routing problems with pickups and deliveries*, European Journal of Operational Research, 162 (2005), pp. 126–141.

[101] Z. NAJI-AZIMI AND M. SALARI, *A complementary tool to enhance the effectiveness of existing methods for heterogeneous fixed fleet vehicle routing problem*, Applied Mathematical Modelling, 37 (2013), pp. 4316–4324.

[102] P. K. NGUYEN, T. G. CRAINIC, AND M. TOULOUSE, *A hybrid genetic algorithm for the periodic vehicle routing problem with time windows*, Technical Report 2011-25, CIRRELT, Montréal, Canada, 2011.

[103] T. NUORTIO, J. KYTÖJOKI, H. NISKA, AND O. BRÄYSY, *Improved route planning and scheduling of waste collection and transport*, Expert Systems with Applications, 30 (2006), pp. 223–232.

[104] J. OPPEN, A. LØKKETANGEN, AND J. DESROSIERS, *Solving a rich vehicle routing and inventory problem using column generation*, Computers & Operations Research, 37 (2010), pp. 1308–1317.

[105] I. OR AND W. P. PIERSKALLA, *A transportation location-allocation model for regional blood banking*, IIE Transactions, 11 (1979), pp. 86–95.

[106] I. H. OSMAN AND N. A. WASSAN, *A reactive tabu search meta-heuristic for the vehicle routing problem with back-hauls*, Journal of Scheduling, 5 (2002), pp. 263–285.

[107] J. PACHECO, A. ALVAREZ, I. GARCIA, AND F. ANGEL-BELLO, *Optimizing vehicle routes in a bakery company allowing flexibility in delivery dates*, Journal of the Operational Research Society, 63 (2012), pp. 569–581.

[108] D. C. PARASKEVOPOULOS, P. P. REPOUSSIS, C. D. TARANTILIS, G. IOANNOU, AND G. P. PRASTACOS, *A reactive variable neighborhood tabu search for the heterogeneous fleet vehicle routing problem with time windows*, Journal of Heuristics, 14 (2007), pp. 247–254.

[109] P. PARTHANADEE AND R. LOGENDRAN, *Periodic product distribution from multi-depots under limited supplies*, IIE Transactions, 38 (2006), pp. 1009–1026.

[110] A. PESSOA, M. POGGI, AND E. UCHOA, *A Robust Branch-Cut-and-Price Algorithm for the Heterogeneous Fleet Vehicle Routing Problem*, vol. 4525 of Lecture Notes in Computer Science, Springer, Berlin, Heidelberg, 2007, pp. 150–160.

[111] S. PIRKWIESER AND G. R. RAIDL, *A variable neighborhood search for the periodic vehicle routing problem with time windows*, in Proceedings of the 9th EU/Meeting on Metaheuristics for Logistics and Vehicle Routing, Troyes, France, 2008.

[112] ——, *Multiple variable neighborhood search enriched with ILP techniques for the periodic vehicle routing problem with time windows*, in Hybrid Metaheuristics, M. J. Blesa, C. Blum, L. Gaspero, A. Roli, M. Sampels, and A. Schaerf, eds., vol. 5818 of Lecture Notes in Computer Science, Springer, Berlin, Heidelberg, 2009, pp. 45–59.

[113] ——, *Boosting a variable neighborhood search for the periodic vehicle routing problem with time windows by ILP techniques*, in Proceedings of the 8th Metaheuristic International Conference (MIC 2009), S. Voss and M. Caserta, eds., Hamburg, Germany, 2009.

[114] ——, *A column generation approach for the periodic vehicle routing problem with time windows.*, in Proceedings of the International Network Optimization Conference 2009, Pisa, Italy, 2009.

[115] ——, *Multilevel variable neighborhood search for periodic routing problems*, in Evolutionary Computation in Combinatorial Optimization, P. Cowling and P. Merz, eds., vol. 6022 of Lecture Notes in Computer Science, Springer, Berlin, Heidelberg, 2010, pp. 226–238.

[116] ——, *Matheuristics for the periodic vehicle routing problem with time windows*, in Proceedings of Matheuristics, Vienna, Austria, 2010.

[117] C. PRINS, *Two memetic algorithms for heterogeneous fleet vehicle routing problems*, Engineering Applications of Artificial Intelligence, 22 (2009), pp. 916–928.

[118] A. RAHIMI-VAHED, T. G. CRAINIC, M. GENDREAU, AND W. REI, *A path relinking algorithm for a multi-depot periodic vehicle routing problem*, Journal of Heuristics, 19 (2013), pp. 497–524.

[119] P. P. REPOUSSIS AND C. D. TARANTILIS, *Solving the fleet size and mix vehicle routing problem with time windows via adaptive memory programming*, Transportation Research Part C: Emerging Technologies, 18 (2010), pp. 695–712.

[120] D. RONEN AND C. A. GOODHART, *Tactical store delivery planning*, Journal of the Operational Research Society, 59 (2007), pp. 1047–1054.

[121] S. ROPKE AND D. PISINGER, *A unified heuristic for a large class of vehicle routing problems with backhauls*, European Journal of Operational Research, 171 (2006), pp. 750–775.

[122] R. A. RUSSELL AND D. GRIBBIN, *A multiphase approach to the period routing problem*, Networks, 21 (1991), pp. 747–765.

[123] R. A. RUSSELL AND W. IGO, *An assignment routing problem*, Networks, 9 (1979), pp. 1–17.

[124] M. SALANI AND I. VACCA, *Branch and price for the vehicle routing problem with discrete split deliveries and time windows*, European Journal of Operational Research, 213 (2011), pp. 470–477.

[125] S. SALHI AND G. NAGY, *A cluster insertion heuristic for single and multiple depot vehicle routing problems with backhauling*, Journal of the Operational Research Society, 50 (1999), pp. 1034–1042.

[126] S. SALHI AND M. SARI, *A multi-level composite heuristic for the multi-depot vehicle fleet mix problem*, European Journal of Operational Research, 103 (1997), pp. 95–112.

[127] M. W. P. SAVELSBERGH, *The vehicle routing problem with time windows: Minimizing route duration*, ORSA Journal on Computing, 4 (1992), pp. 146–154.

[128] Y. SHAO, J. F. BARD, AND A. I. JARRAH, *The therapist routing and scheduling problem*, IIE Transactions, 44 (2012), pp. 868–893.

[129] K. SMILOWITZ, M. NOWAK, AND T. JIANG, *Workforce management in periodic delivery operations*, Transportation Science, 47 (2013), pp. 214–230.

[130] M. M. SOLOMON, *Algorithms for the vehicle routing and scheduling problems with time window constraints*, Operations Research, 35 (1987), pp. 254–265.

[131] A. SUBRAMANIAN, P. H. V. PENNA, E. UCHOA, AND L. S. OCHI, *A hybrid algorithm for the heterogeneous fleet vehicle routing problem*, European Journal of Operational Research, 221 (2012), pp. 285–295.

[132] É. D. TAILLARD, *A heuristic column generation method for the heterogeneous fleet VRP*, RAIRO - Recherche Opérationnelle, 33 (1999), pp. 1–14.

[133] É. D. TAILLARD, P. BADEAU, M. GENDREAU, F. GUERTIN, AND J.-Y. POTVIN, *A tabu search heuristic for the vehicle routing problem with soft time windows*, Transportation Science, 31 (1997), pp. 170–186.

[134] C. C. R. TAN AND J. E. BEASLEY, *A heuristic algorithm for the period vehicle routing problem*, Omega, 12 (1984), pp. 497–504.

[135] C. D. TARANTILIS AND C. T. KIRANOUDIS, *A flexible adaptive memory-based algorithm for real-life transportation operations: Two case studies from dairy and construction sector*, European Journal of Operational Research, 179 (2007), pp. 806–822.

[136] R. TAVAKKOLI-MOGHADDAM, A. R. SAREMI, AND M. S. ZIAEE, *A memetic algorithm for a vehicle routing problem with backhauls*, Applied Mathematics and Computation, 181 (2006), pp. 1049–1060.

[137] S. R. THANGIAH, J.-Y. POTVIN, AND T. SUN, *Heuristic approaches to vehicle routing with backhauls and time windows*, Computers & Operations Research, 23 (1996), pp. 1043–1057.

[138] P. TOTH AND D. VIGO, *Heuristic algorithms for the vehicle routing problem with backhauls*, in Advanced Methods in Transportation Analysis, L. Bianco and P. Toth, eds., Springer-Verlag, Berlin, 1996, pp. 585–608.

[139] ——, *An exact algorithm for the vehicle routing problem with backhauls*, Transportation Science, 31 (1997), pp. 372–385.

[140] ——, *A heuristic algorithm for the symmetric and asymmetric vehicle routing problems with backhauls*, European Journal of Operational Research, 113 (1999), pp. 528–543.

[141] ——, *VRP with backhauls*, in The Vehicle Routing Problem, P. Toth and D. Vigo, eds., SIAM, Philadelphia, 2002, ch. 8, pp. 195–224.

[142] ——, *The granular tabu search and its application to the vehicle-routing problem*, INFORMS Journal on Computing, 15 (2003), pp. 333–346.

[143] Y. TÜTÜNCÜ, *An interactive GRAMPS algorithm for the heterogeneous fixed fleet vehicle routing problem with and without backhauls*, European Journal of Operational Research, 201 (2010), pp. 593–600.

[144] T. VIDAL, T. G. CRAINIC, M. GENDREAU, N. LAHRICHI, AND W. REI, *A hybrid genetic algorithm for multidepot and periodic vehicle routing problems*, Operations Research, 60 (2012), pp. 611–624.

[145] T. VIDAL, T. G. CRAINIC, M. GENDREAU, AND C. PRINS, *A unified solution framework for multi-attribute vehicle routing problems*, Technical Report 2012-23, CIRRELT, Montréal, Canada, 2012.

[146] ——, *A hybrid genetic algorithm with adaptive diversity management for a large class of vehicle routing problems with time-windows*, Computers & Operations Research, 40 (2013), pp. 475–489.

[147] A. C. WADE AND S. SALHI, *An investigation into a new class of vehicle routing problem with backhauls*, Omega, 30 (2002), pp. 479–487.

[148] Z. WANG AND Z. WANG, *A novel two-phase heuristic method for vehicle routing problem with backhauls*, Computers & Mathematics with Applications, 57 (2009), pp. 1923–1928.

[149] N. A. WASSAN, *Reactive tabu adaptive memory programming search for the vehicle routing problem with backhauls*, Journal of the Operational Research Society, 58 (2007), pp. 1630–1641.

[150] J. H. WILCK IV AND T. M. CAVALIER, *A construction heuristic for the split delivery vehicle routing problem*, American Journal of Operations Research, 2 (2012), pp. 153–162.

[151] ——, *A genetic algorithm for the split delivery vehicle routing problem*, American Journal of Operations Research, 2 (2012), pp. 207–216.

[152] Y. XIONG, D. GULCZYNSKI, D. KLEITMAN, B. L. GOLDEN, AND E. A. WASIL, *A worst-case analysis for the split delivery vehicle routing problem with minimum delivery amounts*, Optimization Letters, 7 (2013), pp. 1597–1609.

[153] B. YU AND Z. Z. YANG, *An ant colony optimization model: The period vehicle routing problem with time windows*, Transportation Research Part E: Logistics and Transportation Review, 47 (2011), pp. 166–181.

[154] E. E. ZACHARIADIS AND C. T. KIRANOUDIS, *An effective local search approach for the vehicle routing problem with backhauls*, Expert Systems with Applications, 39 (2012), pp. 3174–3184.

[155] Y. ZHONG AND M. H. COLE, *A vehicle routing problem with backhauls and time windows: A guided local search solution*, Transportation Research Part E: Logistics and Transportation Review, 41 (2005), pp. 131–144.

Chapter 10

Vehicle Routing Problems with Profits

Claudia Archetti
M. Grazia Speranza
Daniele Vigo

10.1 • Introduction

The key characteristic of the class of *Vehicle Routing Problems with Profits* (VRPPs) is that, contrary to what happens for the most classical vehicle routing problems, the set of customers to serve is not given. Therefore, two different decisions have to be taken: (i) which customers to serve, and (ii) how to cluster the customers to be served in different routes (if more than one) and order the visits in each route. In general, a *profit* is associated with each customer that makes such a customer more or less attractive. Thus, any route or set of routes, starting and ending at a given depot, can be measured both in terms of cost and in terms of profit. The difference between route profit and cost may be maximized, or the profit or the cost optimized with the other measure bounded in a constraint.

There are several types of applications that can be modeled by means of a problem of this class:

- scheduling of the daily operations of a steel rolling mill (see, e.g., Balas [13, 16]);

- design of tourist trips to maximize the value of the visited attractions in a limited period (see, e.g., Vansteenwegen and Van Oudheusden [109]);

- identification of suppliers to visit to maximize the recovered claims with a limited number of auditors (see Ilhan, Iravani, and Daskin [61]);

- recruiting of athletes from high schools for a college team (see Butt and Cavalier [24]);

- planning of the visits of a salesperson (see, e.g., Ramesh and Brown [89]);

- routing of oil tankers to serve ships at different locations (see Golden, Levy, and Vohra [57]);

- delivery of home heating fuel, where the urgency of a customer request for fuel is treated as a score (see Golden, Assad, and Dahl [56]);

- reverse logistics problem of a firm that aims to collect used products from its dealers (see Aras, Aksen, and Tekin [2]);

- customers selection in less-than-truckload transportation (see Archetti et al. [8]);

- service outsourcing of unprofitable customers (see Chu [31]).

Despite the practical interest in considering profits in vehicle routing, most of the existing literature concentrates on the single-vehicle case of the problem. Clearly, these findings form a valuable starting point also for more general VRPPs, and we therefore examine them in detail in the first part of this chapter.

The most basic problems of this class with only one route are often presented as variants of the *Traveling Salesman Problem* (TSP) (see Fischetti, Salazar González, and Toth [47]). The *Orienteering Problem* (OP) was first studied by Tsiligirides [105] and Golden, Levy, and Vohra [57]. The name is derived from the orienteering sport where each participant has to maximize the total collected prize associated with visited points while returning to the starting point within a given time limit. The OP is also known with the names *Selective Traveling Salesperson Problem* (see, e.g., Laporte and Martello [74], Gendreau, Laporte, and Semet [53], and Thomadsen and Stidsen [102]), *Maximum Collection Problem* (see, e.g., Kataoka and Morito [65] and Butt and Cavalier [24]), and *Bank Robber Problem* (see Awerbuch et al. [12]). The other two basic routing problems with profits and one vehicle only are the *Profitable Tour Problem* (PTP) and the *Prize-Collecting Traveling Salesman Problem* (PCTSP). In the PTP the objective is to maximize the difference between total collected profit and traveling cost with no constraint on the route. In the PCTSP the objective is to minimize the traveling cost with the constraint of collecting a total profit which is at least equal to a given threshold. In the literature, there is no consistent definition of these two problems. In this chapter we will follow the definitions given by Feillet, Dejax, and Gendreau [45] and explain, case by case, the different contributions.

Among the routing problems with profits and multiple vehicles the only one that has been studied in depth is the so-called *Team Orienteering Problem* (TOP). The TOP was introduced by Butt and Cavalier [24] with the name *Multiple Tour Maximum Collection Problem*. The first paper where the name TOP appears is due to Chao, Golden, and Wasil [28].

Previous surveys appeared covering parts of the literature on routing problems with profits. The survey by Feillet, Dejax, and Gendreau [45] is focused on the routing problems with profits with one vehicle only, whereas the more recent survey by Vansteenwegen, Souffriau, and Van Oudheusden [106] covers the OP and the TOP. Finally, the review by Gavalas et al. [51] is focused on the problems arising in the context of tourist trip design.

All the above-mentioned papers consider problems where customers are represented as vertices of a graph. We call these problems *VRP with Profits* (VRPPs). The problems where customers are represented as edges or arcs of a graph are called *arc routing problems with profits* (see Archetti and Speranza [10] for a recent survey).

In this chapter we cover all the VRPPs, from the problems considering a single vehicle to their extensions with multiple vehicles. We cover also the variants, e.g., those considering time windows and those inspired by applications.. We start with problems with one vehicle (the OP in Section 10.2.1, PTP in Section 10.2.2, PCTSP in Section 10.2.3, and variants in Section 10.2.4). We then present the TOP in Section 10.3.1 and its variants in

10.3.2. Finally, Section 10.3.3 is devoted to the Vehicle Routing Problems with Private Fleet and Common Carrier (VRPPC), a specific class of VRPPs that arise in distribution where subcontracting options are considered.

Following Vansteenwegen, Souffriau, and Van Oudheusden [106] we formulate the problems on a directed graph. We will show how the models change in the case where the graph is undirected. Although different papers often use different notations, for the ease of presentation we have chosen a standard notation that we will use throughout the chapter to present the various problems. A complete graph $G = (V,A)$ is given, where $V = \{0,\dots,n\}$ is the set of vertices and A is the set of arcs. Vertices in $N = V \setminus \{0\} = \{1,\dots,n\}$ correspond to the customers, and vertex 0 corresponds to the depot where the routes start and end. For any subset of vertices $S \subset V$, we define $\delta^+(S) = \{(i,j) \in A : i \in S, j \notin S\}$ and $\delta^-(S) = \{(i,j) \in A : i \notin S, j \in S\}$. For the ease of presentation, in the following we will use the notation $\delta^+(i)$ and $\delta^-(i)$ when $S = \{i\}$. Two nonnegative values may be associated with each arc $(i,j) \in A$: a traveling cost c_{ij} and a traveling time t_{ij}. In most of the problems only one of these two values is relevant. A nonnegative profit p_i is associated with each customer i, while a profit $p_0 = 0$ is associated with the depot. The profit of each customer can be collected at most once. One or more vehicles are available to collect the profit of a subset of customers. Each vehicle route starts from and ends at the depot. While in some problems a time limit $T_{\max}$ is set on the time duration of a route, in some others a minimum value $p_{\min}$ is imposed on the total profit to be collected.

In Table 10.1 we summarize the basic VRPPs and their characteristics. In the first column we indicate whether it is a single- or multiple-vehicle problem; then we give the name we use for the problem and, in parentheses, the other names used in the literature, if any. In the following columns, we show the objective and the kind of constraints. For short we write "max profit" to mean "maximize profit collected". When we simply write "min cost" we mean "minimize the routing cost". While the first three problems

Table 10.1. *Summary of VRPPs.*

Vehicles	Problem name	Objective	Constraints
single	Orienteering Problem (OP) (Selective TSP, Maximum Collection Problem, Bank Robber Problem)	max profit	route duration
single	Profitable Tour Problem (PTP)	max (profit - cost)	–
single	Prize Collecting TSP (PCTSP)	min cost	route profit
multiple	Team Orienteering Problem (TOP) (Multiple Tour Maximum Collection Problem)	max profit	route duration
multiple	Capacitated PTP (CPTP)	max (profit - cost)	capacity
multiple	Capacitated Prize Collecting VRP (CPCVRP)	min cost	route profit capacity
multiple	VRP with Private Fleet and Common Carrier (VRPPC)	min cost	limited fleet capacity

with multiple vehicles are uniquely defined problems, extensions of their single vehicle counterparts listed in the upper part of the table, the last problem, the VRPPC, refers to a class of distribution problems where customers are considered for outsourcing.

10.2 ▪ Single-Vehicle Case

In this section we survey the literature dedicated to the three basic VRPPs with a single vehicle: the *Orienteering Problem* (OP, Section 10.2.1), the *Profitable Tour Problem* (PTP, Section 10.2.2), and the *Prize-Collecting Traveling Salesman Problem* (PCTSP, Section 10.2.3). We first present a general formulation for the VRPP with a single vehicle and then specify, in each of the following sections, how each basic problem can be obtained from it.

We introduce the following variables:

- y_i = binary variable equal to 1 if vertex $i \in V$ is visited by the vehicle route, and 0 otherwise;

- x_{ij} = binary variable equal to 1 if arc $(i,j) \in A$ is traversed by the vehicle, and 0 otherwise.

The mathematical programming formulation of the directed version of the VRPP with a single vehicle is the following, where α is a nonnegative value:

$$(10.1) \qquad \text{maximize } \alpha \sum_{i \in V} p_i y_i - \sum_{(i,j) \in A} c_{ij} x_{ij}$$

$$(10.2) \qquad \text{s.t.} \quad \sum_{(i,j) \in \delta^+(i)} x_{ij} = y_i \qquad \forall\, i \in V,$$

$$(10.3) \qquad \sum_{(j,i) \in \delta^-(i)} x_{ji} = y_i \qquad \forall\, i \in V,$$

$$(10.4) \qquad \sum_{(i,j) \in \delta^+(S)} x_{ij} \geq y_h \qquad \forall\, S \subseteq V \setminus \{0\},\ h \in S,$$

$$(10.5) \qquad \sum_{(i,j) \in A} t_{ij} x_{ij} \leq T_{\max},$$

$$(10.6) \qquad \sum_{i \in V} p_i y_i \geq p_{\min},$$

$$(10.7) \qquad y_i \in \{0,1\} \qquad \forall\, i \in V,$$

$$(10.8) \qquad x_{ij} \in \{0,1\} \qquad \forall\, (i,j) \in A.$$

The objective function (10.1) maximizes the difference between collected profit, multiplied by α, and traveling cost. Constraints (10.2) and (10.3) ensure that one arc enters and one arc leaves each visited vertex. Subtours are eliminated through (10.4). Constraint (10.5) is the maximum duration constraint on the route, while (10.6) imposes collecting a profit not smaller than $p_{\min}$. Finally, (10.7) and (10.8) are variable definitions.

In case of an undirected graph $G = (V, E)$, a variable x_e is defined for each edge $e = (i,j) \in E$. In this case we define c_e and t_e as the traveling cost and the traveling time associated with each edge $e \in E$, respectively. Moreover, let us define $\delta(S) = \{e = (i,j) \in$

$E : (i \in S, j \notin S)$ or $(i \notin S, j \in S)\}$ and $\delta(i)$ when $S = \{i\}$. The formulation changes as follows:

$$(10.9) \qquad \text{maximize } \alpha \sum_{i \in V} p_i y_i - \sum_{e \in E} c_e x_e$$

$$(10.10) \qquad \text{s.t.} \quad \sum_{e \in \delta(i)} x_e = 2y_i \qquad\qquad \forall\, i \in V,$$

$$(10.11) \qquad\qquad \sum_{e \in \delta(S)} x_e \geq 2y_h \qquad\qquad \forall\, S \subseteq V \setminus \{0\},\ h \in S,$$

$$(10.12) \qquad\qquad \sum_{e \in E} t_e x_e \leq T_{\max},$$

$$(10.13) \qquad\qquad \sum_{i \in V} p_i y_i \geq p_{\min},$$

$$(10.14) \qquad\qquad y_i \in \{0,1\} \qquad\qquad \forall\, i \in V,$$

$$(10.15) \qquad\qquad x_e \in \{0,1\} \qquad\qquad \forall\, e \in E \setminus \delta(0),$$

$$(10.16) \qquad\qquad x_e \in \{0,1,2\} \qquad\qquad \forall\, e \in \delta(0).$$

The meaning of the objective function and of the constraints is analogous to the case of a directed graph.

10.2.1 ▪ The Orienteering Problem

In this section we introduce the OP and we overview the exact, heuristic, and approximation algorithms proposed in the literature for its solution.

In the OP, one vehicle is available at the depot with maximum route duration $T_{\max}$. The OP is the problem of finding a vehicle route that maximizes the total collected profit while satisfying the maximum duration constraint on the route. The formulation for the OP is obtained from (10.1)–(10.8) by setting $\alpha = 1$, $c_{ij} = 0$ for each $(i,j) \in A$, and $p_{\min} = 0$.

We now present the solution methods described in the literature to solve the OP. We first focus on exact algorithms, then on heuristics, and finally on approximation algorithms.

Exact Algorithms. Hayes and Norman [59] proposed the first exact algorithm to solve the OP which is based on dynamic programming. The algorithm was tested on a real problem related to the sport of orienteering in a mountain area where traveling times were approximated through an estimation which takes into account uphill and downhill steps.

Laporte and Martello [74] defined a scheme to obtain an upper bound on the optimal value of the OP based on the solution of the following knapsack problem:

$$(10.17) \qquad \text{maximize } \sum_{i \in V} p_i y_i$$

$$(10.18) \qquad \text{s.t.} \quad \sum_{i \in V \setminus \{0\}} w_i y_i \leq T_{\max} - w_0,$$

$$(10.19) \qquad\qquad y_i \in \{0,1\} \qquad\qquad \forall\, i \in V \setminus \{0\},$$

where $w_i = \alpha \min_{j \neq i} \{t_{ij}\} + (1-\alpha) \min_{k \neq i} \{t_{ki}\}$. This upper bound was used in an enumerative algorithm based on the idea of extending a simple path emanating from the depot through a *Branch-and-Bound* (BB) scheme. The authors presented also two simple heuristic algorithms to obtain lower bounds based on the application of the classical nearest neighbor and cheapest insertion algorithms for the TSP. These algorithms were adapted to the OP by inserting a stopping criterion which checks the violation of the maximum route duration constraint. Instances with a number of vertices between 10 and 90 were solved to optimality. Millar and Kiragu [80] presented a mathematical formulation with a polynomial number of constraints. The formulation is enhanced by introducing an upper bound on the number of vertices visited in an optimal solution. This bound is obtained through a modification of the bounding scheme developed by Laporte and Martello [74]. Instances with 10 vertices were solved through the use of CPLEX. Leifer and Rosenwein [75] proposed new valid inequalities that were added to the formulation presented in [74]. They developed a procedure to obtain upper bounds by solving three successive linear programs. Tests were made on instances introduced in Tsiligirides [105] and showed that the deviation between the value of the upper bound and the best known solution was always lower than 5%.

Ramesh, Yoon, and Karwan [90] developed an exact algorithm using a Lagrangian relaxation where flow conservation and maximum route duration constraints were relaxed. The solution algorithm is a BB where the problem is iteratively solved improving the value of Lagrangian multipliers in the first phase while branching is performed in the second phase. They solved instances with up to 150 vertices. Kataoka, Yamada, and Morito [66] proposed a new relaxation scheme to the OP based on the *Minimum Directed* 1-*Subtree Problem* (MD1SP). Two methods were developed to improve the lower bound given by the solution of the MD1SP: the first one is based on the iterative introduction of violated valid inequalities, while the second one is based on Lagrangian relaxation. The two methods were combined together in a solution procedure which was tested on instances with up to 1000 customers. Results showed that the proposed relaxation was superior to the ordinary assignment relaxation. Fischetti, Salazar González, and Toth [46] presented a *Branch-and-Cut* (BC) algorithm using several families of valid inequalities. They solved instances with up to 500 vertices.

Heuristic Algorithms. The literature on heuristic algorithms for the OP is broader than the one on exact approaches. Tsiligirides [105] presented the first heuristic algorithms for the OP, which were a deterministic and a stochastic one. In the stochastic algorithm, different routes are generated using a Monte Carlo process and the best one is chosen. Each route is constructed by inserting customers sequentially on the basis of a desirability measure, given by the ratio between the profit and the closeness to the last inserted customer. The deterministic algorithm builds the route by dividing the Euclidean space in different areas. All customers in a given area are inserted, if feasible, before moving to a different area. Tests were made on instances with up to 33 vertices and showed that the stochastic algorithm is superior. Golden, Levy, and Vohra [57] used a center of gravity heuristic composed by three steps: construction, improvement, and calculation of the center of gravity. In the construction phase, a solution is built by inserting customers sequentially following a given order. In the improvement phase, the 2-opt procedure, proposed by Lin [77] for the TSP, is applied followed by cheapest insertion. Finally, the center of gravity of the route is calculated. Customers are then ordered on the basis of their distance with respect to the center of gravity and the procedure is repeated. Computational tests were made on the instances introduced by Tsiligirides [105] and showed that the center of gravity heuristic was better than both algorithms given in [105]. Keller [69] proposed a new construction heuristic which inserts a customer at each iteration. The customer is chosen on the basis of the ratio between the profit and the distance with respect to the last customer inserted. The algorithm was compared with those by Tsiligirides [105] and by Golden, Levy, and Vohra [57]. Computational results showed that the new algorithm is the best among the competitors. Golden, Wang, and Liu [58] introduced a heuristic that combines previous algorithmic concepts, along with learning capabilities. The heuristic was compared with the algorithms proposed in [105] and [57] on the instances from [105]. The results showed that it beat both previous approaches and it was much faster than the center of gravity algorithm in [57]. Ramesh and Brown [89] developed a four-phase algorithm where the phases are vertex insertion, improvement, vertex deletion, and maximal insertions. The insertion of a vertex is made with a rule which is similar to the one by Tsiligirides [105]. The improvement phase uses the k-opt algorithm for the TSP. The algorithm was shown to beat the heuristics given in [105] and [57]. Chao, Golden, and Wasil [29] introduced a new heuristic composed by two phases: initialization and improvement. The initialization phase constructs different routes with the cheapest insertion method where the first customer inserted is changed from one solution to another. Then, in the improvement phase customer exchanges between different routes or intra-route moves are performed. Tests were made on the instances given in [105] and on new instances. Despite its simplicity, the heuristic compared favorably with respect to all previous heuristic algorithms.

All the heuristic methods cited above use classical local ascent schemes and thus tend to become trapped in local optima. In the following years researchers concentrated on metaheuristic schemes for the OP which could overcome this problem. A neural network based heuristic was proposed by Wang et al. [111], where a state matrix was used to generate solutions and perturbed at each iteration in order to obtain different solutions. Tests were made on the instances from Tsiligirides [105]. Results showed that the neural network heuristic performed much better than the stochastic heuristic in [105] and slightly worse than that of Chao, Golden, and Wasil [29]. The first tabu search heuristic for the OP was presented by Gendreau, Laporte, and Semet [52]. The algorithm iteratively inserts clusters of customers or removes chains of vertices. The algorithm was tested on randomly generated instances with up to 300 vertices. Results were compared with optimal solutions provided by the BC algorithm described in Gendreau, Laporte, and Semet [53] and proved that the tabu search was able to find optimal or near optimal solutions in a very short computing time. The gap with respect to the optimum was typically less than 1%. Liang, Kulturel-Konak, and Smith [76] proposed an ant colony and a tabu search heuristic. Both algorithms are based on standard schemes for ant colony optimization and tabu search. Tests were made on the instances of Tsiligirides [105]. Results showed that the solution quality was comparable to the one produced by the heuristic of Chao, Golden, and Wasil [29]. A heuristic, based on the *Greedy Randomized Adapted Search Procedure* (GRASP) and the path relinking methods, was proposed by Campos et al. [27]. Computational results show good quality solutions obtained in short computing times.

Approximation Algorithms. The first constant-factor approximation algorithm for the OP can be found in Blum et al. [20]. The approximation ratio depends on the approximation factor for the min-excess path problem, for which a constant factor approximation algorithm was proposed in the same paper. Awerbuch et al. [12] introduced an approximation algorithm for the *unrooted* OP, i.e., the problem where no starting and ending vertex is defined. The authors proposed an algorithm which is able to find a route of length $O(T_{\max} \log^2(\min(n, R)))$, where R is the desired value of the total profit to be collected. An approximation result for the directed OP is due to Nagarajan and Ravi [82], who presented an $O(\frac{\log^2(n)}{\log\log n})$ algorithm. Approximation results both for the directed and the undirected case of the OP can be found in Chekuri, Korula, and Pál [30]. A $(2 + \epsilon)$-approximation algorithm was presented for undirected graphs while a $O(\log^2(OPT))$ algorithm was described for the directed case, where OPT is the number of nodes in the optimal solution. A more detailed discussion of these approximation results for the OP can be found in Gavalas et al. [51].

Angelelli et al. [1] studied the undirected OP together with the variant where a service time is associated with the visit of a customer. They studied the complexity of classes of instances corresponding to special structures of the underlying graph, like paths, cycles, stars, and trees. They proved that the problem with service time is always NP-hard, while, when no service time is considered, the problem is polynomially solvable on cycles and on paths and it is NP-hard on stars and trees. Polynomial algorithms for the polynomially solvable cases and fully polynomial-time approximation schemes for some NP-hard cases were presented.

10.2.2 ▪ The Profitable Tour Problem

In this section we present the PTP and overview the solution algorithms proposed in the literature.

The PTP is the problem of finding a vehicle route that maximizes the difference between the total collected profit and the total traveling cost. No constraint is imposed on the vehicle route. The formulation for the PTP is obtained from (10.1)–(10.8) by setting $\alpha = 1$, $p_{\min} = 0$, and $T_{\max} = +\infty$.

Note that

$$\text{maximize} \sum_{i \in V} p_i y_i - \sum_{(i,j) \in A} c_{ij} x_{ij}$$

is equivalent to

$$(10.20) \qquad -\text{minimize} \left(\sum_{(i,j) \in A} c_{ij} x_{ij} - \sum_{i \in V} p_i y_i + \sum_{i \in V} p_i \right)$$

$$= -\text{minimize} \left(\sum_{(i,j) \in A} c_{ij} x_{ij} + \sum_{i \in V} p_i (1 - y_i) \right).$$

Thus, maximizing the difference between the total collected profit and the total traveling cost is equivalent to minimizing the sum of the total traveling cost and the total uncollected profit. Moreover, note that, by subtracting the profit of a customer to each arc outgoing (or ingoing) from each customer, the objective function of the PTP becomes

$$-\text{minimize} \sum_{i \in V} \tilde{c}_{ij} x_{ij},$$

where $\tilde{c}_{ij} = c_{ij} - p_i$, $i \in V$. With the objective function (10.21), the formulation of the PTP corresponds to the one of the *Elementary Shortest Path Problem* (ESPP), as discussed in Drexl and Irnich [40].

To the best of our knowledge, neither exact approaches nor computational analysis of heuristic algorithms were specifically proposed for the PTP. The only exception is the work by Dell'Amico, Maffioli, and Väbrand [37], where a bounding procedure based on Lagrangian relaxation was introduced. Tests were made on instances with up to 500 vertices.

The PTP received, instead, attention in terms of approximation results, probably due to its simple structure. Several approximation results appeared in the literature under the assumption that the c_{ij} satisfy the triangle inequality. Bienstock et al. [19] provided a 2.5-approximation algorithm which was then improved by Goemans and Williamson [55], where a 2-approximation algorithm was provided. Archer et al. [3] were the first to break the barrier of 2 and provide an improvement of the primal-dual algorithm of Goemans and Williamson [55]; their performance guarantee was $2 - \epsilon$. Goemans [54] showed that, by combining the rounding algorithm given in [19] and the primal-dual algorithm in [55], it was possible to obtain a guarantee $\frac{1}{1-\frac{2}{3}e^{-1/3}}$. A $\lceil \log(n) \rceil$-approximation algorithm for the directed version of the PTP was recently presented by Nguyen [83] which improved a previous result proposed by Nguyen and Nguyen [84]. Angelelli et al. [1] studied the undirected PTP together with the variant where a service time is associated with the visit of a customer. Polynomially solvable classes of instances, corresponding to special structures of the underlying graph, were also identified and studied.

10.2.3 ▪ The Prize-Collecting Traveling Salesman Problem

In this section we introduce the PCTSP and overview the algorithms developed for this problem. We adopt the definition of PCTSP used by Feillet, Dejax, and Gendreau [45] and in Vansteenwegen, Souffriau, and Van Oudheusden [106].

The PCTSP is the problem of finding a vehicle route that minimizes the total traveling cost with the constraint that the total collected profit is at least $p_{\min}$. The formulation for the PCTSP is obtained from (10.1)–(10.8) by setting $\alpha = 0$ and $T_{\max} = +\infty$.

In the paper which first introduced the name PCTSP, due to Balas [13], the objective function is the sum of the total traveling cost and a total penalty for the non-visited customers. Denoting by γ_i the penalty associated with customer i, $i \in V \setminus \{0\}$, to distinguish this more general problem we call it PCTSP *with penalties*. The objective function becomes

$$(10.21) \qquad \text{minimize} \sum_{(i,j)\in A} c_{ij}x_{ij} + \sum_{i\in V\setminus\{0\}} \gamma_i(1-y_i),$$

while the constraints remain the same.

The objective function (10.21) is equivalent to the maximization of the difference between the penalties (that can be interpreted now as profits) of the visited customers and the total distance traveled (see the transformation (10.20)). Then, the problem turns out to be a generalization of the PTP, obtained when $p_{\min} = 0$, and of the PCTSP, obtained when $\gamma_i = 0$ for all i.

As in Balas [13], also in Dell'Amico, Maffioli, and Väbrand [37] two different values were associated with each customer i: a profit used in constraint (10.6), and a penalty used in the objective function.

Balas [13] studied the structural properties of the PCTSP with penalties, while in Balas [14] the same author provided a polyhedral study of the problem. He derived different classes of facet-defining inequalities from facet-defining inequalities for the asymmetric TSP. Balas [15] studied the complexity of the PCTSP with penalties and additional constraints on the structure of the route. In particular, he studied different precedence constraints among customers and showed that in all cases the problem can be solved in polynomial time.

We now review the solution approaches proposed in the literature.

Exact Algorithms.   Fischetti and Toth [48] introduced different mathematical formulations for the problem and an additive approach to obtain lower bounds from various problem relaxations. These bounds were then embedded in a BB algorithm which was used to solve instances with up to 200 customers. Dell'Amico, Maffioli, and Väbrand [37] introduced the same relaxation illustrated above for the PTP to obtain a bound also for the PCTSP and tested it on the same set of instances. Bérubé, Gendreau, and Potvin [17] proposed a BC algorithm for the PCTSP where different classes of valid inequalities were used to strengthen the formulation. Tests were made on instances generated from standard VRP and TSP benchmark instances with up to 532 vertices. Computational results showed that the solution time was strictly related to the value of the minimum value of the profit required in constraint (10.6).

Heuristic Algorithms.   To the best of our knowledge, the only heuristic algorithm available in the literature for the PCTSP with penalties is a Lagrangian heuristic proposed by Dell'Amico, Maffioli, and Sciomachen [36]. It starts from a lower bound to the problem and makes the solution feasible. The heuristic was evaluated on randomly generated instances and real-world ones with up to 500 vertices. The solution was compared with the optimal one, when available, or with an upper bound. The performance of the heuristic improved when the minimum prize required in constraint (10.6) increased.

Approximation Algorithms.   An approximation algorithm for the undirected PCTSP with penalties was presented by Awerbuch et al. [12]. The result was derived from previous results for the minimum-weight k-tree problem presented in Ravi et al. [91]. The authors, as done for the OP, proposed an approximation algorithm for the unrooted OP, i.e., the problem where no starting and ending vertex is defined. The algorithm is able to find a route of length $O(T_{\max} \log^2(\min(n, R)))$, where R is the desired value of the total profit to be collected. Angelelli et al. [1] studied the undirected PCTSP together with the variant where a service time is associated with the visit of a customer. They studied the complexity of classes of instances corresponding to special structures of the underlying graph, like paths, cycles, stars, and trees. They proved that the problem with service time is always NP-hard, while, when no service time is considered, the problem is polynomially solvable on cycles and on paths and it is NP-hard on stars and trees. Polynomial algorithms for the polynomially solvable cases and fully polynomial-time approximation schemes for some NP-hard cases were presented.

Ausiello, Bonifaci, and Laura [11] studied the online version of the PCTSP with penalties where customers are disclosed over time. For this problem they derived a 7/3 approximation algorithm which is not far from the best possible one since they proved that the competitive ratio of any approximation algorithm for the online PCTSP is at least two.

10.2.4 ▪ Variants

Wang, Golden, and Wasil [112] and Silberholz and Golden [95] considered a variant of the OP that differs from the basic problem mainly in the objective function that is a nonlinear function of the profits collected from the visited vertices. In this problem each city is assigned a number of scores for different attributes and the overall function to optimize is a function of these attribute scores. A different variant of the OP, where a limited resource is consumed on the traversed arcs and on the visited nodes, was studied by Pietz and Royset [86], who presented a specialized BB algorithm. Two variants of the OP with variable profits were introduced by Erdoğan and Laporte [42]. In the first variant part of the profit of a vertex is collected at each visit if a predefined amount of time is spent at the vertex, and in the second one the collected profit is a function of the time spent at the vertex. They proposed a BC algorithm that was tested on randomly generated instances obtained from benchmark test problems.

Kantor and Rosenwein [64] studied the *OP with Time Windows* (OPTW) and proposed a heuristic. Righini and Salani [92] introduced an exact algorithm for the OPTW which is based on a bi-directional and bounded dynamic programming algorithm with state space relaxation. Fomin and Lingas [49] studied a different generalization of the OP, called *Time-Dependent OP*, which is the problem where the traveling time between two locations depends on the starting time. They proposed an approximation algorithm. The same problem was considered by Verbeeck et al. [110], who proposed a heuristic that combines the principle of an ant colony system with a time-dependent local search procedure. Erkut and Zhang [43] developed a penalty-based greedy heuristic and a BB algorithm for a variant of the OP where the profit is a time-dependent decreasing linear function. Deitch and Ladany [35] studied a more general problem than the OP, called the *One-Period Bus Touring Problem*, where an attractiveness, that we can also see as a profit, is associated also with edges. They presented a transformation of the problem into the OP together with a heuristic that was compared with one of the heuristics proposed by Tsiligirides [105]. The OP with a given set of compulsory vertices is studied by Gendreau, Laporte, and Semet [53]. A BC algorithm based on several families of valid inequalities was proposed that can solve to optimality instances with up to 300 vertices.

The *OP with stochastic profits* is the problem of finding a route that visits a subset of vertices within a pre-specified time limit and maximizes the probability of collecting more than a pre-specified target profit level. This variant of the OP was studied by Ilhan, Iravani, and Daskin [61]. Campbell, Gendreau, and Thomas [26] introduced a variant of the OP where travel and service times are stochastic. Special cases of the problem were solved exactly and heuristics presented for the general problem. Different ways to approximate the objective function are compared by Papapanagiotou, Montemanni, and Gambardella [85]. Stochastic travel times were also considered by Evers et al. [44], who formulated a two-stage stochastic model with recourse.

The only studied variant of the PTP is the *Capacitated PTP* (CPTP), where each customer has a demand and the vehicle has a prefixed capacity which must not be exceeded by the route. Jepsen [62] proposed a BC algorithm for the solution of the undirected version of the CPTP where instances with up to 800 vertices were solved to optimality. Tang and Wang [100] recently introduced an iterated local search heuristic for the Capacitated PCTSP with penalties to model a real-world application related to the optimal scheduling of rolling mills in the steel industry.

Finally, some authors considered variants with multiple objectives. The bi-objective version of the OP was studied by Jozefowiez, Glover, and Laguna [63], who developed a heuristic to find the efficient frontier. The same problem was considered in Bérubé, Gendreau, and Potvin [18]. Schilde et al. [94] studied a multi-objective version of the OP where each vertex may provide different benefits belonging to different categories (e.g., culture, history, leisure).

10.3 ▪ Multiple-Vehicle Case

As mentioned in the introduction, the study of multiple VRPs with profits has started relatively recently and the area is still open for research. We overview in this section the main results presented in the literature which mainly focus on the extension to the case with multiple vehicles of the OP and some of its variants and on the *VRP with Private Fleet and Common Carrier* (VRPPC). The only contributions that do not fall in these categories concern the capacitated and multiple-vehicle version of the PTP (CPTP) that was considered by Archetti et al. [8] and, more recently, by Archetti, Bianchessi, and Speranza [6]. In both papers exact BP algorithms were presented and tested on benchmark instances derived from instances of the VRP with up to 200 vertices. To the best of our knowledge, no work has been done on the *Capacitated Prize Collecting VRP* (CPCVRP), which is the extension to the multiple-vehicle case of the PCTSP. In the CPCVRP a demand is associated with each customer, vehicles are capacitated, and a minimum profit must be collected, either by each route or in total.

10.3.1 ▪ The Team Orienteering Problem

The multiple-vehicle extension of the OP was first introduced by Butt and Cavalier [24], who called it the *Multiple Tour Maximum Collection Problem*. The name *Team Orienteering Problem* (TOP) was coined by Chao, Golden, and Wasil [28] to highlight the connection with the more widely studied single-vehicle case. More precisely, given a set K of vehicles, the TOP calls for the determination of at most $|K|$ vehicle routes that maximize the total collected profit, while satisfying a maximum duration constraint for each route. Several real-world applications of the TOP are mentioned in the literature, such as athlete recruiting (Chao, Golden, and Wasil [28]), technician routing (Tang and Miller-Hooks [99]), and tourist trip planning (Vansteenwegen and Van Oudheusden [109]).

Vehicle flow models, extending those presented in Section 10.2.1, are proposed in the literature (see, e.g., Tang and Miller-Hooks [99]). We present here a formulation for the directed case of the TOP that is an extension of the one presented for the directed OP in Section 10.2.1 and uses the following decision variables:

- y_{ik} = binary variable equal to 1 if vertex $i \in V$ is visited by vehicle route $k \in K$, and 0 otherwise;

- x_{ijk} = binary variable equal to 1 if arc $(i,j) \in A$ is traversed by vehicle route $k \in K$, and 0 otherwise.

The mathematical programming formulation for the directed TOP is as follows:

(10.22) (TOP1) maximize $\sum_{i \in V} p_i \sum_{k \in K} y_{ik}$

(10.23) s.t. $\sum_{j \in V} x_{ijk} = y_{ik}$ $\forall\, i \in V,\, k \in K$,

(10.24) $\sum_{j \in V} x_{jik} = y_{ik}$ $\forall\, i \in V,\, k \in K$,

(10.25) $\sum_{k \in K} y_{0k} \leq |K|$,

(10.26) $\sum_{k \in K} y_{ik} \leq 1$ $i \in V \setminus \{0\}$,

(10.27) $\sum_{(i,j) \in \delta^+(S)} x_{ijk} \geq y_{hk}$ $\forall\, S \subseteq V \setminus \{0\},\, h \in S,\, k \in K$,

(10.28) $\sum_{(i,j) \in A} t_{ij} x_{ijk} \leq T_{\max}$ $\forall\, k \in K$,

(10.29) $y_{ik} \in \{0,1\}$ $\forall\, i \in V,\, k \in K$,

(10.30) $x_{ijk} \in \{0,1\}$ $\forall\, (i,j) \in A,\, k \in K$.

The objective function and most of the constraints are extensions to the multiple-vehicle case of those presented for the OP. Constraint (10.25) limits the number of routes to be at most $|K|$, while constraints (10.26) impose that each customer be visited at most once. Finally, the constraints (10.27) impose that each route be connected and (10.28) limit the maximum distance for each route.

Several authors based their exact methods on route-based formulations, as an alternative to vehicle flow formulations. Let Ω be the set of all feasible routes, i.e., elementary routes with length at most T_{max}, and let A be a matrix of binary coefficients a_{ir} describing such routes. More precisely, for each route $r \in \Omega$ the coefficient a_{ir} takes value 1 if vertex $i \in N$ is included in route r, and 0 otherwise. Moreover, let P_r be the total profit of route r, defined as the sum of the profits of its customers. The set-packing model for the TOP (denoted as TOP2), proposed by Butt and Ryan [25], uses a binary variable λ_r for each route $r \in \Omega$ which takes value 1 if the route is selected in the optimal solution, and 0 otherwise:

$$(10.31) \qquad (\text{TOP2}) \text{ maximize} \sum_{r \in \Omega} P_r \lambda_r$$

$$(10.32) \qquad \text{s.t.} \quad \sum_{r \in \Omega} a_{ir} \lambda_r \le 1 \qquad \forall i \in N,$$

$$(10.33) \qquad \qquad \sum_{r \in \Omega} \lambda_r \le |K|,$$

$$(10.34) \qquad \qquad \lambda_r \in \{0, 1\} \qquad \forall r \in \Omega.$$

The objective function (10.31) maximizes the total profit. Constraints (10.32) ensure that each vertex is visited by at most one route, and constraint (10.33) limits the number of used routes.

Testing of algorithms is generally performed by using a large benchmark set proposed by Chao, Golden, and Wasil [28] (called CGW hereafter) and obtained by adapting instances originally proposed for the OP. The original set contained 353 instances with 21 to 102 customers and up to four vehicles, but was later enlarged to 387 instances in total (see, e.g., Boussier, Feillet, and Gendreau [23]) by considering additional values for T_{max}.

Exact Algorithms. The first exact solution procedure for the TOP was proposed by Butt and Ryan [25] for a heterogeneous fleet variant of the problem. They started from the set-packing formulation TOP2, and their algorithm makes use of both column generation and constraint branching. Pricing in the column generation step is performed through complete enumeration of the routes. Therefore, the algorithm is able to solve instances with up to 100 potential customers only when routes include a few customers each. A more effective BP algorithm was introduced by Boussier, Feillet, and Gendreau [23]. They again used as a base model TOP2 but defined the pricing subproblem as a *Resource Constrained Elementary Shortest Path Problem* (RCESPP) and then solved it by dynamic programming. Using various acceleration procedures in the column generation step, the algorithm is able to solve 270 instances with up to 100 potential customers from the set of 387 CGW benchmark instances. An effective bi-directional dynamic programming procedure and relaxation-based dominance procedures are the base of an enhanced BP approach by Keshtkaran et al. [70] which solves 10% more instances than the previous approaches (i.e., 301 out of the 387 instances of the benchmark).

Poggi de Aragão, Viana de Freitas, and Uchoa [87] defined a *Branch-and-Cut-and-Price* (BCP) algorithm based on a pseudo-polynomial formulation in which binary variables are associated with the visit of vertices and the traversal of an arc at a given time instant between 0 and $T_{\max}$. A Dantzig–Wolfe decomposition scheme was adopted to handle the potentially huge number of arc variables, and the resulting column generation algorithm was enhanced through the addition of min-cut and clique cuts. The computational tests showed that the root node upper bounds computed with this approach are generally better than those of Boussier, Feillet, and Gendreau [23] and that the additional cuts help in reducing the gap with respect to the BP bound. This is confirmed by the results obtained by more recent BC and BCP approaches. In particular, Dang, El-Hajj, and Moukrim [33] defined a BC approach based on model TOP1 which uses Generalized Subtour Elimination and Clique cuts to strengthen the linear relaxation and is able to close some open instances that were hard for BP. The subset-row inequalities are instead used by the BCP of Keshtkaran et al. [70] which, although on average is inferior with respect to the pure BP, is capable of solving some instances that were not solvable by BP.

The complementarity of the different exact approaches available for TOP is further illustrated by Table 10.2. For each existing algorithm we report in the last line the total number of solved instances within 2 hours of computing time, out of the 387 of the extended CGW benchmark set. The BCP of Poggi de Aragão, Viana de Freitas, and Uchoa [87] is not included in the table since the paper does not report complete results on the benchmark set. In the first four lines of the table we make a direct comparison of the algorithms by providing, for each ordered pair, the number of instances that the first algorithm is able to solve and the other is not. For example, the BC by [33] can solve 13 test instances the BP by [70] cannot solve, whereas this latter solves 36 instances unsolved by the BC. By observing the table we note that even though BCP approaches are not currently the best available ones, it is likely that a more effective integration of cuts within the BP may be the path to follow for future research, as happened for other variants of the VRP.

Table 10.2. *Comparison between exact approaches for TOP on the benchmark of Chao, Golden, and Wasil* [28] *including 387 instances.*

	BP [23]	BP [70]	BC [33]	BCP [70]
BP [23]	-	+8	+21	+6
BP [70]	+39	-	+36	+16
BC [33]	+29	+13	-	+19
BCP [70]	+27	+6	+32	-
Total solved instances	270	301	278	291

Heuristic Algorithms. The first heuristic proposed for the TOP is a simple construction algorithm introduced by Butt and Cavalier [24] and tested on small-sized instances with up to 15 vertices by comparing the results with a mathematical programming model similar to TOP1. Chao, Golden, and Wasil [28] developed a more sophisticated construction heuristic in which the initial solution was refined through customer moves and exchanges and various restart strategies. The resulting algorithm was tested on the original CGW set of 353 instances with up to 102 customers and four vehicles.

More recently, several metaheuristics were applied to the TOP, starting from the tabu search algorithm introduced by Tang and Miller-Hooks [99] which is embedded in an adaptive memory procedure that alternates between small and large neighborhoods during the search and outperforms previous heuristics. Archetti, Hertz, and Speranza [9] proposed two variants of a generalized tabu search algorithm and a variable neighborhood search algorithm. Ke, Archetti, and Feng [67] used an ant colony optimization approach which uses four different methods to construct candidate solutions. Other metaheuristic paradigms were successfully applied to the TOP, such as guided local search (Vansteenwegen et al. [107]), path relinking (Souffriau et al. [96]), memetic algorithms (Bouly, Dang, and Moukrim [22]), particle swarm optimization-based memetic algorithm (Dang, Guibadj, and Moukrim [34]), and augmented large neighborhood search (Kim, Li, and Johnson [71]), the latter two being the current best in class.

10.3.2 ▪ Variants of the Team Orienteering Problem

Archetti et al. [8] studied the capacitated version of the TOP, called *Capacitated TOP* (CTOP). In this problem a demand is associated with each customer and each vehicle has a maximum capacity. The objective is to maximize the total collected profit while satisfying the capacity and duration constraint for each route. As previously mentioned in this paper also the CPTP was studied. The authors proposed a BP algorithm adapted from the one defined by Boussier, Feillet, and Gendreau [23] for the TOP. The algorithm is able to solve instances derived from classical VRP instances and including up to 200 customers. Heuristic algorithms were also proposed for both problems adapted from the heuristics described by Archetti, Hertz, and Speranza [9] for the TOP. An improved BP for both CTOP and CPTP was developed by Archetti, Bianchessi, and Speranza [6]. The algorithm implements the bi-directional dynamic programming and decremental state space relaxation of Righini and Salani [92] and clearly obtains better results than those of Archetti et al. [8]. Two new heuristic algorithms for the CTOP were simultaneously proposed by Luo et al. [79] and Tarantilis, Stavropoulou, and Repoussis [101] and compared with the heuristics presented in Archetti et al. [8]. In Luo et al. [79] the algorithm uses the ejection pool framework with an adapted strategy and a diversification mechanism and finds 16 new best solutions among 120 benchmark instances. Tarantilis, Stavropoulou, and Repoussis [101] proposed a method that adopts a hierarchical bi-level search framework. The best tested version of the method finds 18 new best solutions on benchmark instances.

Archetti, Bianchessi, and Speranza [7] studied a variant of the CTOP where incomplete service to the customers is allowed and developed a BP algorithm for its exact solution. In Archetti et al. [4] split deliveries are allowed in CTOP, but each customer has to be either entirely served or not served, while in Archetti et al. [5] incomplete service is also allowed. A BP exact algorithm and heuristic algorithms were proposed for the solution of both problems.

The *TOP with Time Windows* (TOPTW) received considerable attention from the heuristic community in the last few years. Vansteenwegen et al. [108] proposed an iterated local search which quickly produces good solutions on large instances derived from VRPTW ones. Montemanni and Gambardella [81] developed an ant colony optimization approach which was later improved by Gambardella, Montemanni, and Weyland [50]. Other metaheuristic approaches for the TOPTW were recently introduced and overall obtained very good average results on benchmark instances. Tricoire et al. [104] applied a *Variable Neighborhood Search* (VNS) developed for the multi-period variant of the problem. Labadie, Melechovský, and Wolfler Calvo [73] used a hybrid approach which combines a greedy randomized adaptive search procedure and an evolutionary search, whereas Labadie et al. [72] presented a VNS and a hybrid approach which combines VNS with the granular search by Toth and Vigo [103]. Lin and Yu [78] used two heuristics based on the simulated annealing paradigm. The most recent heuristic for the solution of the TOPTW is due to Hu and Lim [60], who proposed a three-component heuristic. The first two components are a local search procedure and a simulated annealing procedure, whereas the third component recombines the routes. The computational results show that their algorithm outperforms previous approaches. They obtained 35 new best solutions.

A variant of the TOP called *Multi-District TOP* (MDTOP) was studied by Salazar-Aguilar, Langevin, and Laporte [93]. In this problem a set of mandatory and optional tasks located in several districts must be scheduled over a planning horizon to maximize the total collected profit. A sequence of tasks assigned to a group of workers cannot have a duration that exceeds a time limit. When interpreted in terms of routing problem with profits, in the MDTOP a set of customers is mandatory and incompatibility constraints are present among customers.

A variant of the TOP where rewards are linearly decreasing over time was studied by Ekici and Retharekar [41]. A heuristic, called cluster-and-route algorithm, was proposed and tested on randomly generated instances obtained by modifying TOP instances.

Vansteenwegen and Van Oudheusden [109] introduced a class of problems, called *Tourist Trip Design Problems* (TTDPs), that is associated with the application of creating a feasible plan for tourists to visit attractions within the available time span. Such a problem class, also considered in Souffriau et al. [97], generalizes the TOP by considering time windows and other practical constraints. The OP with hotel selection introduced by Divsalar, Vansteenwegen, and Cattrysse [38] belongs to this class and is aimed at determining a fixed number of connected trips starting and ending in one of the hotels. A memetic algorithm was proposed for this problem by Divsalar et al. [39]. An extended survey on the class of TTDPs is presented in Gavalas et al. [51].

Uncertainty in the TOP was modeled through robust optimization by Ke et al. [68]. They allowed all data to vary in an interval and proposed a BP algorithm to solve the resulting robust counterpart of the TOP.

10.3.3 ▪ The VRP with Private Fleet and Common Carrier

An important application area of VRPs with profits arises in the context of the *Small Package Shipping* (SPS) industry and is related to service outsourcing of unprofitable customers. In fact, large SPS companies outsource depot operations and last-mile deliveries of unprofitable areas to small regional suppliers, called subcontractors. Subcontracted services are generally efficient in rural areas characterized by few customers and long distances, where a subcontractor may operate profitably by bundling deliveries. In general, SPS companies pay subcontractors per parcel delivered; hence the cost for the SPS company is independent from routing decisions of subcontractors.

The minimization of the overall cost of an SPS company using subcontractors must consider two elements at a time. The first one is the identification of the unprofitable customers that are outsourced at a fixed cost, and the second is the routing for the private fleet required to serve the remaining customers. Despite the practical interest, the literature on routing with outsourcing decision is still relatively scarce. A single-depot routing problem with outsourcing options was first introduced by Chu [31]. The problem, which was later named *VRP with Private Fleet and Common Carrier* (VRPPC), considers a private fleet of vehicles with limited capacity and fixed cost per use. A set of customers with known demand can be served by the private fleet which then incurs travel costs as in standard VRP. As an alternative, customers may be outsourced to a common carrier, and in such a case only fixed service costs must be paid. The objective is to minimize the total cost involving fixed costs for vehicles, variable travel costs, and fixed costs for orders performed by the common carrier. The VRPPC can be actually considered as a class of problems since some variants were also studied in the literature. Chu [31] proposed a simple heuristic based on a modified savings algorithm that was tested on five instances. Bolduc et al. [21] showed that the VRPPC can be modeled as a heterogeneous VRP and presented a metaheuristic based on a perturbation procedure. The algorithm greatly improved the results of [31] and was also tested on two new benchmark sets, one with homogeneous and one with heterogeneous fleet, with up to 483 customers. Two tabu search algorithms for VRPPC were proposed by Côté and Potvin [32] and Potvin and Naud [88]: the latter is based on ejection chain neighborhoods and is able to obtain very good results on both the homogeneous and the heterogeneous versions of the problem within a quite large computing time with respect to that required by Bolduc et al. [21].

Considering a single depot may be not realistic in the case of many SPS companies which manage several depots to serve huge areas including large, medium, and small towns. To optimize the whole delivery network, interdependencies between customer assignment and routing decisions of different self-owned depots have to be considered. Furthermore, instead of simply assuming the existence of a common carrier it is more appropriate to consider the location of potential subcontractors involved which are normally small regional carriers that have a small established depot characterized by a restricted delivery radius and a limited capacity. To this aim Stenger et al. [98] introduced a multiple-depot version of the problem denoted as MDVRPPC. For such a problem they defined a variable neighborhood search algorithm which implemented an effective adaptive mechanism to select routes and customers involved in the shaking step. The resulting algorithm was tested on a benchmark set of instances derived from *Multi-Depot VRP* (MD-VRP) ones showing the potential benefits associated with subcontracting. The algorithm is also capable of obtaining state-of-the-art results on both the single depot VRPPC and on the MDVRP.

10.4 ▪ Conclusions and Future Research Directions

In this chapter we reviewed the large family of vehicle routing problems with profits. These problems are widely studied due to the practical relevance of the applications they model and the scientific interest of their structure in which not all customers have to be served. We provided a homogeneous description of the main problems of the family since in the literature some discrepancies in problem definitions and naming are encountered. Then for each problem and variants we reviewed the most relevant results with particular attention to the computational testing of the proposed methods.

In this area most of the research has been devoted to the single-vehicle case and to the so-called orienteering variants, and relatively little attention was given to multiple-vehicle extensions. Therefore, there is still considerable room for valuable and systematic research in this field, particularly for unified approaches capable of successfully tackling several variants. Furthermore, many additional characteristics, such as multiple depots, heterogeneous fleet, and customer clustering may be added to the basic problems to model specific real-world applications such as SPS and tourist trip determination.

As a general comment on the solution approaches proposed in the literature for vehicle routing problems with profits, we can say that, concerning the heuristic algorithms, the most successful ones combine classical procedures for the TSP and the VRP, aimed at optimizing the route length, with the use of neighborhoods which determine the insertion or the removal of customers from the routes. If we instead focus on exact approaches, BC is the leading methodology for the solution of single-vehicle problems, while BP is the most applied procedure for the multiple-vehicle case. In all the BP algorithms described in the literature, the choice of the set of customers to be served is made in the master problem, while the subproblem turns out to have the same structure as the subproblem obtained in column generation for the VRP; i.e., it is an elementary shortest path problem with resource constraints. Recently, these approaches turned out to be competitive when cuts were added.

Acknowledgments

The authors wish to express their gratitude to Bruce Golden, Stefan Irnich, and an anonymous referee that helped them improve an earlier version of this chapter.

Bibliography

[1] E. ANGELELLI, C. BAZGAN, Z. TUZA, AND M. G. SPERANZA, *Complexity and approximation for traveling salesman problems with profits*, Theoretical Computer Science, 531 (2014), pp. 54–65.

[2] N. ARAS, D. AKSEN, AND M. T. TEKIN, *Selective multi-depot vehicle routing problem with pricing*, Transportation Research Part C: Emerging Technologies, 19 (2011), pp. 866–884.

[3] A. ARCHER, M. BATENI, M. HAJIAGHAYI, AND H. KARLOFF, *Improved approximation algorithms for prize-collecting Steiner tree and TSP*, SIAM Journal on Computing, 40 (2011), pp. 309–332.

[4] C. ARCHETTI, N. BIANCHESSI, A. HERTZ, AND M. G. SPERANZA, *The split delivery capacitated team orienteering problem*, Networks, 63 (2014), pp. 16–33.

[5] ——, *Incomplete service and split deliveries in a routing problem with profits*, Networks, 63 (2014), pp. 135–145.

[6] C. ARCHETTI, N. BIANCHESSI, AND M. G. SPERANZA, *Optimal solutions for routing problems with profits*, Discrete Applied Mathematics, 161 (2013), pp. 547–557.

[7] ——, *The capacitated team orienteering problem with incomplete service*, Optimization Letters, 7 (2013), pp. 1405–1417.

[8] C. ARCHETTI, D. FEILLET, A. HERTZ, AND M. G. SPERANZA, *The capacitated team orienteering and profitable tour problems*, Journal of the Operational Research Society, 60 (2009), pp. 831–842.

[9] C. ARCHETTI, A. HERTZ, AND M. G. SPERANZA, *Metaheuristics for the team orienteering problem*, Journal of Heuristics, 13 (2007), pp. 49–76.

[10] C. ARCHETTI AND M. G. SPERANZA, *Arc routing problems with profits*, Technical Report WPDEM 2013/2, Department of Economics and Management, University of Brescia, Italy, 2013.

[11] G. AUSIELLO, V. BONIFACI, AND L. LAURA, *The online prize-collecting traveling salesman problem*, Information Processing Letters, 107 (2008), pp. 199–204.

[12] B. AWERBUCH, Y. AZAR, A. BLUM, AND S. VEMPALA, *New approximation guarantees for minimum-weight k-trees and prize-collecting salesmen*, SIAM Journal on Computing, 28 (1998), pp. 254–262.

[13] E. BALAS, *The prize collecting traveling salesman problem*, Networks, 19 (1989), pp. 621–636.

[14] ——, *The prize collecting traveling salesman problem. II: Polyhedral results*, Networks, 25 (1995), pp. 199–216.

[15] ——, *New classes of efficiently solvable generalized travelling salesman problems*, Annals of Operations Research, 86 (1999), pp. 529–558.

[16] ——, *The prize collecting traveling salesman problem and its applications*, in Traveling Salesman Problem and its Variations, G. Gutin and A. Punnen, eds., Kluwer Academic Publishers, Dordrecht, 2002, pp. 663–695.

[17] J.-F. BÉRUBÉ, M. GENDREAU, AND J.-Y. POTVIN, *A branch-and-cut algorithm for the undirected prize collecting traveling salesman problem*, Networks, 54 (2009), pp. 56–67.

[18] ——, *An exact ϵ-constraint method for bi-objective combinatorial optimization problems: Application to the travelling salesman problem with profits*, European Journal of Operational Research, 194 (2009), pp. 39–50.

[19] D. BIENSTOCK, M. X. GOEMANS, D. SIMCHI-LEVI, AND D. WILLIAMSON, *A note on the prize collecting traveling salesman problem*, Mathematical Programming, 59 (1993), pp. 413–420.

[20] A. BLUM, S. CHAWLA, D. R. KARGER, T. LANE, A. MEYERSON, AND M. MINKOFF, *Approximation algorithms for orienteering and discounted-reward TSP*, SIAM Journal on Computing, 37 (2007), pp. 653–670.

[21] M.-C. BOLDUC, J. RENAUD, F. BOCTOR, AND G. LAPORTE, *A perturbation metaheuristic for the vehicle routing problem with private fleet and common carriers*, Journal of the Operational Research Society, 59 (2008), pp. 776–787.

[22] H. BOULY, D.-C. DANG, AND A. MOUKRIM, *A memetic algorithm for the team orienteering problem*, 4OR, 8 (2010), pp. 49–70.

[23] S. BOUSSIER, D. FEILLET, AND M. GENDREAU, *An exact algorithm for the team orienteering problem*, 4OR, 5 (2007), pp. 211–230.

[24] S. E. BUTT AND T. M. CAVALIER, *A heuristic for the multiple tour maximum collection problem*, Computers & Operations Research, 21 (1994), pp. 101–111.

[25] S. E. BUTT AND D. M. RYAN, *An optimal solution procedure for the multiple tour maximum collection problem using column generation*, Computers & Operations Research, 26 (1999), pp. 427–441.

[26] A. M. CAMPBELL, M. GENDREAU, AND B. W. THOMAS, *The orienteering problem with stochastic travel and service times*, Annals of Operations Research, 186 (2011), pp. 61–81.

[27] V. CAMPOS, R. MARTÍ, J. SÁNCHEZ-ORO, AND A. DUARTE, *GRASP with path relinking for the orienteering problem*, Journal of the Operational Research Society, doi:10.1957/jors.2013.156, 2013.

[28] I.-M. CHAO, B. L. GOLDEN, AND E. A. WASIL, *The team orienteering problem*, European Journal of Operational Research, 88 (1996), pp. 464–474.

[29] ———, *A fast and effective heuristic for the orienteering problem*, European Journal of Operational Research, 88 (1996), pp. 475–489.

[30] C. CHEKURI, N. KORULA, AND M. PÁL, *Improved algorithms for orienteering and related problems*, ACM Transactions on Algorithms, 8 (2012), pp. 661–670.

[31] C.-W. CHU, *A heuristic algorithm for the truckload and less-than-truckload problem*, European Journal of Operational Research, 165 (2005), pp. 657–667.

[32] J.-F. CÔTÉ AND J.-Y. POTVIN, *A tabu search heuristic for the vehicle routing problem with private fleet and common carrier*, European Journal of Operational Research, 198 (2009), pp. 464 – 469.

[33] D.-C. DANG, R. EL-HAJJ, AND A. MOUKRIM, *A branch-and-cut algorithm for solving the team orienteering problem*, in Integration of AI and OR Techniques in Constraint Programming for Combinatorial Optimization Problems, C. Gomes and M. Sellmann, eds., vol. 7874 of Lecture Notes in Computer Science, Springer, Berlin, Heidelberg, 2013, pp. 332–339.

[34] D.-C. DANG, R. N. GUIBADJ, AND A. MOUKRIM, *An effective PSO-inspired algorithm for the team orienteering problem*, European Journal of Operational Research, 229 (2013), pp. 332–344.

[35] R. DEITCH AND S. P. LADANY, *The one-period bus routing problem: Solved by an effective heuristic for the orienteering tour problem and improvement algorithm*, European Journal of Operational Research, 127 (2000), pp. 69–77.

[36] M. DELL'AMICO, F. MAFFIOLI, AND A. SCIOMACHEN, *A Lagrangian heuristic for the prize collecting travelling salesman problem*, Annals of Operations Research, 81 (1998), pp. 289–306.

[37] M. DELL'AMICO, F. MAFFIOLI, AND P. VÄBRAND, *On prize-collecting tours and the asymmetric travelling salesman problem*, International Transactions in Operational Research, 2 (1995), pp. 297–308.

[38] A. DIVSALAR, P. VANSTEENWEGEN, AND D. CATTRYSSE, *A variable neighborhood search method for the orienteering problem with hotel selection*, International Journal of Production Economics, 145 (2013), pp. 150–160.

[39] A. DIVSALAR, P. VANSTEENWEGEN, K. SÖRENSEN, AND D. CATTRYSSE, *A memetic algorithm for the orienteering problem with hotel selection*, European Journal of Operational Research, 237 (2014), pp. 29–49.

[40] M. DREXL AND S. IRNICH, *Solving elementary shortest-path problems as mixed-integer programs*, OR Spectrum, 36 (2014), pp. 281–296.

[41] A. EKICI AND A. RETHAREKAR, *Multiple agents maximum collection problem with time dependent rewards*, Computers & Industrial Engineering, 64 (2013), pp. 1009–1018.

[42] G. ERDOGĂN AND G. LAPORTE, *The orienteering problem with variable profits*, Networks, 61 (2013), pp. 104–116.

[43] E. ERKUT AND J. ZHANG, *The maximum collection problem with time-dependent rewards*, Naval Research Logistics, 43 (1996), pp. 749–763.

[44] L. EVERS, K. GLORIE, S. VAN DER STER, A. I. BARROS, AND H. MONSUUR, *A two-stage approach to the orienteering problem with stochastic weights*, Computers & Operations Research, 43 (2014), pp. 248–260.

[45] D. FEILLET, P. DEJAX, AND M. GENDREAU, *Travelling salesman problems with profits*, Transportation Science, 39 (2005), pp. 188–205.

[46] M. FISCHETTI, J. J. SALAZAR GONZÁLEZ, AND P. TOTH, *Solving the orienteering problem through branch-and-cut*, INFORMS Journal on Computing, 10 (1998), pp. 133–148.

[47] ——, *The generalized traveling salesman and orienteering problems*, in Traveling Salesman Problem and Its Variations, G. Gutin and A. Punnen, eds., Kluwer Academic Publishers, Dordrecht, 2002, pp. 609–662.

[48] M. FISCHETTI AND P. TOTH, *An additive approach for the optimal solution of the prize-collecting traveling salesman problem*, in Vehicle Routing: Methods and Studies, B. L. Golden and A. A. Assad, eds., North–Holland, Amsterdam, 1988, pp. 319–343.

[49] F. V. FOMIN AND A. LINGAS, *Approximation algorithms for time-dependent orienteering*, Information Processing Letters, 83 (2002), pp. 57–62.

[50] L. M. GAMBARDELLA, R. MONTEMANNI, AND D. WEYLAND, *Coupling ant colony systems with strong local searches*, European Journal of Operational Research, 220 (2012), pp. 831–843.

[51] D. GAVALAS, C. KONSTANTOPOULOS, K. MASTAKAS, AND G. PANTZIOU, *A survey on algorithmic approaches for solving tourist trip design problems*, Journal of Heuristics, 20 (2014), pp. 291–328.

[52] M. GENDREAU, G. LAPORTE, AND F. SEMET, *A tabu search heuristic for the undirected selective travelling salesman problem*, European Journal of Operational Research, 106 (1998), pp. 539–545.

[53] ——, *A branch-and-cut algorithm for the undirected selective travelling salesman problem*, Networks, 32 (1998), pp. 263–273.

[54] M. X. GOEMANS, *Combining approximation algorithms for the prize-collecting TSP*, in Proceedings of CoRR, 2009.

[55] M. X. GOEMANS AND D. P. WILLIAMSON, *A general approximation technique for constrained forest problems*, SIAM Journal on Computing, 24 (1995), pp. 296–317.

[56] B. L. GOLDEN, A. A. ASSAD, AND R. DAHL, *Analysis of a large scale vehicle routing problem with an inventory component*, Large Scale Systems, 7 (1984), pp. 181–190.

[57] B. L. GOLDEN, L. LEVY, AND R. VOHRA, *The orienteering problem*, Naval Research Logistics, 34 (1987), pp. 307–318.

[58] B. L. GOLDEN, Q. WANG, AND L. LIU, *A multifaceted heuristic for the orienteering problem*, Naval Research Logistics, 35 (1988), pp. 359–366.

[59] M. HAYES AND J. M. NORMAN, *Dynamic programming in orienteering: Route choice and siting of controls*, Journal of the Operational Research Society, 35 (1984), pp. 791–796.

[60] Q. HU AND A. LIM, *An iterative three-component heuristic for the team orienteering problem with time windows*, European Journal of Operational Research, 232 (2014), pp. 276–286.

[61] T. ILHAN, S. M. R. IRAVANI, AND M. S. DASKIN, *The orienteering problem with stochastic profits*, IIE Transactions, 40 (2008), pp. 406–421.

[62] M. K. JEPSEN, *Branch-and-cut and branch-and-cut-and-price algorithms for solving vehicle routing problems*, PhD thesis, Technical University of Denmark, Lyngby, Denmark, 2011.

[63] N. JOZEFOWIEZ, F. GLOVER, AND M. LAGUNA, *Multi-objective meta-heuristics for the traveling salesman problem with profits*, Journal of Mathematical Modelling and Algorithms, 7 (2008), pp. 177–195.

[64] M. G. KANTOR AND M. B. ROSENWEIN, *The orienteering problem with time windows*, Journal of the Operational Research Society, 43 (1992), pp. 629–635.

[65] S. KATAOKA AND S. MORITO, *An algorithm for single constraint maximum collection problem*, Journal of the Operations Research Society of Japan, 31 (1988), pp. 515–530.

[66] S. KATAOKA, T. YAMADA, AND S. MORITO, *Minimum directed 1-subtree relaxation for score orienteering problem*, European Journal of Operational Research, 104 (1998), pp. 139–153.

[67] L. KE, C. ARCHETTI, AND Z. FENG, *Ants can solve the team orienteering problem*, Computers & Industrial Engineering, 54 (2008), pp. 648–665.

[68] L. KE, Z. XU, Z. FENG, K. SHANG, AND X. QIAN, *Proportion-based robust optimization and team orienteering problem with interval data*, European Journal of Operational Research, 226 (2013), pp. 19–31.

[69] C. P. KELLER, *Algorithms to solve the orienteering problem: A comparison*, European Journal of Operational Research, 41 (1989), pp. 224–231.

[70] M. KESHTKARAN, K. ZIARATI, A. BETTINELLI, AND D. VIGO, *Enhanced exact solution methods for the team orienteering problem*, Technical Report, DEI University of Bologna, Italy, 2014.

[71] B.-I. KIM, H. LI, AND A. L. JOHNSON, *An augmented large neighborhood search method for solving the team orienteering problem*, Expert Systems with Applications, 40 (2013), pp. 3065–3072.

[72] N. LABADIE, R. MANSINI, J. MELECHOVSKÝ, AND R. WOLFLER CALVO, *The team orienteering problem with time windows: An LP-based granular variable neighborhood search*, European Journal of Operational Research, 220 (2012), pp. 15–27.

[73] N. LABADIE, J. MELECHOVSKÝ, AND R. WOLFLER CALVO, *Hybridized evolutionary local search algorithm for the team orienteering problem with time windows*, Journal of Heuristics, 17 (2011), pp. 729–753.

[74] G. LAPORTE AND S. MARTELLO, *The selective travelling salesman problem*, Discrete Applied Mathematics, 26 (1990), pp. 193–207.

[75] A. C. LEIFER AND M. B. ROSENWEIN, *Strong linear programming relaxations for the orienteering problem*, European Journal of Operational Research, 73 (1994), pp. 517–523.

[76] Y.-C. LIANG, S. KULTUREL-KONAK, AND A. E. SMITH, *Meta heuristics for the orienteering problem*, in Proceedings of the 2002 Congress on Evolutionary Computation, Honolulu, HI, 2002, pp. 384–389.

[77] S. LIN, *Computer solutions of the traveling salesman problem*, Bell System Technical Journal, 44 (1965), pp. 2245–2269.

[78] S.-W. LIN AND V. F. YU, *A simulated annealing heuristic for the team orienteering problem with time windows*, European Journal of Operational Research, 217 (2012), pp. 94–107.

[79] Z. LUO, B. CHEANG, A. LIM, AND W. ZHU, *An adaptive ejection pool with toggle-rule diversification approach for the capacitated team orienteering problem*, European Journal of Operational Research, 229 (2013), pp. 673–682.

[80] H. H. MILLAR AND M. KIRAGU, *A time-based formulation and upper bounding scheme for the selective traveling salesperson problem*, Journal of the Operational Research Society, 48 (1997), pp. 511–518.

[81] R. MONTEMANNI AND L. M. GAMBARDELLA, *Ant colony system for team orienteering problems with time windows*, Foundations of Computing and Decision Sciences, 34 (2009), pp. 287–306.

[82] V. NAGARAJAN AND R. RAVI, *The directed orienteering problem*, Algorithmica, 60 (2011), pp. 1017–1030.

[83] V. H. NGUYEN, *A primal-dual approximation algorithm for the asymmetric prize-collecting TSP*, in Combinatorial Optimization and Applications, Lecture Notes in Computer Science, W. Wu and O. Daescu, eds., Springer, Berlin, 2010, pp. 260–269.

[84] V. H. NGUYEN AND T. T. T. NGUYEN, *Approximating the asymmetric profitable tour*, Electronic Notes in Discrete Mathematics, 36 (2010), pp. 907–914.

[85] V. PAPAPANAGIOTOU, R. MONTEMANNI, AND L. M. GAMBARDELLA, *Objective function evaluation methods for the orienteering problem with stochastic travel and service times*, Journal of Applied Operational Research, 6 (2014), pp. 16–29.

[86] J. PIETZ AND J. O. ROYSET, *Generalized orienteering problem with resource dependent rewards*, Naval Research Logistics, 60 (2013), pp. 294–312.

[87] M. V. POGGI DE ARAGÃO, F. H. VIANA DE FREITAS, AND E. UCHOA, *Team orienteering problem: Formulations and branch-cut and price*, Monografias em Ciência da Computação, 10/13 (2010), pp. 1–15.

[88] J.-Y. POTVIN AND M.-A. NAUD, *Tabu search with ejection chains for the vehicle routing problem with private fleet and common carrier*, Journal of the Operational Research Society, 62 (2011), pp. 326–336.

[89] R. RAMESH AND K. M. BROWN, *An efficient four-phase heuristic for the generalized orienteering problem*, Computers & Operations Research, 18 (1991), pp. 151–165.

[90] R. RAMESH, Y.-S. YOON, AND M. H. KARWAN, *An optimal algorithm for the orienteering tour problem*, ORSA Journal on Computing, 4 (1992), pp. 155–165.

[91] R. RAVI, R. SUNDARAM, M. MARATHE, D. ROSENKRANTZ, AND S. RAVI, *Spanning trees short and small*, in Proceedings of the 5th Annual ACM-SIAM Symposium on Discrete Algorithms, 1994, pp. 546–555.

[92] G. RIGHINI AND M. SALANI, *Decremental state space relaxation strategies and initialization heuristics for solving the orienteering problem with time windows with dynamic programming*, Computers & Operations Research, 36 (2009), pp. 1191–1203.

[93] M. A. SALAZAR-AGUILAR, A. LANGEVIN, AND G. LAPORTE, *The multi-district team orienteering problem*, Computers & Operations Research, 41 (2014), pp. 76–82.

[94] M. SCHILDE, K. F. DOERNER, R. F. HARTL, AND G. KIECHLE, *Metaheuristics for the bi-objective orienteering problem*, Swarm Intelligence, 3 (2009), pp. 179–201.

[95] J. SILBERHOLZ AND B. L. GOLDEN, *The effective application of a new approach to the generalized orienteering problem*, Journal of Heuristics, 16 (2010), pp. 393–415.

[96] W. SOUFFRIAU, P. VANSTEENWEGEN, G. VANDEN BERGHE, AND D. VAN OUDHEUSDEN, *A path relinking approach for the team orienteering problem*, Computers & Operations Research, 37 (2010), pp. 1853–1859.

[97] ——, *The multiconstraint team orienteering problem with multiple time windows*, Transportation Science, 47 (2013), pp. 53–63.

[98] A. STENGER, D. VIGO, S. ENZ, AND M. SCHWIND, *A variable neighborhood search algorithm for a vehicle routing problem arising in small package shipping*, Transportation Science, 47 (2013), pp. 64–80.

[99] H. TANG AND E. MILLER-HOOKS, *A tabu search heuristic for the team orienteering problem*, Computers & Operations Research, 32 (2005), pp. 1379–1407.

[100] L. TANG AND X. WANG, *An iterated local search heuristic for the capacitated prize-collecting travelling salesman problem*, Journal of the Operational Research Society, 59 (2008), pp. 590–599.

[101] C. D. TARANTILIS, F. STAVROPOULOU, AND P. P. REPOUSSIS, *The capacitated team orienteering problem: A bi-level filter-and-fan method*, European Journal of Operational Research, 224 (2013), pp. 65–78.

[102] T. THOMADSEN AND T. STIDSEN, *The quadratic selective travelling salesman problem*, IMM-Technical Report-2003-17, Technical University of Denmark, Lyngby, Denmark, 2003.

[103] P. TOTH AND D. VIGO, *The granular tabu search and its application to the vehicle routing problem*, INFORMS Journal on Computing, 15 (2003), pp. 333–346.

[104] F. TRICOIRE, M. ROMAUCH, K. F. DOERNER, AND R. F. HARTL, *Heuristics for the multi-period orienteering problem with multiple time windows*, Computers & Operations Research, 37 (2010), pp. 351–367.

[105] T. TSILIGIRIDES, *Heuristic methods applied to orienteering*, Journal of the Operational Research Society, 35 (1984), pp. 797–809.

[106] P. VANSTEENWEGEN, W. SOUFFRIAU, AND D. VAN OUDHEUSDEN, *The orienteering problem: A survey*, European Journal of Operational Research, 209 (2011), pp. 1–10.

[107] P. VANSTEENWEGEN, W. SOUFFRIAU, G. VANDEN BERGHE, AND D. VAN OUDHEUSDEN, *A guided local search metaheuristic for the team orienteering problem*, European Journal of Operational Research, 196 (2009), pp. 118–127.

[108] ——, *Iterated local search for the team orienteering problem with time windows*, Computers & Operations Research, 36 (2009), pp. 3281–3290.

[109] P. VANSTEENWEGEN AND D. VAN OUDHEUSDEN, *The mobile tourist guide: An OR opportunity*, OR Insight, 20 (2007), pp. 21–27.

[110] C. VERBEECK, K. SÖRENSEN, P. AGHEZZAF, AND E.-H. VANSTEENWEGEN, *A fast solution method for the time-dependent orienteering problem*, European Journal of Operational Research, 236 (2014), pp. 419–432.

[111] Q. WANG, X. SUN, B. L. GOLDEN, AND J. JIA, *Using artificial neural networks to solve the orienteering problem*, Annals of Operations Research, 61 (1995), pp. 111–120.

[112] X. WANG, B. L. GOLDEN, AND E. A. WASIL, *Using a genetic algorithm to solve the generalized orienteering problem*, in The Vehicle Routing Problem: Latest Advances and New Challenges, B. L. Golden, S. Raghavan, and E. A. Wasil, eds., Springer, New York, 2008, pp. 263–274.

Chapter 11

Dynamic Vehicle Routing Problems

Tolga Bektaş
Panagiotis P. Repoussis
Christos D. Tarantilis

11.1 ▪ Introduction

When a vehicle routing model is cast and solved, it is normally assumed that the values of all input parameters are known with certainty. However, this is hardly the case in real-life applications where parameters such as customer demands, travel and service times, or even the information of whether a particular customer will require service or not are often incomplete, uncertain, or unknown during the route design phase (see Gounaris et al. [60]). As pointed out by Psaraftis [123], there are two important dimensions of input data, namely *evolution* and *quality of information*. The former implies that the available information is subject to change even after the routing plan is realized, while the latter reflects the possible uncertainties in the available data. What is common in both cases is that the partially known, uncertain, or unknown input parameters are revealed or updated concurrently with the execution of the routing process.

Depending on the availability and quality of a priori information and other characteristics of the problem under consideration, two alternatives emerge for solving the routing problem. Assuming that sufficient information is available (e.g., all input data are known in advance with a predefined degree of uncertainty), the first option is to treat the problem as static and solve it once during the design phase. The goal in this case is to obtain a robust routing plan that will possibly be subject to relatively small changes during the actual execution. This option is named *a priori optimization*, for which anticipating uncertainty is crucial in order to find realizable routing plans, and to avoid hefty penalties, both economic and reputational, when one fails to provide the required level of service (see Gounaris et al. [60]).

The second option is to address the problem in an ongoing and dynamic fashion as new input data arrive or are revealed in real time. While information evolves and decisions must be continuously made in a changing environment, the goal is to react to the new events as well as to anticipate future events, particularly if exploitable stochastic information is available or can be derived from past data. This approach, namely *dynamic optimization*, recognizes the additional decisions that become available during the execution of the routing plan and attempts to handle uncertainty in real time. However, it also requires advanced technological support and real-time communication between the vehicle and the dispatcher.

Based on the above described optimization frameworks, and the quality of advanced information, a possible taxonomy for VRPs can be stated as follows:

- *Static and Deterministic*. All input parameters are known in advance and with certainty, and are assumed not to change during planning and execution. This class of problems can be solved once and before the beginning of the planning horizon, and includes the traditional VRPs discussed in other chapters.

- *Static and Stochastic*. Part of the input parameters are known as random or stochastic variables, for which the actual values are revealed during the execution of the routing process. Unlike static and deterministic problems that have been studied thoroughly, VRPs under uncertainty have received less attention. Most papers address the stochastic *Capacitated VRP* (CVRP) through recourse and chance-constrained models, considering stochastic customers (Cheung, Xu, and Guan [28]), demands (Christiansen and Lysgaard [29]), and/or travel times (Kenyon and Morton [82]). An alternative to stochastic modeling is the use of robust optimization techniques (see Gounaris, Wiesemann, and Floudas [59], Gounaris et al. [60], and Sungur, Ordonez, and Dessouky [138]). For reviews of the literature on stochastic VRPs, we refer the reader to Cordeau et al. [34], Gendreau, Laporte, and Séguin [49], Toth and Vigo [145], and Chapter 8.

- *Dynamic and Deterministic*. As opposed to the above static groups of problems, dynamic problems assume that either part or all input data are not known prior to the execution of the plan, but only become available incrementally over time. Therefore, this group of problems is characterized by total uncertainty, since only probabilistic information is available for future events, and optimization can only be performed as new information arrives. The terms *real-time* and *online* typically refer to this group of problems.

- *Dynamic and Stochastic*. This group can be seen as dynamic problems that cannot be solved once and before the realization of the routing process; however, part of the unknown input data is in the form of stochastic information (e.g., forecasts, range values, and prescribed distributions). In contrast to pure dynamic and deterministic problems, there is a strong incentive to exploit and integrate all available information on foreseen future events in the solution process. This group of problems is usually referred to as *partially dynamic* and *mix dynamic and stochastic*.

This chapter focuses on the latter two groups of the above taxonomy, namely dynamic VRPs either with deterministic and/or stochastic data, in which *dynamic data* or *interaction of activities* over time are considered explicitly. The term dynamic data refers to one or more problem parameters (e.g., customer locations, demands, and travel times) that can be expressed as a function of time, including time-dependent travel times that are known in advance (see Powell, Jaillet, and Odoni [118]). An interacting activity refers to a dynamic event that affects the execution of the routing plan (e.g., service cancellations and vehicle availabilities). In the chapter, emphasis is given to dynamic VRPs where consolidation of requests is allowed (i.e., many customer requests can be served by the same vehicle); however, dynamic vehicle dispatching problems without consolidation (i.e., a vehicle is dispatched to serve a single customer) are also discussed.

Dynamic vehicle routing and dispatching problems have received significant attention in the literature. The early works of Psaraftis [123, 124] provide formal definitions and discuss the differences between dynamic and ordinary static VRPs. Bianchi [18] reviews early solution approaches. Ghiani et al. [53] differentiate between sequential and parallel solution approaches for both single- and multi-vehicle problems, and discuss key implementation features. An excellent review of the more recent approaches, frameworks, and routing strategies is provided by Gendreau and Potvin [52] and Ichoua, Gendreau, and Potvin [77]. The survey papers by Larsen, Madsen, and Solomon [89, 90] discuss technological advances, analyze the degrees of dynamism, different objectives, and performance measures, and suggest a three-echelon classification scheme for real-life applications. The works by Berbeglia, Cordeau, and Laporte [13] and Cordeau et al. [33] focus on dynamic pickup-and-delivery problems. Pillac et al. [111] review recent approaches for both dynamic and partially dynamic problems where some form of advance information is provided for near and long-term future events. Finally, Ritzinger, Puchinger, and Hartl [128] discuss solutions approaches proposed for stochastic VRPs as well as for VRPs that combine dynamic and stochastic information.

The aim of this chapter is to present the latest advances and research trends in the field of dynamic vehicle routing and dispatching. The goal is not only to provide an overview of the relevant literature but also to present the state of the art in frameworks and strategies, to provide useful insights, and to identify directions for further research in the area.

While a chapter on the dynamic VRPs can be organized in a number of ways, we have decided to structure it according to the following outline, as illustrated in Figure 11.1. According to this structure, we start by introducing the concept of degrees of dynamism, discuss the different sources of dynamism, namely *requests*, *travel times*, and *vehicle availability*, present the objectives that are often encountered in dynamic problems, and provide an overview of important problem variants and applications in Section 11.2.

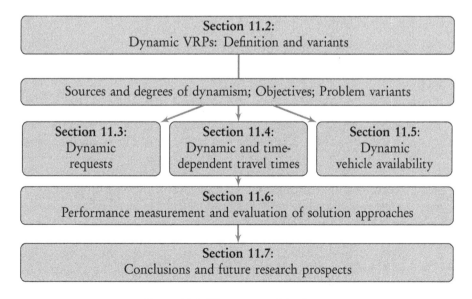

Figure 11.1. *The structure of the chapter.*

The three subsequent sections correspond to the three sources of dynamism. In particular, dynamic requests are discussed in Section 11.3, where we differentiate between pure dynamic problems (a.k.a. dynamic and deterministic) and dynamic problems that incorporate stochastic information (a.k.a. dynamic and stochastic). For the former, we present a review of routing policies, heuristics, reoptimization, and multiple plan approaches, as well as exact techniques based on dynamic and linear programming. As for the latter, we discuss anticipatory and predictive approaches, and review algorithms based on sampling, scenario analysis, and stochastic programming. Problems with dynamic and time-dependent travel times are presented in Section 11.4, and problems with dynamic vehicle availabilities are the subject of Section 11.5. Finally, Section 11.6 discusses performance measurement issues, and Section 11.7 presents the conclusions and offers directions for future research.

11.2 ▪ Definitions, Objectives, and Overview of Problem Variants

11.2.1 ▪ Sources of Dynamism

Based on the definition provided by Psaraftis [123], dynamic VRPs deal with the evolution and manipulation of routes under various operational constraints (e.g., time windows) performed by a fleet of vehicles on the move to serve future or immediate (customer) requests as a function of those inputs (e.g., customer demands, travel times, and on-site service times) that evolve in real time. The term *advance requests* refers to static service requests received before the realization of the routing process, whereas *immediate requests* refers to dynamic service requests revealed over time during the execution of the routing process.

As opposed to a static problem setting, the solution of a dynamic VRP seeks to handle and respond to all dynamic elements of the problem at hand as well as to exploit and integrate the available information on future events (e.g., customer demand forecasts) in an on-going fashion. Hence, the decision making process is executed in a changing and continuously evolving environment. Such an approach allows planners to react to external events and to anticipate the future events and handle uncertainty in real time. It also provides opportunities to reduce costs, improve customer service, and reduce the environmental impacts (Pillac et al. [111]).

Figure 11.2 illustrates the evolution and execution process of a single vehicle with the occurrence of an immediate request. Time t_0 refers to the beginning of the planning horizon, i.e., the time at which the vehicle leaves the depot. At this point, the initial planned route shown by the dotted arrows only contains the requests known a priori, shown by A, B, C, and D. During the execution of the route by the vehicle, one new request denoted by X appears at time t_r and the planned route is reconfigured based on the new input data. In particular, the route is divided into three parts: (i) the part of the route that is already executed and that cannot be modified, (ii) the current movement (position) of the vehicle to reach the next customer, and (iii) the remaining part of the route that will possibly be executed in the future and which can be modified. In this example, the route executed by the end of the planning horizon (t_T) is shown by the sequence A, B, C, X, and D. Note that the planned route can be used not only to decide on the next destination but also to accept or reject an immediate request (Ichoua, Gendreau, and Potvin [77]).

Based on the above dynamic routing scenario, Figure 11.3 describes the timeline of events, the interaction and communication points between the vehicle and the dispatcher, and the type of information and decisions that are exchanged between them.

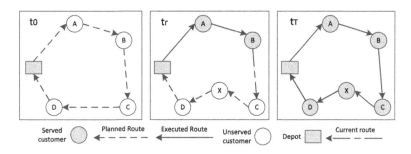

Figure 11.2. *Vehicle routing scenario with advance and immediate requests.*

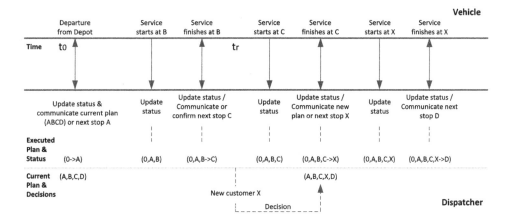

Figure 11.3. *Timeline of events and real-time communication between the vehicle and the dispatcher.*

There is a significant body of literature on problems dealing with dynamic requests. The requests can be a demand for goods (see Azi, Gendreau, and Potvin [7], Hvattum, Løkketangen, and Laporte [73], Khouadjia et al. [84], Mes, Van der Heijden, and Van Harten [101], and van Hemert and La Poutré [146]) or services (see Bertsimas and Van Ryzin [16], Gendreau et al. [48], and Mitrović-Minić and Laporte [104]), including the cases of on-demand transportation (see Attanasio et al. [6] and Beaudry et al. [9]). A special case is to dynamically consider revealed demands for a given set of customers (see Novoa and Storer [108], Secomandi [133], Secomandi and Margot [134], and Wang and Cao [149]). Besides (and in addition to) dynamic requests, variability in travel times has also been taken into account in various studies (see Fleischmann, Gnutzmann, and Sandvoss [44], Haghani and Jung [61], Jung and Haghani [81], Lorini, Potvin, and Zufferey [96], and Potvin, Xu, and Benyahia [115]). The literature is also rich on time-dependent travel times (see Chen, Hsueh, and Chang [25]). In contrast, only a few papers consider variable service times. However, in most cases, this source of dynamism can be treated as part of the dynamic travel times. The literature is also scarce concerning less predictable events, such as service cancellations, vehicle breakdowns, unexpected congestion and accidents, cargo damages, and unexpected changes in customer locations and demands (see Wang et al. [151]). Such events may disrupt the routing plan significantly, and therefore require a special recovery treatment and recourse strategies to minimize the potential negative effects (see Ichoua, Gendreau, and Potvin [77], and Wang et al. [151]).

Given that routing plans are readjusted over time in dynamic VRPs, this requires the exploitation of advanced telematic technology to support the real-time communication between the vehicle and the dispatcher. Recent advances and the rapid growth in information and communication technologies, such as *Radio Frequency Identification* (RFID), *Global Positioning Systems* (GPS), *Geographical Information Systems* (GIS), *General Packet Radio Services* (GPRS), and 3G/4G cellular networks, have overcome the barriers to accessing real-time data and information sharing, and have allowed for a real-time monitoring of vehicle fleets. Larsen, Madsen, and Solomon [89] highlight that nowadays even medium-sized companies have adopted and implemented advanced GPS/GIS systems coupled with wireless telecommunication facilities and mobile equipment, using them to be able to track the status and current position of their vehicles in real time. For descriptions of and discussions on the technological environment required for dynamic vehicle routing and dispatching problems, we refer the reader to Larsen, Madsen, and Solomon [90] and Giaglis et al. [55] as well as to the books edited by Goel [56] and Zeimpekis et al. [160].

Figure 11.4 sketches the basic flow and exchange of information between a dispatcher and a vehicle, and shows various sources of historical and real-time data. Obviously, the type of technology adopted for transmitting and storing data and the nature of the equipment used (e.g., the communication devices to transfer information between the dispatcher and the vehicles) will determine the quality of not only the data obtained, the frequency of interactions, and the system updates but also of the structure of the optimization system itself. For example, if up-to-date information is provided at any time to the dispatcher about the current positions of the vehicles, then vehicle diversion options become available.

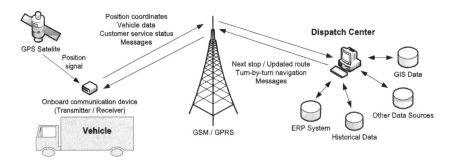

Figure 11.4. *Information flow of a typical GPS-based vehicle routing and monitoring system.*

Finally, access to historical data pertaining to certain problem attributes and/or real-time information about traffic, weather conditions, etc., will allow the use of stochastic methodologies. This is a key element for successfully solving dynamic and stochastic VRPs. As mentioned earlier, the integration of stochastic information about future events (e.g., location of requests, customer demands, travel times, cancellation of requests, and vehicle breakdowns) can significantly increase the look-ahead capability, reliability, and robustness of an optimization system. However, data analysis, data provision, and development of predictive models for estimating future levels and trends for specific problem parameters are often very challenging.

11.2.2 ▪ Degrees of Dynamism

The number of dynamic events as well as the time they actually take place during the planning horizon determine, to a large extent, the *degrees of dynamism*. For problem instances with dynamic requests, the *frequency of changes* and the *urgency of requests* can be used to measure the dynamism of the environment (see Ichoua, Gendreau, and Potvin [77]). The former refers to the rate at which new service requests or their attributes (e.g., demand and time windows) become available or are updated over time. The latter refers to the available *response time*, which can be seen as the time gap between the arrival time of a new request and the latest allowable time at which service begins.

As described in Larsen, Madsen, and Solomon [89], four factors contribute to the dynamism of a VRP with dynamic requests, namely the number of advance requests, the number of immediate requests, the arrival time, and the response time. Looking myopically at the ratio between the number of immediate requests n_{dr} and the total number of requests n_{tr} (see Lund, Madsen, and Rygaard [97]), the degree of dynamism can be expressed as

$$(11.1) \qquad \alpha = \frac{n_{dr}}{n_{tr}}.$$

Metric (11.1) is indicative of the extent of the information received in real time in relation to the overall system information and provides an indication of how dynamic the system actually is (see Larsen, Madsen, and Solomon [89]. Depending on the length T of the planning horizon and the distribution of immediate requests over time, it is also important to consider the arrival times of requests. Clearly, the later the immediate requests are received (e.g., close to T), the more difficult it is to react accordingly. To capture this aspect, Larsen, Madsen, and Solomon [89] proposed the so-called *effective degree of dynamism* α^e. If the arrival time of the ith immediate request is denoted by τ_i, i.e., $0 < \tau_i \leq T$, α^e can then be calculated as the normalized average of the disclosure time (temporal distribution) as follows:

$$(11.2) \qquad \alpha^e = \frac{\sum_{i=1}^{n_{dr}} \frac{\tau_i}{T}}{n_{tr}}.$$

For problem instances with service time windows, it is also important to take into account the level of urgency, i.e., the temporal distance between the request arrival time and the latest requested service time. One may expect that the degree of dynamism is not only related to the length of the time windows but also to the available reaction times and the remaining planning time. For this reason, the latest possible service starting time should be also considered. To that end, the effective degree of dynamism, as described in [89], can be extended as follows:

$$(11.3) \qquad \alpha^e_{TW} = \frac{1}{n_{tr}} \sum_{i=1}^{n_{dr}} \left(1 - \frac{r_i}{T}\right),$$

where r_i denotes the reaction time that corresponds to the difference between the arrival time τ_i and the latest allowable service time l_i of the ith immediate request, with $l_i - \tau_i \leq T$ for all $i = 1, 2, \ldots, n_{dr}$. Both α^e and α^e_{TW} take values in the interval $[0,1]$, where 0 corresponds to the static case (all requests are received at time 0).

The effective degree of dynamism captures the volume and the temporal composition of immediate requests, and can be used to classify the problem as being either weakly, moderately, or strongly dynamic. As reported in [89] and [90], a VRP is *weakly dynamic* if α^e is below 30%, *moderately dynamic* if between 30% and 80%, and *strongly dynamic* if higher than 80%. This information is useful not only to adopt the right methods but also to select which and how different decision strategies can be applied for obtaining high-quality routing plans. In weakly dynamic environments, more time-consuming and accurate methods can be used to obtain optimal or near-optimal solutions, since interruptions are scarce and such methods can be repeatedly applied for a long period of time. On the other hand, methods able to deliver high quality solutions in short computational times are better suited for strongly dynamic environments where the quality of advance information is low.

Bent and Van Hentenryck [10] and Larsen, Madsen, and Solomon [87, 88] have examined the performance of different solution approaches for varying degrees of dynamism. In particular, they focused on problem instances with different mixes of immediate and future requests, without considering other sources of dynamism. However, they did not consider the spatial distribution of requests, demands and the traveling times, frequency of updates, or the availability of a priori information for future events. An extension to a_{TW}^e that takes into account travel times and the required capacity for the demand is proposed by Wohlgemuth, Oloruntoba, and Clausen [156].

Recently, Ferrucci, Bock, and Gendreau [41] have introduced the *degree of structural diversity* as a measure of the existing variability in the requests. Having divided a given service area into subareas and segments with a spatial and a temporal dimension, the degree of structural diversity is calculated as the sum of the distances between the barycenters of consecutive time periods weighted with the number of arrived requests in the respective time period, divided by the total number of requests times the maximum distance between subareas. Their results show that as structural diversity increases (i.e., larger distances between barycenters and more requests occur) the more valuable is the utilization of (available or derived) stochastic knowledge about future request arrivals.

11.2.2.1 ▪ Objectives

In static VRPs the objective is to minimize the transportation cost, expressed in terms of one-off costs (e.g., fixed fleet acquisition and depreciation costs) and/or recurring costs (e.g., variable costs related to the distance traveled). In cases where the fleet size is not prescribed, a hierarchical objective is generally used, which is to minimize the total number of vehicles, followed by minimizing the total distance traveled for the fixed number of vehicles. In contrast, objective functions used for dynamic VRPs include various fitness criteria and performance measures, relevant to the degrees of dynamism. Furthermore, the objective(s) of dynamic VRPs often differ from one application to another, since various factors are considered to determine the quality of a solution including, among others, routing costs, customer requirements, and company policies.

Below, important attributes when defining the objective(s) are discussed:

- *Transportation Costs.* A successful routing plan must be characterized by short distances between the customers, since the distances traveled often reflect the expected transportation costs (see Montemanni et al. [105]). For this reason, transportation costs are usually included in the objective function, especially for weakly dynamic problems.

- *Service Level.* The service level can be seen as a measure of the response time to changes. From a customer service perspective, this can be translated to minimizing the waiting time and lateness to maximize customer satisfaction, i.e., the delay between the arrival of a request and the time at which it is serviced. However, a fast response to an immediate request is often conflicting with the objective of minimizing transportation costs, as it might result in long detours. In moderately and strongly dynamic environments, the objective is often a weighted sum (or a hierarchical ordering) of routing costs and lateness (see Branchini, Armentato, and Lokketangen [20], Chang, Chen, and Hsueh [24], Ichoua, Gendreau, and Potvin [74], and Chen and Xu [26]). Ferrucci, Bock, and Gendreau [41] considered the objective of minimizing customer inconvenience, calculated as a function of request response time, and studied the effects of using a linear and a quadratic dependency between response time and customer inconvenience. Ghiani et al. [54] seek to minimize the expected inconvenience cost, which is expressed via a non-decreasing convex function.

- *Throughput.* Throughput is typically related to revenue, utilization levels (e.g., minimum idle times for the vehicles), and the ability of serving as many customers as possible. In highly dynamic environments and time-constraint problem settings, such as emergency and on-demand transportation services, maximizing the total number of served requests (or similarly the minimization of the number of rejected requests) is the primary objective. This happens because new customer requests can be either accepted or rejected, and to serve them is either feasible or very costly (see Ichoua, Gendreau, and Potvin [75]). Van Hentenryck and Bent [147] refer to this decision process as *service guarantee.*

A combination of the objectives described above can also be considered for dynamic VRPs. However, multiple objectives are often conflicting, for example in terms of speed and flexibility or reliability and robustness. Therefore, a compromise or different hierarchy levels among them might be considered. Additional objectives or a different mix are usually adopted for solution approaches specifically designed for dealing with less predictable events, such as vehicle breakdowns and accidents. The goal of handling such disruption events is to alleviate the negative impact. To this end, the cost of delaying a service request, the deviation from the original plan, and the inconvenience to the customers are often adopted as primary or secondary objectives.

As mentioned by Pillac et al. [111] the requirement for real-time decision making within an evolving environment often compromises solution quality to better react to changes in the input data. Therefore, short-term decisions should maintain a reasonable tradeoff between accuracy and speed, while long-term decisions need to ensure an adequate level of solution quality. To that end, important elements for defining the objectives properly are the size of the dispatching area and the length of the planning horizon. For example, local area dispatching systems (e.g., repair and courier services) often exhibit highly dynamic environments that evolve over narrow planning horizons, requiring prioritization of high service levels and quick response times over, for example, transportation costs (see Gendreau and Potvin [52]).

11.2.3 ▪ Overview of Problem Variants and Important Applications

Dynamic vehicle routing and dispatching problems have been studied extensively and linked with numerous important applications, encountered in a large variety of practical contexts including, among others, maintenance operations, courier services, on-demand dial-a-ride systems, emergency services, and pickup and/or delivery of goods. In the literature, several problem variants and models have been defined with various combinations of operational requirements and constraints. Below, we briefly discuss the various problem classes that have emerged. The main goal here is to provide a link between problem settings and practical applications, and also to identify key structural characteristics.

Initially, one may distinguish between problems with and without consolidation, as in, for example, less-than-truckload trucking and truckload trucking, respectively. In the former, where many customer requests are served by a single vehicle, the main challenge is to determine the "right" sequence of visits. In the latter, one vehicle is dispatched to visit a single customer. These problems are of an assignment nature, where the main challenge is the repositioning of idle vehicles in anticipation of future requests, which are particularly relevant in the planning of emergency services.

Dynamic VRPs with consolidation are often the most difficult to solve, and they have been studied assuming various operational settings and constraints pertaining to, among others, time windows, capacity, and route durations. The typical setting involves a set of capacitated or uncapacitated vehicles which all start from and end at a depot, and have to visit a set of geographically scattered locations. These problems can be classified into four groups as follows:

- *One-to-many* and *many-to-one* problems. A single location is associated with each request. A request can be either for collecting or delivering a product or providing a service.

- *One-to-one* problems. A pickup and a delivery location are associated with each request; i.e., products or people are moved from a given origin to a given destination. In this case, a pickup must precede a delivery for each request.

- *One-to-many-to-one* problems. All delivery demands (shipments to linehaul customers) are initially located at the depot, and all pickup demands (collections from backhaul customers) are returned to the depot.

- *Many-to-many* problems. Any node can serve as a destination or as a source for any product.

Collection or delivery systems of various products, including variants of the dynamic *Traveling Salesman Problem* (TSP), are typical examples of one-to-many and many-to-one problems (see Hvattum, Løkketangen, and Laporte [73], Khouadjia et al. [84], and Montemanni et al. [105]). Maintenance operations and repair services that respond to immediate requests for providing service (e.g., equipment and facility maintenance or providing utility services) at the customer's premises (see Madsen, Tosti, and Voelds [99]) also fall into this category. A typical example is the dynamic *Traveling Repairman Problem* (TRP) (see Bertsimas and Van Ryzin [15] and Larsen, Madsen, and Solomon [87]). Note that in some of these problems the only requirement is to visit the customer location, possibly within a time window, without any constraints on vehicle capacity. Time windows can be either hard or soft in the sense that they can be violated to allow either early or late arrivals. Newspaper delivery (see Ferrucci, Bock, and Gendreau [41]) and courier services (see Gendreau et al. [48]), where parcels and mail are consolidated in a central location for further processing, can be also classified as one-to-many or many-to-one problems. As described by Kilby, Prosser, and Shaw [85], it is important to highlight the difference between one-to-many and many-to-one problems in a dynamic environment. In particular, it is physically impossible to add a delivery service to a vehicle that has already left the depot, whereas a pickup service can be added to an existing route (see Azi, Gendreau, and Potvin [7]). Therefore, one-to-many problems can be much harder to solve compared to many-to-one problems, where service guarantee issues often prevail. On the other hand, dynamic multi-period one-to-many and many-to-one problem settings appear in the works of Wen et al. [155] and Angelelli et al. [1], respectively.

One-to-one problems include virtually all variants of the dynamic *Dial-a-Ride Problems* (DARP) for on-demand transportation and door-to-door services (see Attanasio et al. [6] and Psaraftis [122]), such as transportation of elderly, handicapped, or disabled people (see Madsen, Ravn, and Rygaard [98]). All dynamic *Pickup-and-Delivery Problems* (PDPs) arising in local less-than-truckload applications, such as local express mail delivery services in urban areas (see Gendreau et al. [47]), pickup-and-delivery systems for different kinds of products and goods (see Wohlgemuth, Oloruntoba, and Clausen [156]), and multi-cab metropolitan transportation services (see Caramia et al. [23]), share the same properties. Where the problem concerns transportation of people, additional constraints might need to be taken into account relevant to passenger waiting and travel times.

In contrast to the two groups of problems discussed above, only few papers have looked at dynamic one-to-many-to-one problems. Wang and Cao [149] studied dynamic requests and service cancellations in the context of a VRP with Time Windows and Backhauls (VRPBTW). Chang, Chen, and Hsueh [24] investigated the dynamic *VRP with Simultaneous Pickups and Deliveries* (VRPSPD). Finally, we are not aware of any work addressing dynamic many-to-many problems.

The dynamic VRPs described above seek to determine the assignment of customer requests to the vehicles as well as the sequencing of visits over the planning horizon. The possibility of serving several customers with a single vehicle implies consolidation of goods or services. In contrast, problems without consolidation where a vehicle serves only one request at a time are also referred to as *resource allocation*. Dynamic fleet management problems as well as dynamic vehicle scheduling and rescheduling problems also fall into this category. Throughout this chapter, we will refer to this class of problems as dynamic vehicle dispatching problems.

To our knowledge, Powell [116] is among the first to address dynamic vehicle dispatching problems in the context of long-haul truckload trucking. In particular, this work deals with a dynamic fleet management problem for truckload motor carriers and seeks to dynamically assign incoming requests to a fleet of vehicles. Each request is characterized by its start and ending time window, origin and destination, duration (including travel time if origin and destination are different), and requirements on the driver for dealing with a particular request. Drivers are, in turn, characterized by their time of availability, location, and factors pertaining to hours of service, desired time off, etc.

Besides truckload trucking applications, various dynamic vehicle dispatching problems have been defined in other application areas with similar features and structure. One example concerns the emergency vehicle dispatching problems, including ambulance location and relocation models for which a review is provided by Brotcorne, Laporte, and Semet [22]. Here, the existing models are differentiated with respect to their deterministic, probabilistic, and dynamic natures. For other applications arising in the context of emergency vehicles, the reader is referred to Gendreau, Laporte, and Semet [50] and Haghani and Yang [63]. One further example arises in the context of real-time scheduling of automated guided vehicles in airports, for which an agent-based approach is described in Mes, Van der Heijden, and Van Harten [101].

11.3 ▪ Dynamic Requests

This section discusses methods and algorithms proposed for solving dynamic vehicle routing and dispatching with dynamic requests. We distinguish between *dynamic and deterministic* and *dynamic and stochastic* problems, and have divided the rest of this section into two parts, one for each class of problems. The former class deals with pure dynamic problem settings, while the latter is characterized by dynamic requests with stochastic information. From a methodological perspective, the treatment of these two classes of problems differs significantly, since the advance knowledge for future events leads to distinct solution frameworks and routing strategies.

11.3.1 ▪ Dynamic and Deterministic Problems

Solution approaches for problems with dynamic requests must follow the online routing process where, at the beginning of the planning horizon, an initial plan is constructed based on the advanced requests. This base plan can be followed without any modifications, until a new customer request is received. In this case, there is always a chance that the new customer can be inserted into the existing planned routes without affecting the order of subsequent customers and with minimal delay. However, it is more likely that the insertion of new requests into the existing route will require either partial or full rescheduling of the vehicle route.

The common practice for generating a base routing plan is to use exact or metaheuristic algorithms already developed for the corresponding static problem. These algorithms can be applied in a *rolling horizon* basis to reoptimize the existing solution when there is a new (immediate) request or an update to the input data. In this case, optimization is performed at discrete time intervals called *decision epochs*. Reoptimization approaches have the drawback of repeatedly solving difficult optimization problems, which may require excessive computational times. Note also that exact approaches can provide optimal solutions for the current state only (unless information is available over the entire planning horizon in advance). In this case, any solution at hand may be suboptimal once new data arrives. We refer the reader to Psaraftis [123] for an example where insertion of a new customer into an existing optimal tour renders the tour suboptimal. Despite this drawback, empirical evidence shows that the use of exact approaches often results in better overall solutions compared to using heuristics for the same purpose (see Yang, Jaillet, and Mahmassani [158] and Chen and Xu [26]).

Reoptimization approaches solve and resolve either part or the entire problem at each decision epoch. Depending on the way the time intervals for reoptimization are defined, one may distinguish between *instant* and *periodic* approaches. Instant reoptimization is performed whenever there is an update to the input data. This scheme is suitable for weakly dynamic systems, given that the time elapsed between subsequent updates determines the time available for reoptimization. On the other hand, periodic reoptimization schemes incorporate the latest updates only after a predefined time period, and any new requests (or other changes in the input data) are kept until the beginning of the next decision epoch. An advantage of this scheme is that the time available for optimization can be predefined. However, it is less suitable for settings where immediate actions are required (e.g., emergency services; see Hvattum, Løkketangen, and Laporte [73]).

As mentioned above, heavy computational requirements might hinder the use of reoptimization procedures, especially in highly dynamic environments where the problem quickly becomes more complex with arrivals of new information (see Ichoua, Gendreau, and Potvin [77]). In this case, an alternative approach is to locally *update* the existing solution. For this purpose, a wide variety of local update and instant reaction heuristic methods (e.g., insertion heuristics) have been proposed. Although these heuristic methods are likely to run in short computational times, the solution quality is often poor and necessitates the additional use of more enhanced reoptimization methods. For this reason, the majority of reoptimization approaches found in the literature are hybrid in the sense that they utilize local update techniques to react to the incoming information, while reoptimization is periodically applied to obtain further improvement.

Instead of reacting to the changes in the environment myopically as done by the above schemes, one may resort to routing strategies that aim to anticipate future events. One way to follow is to use relocation strategies with the aim of (re)positioning idle vehicles to strategic locations. Another effective approach is to make use of waiting strategies, which look at delaying commitments in an attempt to take advantage of future opportunities. For example, it might be beneficial for a vehicle to wait at its current location, as opposed to returning to the depot or to commit early, if there is an expected wait at its next visit dictated by, e.g., time-window restrictions (see Branchini, Armentato, and Lokketangen [20], Chang, Chen, and Hsueh [24], Gendreau et al. [48], and Kilby, Prosser, and Shaw [85]).

Another issue with local update heuristics and myopic reoptimization approaches is guaranteeing robustness, especially for dynamic problems. A methodological framework that seeks to provide reliability and stability as well as a proactive way of avoiding infeasibility is the so-called multiple plan approach. The main idea is to generate and maintain throughout the planning horizon a pool of solutions that correspond to alternative plans. These solutions are coherent with the current state, and one of them can be selected as the current plan at each epoch through a consensus function.

The flexibility of selecting and altering the current master routing plan is a key methodological differentiator between dynamic and a priori routing. In the latter, the base routing plan (a.k.a. skeleton routes, backbones routes, semi-fixed routes, standard routes, master plan, etc.) is typically used to determine the customer assignment and sequencing decisions at the first stage so as to avoid the complexity of full reoptimization in later stages. Therefore, a priori routing approaches can be effective in multi-period problem settings only if rich stochastic information is available and the customer demand patterns are relatively stable.

In the following sections we discuss various methods and algorithms that are proposed for pure dynamic problems. Most sections include a tabulated summary of the papers reviewed. These tables are not intended to serve as an exhaustive coverage of all publications on each topic, but are rather provided to describe the landscape of research pertaining both to problem variants and to solution algorithms.

11.3.1.1 ▪ Routing Policies and Local Update Heuristics

Local update heuristics refer to a set of heuristic rules, which specify what actions should be taken with respect to the current state and the new information at hand. In the literature, various *routing policies* (a.k.a. dispatching policies, online algorithms) have been proposed, ranging from local assignment rules, such as First-Come-First-Served, Nearest Neighbor, Stochastic Queue Median, and Partitioning Policies, to more elaborate algorithms. Local rules can be used to assign customers to a vehicle by considering a queue of pending requests and to build routes sequentially. Therefore, no explicitly planned routes are constructed a priori, although they can be traced back a posteriori (see Ichoua, Gendreau, and Potvin [77]). Routing policies have often been inspired by queuing theory, and they can perform very well even if the rate of immediate requests is very high and the system is congested. So far, studies that analytically examine routing policies have only been investigated for a limited number of dynamic problems, which are discussed in Section 11.3.2.3.

Jaillet and Wagner [78] examined various routing policies for dynamic TSPs with disclosure dates for the requests. Larsen, Madsen, and Solomon [87] studied the partial dynamic TRP with a mix of advance and immediate requests, and provided an evaluation of several routing policies proposed earlier by Bertsimas and Van Ryzin [15] for different degrees of dynamism. Later, Yang, Jaillet, and Mahmassani [158] addressed the multi-vehicle truckload PDP, and compared five rolling horizon strategies based on reoptimization and simple heuristic rules under varying traffic intensities, degrees of advance information, and degrees of flexibility for rejection decisions. Competitive analysis studies for various routing policies can be found in Ascheuer, Krumke, and Rambau [4], Feuerstein and Stougie [42], Hauptmeier, Krumke, and Rambau [68], and Lipmann et al. [95] for the dynamic DARPs, as well as in Angelelli, Savelsbergh, and Speranza [3] for a dynamic multi-period uncapacitated VRP. We refer the reader to Jaillet and Wagner [79] for a survey on complexity results and competitive and analytical studies for dynamic VRPs.

Instead of using local assignment rules that perform well under specific assumptions, ordinary *insertion procedures* and well-known construction heuristics have often been employed in the literature for various problem settings under several operational constraints. In this case, planned routes are constructed for all known requests. Besides flexibility, one other advantage of using insertion procedures to react to incoming immediate requests is that the planned routes can be also used for later decisions. Furthermore, insertion procedures are sufficiently fast and can also be used in real time to accept or reject a request or to specify a time window for a customer visit (see Ichoua, Gendreau, and Potvin [77]).

From an implementation point of view, whenever an immediate request is received (or at regular time intervals) the effort is initially to find feasible insertion positions, or to dispatch a new vehicle, for the new requests in the existing plan. At this point, *accept* or *reject* decisions also occur (see Gendreau et al. [48] and Ichoua, Gendreau, and Potvin [74]). Subsequently, depending on the objective(s), the best feasible insertion position(s) is selected and the new requests are incorporated into the routing plan. Although various fitness criteria and metrics have been proposed (e.g., related to the geographical proximity, the temporal closeness, the latency, the response times, and the smallest disruption), the most used methods adopt an insertion position that results in the shortest detour over a subset of vehicle routes. Insertion procedures for one-to-many and many-to-one problems are relatively simple, since they look at the insertion of single locations. However, for one-to-one problems, one needs to handle the insertion of pairs of pickup and delivery locations. Readers are referred to Berbeglia, Cordeau, and Laporte [13], Cordeau and Laporte [32], Cordeau et al. [34], and Madsen, Ravn, and Rygaard [98] for insertion procedures as applied in PDPs.

One drawback of the myopic savings-based insertion procedures is that they fail to introduce sufficient slack to accommodate future requests. To that end, Mitrović-Minić, Krishnamurti, and Laporte [103] extended the ordinary rolling horizon approach (where all known requests between decision epochs are treated equally or only those requests that are sufficiently close to the current time are considered) to that of *double horizon*. In particular, they considered both short- and long-term planning horizons. The main effort was to alleviate the adverse long-term effects of good short-term decisions. For this problem, it is preferable to put emphasis on minimizing the traveling distance in the short-term since the later portions of the routes are likely to change in the future, and on maximizing the slack time in the long-term, thus favoring the accommodation of new future requests.

In a dynamic VRP, as mentioned earlier, a vehicle at a given point in time is either serving (or waiting to start service) a customer, idle (waiting) at some location, or moving to serve a customer. Whenever a vehicle finishes serving a customer, there are two decisions to be made: (i) *wait* at the current location, or (ii) *move* towards a known (or dummy) but unserved customer. A wait decision after service or a planned wait at a "strategic" location is very important and can be used as a way of anticipating future requests. Similarly, request buffering (holding) techniques (see Pureza and Laporte [125]) can be applied to prioritize more urgent requests. Section 11.3.1.4 describes various routing strategies for the anticipation of future requests and discusses vehicle diversion strategies as well.

11.3.1.2 ∎ Instant, Periodic, and Continuous Reoptimization Approaches

Various reoptimization approaches have appeared in the literature for solving VRPs with dynamic requests. This section presents an overview of such approaches.

Kilby, Prosser, and Shaw [85] presented a periodic reoptimization approach wherein the time horizon is divided into fixed time slots, the size of which are determined based on the degree of dynamism. A rolling horizon with fixed decision intervals for solving the dynamic VRP with and without time windows is described by Montemanni et al. [105], Gambardella et al. [46], and Rizzoli et al. [129]. These papers assume an event manager who receives the immediate requests and keeps track of the served customers, as well as the position and the remaining capacity of each vehicle. Based on this information, an *Ant Colony Optimization* (ACO) algorithm is used to solve the static VRP instances. A pheromone conservation mechanism is used to determine the good solution parts and to transfer this information from one time slot to another. Later, Euchi, Yassine, and Chabchoub [39] extended the work of Montemanni et al [105] and enhanced the ACO system by using a local search algorithm.

A *Large Neighborhood Search* (LNS) is proposed by Hong [69] for the dynamic *VRP with Time Windows* (VRPTW). LNS approaches are based on the so-called ruin-and-recreate principle. More specifically, part of a given solution is destroyed by removing customers from their current positions; then the feasibility is restored by inserting the removed customers into different positions in the solution. Here, all immediate requests and customers already routed but hitherto unserved are treated as "removed" customers. The removal and reinsertion processes are repeated until the next triggering event. The approach has been shown to be an effective one, with the added advantage that only a part of the corresponding static problem is solved at each LNS iteration.

Angelelli et al. [1] proposed short-term solution strategies for solving an uncapacitated multi-period VRP with dynamic pickup requests. In every period, a fixed fleet of vehicles services a set of advance offline and immediate online requests. Each online request has a service deadline of at most two consecutive periods after its arrival, and depending on the length of the deadline, the online request can be also postponable. To that end, an additional decision occurs, which is whether or not to serve the postponable requests, in addition to the immediate requests that may appear later the same day or the day after. The objective is a hierarchical one; i.e., it requires the maximization of the number of served requests (primary) and the minimization of traveling distance (secondary). A periodic reoptimization scheme is adopted, combined with a *Variable Neighborhood Search* (VNS) algorithm. Various computational experiments are reported by varying the criteria to evaluate reoptimized solutions, the length of the look-ahead period (i.e., the period of time which reoptimization will be applied), and the reoptimization intervals.

Wen et al. [155] proposed a three-phase rolling horizon approach for a dynamic multi-period VRP motivated by a large distributor operating in Sweden. Immediate requests are revealed over a multi-period time horizon. The objectives are to minimize cost and customer waiting times as well as to balance the daily workload. Initially, a subset of requests is selected for a given (current) day and for a number of days ahead using a time-space correlation analysis. A routing plan is then generated by solving the corresponding static Periodic VRP (PVRP) with service frequencies equal to one (and visit combinations made up of consecutive days) using VNS. Finally, a *Tabu Search* (TS) post-optimization procedure is applied to minimize the total travel time of each day.

Angelelli, Mansini, and Speranza [2] developed a VNS algorithm for a courier service application. The service area is divided into zones, each with a central hub. In this problem, customer requests must be served within a time window, which might be either pickup or delivery. When the destination zone of a request is different from its origin, the transhipment between the hubs is performed overnight. Customer requests cannot be rejected, but are allowed to be allocated to future shifts. VNS and LNS local search heuristics have been also applied in the dynamic context by Goel and Gruhn [57] for the so-called Generalized VRP (GVRP).

Chang, Chen, and Hsueh [24] proposed a TS algorithm for the dynamic VRPTW with pickup and delivery demands. The objective is to minimize a weighted function of traveling and waiting times. Insertion-based heuristics are employed for the construction of the starting solution as well as for the insertion of new requests. Reoptimization is interrupted at checkpoints, i.e., either if a new request occurs or if the earliest departure time of the last node being served or scheduled to be served has arrived. Beaudry et al. [9] addressed a patient transportation problem between several locations in a hospital campus. This problem can be seen as a dynamic DARP with various side constraints, such as multiple degrees of urgencies, different equipment requirements, soft time windows, and multiple transportation modes. For solving this problem a TS algorithm has been developed, coupled with a local update insertion procedure. Computational experiments on real data are reported. Attanasio et al. [6] proposed different variations of a parallel TS heuristic for the dynamic DARP, which was described earlier by Cordeau and Laporte [31] for the corresponding static problem. In the proposed implementation, immediate requests are randomly inserted into the existing solution and the TS is used to restore feasibility. If a feasible solution is obtained, the request is accepted and TS is applied again as a post-optimization procedure for further improvement.

Du, Wang, and Lu [38] proposed a local search re-optimization approach, coupled with First-Fit and Best-Fit insertion procedures, for solving a two-level VRP with pickups and deliveries that involves transportation of products between suppliers and customers via a distribution center. Pickup-and-delivery orders are placed dynamically, and vehicles can be used for both pickups and deliveries; however, pickup products cannot be sent to customers before being deposited to the distribution center. The problem includes three types of service time windows, namely hard, soft, and mixed, and the objective seeks to minimize both the service penalty (i.e., delay in service) and traveling costs.

A dynamic PDP motivated by a multi-cab metropolitan transportation service company has been studied by Caramia et al. [23]. A cab is allowed to carry up to six customers. Each customer requests a pickup and delivery location, a pickup time window, and a so-called stretch-factor that denotes the maximum deviation from the shortest travel time the customer accepts. Later, Fabri and Recht [40] extended this work, and replaced the stretch-factor with a delivery time window, and allowed cabs to wait at customer locations. For solving the problem, a hybrid TS heuristic combined with an A^*-algorithm is proposed.

A hybrid reoptimization approach is proposed by Berbeglia, Cordeau, and Laporte [14] for the dynamic DARP combining an exact constraint programming algorithm with a TS heuristic. The primary role of the former is to determine whether it is feasible to insert a new request into the existing plan. The role of the latter is to continuously improve the existing solution as well as to insert new incoming requests. For this purpose, the algorithm is equipped with three scheduling procedures. The constraint programming algorithm is executed in parallel to the TS heuristic, either to find a feasible solution or to prove that no feasible compatible solution exists. Experiments indicated that the proposed hybrid scheme outperformed each of the two algorithms when executed independently.

Evolutionary algorithms have also been applied to this class of problems. These algorithms maintain a population of individuals, but when used to solve dynamic problems the population needs to be consistent with the current state of information, and must be updated at least periodically in line with the changes in the input data and the currently executed plan. One advantage of using a population is that it can provide useful information when the search is restarted. To that end, Khouadjia et al. [84] proposed *Particle Swarm Optimization* (PSO) and VNS algorithms for the dynamic VRP. They report that VNS is more accurate than PSO, but PSO is more stable with respect to the changes in the environment. In contrast, the PSO seemed to work better when the objective is to serve as many customers as possible, for increasing degrees of dynamism.

Table 11.1 provides an overview of reoptimization approaches proposed for various dynamic and deterministic problem classes.

Table 11.1. *Overview of reoptimization approaches.*

References	Problem features	Algorithmic features	Objectives	Data set
Kilby, Prosser, and Shaw [85]	VRP	Periodic reoptimization; Local Search heuristics	Min. distance traveled	Modified CVRP benchmark data sets (up to 385 visits)
Montemanni et al. [105]; Euchi, Yassine, and Chabchoub [39]	VRP	Periodic reoptimization; ACO	Min. distance traveled	Modified CVRP benchmark data sets (see [85])
Gambardella et al. [46]; Rizzoli et al. [129]	VRP; VRPTW	Periodic reoptimization; ACO	Min. n. of routes; Min. distance traveled	Data from real-life applications
Hong [69]	VRPTW	Continuous reoptimization; LNS	Min. n. of routes; Min. distance traveled	Based on Solomon VRPTW benchmark data sets
Angelelli et al. [1]	Multi-period VRP; Postponable requests	Short-term strategies; VNS	Max. n. of served requests; Min. distance traveled	Based on Solomon VRPTW benchmark data sets
Wen et al. [155]	Multi-period VRP; Service frequencies	Three-phase rolling horizon heuristic; VNS	Min. distance traveled and customer waiting; Balance daily workloads	Data from real-life application (15-day planning period; up to 80 orders)
Angelelli, Mansini, and Speranza [2]	PDP; Courier Services; Multiple Shifts	VNS; Insertion and Local Update heuristics	Min. operational costs; Max. value of requests served	Numerical example
Chang, Chen, and Hsueh [24]	VRPTW with PDP	Instant reoptimization; TS	Min. weighted function of traveling and waiting times	Based on Solomon VRPTW benchmark data sets
Beaudry et al. [9]	DARP with various side constraints	TS; Local update heuristics	Min. fleet operating costs and patient inconvenience	Real data from a large German hospital
Attanasio et al. [6]	DARP	Parallel TS	Max. n. of served requests; Min. distance traveled	Real world and randomly generated instances
Du, Wang, and Lu [38]	VRPTW with PDP	Local search and local update heuristics	Min. service penalty and distance traveled	Based on Solomon VRPTW benchmark data sets
Fabri and Recht [40]	PDP with several time windows	TS; A^*-algorithm	Max. n. of served requests; Min. distance traveled	Randomly generated problem instances
Berbeglia, Cordeau, and Laporte [14]	DARP	TS; Constraint Programming	Min. distance traveled	Modified DARP benchmark data sets (see Ropke, Cordeau, and Laporte [130])
Khouadjia et al. [84]	VRP with route duration constraints	Continuous reoptimization; PSO; VNS	Min. distance traveled; Max. n. served requests	Modified CVRP benchmark data sets

11.3.1.3 ▪ Multiple Plan Approaches

Motivated by the local operation of long-distance express courier services, Gendreau et al. [48] proposed a parallel TS method for the dynamic VRPTW. This work assumes that the only information provided to the drivers is their next destination. The proposed approach utilizes an *adaptive memory* that is used to maintain a pool of elite solutions. Whenever an immediate request occurs, insertion procedures are applied to all elite solutions to check whether feasible insertion positions exist. If the request is accepted, then a new starting solution that includes the new request is generated by combining the routes of the elite solutions. The TS method is resumed to improve this new solution and terminates when there is an immediate request or the service of a known request is completed. The solution space is explored on the basis of a cross-exchange neighborhood, while a two-level parallelization scheme is also adopted. Ichoua, Gendreau, and Potvin [75, 76] applied the same algorithm for the dynamic VRP with time-dependent travel times and for the dynamic PDP, respectively. Gendreau et al. [47] proposed a TS heuristic for solving a local express courier problem with pickups and deliveries. The neighborhood structure is based on node ejection chains. The problem of determining the best chain or cycle of ejection/insertion moves (of any length) over the existing set of routes is modeled as a constrained shortest path problem. The latter is solved via an adaptation of the all-pairs Floyd–Warshall algorithm. Furthermore, a master-slave parallelization scheme is employed. Recently, Kergosien et al. [83] have adopted a similar TS method for solving a patient transportation problem related to the dynamic DARP.

Bent and Van Hentenryck [10] generalized the "adaptive memory" framework of Gendreau et al. [48] and proposed a multiple plan approach. The idea is to maintain a set of routing plans that are coherent with the plan being executed, as well as with the current state of the vehicles and customers. At each iteration, a so-called distinguished plan is generated to serve known requests, and this plan is followed until the next event occurs. In an effort to limit the amount of modification to the existing plan, the distinguished plan selected is not necessarily one of minimum cost but one that is most similar to the other plans. When an immediate request comes in, a local update procedure is applied to check whether it can be accommodated without destroying feasibility. If at least one feasible plan is found, the request is accepted and incompatible plans are discarded.

Multiple plan approaches equipped with a parallel Adaptive LNS algorithm have recently been developed by Pillac, Guéret, and Medaglia [113, 114]. The first paper studies the dynamic TRP and examines different objectives. The authors demonstrated that the minimization of the total working time as an objective is not well suited to a dynamic environment. Instead, they suggest that minimizing the total distance leads to solutions that are better both in terms of total distance and duration. The second paper studies a dynamic VRP and introduces the notion of driver inconvenience, indicative of the consistency between an updated routing plan and the initial reference plan handed out to the drivers. To that end, a bi-objective optimization problem is defined that minimizes the cost, while maintaining consistency throughout the day.

The concept of maintaining a repository of feasible as well as compatible solutions according to partial executed routes is also utilized by Coslovich, Pesenti, and Ukovich [37]. They proposed a two-phase insertion algorithm for the DARP. While the vehicle is moving between two successive stops (first phase), a set of feasible neighboring solutions is generated and maintained. Whenever an immediate request occurs, i.e., a trip demand by a person located at a stop, an insertion algorithm is used to see whether it could be inserted into the existing route by checking all solutions in the repository (second phase).

Table 11.2 provides an overview of multiple plan approaches proposed for various dynamic and deterministic problem classes.

Table 11.2. *Overview of multiple plan approaches.*

References	Problem features	Algorithmic features	Objectives	Data set
Gendreau et al. [48]; Gendreau et al. [47]	VRPTW; PDP; Courier Services	Parallel TS	Min. weighted sum of distance traveled, lateness and overtime	Randomly generated instances (33 and 24 average request per hour)
Ichoua, Gendreau, and Potvin [75]	VRP with time-dependent travel times	Parallel TS	Min. weighted sum of distance traveled and lateness	Based on Solomon VRPTW benchmark data sets
Kergosien et al. [83]	DARP	TS	Min. transportation costs and tardiness	Randomly generated instances (130 requests per day)
Bent and van Hentenryck [10]	VRPTW	Scenario-based planning; Local update heuristics; Consensus function	Max. n. of served customers	Based on Solomon VRPTW benchmark data sets
Pillac, Guéret, and Medaglia [113]	TRP	Parallel Adaptive LNS	Min. distance traveled and working time	Based on Solomon VRPTW benchmark data sets
Pillac, Guéret, and Medaglia [114]	Bi-objective VRP	Parallel Adaptive LNS	Min. distance traveled and driver inconvenience	Based on Solomon VRPTW benchmark data sets
Coslovich, Pesenti, and Ukovich [37]	DARP	Two-phase insertion algorithm; Solution repository	Min. distance traveled; Max. no of served customers	Randomly generated instances (up to 50 customers)

11.3.1.4 ▪ Routing Strategies for the Anticipation of Future Requests

In contrast to local update heuristics, look-ahead routing strategies can be applied to smooth future perplexities and enhance the performance of the solution methods by anticipating future events. Mitrović-Minić and Laporte [104] proposed a TS algorithm for the dynamic uncapacitated Pickup-and-Delivery Problem with Time Windows (PDPTW). The objective is to serve all requests, while minimizing the total distance traveled. A cheapest insertion algorithm is used to accommodate immediate requests. In addition, a TS algorithm that is similar to the one proposed earlier by Gendreau et al. [47] is applied for further improvements. Four waiting strategies are examined, namely drive-first, wait-first, dynamic waiting, and advanced dynamic waiting. If the dispatcher is forced to postpone the decision until the next destination (wait-first), this reduces the total detour but increases the number of vehicles. In contrast, if waiting times are used (drive-first), the number of vehicles is reduced, but this comes at the expense of increased traveling distance. Dynamic and advance dynamic waiting strategies produced the best results, which seek to distribute the waiting times along routes. For this purpose, the service area is divided into zones, and waiting time is allocated proportionally to the time needed to serve them.

Branke et al. [21] studied the dynamic VRP and examined alternative waiting strategies. The objective is to maximize the probability that immediate requests can be inserted into the fixed routing plan and to minimize the average distance traveled to visit them without violating time constraints. They developed an evolutionary algorithm for solving the corresponding waiting drivers problem with known request arrival times and performed a comparative analysis for different heuristic waiting strategies. Additionally, optimal waiting strategies are proposed for special variants with limited fleet size. In particular, the optimal policy is not to wait for the single-vehicle case. If two vehicles are available and they travel from opposite directions to the depot, the best waiting positions are about half the total distance they have to travel. The waiting strategies proposed by Branke et al. [21] can be summarized as follows:

- *Depot strategy* that forces a vehicle to wait at the depot as long as possible before visiting any customer. Here, the total available waiting time of a vehicle is spent at the beginning, which is the opposite of the no wait strategy.

- *Current location strategy* where each vehicle is forced to wait at a customer location and the waiting time is equally distributed among the customers

- *Proportional current location strategy* where each vehicle is forced to wait after service completion at customers and the waiting time is distributed to each customer proportionally to the distance traveled. In other words, this strategy assumes that the more distant a customer is, the more a vehicle should wait at this location.

- *Distant customer strategy* where the vehicle waits only at the most distant customer; therefore, it is less costly to service new nearby isolated requests.

- *Variable strategy* where a vehicle waits following the proportional current location strategy but only after visiting a number of customers, and where the residual travel distance to the depot is equal to the total waiting time.

Of the strategies above, the variable strategy proved to be the best. The overall performance of other strategies is consistent with the results reported by Mitrović-Minić and Laporte [104].

Pureza and Laporte [125] described a method for solving the dynamic PDPTW with random travel times using a construction-destruction heuristic. They examined the impact of waiting and request buffering strategies for different problem sizes and degrees of dynamism. The former is based on fastest paths with random travel times, whether these are time-dependent or not. The main effort is to take advantage of faster paths in order to wait at given location and arrive no earlier than, but as close as possible to. the time at which service may start at the next destination. The latter seeks to postpone the insertion of non-urgent requests to a later time. Request buffering strategies are rarely studied in the literature. The key issue is to define rules that can provide a quick and correct assessment of the viability of inserting new requests in future route adjustments.

Instead of waiting at customer locations, more advanced strategies may suggest performing anticipatory moves and waiting at promising areas (e.g., moving to "dummy" customer locations) that exhibit a higher probability for a new request to appear nearby. However, stochastic information about future requests is needed to identify promising areas in this case. This information can be either provided in advance or it can be generated from historical request information without assuming any existing distributions. Relocation strategies that exploit available stochastic information are discussed in Section 11.3.2.1. It is worth mentioning that even if a vehicle is waiting at a promising area, sometimes it is not feasible to visit a new customer due to capacity constraints, even if an extra travel cost has already been paid. Therefore, relocation strategies need careful design and validation.

A proactive reoptimization approach for the dynamic VRP is proposed by Ferrucci, Bock, and Gendreau [41] with an objective of minimizing customer inconvenience. Without assuming any prior information, stochastic knowledge about future events is generated using historical information on the requests. This information is used to coordinate the utilization of vehicles and to guide them into request-likely areas by integrating dummy customers. The process is controlled by a TS algorithm that switches between

Table 11.3. *Overview of routing strategies for anticipating future requests.*

References	Problem features	Algorithmic features	Objectives	Data set
Mitrović-Minić and Laporte [104]	PDPTW (uncapacitated)	TS; Various waiting strategies	Min. distance traveled	Randomly generated instances (10–1000 requests)
Branke et al. [21]	VRP	Evolutionary algorithm; Various waiting strategies	Max. n. of served requests; Min. distance traveled	Based on Beasley's ORLib VRP instances
Pureza and Laporte [125]	PDPTW with random travel times	Insertion heuristics; Buffering strategy; Waiting strategy	Hierarchical; Min. n. of lost requests, n. of routes and distance traveled	Randomly generated instances (up to 100 requests)
Ferrucci, Bock, and Gendreau [41]	VRP with customer inconvenience	Proactive reoptimization; TS; Relocation strategies	Min. customer inconvenience	Random instances based on the road network of Dortmund (150 requests on average)

different stages (depending on previously explored solutions) to control the neighborhood operators. Computational experiments demonstrated that the integration of derived stochastic knowledge may lead to considerable improvements, especially when the request data has high structural diversity.

Table 11.3 provides an overview of selected papers that discuss and demonstrate the application of various routing strategies for anticipating future requests.

Another issue which may arise during the solution update process is *vehicle diversion*. Diversion allows a moving vehicle to change its current destination to serve an immediate request that is geographically close to its location. Using this option increases the flexibility to manipulate the current routing plan to a large extent and may yield considerable improvements. Recent technological advances allow the dispatcher to track vehicle positions and speeds in real time and allow exploiting diversion opportunities. However, the use of vehicle diversion also makes the driver operations more complex (see Berbeglia, Cordeau, and Laporte [13]) and a number of issues must be taken into consideration, especially when applied in highly dynamic environments.

Regan, Mahmassani, and Jaillet [126, 127] were the first to study vehicle diversion issues for dynamic full truckload PDPs. They assessed the benefits of diversion assuming different demand patterns and dispatching rules. Later, Ichoua, Gendreau, and Potvin [74] studied the effect of a broader diversion strategy for a dynamic uncapacitated VRP in the context of long-distance courier services. The authors employ a parallel TS heuristic of Gendreau et al. [48] and allow vehicle redirections between their current and planned destinations. This, however, might result in destinations of one or more vehicles being changed. A critical factor is the amount of time available for reoptimization, since the environment constantly changes and diversion opportunities may no longer be valid. Therefore, it is important to keep a balance between the solution quality (as a result of the amount of time invested for reoptimization) and the response time. Vehicle diversion strategies have been also considered in Angelelli et al. [1], Branchini, Armentato, and Lokketangen [20], Chen and Xu [26], and Lorini, Potvin, and Zufferey [96], among others.

Finally, one alternative approach that allows one to reduce the complexity of reoptimization for dynamic multi-period VRPs is districting, i.e., dividing the service region into smaller districts. In this setting, all customer requests that appear within the district are served by a single vehicle or by a predefined group of vehicles that service this region. Territory shaping and sharing schemes applied to multi-period problems can be found in the works of Haughton [64, 65, 66, 67].

11.3.1.5 ▪ Approaches Based on Dynamic Programming and Integer Programming

This section focuses on exact approaches that are based on adaptations of static algorithms. The literature in this field is scarce, and the existing methods are either based on dynamic programming or linear (mixed) integer programming. Table 11.4 provides an overview of exact approaches proposed for pure dynamic problems.

Table 11.4. *Overview of exact approaches.*

References	Problem features	Algorithmic features	Objectives	Data set
Psaraftis [122]	DARP	DP	Min. time and ride-time	Numerical example
Savelsbergh and Sol [132]	PDP (general)	Branch-and-Price	Min. n. of vehicles; Min. route costs	Random generated data
Yang, Jaillet, and Mahmassani [158]	Multi-vehicle truckload PDP	Mathematical Programming	Max. Net revenue	Random generated data
Chen and Xu [26]	VRPTW	Column generation	Min. distance traveled	Based on Solomon's VRPTW benchmark data sets

An application of *Dynamic Programming* (DP) for static and dynamic variants of the DARP, where a single vehicle serves incoming requests from customers wishing to be picked up from a given origin and be transported to a specified destination, is presented by Psaraftis [122]. The features of this problem are that (i) the vehicle routes should be in the form of open paths (as opposed to tours); (ii) each customer should be picked up before being dropped off; (iii) the vehicle capacity, defined as the total number of passengers that can be transported, should be respected; and (iv) there are special priority constraints preventing a customer's request from being deferred indefinitely. DP recursions, for both the static and the dynamic cases, are described, where the latter is based on the observation that there is no need to reoptimize an existing (optimized) route unless new customer requests are introduced into the problem. If there are updates, then the objective function can be modified by excluding customers that are no longer part of the input.

Linear integer programming stands as another exact solution technique on which solution algorithms for the dynamic VRP are based. One of the earlier approaches developed for a dynamic *General Pickup and Delivery Problem* (GPDP) arising in shipping cargo is presented by Savelsbergh and Sol [132]. The approach is based on repeated solution of the GPDP using a Branch-and-Price algorithm whereby new incoming requests are taken into account through algorithmic adjustments. More specifically, all routes obtained through solving the GPDP are split into two parts, head and tail, where head tours are fixed and implemented as short-term decisions, and tail tours can be changed in the future depending on the new requests. The overall approach has been embedded into a decision support system, and results are reported on randomly generated test problems as well as on a case study which involves simulating the dynamic planning environment with real-life data.

Another mathematical programming–based approach is used to model an offline version of a real-time multi-vehicle truckload PDP in Yang, Jaillet, and Mahmassani [158], where a fleet of trucks is available to serve a number of requests, each identified by its pickup location, the delivery location, the earliest pickup time, and the latest delivery time. A request can be either rejected or accepted, and in the latter case the revenue generated is proportional to the distance between its pickup-and-delivery locations. Each truck can serve only one request at a time. This problem is modeled as a variant of the classical assignment problem, which is used to find a least-cost set of cycles going through all the nodes. In an "online" setting, the model includes the requests that are known at a particular decision epoch which have not yet been served or rejected. Yang, Jaillet, and

Mahmassani [158] use such a model in combination with several real-time policies to solve the problem.

Another application of linear (mixed) integer programming is presented by Chen and Xu [26] to solve the dynamic variant of the VRPTW. This problem is defined on a planning horizon divided into evenly spaced epochs. A set of homogeneous vehicles is located at a depot to serve customers with incoming requests, each with a time window that cannot be violated. Each request has a weight, and the total weight carried on a vehicle cannot exceed the vehicle's capacity. Chen and Xu [26] describe a formulation for the static version of the problem where a binary variable is associated with every possible trip to indicate whether it is used by a vehicle or not. To deal with the dynamic nature of the problem, column generation is used to model incoming requests in future epochs. These requests are incorporated into the existing formulation by either (i) modifying the existing routes by inserting or deleting requests, or (ii) creating new routes to serve new requests. New columns are generated by using fast local-search-based heuristics. The proposed approach allows tackling problems of up to 100 requests with planning horizons of up to 2400 seconds.

11.3.1.6 ▪ Miscellaneous Approaches

An application of genetic programming for a dynamic PDP was described by Benyahia and Potvin [12]. The problem was motivated by a local courier service company that receives calls for the pickup and delivery of express mail in an urban area. The goal was to model and approximate the decision process of an expert vehicle dispatcher via a utility function. Computational results are reported on real data sets, and a comparative analysis is carried out with a neural network model and a simple dispatching policy.

Jemai and Mellouli [80] proposed a neural TS approach for the dynamic VRPTW. The proposed approach is composed of two parts. The first part consists of learning and reproducing previous routing decisions using a feed forward neural network. For this purpose, different problem instances are simulated, and the learning sets are constructed using the exact solver GLPK. The second part consists of a post-optimization TS algorithm that tries to improve the initial solution generated based on the assignment provided earlier by the neural network. Computational results demonstrated better performance compared to well-known local update heuristics.

A disruption recovery model is proposed by Wang and Cao [149] for the VRPBTW with new pickup requests, service cancellations, and increase/decrease in the pickup quantities from the backhaul customers. The objective is to find a recovered solution that minimizes the deviation from the original plan after the occurrence of a disruption. Disruption recovery models are also proposed for the multi-objective VRPTW by Wang, Xu, and Yang [153] and for the VRP with fuzzy time windows by Wang, Zhang, and Yang [154]. A more enhanced combinatorial disruption recovery model for the VRPTW is described by Wang et al. [151] that takes into account multiple types of customer disruption events, such as changes of time windows, changes of customer locations, removal of requests, and combinatorial disruption events. The term combinatorial, in this context, indicates that some disruption events occur simultaneously at one or more customers.

11.3.2 ▪ Dynamic and Stochastic Problems

So far, it was assumed that no exploitable stochastic information is available in advance. However, it is often the case in practice that a dispatcher has some valuable a priori knowledge regarding the demand patterns, not only in terms of time (e.g., "peak" time periods) but also in terms of locations (e.g., intense geographical areas; see Ichoua, Gendreau, and Potvin [77]). This information can be used to anticipate forecasted needs.

A common scenario is that the locations of the customers are known in advance; however, the demand of each customer is a random variable following a known probability distribution and is revealed only after the vehicle visits the customer location. This problem variant is known as the dynamic VRP with *stochastic demand*. Another scenario arises when the only stochastic information available is on the expected number of customer requests. This problem variant is known as the dynamic VRP with *stochastic customers*. Additionally, other forms of stochastic information can also be provided and/or extracted from historical data, for example, service cancellations, absence of customers, and service times.

This section reviews models and algorithms proposed for addressing dynamic vehicle routing and dispatching problems with various forms and combinations of available stochastic information. In particular, Section 11.3.2.1 discusses anticipatory algorithms and predictive routing strategies that exploit stochastic information about customer locations, demand patterns, and/or the occurrence time of immediate requests. More advanced sampling algorithms and multiple scenario approaches are described in Section 11.3.2.2. On the other hand, Section 11.3.2.3 presents analytical studies of various routing policies and problem settings. Finally, stochastic programming as well as other stochastic models and algorithms are described in Section 11.3.2.4 for a wide variety of dynamic problems with stochastic information.

11.3.2.1 ▪ Anticipatory Algorithms and Predictive Routing Strategies

Anticipatory algorithms in the context of vehicle dispatching were first introduced by Powell et al. [119] for long-haul truckload trucking applications. This problem consists of assigning drivers to pickup-and-delivery requests that arise randomly over time within a given time window. A vehicle relocation strategy in anticipation of future demands is introduced. Future demand forecasts are used, and time-space graphs are proposed where nodes correspond to actual and forecasted demands at different regions and time periods. The deterministic cost and the expected cost of forecasted capacity needs of future periods are minimized at each period. Computational results show that it is always advantageous to take forecasted demands into account, as compared to a model that only reacts whenever a new load appears. Section 11.3.2.4 reviews works based on stochastic modeling, including approaches based on *Markov Decision Processes* (MDPs) and *Approximate Dynamic Programming* (ADP) methods, which integrate stochastic information by evaluating expected values to take predictive or preventing actions.

Larsen, Madsen, and Solomon [88] studied the so-called partially dynamic TSP with soft time windows, and they define waiting and relocation points based on a priori information regarding the distribution and clustering of customers. The requests are revealed over time in a number of subregions following a Poisson process with different arrival rates. Starting from the basic latest departure policy, they have developed more enhanced relocation policies that make use of advance information to reposition the vehicle at idle points. The set of idle points are defined as locations with a high probability of generating new requests, and they are selected based on different criteria, i.e, the nearest idle point, the busiest idle point (highest arrival intensity), and the idle point with the highest expected number of new requests. Among them, the policy to reposition the vehicle at the nearest location performed best in terms of average lateness and number of late customers.

An evolutionary algorithm applied over a rolling horizon basis is proposed by van Hemert and La Poutré [146] for a VRP with dynamic pickup loads. The objective is to serve as many customers as possible. They considered that advanced knowledge regarding the probability distributions for the occurrence of a new service request is provided, but in the form of regions. To that end, the concept of fruitful regions is introduced. From the implementation viewpoint, a self-adaptive mechanism is employed that allows the evolutionary algorithm to explore the possibility of vehicles to visit nodes that have not yet requested service. The fitness function is extended with anticipated moves towards these fruitful regions. They also examine the conditions under which such moves improve effectiveness.

Ichoua, Gendreau, and Potvin [76] proposed a threshold-based waiting strategy. The latter exploits probabilistic knowledge about future requests in an effort to better manage the fleet of vehicles and to provide better area coverage. In particular, a vehicle waits at its current location if (a) its next destination is "far enough"; (b) the probability of an immediate request to occur in the vehicle's neighborhood in the near future is "high enough"; and (c) there are not "too many" vehicles in the current zone. The proposed strategy is employed within an adaptive parallel TS algorithm, and significant improvements are observed.

Branchini, Armentano, and Løkketangen [20] proposed an adaptive granular local search heuristic for the dynamic VRPTW. Time windows can be violated by assuming a lateness cost, and the objective is to maximize the expected profit, i.e., the difference between the total revenue and the sum of lateness and travel costs. Initially, a construction heuristic is applied that scatters vehicles over the service area. The number of vehicles deployed is based not only on advanced requests but also on the expected required capacity based on historical data. The local search heuristic employs an adaptive granularity threshold to control the neighborhoods that is relative to the frequency of new customers. Additionally, waiting, repositioning, and vehicle diversion strategies are employed. In particular, a waiting-first scheme is followed. The option of repositioning a vehicle to wait at a strategic location (with a higher probability of new customers to appear) depends on the amount of time needed to reach the location and return before the depot closing time (spare time) and the number of vehicles currently repositioned at each location.

Repositioning strategies have long been recognized as key elements, especially in the context of emergency response systems. Vehicles' preparation and travel times directly influence the resulting service level. Thus, it is essential to exploit advance information and move idle vehicles currently siting at a low demand areas to cover higher demand areas. Gendreau, Laporte, and Semet [50] studied the ambulance redeployment problem. The objective is to maximize the proportion of the demand covered by at least two vehicles within a predefined radius minus a relocation cost. A parallel TS algorithm is employed to predefined redeployment scenarios in anticipation of future events. More recently, Gendreau, Laporte, and Semet [51] proposed a dynamic relocation strategy that seek to maximize the expected covered demand. Haghani and Yang [63] described an emergency response fleet deployment system having the capability to look ahead for future demands. Several dispatching policies are examined, and a mathematical model that considers vehicle dispatching and relocation decisions jointly, in a way similar to that of Yang, Hamedi, and Haghani [159], is solved in an exact fashion over a rolling horizon.

An agent-based approach for the dynamic scheduling of full truckload transportation orders with time windows has been proposed by Mes et al. [101], where intelligent vehicle agents schedule their own routes. Multi-agent systems consist of groups of agents that interact with each other. Whenever an immediate request occurs, vehicle agents interact with job agents, who seek to minimize transportation costs using a Vickrey auction. For every new request, the bid consists of a price and the expected arrival and departure times. Based on this information, the job agent assigns the request to the highest bidder, but at the price of the second highest bidder. Later, Mes et al. [100] extended this work and presented new look-ahead strategies for the dynamic PDP. They considered the expected profit of new jobs, where in addition to bids, vehicles also decide about insertion positions for the new jobs and waiting locations. A multi-agent approach for the dynamic VRP is also proposed in Barbucha [8].

Below, Table 11.5 provides an overview of selected anticipatory algorithms and predictive routing strategies.

Table 11.5. *Overview of anticipatory algorithms and predictive routing strategies.*

References	Problem setting	Algorithmic features	Objectives	Data set
Larsen, Madsen, and Solomon [88]	Partial dynamic TSP;	Various relocation strategies	Min. lateness	Randomly generated data; Real world data
van Hemert and Poutré [146]	VRP with dynamic pickup loads	Self-adapted Evolutionary algorithm	Max. n. of served customers	Randomly generated data (up to 50 customers and 5 clusters)
Ichoua, Gendreau, and Potvin [76]	VRP with soft time windows;	Parallel TS	Max. no of served customers; Min. distance traveled and lateness	Randomly generated data (up to 204 requests)
Branchini, Armentano, and Løkketangen [20]	VRP with soft time windows	Adaptive granular local search; Waiting, repositioning and diversion strategies	Max. expected profit	Randomly generated data (up to 1101 customers)
Mes et al. [101]	Full truckload trucking with time windows; stochastic customers	Multi-agent approach; Coordination through auctions	Min. transportation costs	Self-generated instances based on real world data

11.3.2.2 ▪ Sampling-Based Anticipatory Algorithms and Multiple Scenario Approaches

In the previous section, various anticipatory strategies are discussed, including, among others, waiting and repositioning heuristics integrated within reoptimization frameworks, which aim to enhance the response to foreseen upcoming events. Another alternative is to evaluate possible decisions at each time step (or decision epoch), and thus to solve (optimize) future scenarios in order to determine the best course of action. On this basis, several sampling-based anticipatory algorithms have been proposed in the literature that seek to make decisions using samples (scenarios or simulations resulting from the sampling of the input distributions) of the future.

For practical dynamic VRP applications with time-critical decisions, the task of evaluating every possible decision at each time step is extremely challenging, since a large number of samples may be required to effectively reflect the reality (see Pillac et al. [111]). Therefore, most authors resort to approximations, such as consensus and regret, and to sampling/scenario–based approaches that will provide similar benefits at a fraction of the computational cost. Interested readers on sampling-based anticipatory algorithms applied in a variety of applications, including vehicle routing, vehicle dispatching, packet scheduling, and reservation systems, as well as on the relevant theoretical analysis, may refer to Van Hentenryck and Bent [147].

A multiple scenario approach for the dynamic VRPTW with stochastic customers is introduced by Bent and Van Hentenryck [10]. The goal is to maximize the number of serviced customers. The proposed approach is highly flexible and generic, and it extends the multiple plan approach described earlier in Section 11.3.1.3. It maintains a pool of routing plans for scenarios that include both known and possible future requests. The realizations of future requests are obtained by sampling their probability distributions. Once a routing plan is obtained for each scenario, future virtual customers are removed from the plan, leaving room for new immediate requests. At each decision epoch, a distinguished plan is chosen according to a consensus function (based on similarity among plans), and the pool is updated to reflect the current state. In particular, obsolete scenarios are removed as new information is disclosed. Experiments illustrate the effectiveness of the multiple scenario approach, which in many cases outperformed the multiple plan approach.

Vehicle relocation and waiting strategies guided by scenarios are proposed by Bent and Van Hentenryck [11]. In particular, vehicles can wait or can be relocated anywhere at any time. Decisions of when and where to relocate or whether to wait are derived systematically, based on the scenario solutions and not on any prior knowledge or distribution. In addition to the decision of moving to a known or an accepted request, a wait action is considered whenever a sampled request is scheduled after an accepted request. As for relocations, this strategy simply considers moving a vehicle to the location of sampled customers which yields the best evaluation. Computational experiments indicate that both strategies are very effective in maximizing the number of serviced customers, especially for highly dynamic problems with many late requests.

Ghiani et al. [54] proposed anticipatory insertion and local search algorithms for the uncapacitated dynamic VRP with pickups and deliveries. Requests arrive according to a given stochastic process, while the objective is to minimize the expected customer inconvenience. Whenever an immediate request occurs, the request arrival process is sampled and potential solutions are evaluated with respect to the expected inconvenience assuming perfect information. A particular feature is that sampling is only performed for a predefined short-term period (empirically correlated to the problem data). As a result, the computational effort is relatively small. On the other hand, a fully sequential indifference zone selection procedure is employed to determine the number of demand samples which allows one to eliminate clearly inferior solutions determined earlier. Compared to the corresponding pure reactive algorithms, significant improvements are observed.

Azi, Gendreau, and Potvin [7] proposed a scenario-based adaptive LNS approach for a dynamic VRP with multiple delivery routes. Customer requests are received according to an independent time-space Poisson process. Based on this information, multiple scenarios for the occurrence in time and space of future requests are generated and maintained, while they are used to decide whether to include an immediate request based on an opportunity value. Whenever a new request occurs, it is checked whether it can be feasibly inserted into the actual as well as into every scenario solution. Subsequently, an opportunity value is determined as the sum of differences in solution prior to and after the insertion of the new request. If the value is positive, the new request is accepted. On the other hand, the adopted and all scenario solutions are improved via the adaptive LNS, while a repair procedure is applied to ensure the consistency of scenarios with respect to the executed plan.

Flatberg et al. [43] presented a scenario-based anticipatory algorithm and described the relevant architecture of a solver for real-life dynamic and stochastic VRPs. Several practical and implementation issues are also discussed, including the representation of statistical knowledge of events and learning event models building upon both domain knowledge and historical data. Furthermore, two real-life application examples are discussed (transportation of goods and transportation of persons) that illustrate how dynamic and stochastic aspects show up and why proper handling of dynamic events is essential for successful operations. More recently, an event-driven framework for the multiple scenario approach that allows high reactivity to changes occurring in highly dynamic environments is described in Pillac, Guéret, and Medaglia [112].

Below, Table 11.6 provides an overview of sampling algorithms and multiple scenario approaches proposed for the dynamic and stochastic VRPs. Compared to the corresponding myopic counterparts of these methods, in all cases it is demonstrated that incorporating stochastic information about future requests can lead to significant improvements.

Table 11.6. *Overview of sampling algorithms and multiple scenario approaches.*

References	Problem features	Algorithmic features	Objectives	Data set
Bent and Van Hentenryck [10, 11]	VRPTW with stochastic customers	Multiple scenario approach; Consensus function	Max. n. of serviced customers	Based on Solomon VRPTW benchmark data sets
Ghiani et al. [54]	VRP with PDP (uncapacitated)	Sampling; Anticipatory insertion and local search heuristics	Min. expected customer inconvenience	Randomly generated data (up to 600 requests)
Azi, Gendreau, and Potvin [7]	VRP with multiple delivery routes	Scenario-based adaptive LNS	Max. profit	Randomly generated data (up to 144 requests)

11.3.2.3 ▪ Analytical Studies

Stochasticity in the dynamic VRP is often related to the nature of information on customer demands, which are usually described by probability distributions. One of the earliest mentions of a dynamic and stochastic routing problem is in Psaraftis [123]. The considered problem consists of a single (uncapacitated) vehicle that seeks to service a number of nodes, where the arrival of demands at each node is described by a Poisson distribution. This problem is termed the dynamic TSP and concerns finding an optimal policy with respect to a number of objectives, such as maximizing the average expected number of demands serviced per unit time or minimizing the average expected time to serve the demands. No analytical results are derived by Psaraftis [123] for the dynamic TSP, but useful insights are provided.

To the best of our knowledge, the first analytical study on the so-called dynamic TRP, a variant of the stochastic and dynamic VRP, is introduced and studied by Bertsimas and Van Ryzin [15]. This problem arises in practical contexts where the wait for deliveries has priority over travel cost; hence the objective of the problem is to minimize the average system time (or to maximize the level of services provided by the vehicle). Such problems arise in route planning for replenishment of stock, a fleet of taxis, emergency services, etc. Bertsimas and Van Ryzin [15] describe a model for the problem, from which a lower bound on the optimal system time is derived. They also propose a number of policies for the problem and show that the one based on locating the vehicle at the median of a given region, serving customers according to the *First-Call First-Served* (FCFS) principle and returning to the median after each service, is optimal for the light traffic case. These results are extended to the case of multiple capacitated vehicles with objective functions involving system time *and* travel cost in Bertsimas and Van Ryzin [16]. Further generalization to the case where assumptions on the distribution of demand locations and on the arrival process are relaxed are considered in [17]. The modeling framework put forward in [15] has been adopted for the single-vehicle dynamic PDP in Swihart and Papastavrou [139]. While in the dynamic TRP the services are characterized by the time spent at respective locations, and in the dynamic PDP the services require that the vehicle change location during service, as each demand must be transported from its pick-up point to its location of delivery. Swihart and Papastavrou [139] derive bounds for cases where the vehicle has capacity for unit or multiple demands and investigate the performance of a number of policies for both cases.

A stochastic and dynamic version of the VRPTW is addressed by Pavone et al. [110], who looked at two variations where (i) for each customer there is a time at which demand is released and a deadline by which service needs to be made, and (ii) the deadline for each customer is modeled by using a random variable denoting the *impatience* time, and any request not met within this time after it being available will expire. The authors derive lower bounds on the minimum number of vehicles for both cases and computationally test several policies for the problem.

11.3.2.4 ▪ Stochastic Programming Models, Markov Decision Processes, and Other Algorithms

This section describes various stochastic models, MDP-based approaches, and other algorithms for solving dynamic vehicle routing and dispatching problems with stochastic information, including problem settings with stochastic customers and/or stochastic demands and/or other combinations.

Thomas and White [142] studied a single (uncapacitated) vehicle dynamic PDP. The vehicle travels from an origin to a known destination point (that must be reached by some fixed time) and services potential customers at known locations along the route. If the vehicle visits a customer after that customer has requested a pickup, the vehicle gets a reward. The probability that a customer will request a pickup before a particular time is known. The objective is to minimize the expected total cost (travel costs minus pickup rewards). The problem is modeled as a finite-horizon MDP. Based on structural results, an optimal policy is derived, and it is compared with a reactive strategy that ignores potential customer requests. Computational experiments on instances derived from a real-world road network demonstrate that the proposed approach obtains good results, especially when customer requests occur late.

A similar problem with a mix of advance and late customer requests from known locations is studied later by Thomas [141]. The author examined policies for selecting the next customer as well as where and how long to wait. The goal is to maximize the expected number of late request customers. The problem is modeled as a finite-horizon MDP, and the optimal policy is derived for the single dynamic customer. Using the structural results, the author developed a real-time heuristic and compared it with five waiting heuristic algorithms, namely the center-of-gravity longest wait heuristic with and without stochastic information, the center-of-gravity closest heuristic, the wait-at-start heuristic, and the distribute-available-waiting-time heuristic. Overall, the heuristic based on structural results performs better for problems with less than 25% late customer requests. Furthermore, the customer location information seems more valuable compared to the likelihood of occurrence. For problems with many late customer requests, the strategy of distributing waiting time across the already routed customers performs better. Note that this result is consistent with those reported earlier in Branke et al. [21] and Mitrović-Minić and Laporte [104].

Based on a real-world case, Hvattum, Løkkentangen, and Laporte [72] proposed a so-called dynamic sample scenario hedge heuristic for solving a dynamic VRP with pickup and delivery orders and rich stochastic information. The problem is formulated as a multi-stage stochastic programming model with recourse. Customer locations and demands are stochastic variables, and they are revealed at the time of call in, while historical data are used to determine probability distributions of the order attributes. Sample scenarios are generated and used to guide a heuristic method that builds a plan for each predefined time interval. Each sample scenario is solved at each period as a static VRP. When all scenarios are solved, the most common features (based on the assignment frequency) are used to build the routing plan for the upcoming period. This gradual solution construction scheme based on the sample solutions can be seen as progressive hedging. Computational experiments illustrates the effectiveness of the proposed method compared to a pure dynamic heuristic method. Later, Hvattum, Løkketangen, and Laporte [73] extended the problem and considered the possibility of stochastic demands, while a more enhanced sampling-based Branch-and-Regret heuristic is proposed.

MDP-based approaches have been proposed by Secomandi [133], Secomandi and Margot [134], and Novoa and Storer [108] for the single-vehicle dynamic VRP with stochastic demands. Compared to traditional two-stage approaches with recourse that seek to minimize expected routing costs (including the cost of recourse actions at the second stage), routing decisions are made whenever a vehicle arrives at a customer location and demand is revealed. The idea is to generate an initial solution, and adapt it at each decision point so it always terminates from the state the system is currently in. Secomandi [133] formulated the problem as a stochastic shortest path problem and proposed a sequential consistent roll-out algorithm based on a cycling heuristic. Later, Secomandi and Margot [134] provided a partial characterization of the optimal policy and proposed two partial reoptimization heuristics. The idea is to take into account only a subset of all the possible states and compute an optimal policy on this restricted set of states. Computational experiments showed that the partial reoptimization heuristics are highly effective.

Novoa and Storer [108] described a two-stage roll-out algorithm for the single-vehicle dynamic VRP with stochastic demand using ADP approaches (see Powell [117]). As proposed in [133], the problem is formulated as a stochastic shortest path problem; however, a Monte Carlo simulation is used to compute the expected cost-to-go. At each stage, decisions are estimated based on the current and expected future costs. The proposed two-step roll-out algorithm sequentially improves a single or multiple a priori suboptimal base solution, while improved pruning schemes and look-ahead strategies are also introduced. The authors presented results of computational experiments on randomly generated instances, similar to those in [133], and provided solutions close to those obtained with perfect information.

Goodson, Ohlmann, and Thomas [58] propose a family of roll-out policies based on fixed routes to obtain dynamic solutions for the multi-vehicle VRP with stochastic demand and duration limits. Initially, a traditional one-step algorithm is proposed that yields promising results for up to 50 customers. Next, a post-decision roll-out is proposed that looks ahead from each post-decision state that fixes vehicle destinations, capacities, and arrival times at the next decision epoch (but does not observe demand as in one-step roll-out). To that end, a hybrid policy is also introduced that limits the number of post-decision states. In an effort to enhance computational tractability, a dynamic decomposition scheme (i.e., repartition of customers at each decision epoch) is adopted that allows applying roll-out policies to single-vehicle problems and using the resulting policies to select actions for multiple vehicles. Compared to the static version, the proposed dynamic decomposition-based roll-out policy produces much better results.

ADP and other approaches have been successfully applied for a wide range of fleet management problems, including, among others, dynamic vehicle allocation and dynamic resource allocation problems (see Powell and Topaloglu [121], Spivey and Powell [137], Topaloglu [143], and Topaloglu and Powell [144]). In an earlier work Powell [116] studied a long-haul truckload trucking problem and proposed models for static and dynamic cases where the demand could be deterministic or stochastic. One of the ways in which this paper models the stochastic and dynamic assignment problem is using a conditional recourse function, which is approximated through several methods such as scenario aggregation, stochastic gradient methods, and a successive linear approximation procedure. Computational testing showed that the dynamic model outperforms a more standard myopic one. It also provides evidence that the value of having advance information on shipper demands is minimal. A later work by Powell, Snow, and Cheung [120] described two heuristics for the solution of the dynamic problem using static snapshots of data, where the focus is on routing a driver through a sequence of more than one request. Spivey and Powell [137] studied a more general class of dynamic assignment problems and tested the effect of advance information on myopic and nonmyopic models. The tests showed that the latter outperformed the former only when insufficient advance information is available, but the situation is the opposite when sufficient advance information is at hand.

A tabulated summary of algorithms based on stochastic programming models and MDPs is provided in Table 11.7.

Table 11.7. *Overview of algorithms based on stochastic programming models and MDPs.*

References	Problem features	Algorithmic features	Objectives	Data set
Thomas and White [142]; Thomas [141]	single vehicle (uncapacitated) PDP	Finite-horizon MDP; Optimal policies; Reactive strategies; Waiting Strategies	Min. expected travel cost minus pickup rewards; Max. n. of late request customers served	Randomly generated data based on real world road network
Hvattum, Løkkentangen, and Laporte [72]	VRP with PDP	Dynamic sample scenario hedge heuristic	Min. unserved customers, vehicles, travel time	Real and random data (up to 200 orders)
Hvattum, Løkkentangen, and Laporte [73]	VRP with PDP and stochastic demands	Branch-and-Regret heuristic	Min. unserved customers, vehicles, travel time	Real and random data (up to 200 orders)
Secomandi [133]; Secomandi and Margot [134]	single-vehicle VRP with stochastic demands	Roll-out algorithms; Partial reoptimization	Min. travel cost	Randomly generated instances (up to 60 customers)
Novoa and Storer [108]	single-vehicle VRP with stochastic demands	ADP; Two-stage roll-out algorithm	Min. travel cost	Similar to Secomandi [133]
Goodson, Ohlmann, and Thomas [58]	VRP with stochastic demands and duration limits	One-step, Post-decision, hybrid and decomposition-based roll-out policies	Min. travel cost	Based on Solomon VRPTW benchmark data sets

11.4 ▪ Dynamic and Time-Dependent Travel Times

This section discusses treatment and solution approaches for problem variants with dynamic and time-dependent travel times.

Dynamic VRPs have mainly been characterized through arrival process of customer requests over time not known a priori, but assuming that travel times between pairs of customers are fixed based on forecasted or historical information. However, travel times might change over the course of time (i.e., time-dependent) or be dynamic in nature themselves, only revealed in real time as the vehicles travel. Vehicle routing planning that takes into account real-time information on the travel times might not only have benefits to reap in the total cost of travel but also on the reliability of arrivals at customers, particularly when there are time-window restrictions in place. Advantages of such an approach over assuming fixed (forecasted) travel times are shown by Taniguchi and Shimamoto [140], who show through numerical experiments on a test network of 25 nodes and 80 links that incorporating real-time information enables reducing total costs by 1.5–3.7% and penalties for early or delayed arrivals at customers by 22.2–100%. The authors also state that such an approach would also alleviate traffic congestion by reducing the operating time of vehicles.

A more general VRPTW where *both* service requests and travel times are assumed to be dynamic is studied by Haghani and Jung [61], where requests at customers can be either pickup or delivery, assuming no priority order on requests. Travel time on a given arc is modeled as a continuous function changing over time, as was proposed in the same authors' earlier work (see Jung and Haghani [81]), where a formulation and a genetic algorithm to solve the problem are also described. The approach described in [61] adjusts vehicle routes at given points in time within a planning period as new information becomes available. The adjustments are then communicated to the vehicles from a centralized control center. Results of computational experiments on randomly generated problems with up to 10 nodes comparing the genetic algorithm with the exact solution of the formulation solved by CPLEX as well as a lower-bounding solution procedure confirm that the genetic algorithm is able to produce solutions within 7% from the optimal value. The genetic algorithm is then tested on a larger case study network with 382 nodes and 1398 links. The results reveal that route adjustments are worthwhile when travel times fluctuate significantly over a planning horizon even when requests are static. When requests themselves also change with time a static routing strategy is seen to be between 1.43 and 2.21 times more costly than a dynamic strategy, being particularly effective as travel time uncertainty increases.

Potvin, Xu, and Benyahia [115] look at a similar problem with uncapacitated vehicles but with the additional constraint that each customer i and the depot ($i = 0$) be associated with a predefined time window $[a_i, b_i]$ in which arrivals should be made. A further assumption is that updates to travel information can be conveyed to the vehicles only at customer locations. The objective of the problem is to minimize an objective function comprising travel time, late arrivals at customer locations as calculated by $(\tau_i - b_i)^+ = \max\{0, \tau_i - b_i\}$ for each customer i where τ_i is the planned arrival time, and late arrival to the depot. More specifically, if the sequence of customers visited by vehicle $k = 1, \ldots, |K|$ is given by $\{i_0^k, i_1^k, \ldots, i_{m_k}^k\}$ with $i_0^k = i_{m_k}^k = 0$, the objective can be expressed as follows:

$$(11.4) \qquad \sum_{k=1}^{|K|} \left(\alpha_1 \sum_{p=1}^{m_k} t(i_{p-1}^k, i_p^k) + \alpha_2 \sum_{p=1}^{m_k-1} (\tau_{i_p^k} - b_{i_p^k})^+ + \alpha_3 (\tau_0^k - b_0)^+ \right),$$

with $\alpha_1, \alpha_2,$ and α_3 weights and τ_0^k being the time by which vehicle k should return to the depot. One contribution of Potvin, Xu, and Benyahia [115] is the introduction of *tolerance* for delayed arrival times at each customer, which also serves as a threshold value at which "reactive" changes are introduced to existing routes. The authors describe a dispatching algorithm based on local search and a discrete-event simulation scheme to handle the dynamic aspect of the problem, and results on Solomon's 100-node benchmark problems indicate that allowing for a small but positive, rather than zero, tolerance significantly improves the objective function value.

Lorini, Potvin, and Zufferey [96] later extend the approach of Potvin, Xu, and Benyahia [115] to take into account a version of the problem where it is assumed that communications between drivers and the dispatch office are possible even when the vehicles are en-route. This variant of the problem allows for diversion of vehicles from their planned destination and opens up opportunities to serve new requests. Computational results on benchmark instances reported in [96] confirm that allowing for diversion yields savings in the objective function, which increase with the amount of allowable tolerance in the maximum acceptable delay of a vehicle's planned arrival time to a given destination.

As mentioned above, Haghani and Jung [61] do not consider time-window constraints within the context of the dynamic VRP, whereas [115] does not consider capacity constraints on the vehicles. The consideration of both constraints under the time-dependent nature of requests and travel times can be seen in the work by Chen, Hsueh, and Chang [25], who look at the problem with aiming at minimizing an objective function comprising a weighted function of travel times on each arc, as well as pre- and post-service waiting times under time-window and vehicle capacity constraints. Time-dependent travel times for each arc of the network are modeled as a step-wise function, where a different constant travel time is prescribed for each interval of a planning horizon. The authors suggest that the idle waiting after service (i.e., a positive difference between service completion time at and departure time from a customer) allows for flexibility, as such a vehicle is able to receive and respond to new route instructions under real-time dispatching. Chen, Hsueh, and Chang [25] describe a mathematical model and a heuristic procedure for the solution of the problem. The authors also present computational results on Solomon's benchmark problems and a real-life logistics network using the heuristic procedure, which indicate that, in comparison to a method which does not account for time-dependent travel times, the proposed approach not only improves the objective function value from 1% to 22%, but it also reduces the probability of service requests rejected.

Hsueh, Chen, and Chou [70] studied a dynamic VRP for relief logistics in natural disasters and proposed a local search reoptimization approach. They considered in their model dynamic pickup-and-delivery demands for relief goods as well as dynamic travel times. The objective was to minimize the traveling time and the total penalty due to late arrivals. An anticipatory algorithm of a routing and scheduling problem for forwarding agencies handling less-than-truckload freight in disaster reliefs is studied by Wohlgemuth, Oloruntoba, and Clausen [156]. The problem involves dynamic pickup-and-delivery requests that may both receive and send goods and varying travel times. The objective is to avoid delays (minimize travel distance) and increase utilization (minimize the number of vehicles). A TS approach is employed that benefits from the anticipation of the availability of connections and the integration of possible demand regions.

Wang and Zhu [148] proposed an ACO algorithm for the VRP with dynamic requests and travel times. The objective is to minimize the weighted sum of expected travel times, expected waiting times, and expected penalties. Finally, a GA-based reoptimization procedure coupled with effective local update heuristics has been developed by Cheung et al. [27] for the dynamic PDP with capacity constraints, time windows, and dynamic travel times. On the other hand, Haghani, Tian, and Hu [62] developed a simulation model to evaluate vehicle dispatching strategies for emergency response with the objective of minimizing average response times associated with different types of accidents. Various strategies were analyzed, namely the FCFS strategy, the nearest origin assignment strategy, and the flexible assignment strategy, for assigning and dispatching response vehicles under various circumstances (i.e., different accident occurrence rates, route change strategies, and dynamic travel time information).

Cortés et al. [36] modeled the capacitated dynamic PDP as a *Hybrid Adaptive Predictive Control* (HAPC) problem based on state space variables. The system state is defined in terms of departure times and vehicle loads at stops. The inputs are routing decisions, and the outputs are departure times. The demand requests are modeled as disturbances. The proposed model is discrete and considers a variable step size that is equal to the time between successive calls. The objective function takes into account future requests via probabilities computed from historical data, and accounts for both the total expected waiting and travel time for passengers and the idle travel time. Furthermore, it is considered that potential rerouting of vehicles could affect the current dispatch decisions through the extra cost of inserting immediate requests into the current routing plan. A clustering technique (classic zoning) is used to estimate the spatial-temporal probabilities to forecast future request points. The HAPC problem is solved by means of a PSO method. Computational experiments indicated savings in actual waiting times as compared to myopic dispatch decision models. Saez, Cortés, and Nunez [131] proposed a similar HAPC approach, where fuzzy zoning is used to compute probabilities and trip patterns from historical data and a GA is used to solve the HAPC problem. Finally, Cortés, Nunez, and Saez [35] extended the HAPC scheme proposed in [131] to consider network traffic conditions and other types of uncertainties (e.g., accidents).

Attanasio et al. [5] presented a real-time fleet management system for a same-day courier service application. The problem addressed can be seen as a dynamic uncapacitated PDP. An allocation procedure is proposed to assign customer requests to available couriers, and to reposition idle couriers to high demand zones. The latter employs fast insertion heuristics, feasibility procedures, and a TS method. Furthermore, a forecasting module is utilized to provide reliable near future predictions of customer demands via a service territory zoning method and forecasted travel times to the allocation procedure. Travel time forecasts took into account both traffic and real-time information, and they are adjusted using artificial neural networks. Computational experiments based on real-world data are reported.

Finally, another alternative to capture the uncertain nature of travel times is to consider the combination of stochastic and time dependency. The latter is known as stochastic time-dependent networks, and travel times are random variables with time-dependent distributions. We are not aware of any published work addressing stochastic time-dependent travel times for dynamic VRPs; however, interested readers may refer to the works of Fu [45], Kim, Lewis, and White [86], and Lecluyse, Van Wossel, and Peremans [91].

11.5 • Dynamic Vehicle Availability

Besides more or less expected events, for example dynamic customer requests, in practice it is also likely to experience unforeseen disruption events, such as vehicle breakdowns and accidents, traffic congestions, delayed departures from the depot, and so on. Among others, dynamic vehicle availability events have a major impact, since the original or the currently executed routing plan becomes in most cases infeasible, and thus complex recovery actions are needed to restore it. As mentioned earlier, in the context of disruption management the goal is to revise the plan such that the negative effects are minimized and their impact is alleviated. However, it is often the case that conflicting objectives need to be considered (see Mu and Eglese [106]).

So far, few studies have addressed real-time rerouting and rescheduling problems with dynamic vehicle availabilities. Li, Mirchandani, and Borenstein [94] introduced the so-called Real-time Vehicle Rescheduling problem. Both pickup and delivery service functions are examined, while time-window and vehicle capacity constraints are considered. Whenever a vehicle disruption event occurs, the fleet plan needs to be adjusted in real time depending on the current state. A set-covering model is developed for the rerouting problem. The objective is to minimize the weighted sum of operation, service cancellation, and route disruption costs. A Lagrangian relaxation approach is developed that employs an insertion heuristic to obtain a feasible solution for the primal problem, while the shortest path problems with resource constraints are solved via a DP algorithm.

Earlier, Li, Mirchandani, and Borenstein [92, 93] studied the single-depot Vehicle Rescheduling Problem. The problem consists of assigning vehicles to a set of predetermined trips with fixed starting and ending times with an objective of minimizing capital and operating costs. The problem is partially modeled as a sequence of static vehicle scheduling problems, and a pseudo-polynomially auction algorithm is used to solve it. Two assumptions are made, i.e., the scheduled trips, except the disrupted trip, cannot suffer delays, while there are no restrictions on the number of trips that may be reassigned. In [94] this model is extended by taking into account both schedule disruptions and delays of multiple trips. Another related work is that of Huisman, Freling, and Wagelmans [71]. They proposed a cluster-reschedule heuristic for the dynamic multi-depot vehicle scheduling problem to avoid trips starting late in environments characterized by significant traffic jams. The proposed approach consists of solving a series of optimization problems, where different scenarios for travel times are considered. For this purpose, the cluster-reschedule heuristic first assigns trips to depots by solving the corresponding static problem and then dynamically reschedules the trips per depot.

Mu et al. [107] studied the so-called Disrupted CVRP with Vehicle Breakdown. An extra (backup) vehicle with full capacity is assumed available at the depot, and the objective is to minimize the total number of vehicles used and the total travel distance, such that all customers (including those of the immobilized vehicle) are served. Vehicle diversion is not allowed. Two TS algorithms are developed for solving the problem that differ in the neighborhood structure and the neighborhood selection procedure. Both approaches are assessed in relation to an exact algorithm, and they seem adequate to respond under tight time constraints when disruption occurs. Later, Mu and Eglese [106] studied the so-called Disrupted CVRP with Order Release Delay. The disruption event is that the supplies do not arrive at the depot on time; therefore some vehicles cannot be loaded and begin their delivery schedules. Therefore, new plans must be generated to reduce the impact of delays.

More recently, Minis, Mamasis, and Zeimpekis [102] studied a problem in which a vehicle breaks down and the remaining vehicles are rerouted to serve some or all customers of the disrupted route. The problem is modeled as a Capacitated Team Orienteering Problem with vehicle breakdown. Compared to Mu et al. [107] and Li, Mirchandani, and Borenstein [94], they considered that each vehicle can serve only its own customers and the customers of the failed vehicle, while no extra vehicles are available other than those already routed and products cannot be exchanged between customers (as in the case of Mu et al. [107]). The objective is to maximize the profit. A heuristic and a genetic algorithm are proposed.

Zhang and Tang [161] presented a rescheduling model for the VRPTW with vehicle disruptions. The objective is to revise the vehicle routing plan such that the weighted sum of traveling distance and deviations from time windows is minimized. A hybrid metaheuristic approach is proposed. On the other hand, Wang et al. [150] developed a disruption recovery model for the VRPTW with vehicle breakdowns, and proposed two rescue strategies, namely adding vehicles and neighboring rescue. A similar vehicle breakdown problem is also studied by Wang, Wu, and Hu [152].

Finally, Xiang, Chu, and Chen [157] addressed a dynamic DARP on a time-dependent network with many different types of dynamic events, including traveling and service time fluctuations, new requests, no-shows, vehicle breakdowns, cancellations, and traffic jams. Each event has its own priority, and it is placed within an event queue that is sorted in ascending order of priority (a first-come first-served policy is applied for events with the same priority). Furthermore, a prescribed reaction and/or rescuing decision is proposed for each event type. A local search procedure based on inter-trip moves is applied for the insertion of new requests. Additionally, a diversification strategy is adopted that utilizes a secondary objective measuring the idle time of the vehicles.

11.6 • Performance Measurements and Evaluation of Solution Approaches

A well-known method for measuring the performance of algorithms applied to dynamic problems is the so-called *competitive analysis* (see Borodin and El-Yaniv [19] and Sleator and Tarjan [135]). Let I denote an instance of the set of all instances $\mathscr{I}$ of a minimization problem. The competitive ratio can be defined as $c_r = \sup_I \frac{z_A(I)}{z^*(I_s)}$, where $z_A(I)$ is the cost (objective function value) of the solution obtained by an algorithm A for the problem instance I, and $z^*(I_s)$ is the optimal cost found by an offline algorithm that had access to all input data of instance I beforehand. To that end, the competitive ratio can be used to measure the worst-case performance of a routing policy. However, in principle the competitive analysis requires examining all instances of a given problem and to obtain the optimal solution for the corresponding offline (static) instances, which is in most cases a complex task even for deriving tight lower bounds.

The analytical studies discussed in Section 11.3.2.3 used the competitive analysis framework and produced important analytical results and insights for a number of routing policies. Another more flexible alternative is to measure the so-called *value of information* V_A. In this case, the algorithm is applied on both the dynamic instance I (via some form of discrete-event simulation) and the corresponding static instance I_s. To that end, the value of information for an algorithm A can be defined as $V_A = \frac{z_A(I_s) - z_A(I)}{z_A(I_s)}$. Although this is an empirical study that provides only an estimate of the performance of the algorithm, in most cases it captures the impact of the dynamism on the solution yielded by the examined algorithm (assuming that the arrival order of the inputs is not too significant).

Several benchmark data sets appear in the literature that can be used for the evaluation of different solution approaches. Unfortunately, publicly available data sets are scarce even for well-studied dynamic VRPs, while in many cases the performance of different algorithms is assessed in the restricted scope of the problem variant examined at each paper. Furthermore, it is difficult to find a common ground for comparisons due to the wide variety of objective functions. Many authors base their computational studies on adaptations of well-known data sets used for static VRPs, including those of Solomon [136] and Christofides, Mingozzi, and Toth [30]. Interested readers may refer to the procedure proposed by Bent and Van Hentenryck [10] for the modification of Solomon's data sets [136]. A repository of data sets is also maintained by Pankratz and Krypczyk in [109].

11.7 • Conclusions and Future Research Directions

This chapter has presented a review of dynamic vehicle routing and dispatching problems where information evolves over time and decisions must be made in a continuously changing environment. The review has shown that many variants of such problems have been looked at, in varying levels of detail, with the VRP (with or without time windows) and the PDP being the most prominent. The review has indicated that the nature of the solution algorithms developed for dynamic VRPs primarily relies on either (i) testing a variety of policies or (ii) various heuristic or metaheuristic algorithms, some of them making use of parallel implementations. Exact algorithms that have been proposed for this class of problems are highly limited in number, with the most recent reference reviewed in this chapter dating back to 2006, even though interesting ideas have been put forward. Another interesting observation on the existing literature is on the imbalance between the number of studies concerning the deterministic and stochastic versions of the problem, where the amount of research on the former is seen to be considerably greater than that of the latter. One reason behind this discrepancy is the inherent difficulty of the stochastic version.

Although the broader research field has arguably reached some level of maturity, many challenges still stand and new ones emerge. Furthermore, many important problem variants and application areas remain open. Research is still needed for theory building, development of solution methods for moderately and strongly dynamic problem variants, and the study and adaptation of existing approaches for real-life applications. Below is a list of suggestions for further research in this area:

- *Development of Taxonomies and Classification Schemes.* The domain of dynamic VRPs itself is very wide. Although various taxonomies and classification schemes have been proposed in earlier survey papers, the boundaries and similarities among different problem variants as well as links with particular applications need to be clearly defined. As mentioned in Pillac et al. [111] this will foster the development of more generic solution frameworks.

- *Integrating and Exploiting Knowledge about Future Uncertain Events.* Strategies that anticipate future events have been receiving an increasing interest and are incorporated into the recently developed solution methods. However, there is still a lack of formal and general mathematical modeling frameworks and analytical studies. Furthermore, the study of more advanced location analysis and forecasting techniques may yield considerable improvements.

- *Handling Less Predictable Events.* Unforeseen disruption events, such as vehicle failures, accidents, congestions, and others, need further investigation both in terms of robustness considerations as well as novel strategies to alleviate their impact. In the context of emergency services, disaster events may create a highly dynamic problem environment characterized by total uncertainty as well as frequent changes to the input data. Thus, special solution methodologies must be developed to address these challenging problems.

- *Degrees of Freedom, Real-Time Performance Measurements, and Benchmark Data Sets.* There is an evident need to design more enhanced and encompassing measures that will determine the level of dynamism for dynamic vehicle routing and dispatching problems with different sources of uncertainties. The ability to monitor, in real time, the impact of decisions already made or those of alternative future actions is essential for the evaluation of the services provided. Furthermore, in many cases there is a lack of publicly available reference benchmark data sets, even for well-studied dynamic VRP variants.

- *Parallel Algorithms.* As mentioned in Ichoua, Gendreau, and Potvin [77], time pressure remains as the major impediment for the adaptation of well-performing solution approaches proposed for the corresponding static problem instances. The recent emergence of multi-core processors, graphical processing units, and the ever growing computing power of workstations and parallel implementation techniques can help in solving increasingly larger and more realistic problems, even for the time-demanding multi-plan and multi-scenario sampling approaches.

– *Emergency Services.* The field of emergency services, including allocation and re-location problems for emergency vehicles, warrants more research efforts. The seeming randomness of impacts, the uniqueness of emergency incidents, and the uncertain nature of emergency services make them ideally suitable for real-time optimization solution frameworks. The main idea behind integration of real-time information flows within the optimization process is to enable governmental or other organizations to make decisions in real time, to draw policies, and to react to emergency events under tight time constraints.

– *Links with Other Decision Areas.* As highlighted by Larsen, Madsen, and Solomon [90], an important yet unexplored area of research is to study the interface between dynamic vehicle routing and dispatching problems and other parts of the supply chain in which these problems arise, such as warehousing and manufacturing. This is particularly relevant in the context of just-in-time global supply chains.

– *Dynamic Arc Routing Problems.* Modeling and design of algorithms for dynamic arc routing problems is an important research direction that has only received little attention. We refer the reader to Ghiani et al. [53] for a review of applications related to road gritting and snow collection.

For many of the extensions listed above, there is a need to design solution algorithms that can generate high-quality solutions in short time scales to guarantee that new information is reacted to in a timely fashion. Such algorithms should be coupled with look-ahead capabilities that exploit special problem structures and advance knowledge to anticipate future uncertain events and to be able to go beyond the current state of the art and open new scholarly and technological horizons. Given the recent innovations in vehicle tracking technologies, which are able to provide information in real time, the advances in the emerging fields of big data and predictive analytics might further stimulate algorithmic developments in the area of dynamic VRPs. In our opinion, this research topic is still open to a new generation of solution methods that will not only allocate resources optimally but will also integrate real-time decision making, react to expected or unexpected events, anticipate the future, and learn from the past to produce robust solutions.

Acknowledgments

Thanks are due to two anonymous referees for their comments and suggestions on an earlier version of this chapter.

Bibliography

[1] E. ANGELELLI, N. BIANCHESSI, R. MANSINI, AND M. G. SPERANZA, *Short term strategies for a dynamic multi-period routing problem*, Transportation Research Part C: Emerging Technologies, 17 (2009), pp. 106–119.

[2] E. ANGELELLI, R. MANSINI, AND M. G. SPERANZA, *A real-time vehicle routing model for a courier service problem*, in Distribution Logistics - Advanced Solutions to Practical Problems, Lecture Notes in Economics and Mathematical Systems, vol. 544, Springer-Verlag, Berlin, Heidelberg, 2005, pp. 87–104.

[3] E. ANGELELLI, M. W. P. SAVELSBERGH, AND M. G. SPERANZA, *Competitive analysis of a dispatch policy for a dynamic multi-period routing problem*, Operations Research Letters, 35 (2007), pp. 713–721.

[4] N. ASCHEUER, S. O. KRUMKE, AND J. RAMBAU, *Online dial-a-ride problems: Minimizing the completion time*, in Lecture Notes in Computer Science, vol. 1770, Springer-Verlag, Berlin, Heidelberg, 2004, pp. 639–650.

[5] A. ATTANASIO, J. BREGMAN, G. GHIANI, AND E. MANNI, *Real-time fleet management at eCourier Ltd*, in Dynamic Fleet Management: Concepts, Systems, Algorithms and Case Studies, V. Zeimpekis, C. D. Tarantilis, G. M. Giaglis, and I. Minis, eds., vol. 38 of Operations Research/Computer Science Interfaces Series, Springer-Verlag, New York, 2007, pp. 219–238.

[6] A. ATTANASIO, J.-F. CORDEAU, G. GHIANI, AND G. LAPORTE, *Parallel tabu search heuristics for the dynamic multi-vehicle dial-a-ride problem*, Parallel Computing, 30 (2004), pp. 377–387.

[7] N. AZI, M. GENDREAU, AND J.-Y. POTVIN, *A dynamic vehicle routing problem with multiple delivery routes*, Annals of Operations Research, 199 (2012), pp. 103–112.

[8] D. BARBUCHA, *Simulating activities of the transportation company through multi-agent system solving the dynamic vehicle routing problem*, in Lecture Notes in Artificial Intelligence, vol. 5559, Springer-Verlag, Berlin, Heidelberg, 2009, pp. 773–782.

[9] A. BEAUDRY, G. LAPORTE, T. MELO, AND S. NICKEL, *Dynamic transportation of patients in hospitals*, OR Spectrum, 32 (2010), pp. 77–107.

[10] R. W. BENT AND P. VAN HENTENRYCK, *Scenario-based planning for partially dynamic vehicle routing with stochastic customers*, Operations Research, 52 (2004), pp. 977–987.

[11] ——, *Waiting and relocation strategies in online stochastic vehicle routing*, in Proceedings of the Twentieth International Joint Conference on Artificial Intelligence, 2007, pp. 1816–1821.

[12] I. BENYAHIA AND J.-Y. POTVIN, *Decision support for vehicle dispatching using genetic programming*, IEEE Transactions on Systems Man and Cybernetics Part A - Systems and Humans, 28 (1998), pp. 306–314.

[13] G. BERBEGLIA, J.-F. CORDEAU, AND G. LAPORTE, *Dynamic pickup and delivery problems*, European Journal of Operational Research, 202(1) (2010), pp. 8–15.

[14] ——, *A hybrid tabu search and constraint programming algorithm for the dynamic dial-a-ride problem*, INFORMS Journal on Computing, 24 (2012), pp. 343–355.

[15] D. J. BERTSIMAS AND G. VAN RYZIN, *A stochastic and dynamic vehicle routing problem in the Euclidean plane*, Operations Research, 39 (1991), pp. 601–615.

[16] ——, *Stochastic and dynamic vehicle routing in the Euclidean plane with multiple capacitated vehicles*, Operations Research, 41 (1993), pp. 60–76.

[17] ——, *Stochastic and dynamic vehicle routing with general demand and interarrival time distributions*, Advances in Applied Probability, 25 (1993), pp. 947–978.

[18] L. BIANCHI, *Notes on dynamic vehicle routing - the state of the art*, 2000. Technical Report IDSIA-05-01, Instituto Dalle Molle Di Studi Sull Intelligenza Artificiale, Manno, Switzerland.

[19] A. BORODIN AND R. EL-YANIV, *Online Computation and Competitive Analysis*, Cambridge University Press, Cambridge, UK, 2005.

[20] R. M. BRANCHINI, V. A. ARMENTATO, AND A. LOKKETANGEN, *Adaptive granular local search heuristic for a dynamic vehicle routing problem*, Computers & Operations Research, 36 (2009), pp. 2955–2968.

[21] J. BRANKE, M. MIDDENDORF, G. NOETH, AND M. DESSOUKY, *Waiting strategies for dynamic vehicle routing*, Transportation Science, 39(3) (2005), pp. 298–312.

[22] L. BROTCORNE, G. LAPORTE, AND F. SEMET, *Ambulance location and relocation models*, European Journal of Operational Research, 147 (2003), pp. 451–463.

[23] M. CARAMIA, G. ITALIANO, G. ORIOLO, A. PACIFICI, AND A. PERUGIA, *Routing a fleet of vehicles for dynamic combined pick-up and deliveries services*, in Proceedings of the Symposium on Operations Research 2001, Duisburg, Germany, 2001, pp. 3–5.

[24] M.-S. CHANG, S.-R. CHEN, AND C.-F. HSUEH, *Real-time vehicle routing problem with time windows and simultaneous delivery/pickup demands*, Journal of the Eastern Asia Society for Transportation Studies, 5 (2003), pp. 2273–2286.

[25] H.-K. CHEN, C.-F. HSUEH, AND M.-S. CHANG, *The real-time time-dependent vehicle routing problem*, Transportation Research Part E: Logistics and Transportation Review, 42 (2006), pp. 383–408.

[26] Z.-L. CHEN AND H. XU, *Dynamic column generation for dynamic vehicle routing with time windows*, Transportation Science, 40 (2006), pp. 74–88.

[27] B. K.-S. CHEUNG, K. L. CHOY, C.-L. LI, W. SHI, AND J. TANG, *Dynamic routing model and solution methods for fleet management with mobile technologies*, International Journal of Production Economics, 113 (2008), pp. 694–705.

[28] R. K. CHEUNG, D. XU, AND Y. GUAN, *A solution method for a two-dispatch delivery problem with stochastic customers*, Journal of Mathematical Modelling and Algorithms, 6 (2007), pp. 87–107.

[29] C. H. CHRISTIANSEN AND J. LYSGAARD, *A branch-and-price algorithm for the capacitated vehicle routing problem with stochastic demands*, Operations Research Letters, 35 (2007), pp. 773–781.

[30] N. CHRISTOFIDES, A. MINGOZZI, AND P. TOTH, *The vehicle routing problem*, in Combinatorial Optimization, N. Christofides, A. Mingozzi, P. Toth, and C. Sandi, eds., Wiley, Chichester, UK, 1979, pp. 315–338.

[31] J.-F. CORDEAU AND G. LAPORTE, *A tabu search for the static multi-vehicle dial-a-ride problem*, Transportation Research Part B: Methodological, 37 (2003), pp. 579–594.

[32] ——, *The dial-a-ride problem: Models and algorithms*, Annals of Operations Research, 153 (2007), pp. 29–46.

[33] J.-F. CORDEAU, G. LAPORTE, J.-Y. POTVIN, AND M. W. P. SAVELSBERGH, *Transportation on demand*, in Transportation, C. Barnhart and G. Laporte, eds., vol. 14 of Handbooks in Operations Research and Management Science, Elsevier, Amsterdam, 2007, pp. 429–466.

[34] J.-F. CORDEAU, G. LAPORTE, M. W. P. SAVELSBERGH, AND D. VIGO, *Vehicle routing*, in Transportation, C. Barnhart and G. Laporte, eds., vol. 14 of Handbooks in Operations Research and Management Science, Elsevier, Amsterdam, 2007, pp. 367–428.

[35] C. E. CORTÉS, A. NUNEZ, AND D. SAEZ, *Hybrid adaptive predictive control for a dynamic pick-up and delivery problem including traffic congestion*, International Journal of Adaptive Control and Signal Processing, 22 (2008), pp. 103–123.

[36] C. E. CORTÉS, D. SAEZ, A. NUNEZ, AND D. MUNOZ-CARPINTERO, *Hybrid adaptive predictive control for a dynamic pick-up and delivery problem*, Transportation Science, 43 (2009), pp. 27–42.

[37] L. COSLOVICH, R. PESENTI, AND W. UKOVICH, *A two-phase insertion technique of unexpected customers for a dynamic dial-a-ride problem*, European Journal of Operational Research, 175 (2006), pp. 1605–1615.

[38] T. DU, F. K. WANG, AND P.-Y. LU, *A real-time vehicle-dispatching system for consolidating milk runs*, Transportation Research Part E: Logistics and Transportation Review, 45 (2007), pp. 565–577.

[39] J. EUCHI, A. YASSINE, AND H. CHABCHOUB, *Solving the dynamic vehicle routing problem by means of artificial ant colony*, in Proceedings of the International Conference on Metaheuristics and Nature Inspired Computing (META'2010), Tunisia, 2010.

[40] A. FABRI AND P. RECHT, *On dynamic pickup and delivery vehicle routing with several time windows and waiting times*, Transportation Research Part B: Methodological, 40 (2006), pp. 335–350.

[41] F. FERRUCCI, S. BOCK, AND M. GENDREAU, *A pro-active real-time control approach for dynamic vehicle routing problems dealing with the delivery of urgent goods*, European Journal of Operational Research, 225 (2013), pp. 130–141.

[42] E. FEUERSTEIN AND L. STOUGIE, *On-line single-server dial-a-ride problems*, Theoretical Computer Science, 268 (2001), pp. 91–105.

[43] T. FLATBERG, G. HASLE, O. KLOSTER, E. J. NILSSEN, AND E. RIISE, *Dynamic and stochastic vehicle routing in practice*, in Dynamic Fleet Management: Concepts, Systems, Algorithms and Case Studies, V. Zeimpekis, C. D. Tarantilis, G. M. Giaglis, and I. Minis, eds., vol. 38 of Operations Research/Computer Science Interfaces Series, Springer-Verlag, New York, 2007, pp. 42–63.

[44] B. FLEISCHMANN, S. GNUTZMANN, AND E. SANDVOSS, *Dynamic vehicle routing based on online traffic information*, Transportation Science, 38 (2004), pp. 420–433.

[45] L. FU, *Scheduling dial-a-ride paratransit under time-varying, stochastic congestion*, Transportation Research Part B: Methodological, 36 (2004), pp. 485–506.

[46] L. GAMBARDELLA, A. RIZZOLI, F. OLIVERIO, N. CASAGRANDE, A. DONATI, R. MONTEMANNI, AND E. LUCIBELLO, *Ant colony optimization for vehicle routing in advanced logistics systems*, in Proceedings of the International Workshop on Modelling and Applied Simulation (MAS 2003), 2003, pp. 3–9.

[47] M. GENDREAU, F. GUERTIN, J.-Y. POTVIN, AND R. SEGUIN, *Neighborhood search heuristics for a dynamic vehicle dispatching problem with pick-ups and deliveries*, Transportation Research Part C: Emerging Technologies, 14 (2006), pp. 157–174.

[48] M. GENDREAU, F. GUERTIN, J.-Y. POTVIN, AND É. D. TAILLARD, *Parallel tabu search for real-time vehicle routing and dispatching*, Transportation Science, 33 (1999), pp. 381–390.

[49] M. GENDREAU, G. LAPORTE, AND R. SÉGUIN, *Stochastic vehicle routing*, European Journal of Operational Research, 88 (1996), pp. 3–12.

[50] M. GENDREAU, G. LAPORTE, AND F. SEMET, *A dynamic model and parallel tabu search heuristic for real-time ambulance relocation*, Parallel Computing, 27 (2001), pp. 1641–1653.

[51] ——, *The maximal expected coverage relocation problem for emergency vehicles*, Journal of the Operational Research Society, 57 (2006), pp. 22–28.

[52] M. GENDREAU AND J.-Y. POTVIN, *Dynamic vehicle routing and dispatching*, in Fleet Management and Logistics, T. G. Crainic and G. Laporte, eds., Kluwer, Boston, 1998, pp. 115–126.

[53] G. GHIANI, F. GUERRIERO, G. LAPORTE, AND R. MUSMANNO, *Real-time vehicle routing: Solution concepts, algorithms and parallel computing strategies*, European Journal of Operational Research, 151 (2003), pp. 1–11.

[54] G. GHIANI, E. MANNI, A. QUARANTA, AND C. TRIKI, *Anticipatory algorithms for same-day courier dispatching*, Transportation Research Part E: Logistics and Transportation Review, 45 (2009), pp. 96–106.

[55] G. M. GIAGLIS, I. MINIS, A. TATARAKIS, AND V. ZEIMPEKIS, *Minimizing logistics risk through real time vehicle routing and mobile technologies*, International Journal of Physical Distribution and Logistics Management, 34 (2004), pp. 749–764.

[56] A. GOEL, *Fleet telematics: Real-time management and planning of commercial vehicle operations*, vol. 40 of Operations Research and Computer Science Interfaces Series, Springer-Verlag, New York, 2008.

[57] A. GOEL AND V. GRUHN, *A general vehicle routing problem*, European Journal of Operational Research, 191 (2008), pp. 650–660.

[58] J. GOODSON, J. OHLMANN, AND B. THOMAS, *Rollout policies for dynamic solutions to the multivehicle routing problem with stochastic demand and duration limits*, Operations Research, 61 (2013), pp. 138–154.

[59] C. E. GOUNARIS, W. WIESEMANN, AND C. A. FLOUDAS, *The robust capacitated vehicle routing problem under demand uncertainty*, Operations Research, 61 (2013), pp. 677–693.

[60] G. E. GOUNARIS, P. P. REPOUSSIS, C. D. TARANTILIS, W. WIESEMANN, AND C. A. FLOUDAS, *An adaptive memory programming framework for the robust capacitated vehicle routing problem*, Transportation Science, forthcoming.

[61] A. HAGHANI AND S. JUNG, *A dynamic vehicle routing problem with time-dependent travel times*, Computers & Operations Research, 32 (2005), pp. 2959–2986.

[62] A. HAGHANI, Q. TIAN, AND H. HU, *A simulation model for real-time emergency vehicle dispatching and routing*, in Proceedings of the Transportation Research Board Annual Meeting, 2003.

[63] A. HAGHANI AND S. YANG, *Real-time emergency fleet deployment*, in Dynamic Fleet Management: Concepts, Systems, Algorithms and Case Studies, V. Zeimpekis, C. D. Tarantilis, G. M. Giaglis, and I. Minis, eds., vol. 38 of Operations Research/Computer Science Interfaces Series, Springer, New York, 2007, pp. 133–162.

[64] M. A. HAUGHTON, *Quantifying the benefits of route reoptimisation under stochastic customer demands*, Journal of the Operational Research Society, 51 (2000), pp. 320–332.

[65] ———, *Measuring and managing the learning requirements of route reoptimization on delivery vehicle drivers*, Journal of Business Logistics, 23 (2002), pp. 45–66.

[66] ———, *Assigning delivery routes to drivers under variable customer demands*, Transportation Research Part E: Logistics and Transportation Review, 43 (2007), pp. 157–172.

[67] ———, *The efficacy of exclusive territory assignments to delivery vehicle drivers*, European Journal of Operational Research, 184 (2008), pp. 24–38.

[68] D. HAUPTMEIER, S. O. KRUMKE, AND J. RAMBAU, *The online dial-a-ride problem under reasonable load*, in Lecture Notes in Computer Science, vol. 1767, Springer-Verlag, Berlin, Heidelberg, 2000, pp. 125–136.

[69] L. HONG, *An improved LNS algorithm for real-time vehicle routing problem with time windows*, Computers & Operations Research, 39 (2012), pp. 151–163.

[70] C.-F. HSUEH, H.-K. CHEN, AND H.-W. CHOU, *Dynamic vehicle routing for relief logistics in natural disasters*, in Vehicle Routing Problem, T. Caric and H. Gold, eds., I-Tech, Vienna, Austria, 2008, pp. 71–84.

[71] D. HUISMAN, R. FRELING, AND A. P. M. WAGELMANS, *A robust solution approach to the dynamic vehicle scheduling problem*, Transportation Science, 38 (2004), pp. 447–458.

[72] L. M. HVATTUM, A. LØKKETANGEN, AND G. LAPORTE, *Solving a dynamic and stochastic vehicle routing problem with a sample scenario hedging heuristic*, Transportation Science, 40 (2006), pp. 421–438.

[73] ———, *A branch-and-regret heuristic for stochastic and dynamic vehicle routing problems*, Networks, 49 (2007), pp. 330–340.

[74] S. ICHOUA, M. GENDREAU, AND J.-Y. POTVIN, *Diversion issues in real-time vehicle dispatching*, Transportation Science, 34 (2000), pp. 426–438.

[75] ———, *Vehicle dispatching with time-dependent travel times*, European Journal of Operational Research, 114 (2003), pp. 379–396.

[76] ———, *Exploiting knowledge about future demands for real-time vehicle dispatching*, Transportation Science, 40 (2006), pp. 211–225.

[77] ———, *Planned route approaches for real-time vehicle routing*, in Dynamic Fleet Management: Concepts, Systems, Algorithms and Case Studies, V. Zeimpekis, C. D. Tarantilis, G. M. Giaglis, and I. Minis, eds., vol. 38 of Operations Research/Computer Science Interfaces Series, Springer, New York, 2007, pp. 1–18.

[78] P. JAILLET AND M. R. WAGNER, *Online routing problems: Value of advanced information as improved competitive ratios*, Transportation Science, 40 (2006), pp. 200–210.

[79] ———, *Online vehicle routing problems: A survey*, in The Vehicle Routing Problem: Latest Advances and New Challenges, B. L. Golden, S. Raghavan, and E. A. Wasil, eds., vol. 43 of Operations Research/Computer Science Interfaces Series, Springer, New York, 2008, pp. 221–237.

[80] J. JEMAI AND K. MELLOULI, *A neural-tabu search heuristic for the real time vehicle routing problem*, Journal of Mathematical Modelling and Algorithms, 7 (2008), pp. 161–176.

[81] S. JUNG AND A. HAGHANI, *A genetic algorithm for the time dependent vehicle routing problem*, Transportation Research Record: Journal of the Transportation Research Board, 1771 (2001), pp. 161–171.

[82] A. S. KENYON AND D. P. MORTON, *Stochastic vehicle routing with random travel times*, Transportation Science, 37 (2003), pp. 69–82.

[83] Y. KERGOSIEN, C. LENTE, L. PITON, AND J.-C. BILLAUT, *A tabu search heuristic for the dynamic transportation of patients between care units*, European Journal of Operational Research, 214 (2011), pp. 442–452.

[84] M. R. KHOUADJIA, B. SARASOLA, E. ALBA, L. JOURDAN, AND E. TALBI, *A comparative study between dynamic adapted PSO and VNS for the vehicle routing problem with dynamic requests*, Applied Soft Computing, 12 (2012), pp. 1426–1439.

[85] P. KILBY, P. PROSSER, AND P. SHAW, *Dynamic VRPS: A study of scenarios*, APES-06-1998, University of Strathclyde, Glasgow, Scotland, 1998.

[86] S. KIM, M. LEWIS, AND C. C. WHITE III, *Optimal vehicle routing with real-time traffic information*, IEEE Transactions on Intelligent Transportation Systems, 6 (2005), pp. 178–188.

[87] A. LARSEN, O. B. G. MADSEN, AND M. M. SOLOMON, *Partially dynamic vehicle routing—models and algorithms*, Journal of the Operational Research Society, 53 (2002), pp. 637–646.

[88] ———, *The a-priori dynamic travelling salesman problem with time windows*, Transportation Science, 38 (2004), pp. 459–472.

[89] ———, *Classification of dynamic vehicle routing systems*, in Dynamic Fleet Management: Concepts, Systems, Algorithms and Case Studies, V. Zeimpekis, C. D. Tarantilis, G. M. Giaglis, and I. Minis, eds., vol. 38 of Operations Research/Computer Science Interfaces Series, Springer, New York, 2007, pp. 19–40.

[90] ———, *Recent developments in dynamic vehicle routing systems*, in The Vehicle Routing Problem: Latest Advances and New Challenges, B. L. Golden, S. Raghavan, and E. A. Wasil, eds., vol. 43 of Operations Research/Computer Science Interfaces Series, Springer, New York, 2008, pp. 199–218.

[91] C. LECLUYSE, T. VAN WOSSEL, AND H. PEREMANS, *Vehicle routing with stochastic time-dependent travel times*, 4OR, 7 (2009), pp. 363–377.

[92] J.-Q. LI, P. B. MIRCHANDANI, AND D. BORENSTEIN, *A decision support system for the single-depot vehicle rescheduling problem*, Computers & Operations Research, 34 (2007), pp. 1008–1032.

[93] ———, *Vehicle rescheduling problem: Model and algorithms*, Networks, 50 (2007), pp. 211–229.

[94] ———, *Real-time vehicle rerouting problems with time windows*, European Journal of Operational Research, 194 (2009), pp. 711–727.

[95] M. LIPMANN, X. LU, W. E. PAEPE, R. A. SITTERS, AND L. STOUGIE, *Online dial-a-ride problems under restricted information model*, Algorithmica, 40 (2004), pp. 319–329.

[96] S. LORINI, J.-Y. POTVIN, AND N. ZUFFEREY, *Online vehicle routing and scheduling with dynamic travel times*, Computers & Operations Research, 38 (2011), pp. 1086–1090.

[97] K. LUND, O. B. G. MADSEN, AND J. M. RYGAARD, *Vehicle routing problems with varying degrees of dynamism*, Technical Report IMM, Institute of Mathematical Modelling, Technical University of Denmark, Lyngby, Denmark, 1996.

[98] O. B. G. MADSEN, H. F. RAVN, AND J. M. RYGAARD, *A heuristic algorithm for a dial-a-ride problem with time windows, multiple capacities and multiple objectives*, Annals of Operations Research, 60 (1995), pp. 193–208.

[99] O. B. G. MADSEN, K. TOSTI, AND J. VOELDS, *A heuristic method for dispatching repair men*, Annals of Operations Research, 61 (1995), pp. 213–226.

[100] M. MES, M. VAN DER HEIJDEN, AND P. SCHUUR, *Look-ahead strategies for dynamic pickup and delivery problems*, OR Spectrum, 32 (2010), pp. 395–421.

[101] M. MES, M. VAN DER HEIJDEN, AND A. VAN HARTEN, *Comparison of agent-based scheduling to look-ahead heuristics for real-time transportation problems*, European Journal of Operational Research, 181 (2007), pp. 59–75.

[102] I. MINIS, K. MAMASIS, AND V. ZEIMPEKIS, *Real-time management of vehicle breakdowns in urban freight distribution*, Journal of Heuristics, 18 (2012), pp. 375–400.

[103] S. MITROVIĆ-MINIĆ, R. KRISHNAMURTI, AND G. LAPORTE, *Double-horizon based heuristics for the dynamic pickup and delivery problem with time windows*, Transportation Research Part B: Methodological, 38 (2004), pp. 669–685.

[104] S. MITROVIĆ-MINIĆ AND G. LAPORTE, *Waiting strategies for the dynamic pickup and delivery problem with time windows*, Transportation Research Part B: Methodological, 38 (2004), pp. 635–655.

[105] R. MONTEMANNI, L. M. GAMBARDELLA, A. E. RIZZOLI, AND A. V. DONATI, *A new algorithm for a dynamic vehicle routing problem based on ant colony system*, Journal of Combinatorial Optimization, 10 (2005), pp. 327–343.

[106] Q. MU AND R. EGLESE, *Disrupted capacitated vehicle routing problem with order release delay*, Annals of Operations Research, 207 (2013), pp. 201–216.

[107] Q. MU, Z. FU, J. LYSGAARD, AND R. EGLESE, *Disruption management of the vehicle routing problem with vehicle breakdown*, Journal of the Operational Research Society, 62 (2010), pp. 742–749.

[108] C. NOVOA AND R. STORER, *An approximate dynamic programming approach for the vehicle routing problem with stochastic demands*, European Journal of Operational Research, 196 (2009), pp. 509–515.

[109] G. PANKRATZ AND V. KRYPCZYK, *Benchmark data sets for dynamic vehicle routing problems*, 2009. http://www.fernuni-hagen.de/WINF/inhfrm/benchmark_data.htm.

[110] M. PAVONE, N. BISNIK, E. FRAZZOLI, AND V. ISLER, *A stochastic and dynamic vehicle routing problem with time windows and customer impatience*, Mobile Network Applications, 14 (2009), pp. 350–364.

[111] V. PILLAC, M. GENDREAU, C. GUÉRET, AND A. L. MEDAGLIA, *A review of dynamic vehicle routing problems*, European Journal of Operational Research, 225 (2013), pp. 1–11.

[112] V. PILLAC, C. GUÉRET, AND A. L. MEDAGLIA, *An event-driven optimization framework for dynamic vehicle routing*, Technical Report 11/2/AUTO, École des Mines de Nantes, France, 2011.

[113] ——, *On the dynamic technician routing and scheduling problem*, Technical Report 12/5/AUTO, École des Mines de Nantes, France, 2012.

[114] ——, *A fast re-optimization approach for dynamic vehicle routing*, Technical Report 12/6/AUTO, École des Mines de Nantes, France, 2012.

[115] J.-Y. POTVIN, Y. XU, AND I. BENYAHIA, *Vehicle routing and scheduling with dynamic travel times*, Computers & Operations Research, 33 (2006), pp. 1129–1137.

[116] W. B. POWELL, *A stochastic formulation of the dynamic assignment problem, with an application to truckload motor carriers*, Transportation Science, 30 (1996), pp. 195–219.

[117] ——, *What you should know about approximate dynamic programming*, Naval Research Logistics, 56 (2009), pp. 239–249.

[118] W. B. POWELL, P. JAILLET, AND A. ODONI, *Stochastic and dynamic networks and routing*, in Network Routing, M. O. Ball, T. L. Magnanti, C. L. Monma, and G. L. Nemhauser, eds., vol. 8 of Handbooks Operations Research and Management Science, Elsevier Science, Amsterdam, 1995, pp. 141–295.

[119] W. B. POWELL, Y. SHEFFI, K. S. NICKERSON, K. BUTTERBAUGH, AND S. ATHERTON, *Maximizing profits for North American Van Lines' truckload division: A new framework for pricing and operations*, Interfaces, 18 (1988), pp. 21–41.

[120] W. B. POWELL, W. SNOW, AND R. K. CHEUNG, *Adaptive labeling algorithms for the dynamic assignment problem*, Transportation Science, 34 (2000), pp. 50–66.

[121] W. B. POWELL AND H. TOPALOGLU, *Stochastic programming in transportation and logistics*, in Stochastic Programming, Handbooks Operations Research and Management Science, Elsevier Science, Amsterdam, 2003, pp. 555–635.

[122] H. N. PSARAFTIS, *A dynamic programming solution to the single vehicle many-to-many immediate request dial-a-ride problem*, Transportation Science, 14 (1980), pp. 130–154.

[123] ———, *Dynamic vehicle routing problems*, in Vehicle Routing: Methods and Studies, B. L. Golden and A. A. Assad, eds., North–Holland, Amsterdam, 1988, pp. 223–248.

[124] ———, *Dynamic vehicle routing: Status and prospects*, Annals of Operations Research, 61 (1995), pp. 143–164.

[125] V. PUREZA AND G. LAPORTE, *Waiting and buffering strategies for the dynamic pickup and delivery problem with time windows*, INFOR, 46 (2008), pp. 165–175.

[126] A. C. REGAN, H. S. MAHMASSANI, AND P. JAILLET, *Improving efficiency of commercial vehicle operations using real time information: Potential uses and assignment strategies*, Transportation Research Record: Journal of the Transportation Research Board, 1493 (1995), pp. 188–197.

[127] ———, *Evaluation of dynamic fleet management systems*, Transportation Research Record: Journal of the Transportation Research Board, 1645 (1998), pp. 176–184.

[128] U. RITZINGER, J. PUCHINGER, AND R. HARTL, *A survey on dynamic and stochastic vehicle routing problems*, Technical Report, 2012. Available at: http://www.jakobpuchinger.com/publications.

[129] A. RIZZOLI, R. MONTEMANNI, E. LUCIBELLO, AND L. GAMBARDELLA, *Ant colony optimization for real-world vehicle routing problems*, Swarm Intelligence, 1 (2007), pp. 135–151.

[130] S. ROPKE, J.-F. CORDEAU, AND G. LAPORTE, *Models and branch-and-cut algorithms for pickup and delivery with time windows*, Networks, 49 (2007), pp. 258–272.

[131] D. SAEZ, C. E. CORTÉS, AND A. NUNEZ, *Hybrid adaptive predictive control for the multi-vehicle dynamic pick-up and delivery problem based on genetic algorithms and fuzzy clustering*, Computers & Operations Research, 35 (2008), pp. 3412–3438.

[132] M. W. P. SAVELSBERGH AND M. SOL, *DRIVE: Dynamic routing of independent vehicles*, Operations Research, 46 (1998), pp. 474–490.

[133] N. SECOMANDI, *Comparing neuro-dynamic programming algorithms for the vehicle routing problem with stochastic demands*, Computers & Operations Research, 27 (2000), pp. 1201–1225.

[134] N. SECOMANDI AND F. MARGOT, *Reoptimization approaches for the vehicle routing problem with stochastic demands*, Operations Research, 57 (2009), pp. 214–230.

[135] D. SLEATOR AND R. E. TARJAN, *Amortized efficiency of list update and paging rules*, Communications of the ACM, 28 (1985), pp. 202–208.

[136] M. M. SOLOMON, *Algorithms for the vehicle routing and scheduling problems with time window constraints*, Operations Research, 35 (1987), pp. 254–265.

[137] M. Z. SPIVEY AND W. B. POWELL, *The dynamic assignment problem*, Transportation Science, 38 (2004), pp. 399–419.

[138] I. SUNGUR, F. ORDONEZ, AND M. M. DESSOUKY, *A robust optimization approach for the capacitated vehicle routing problem with demand uncertainty*, IIE Transactions, 40 (2008), pp. 509–523.

[139] M. R. SWIHART AND J. D. PAPASTAVROU, *A stochastic and dynamic model for the single-vehicle pick-up and delivery problem*, European Journal of Operational Research, 114 (1999), pp. 447–464.

[140] E. TANIGUCHI AND H. SHIMAMOTO, *Intelligent transportation system based dynamic vehicle routing and scheduling with variable travel times*, Transportation Research Part C: Emerging Technologies, 12 (2004), pp. 235–250.

[141] B. W. THOMAS, *Waiting strategies for anticipating service requests from known customer locations*, Transportation Science, 41 (2007), pp. 319–331.

[142] B. W. THOMAS AND C. C. WHITE III, *Anticipatory route selection*, Transportation Science, 38 (2004), pp. 473–487.

[143] H. TOPALOGLU, *A parallelizable and approximate dynamic programming-based dynamic fleet management model with random travel times and multiple vehicle types*, in Dynamic Fleet Management: Concepts, Systems, Algorithms and Case Studies, V. Zeimpekis, C. Tarantilis, G. Giaglis, and I. Minis, eds., vol. 38 of Operations Research/Computer Science Interfaces Series, Springer, New York, 2007, pp. 65–93.

[144] H. TOPALOGLU AND W. P. POWELL, *Dynamic-programming approximations for stochastic time-staged integer multicommodity-flow problems*, INFORMS Journal on Computing, 18 (2006), pp. 31–42.

[145] P. TOTH AND D. VIGO, *The Vehicle Routing Problem*, SIAM, Philadelphia, 2002.

[146] J. I. VAN HEMERT AND J. A. LA POUTRÉ, *Dynamic routing problems with fruitful regions: Models and evolutionary computation*, in Lecture Notes in Computer Science, vol. 3242, X. Yao, E. Burke, and J. A. Lozano, eds., Springer-Verlag, Berlin, Heidelberg, 2004, pp. 692–701.

[147] P. VAN HENTENRYCK AND R. BENT, *Online Stochastic Combinatorial Optimization*, MIT Press, Cambridge, MA, 2006.

[148] J. WANG AND R. ZHU, *Efficient intelligent optimized algorithm for dynamic vehicle routing problem*, Journal of Software, 6 (2011), pp. 2201–2208.

[149] X. WANG AND H. CAO, *A dynamic vehicle routing problem with backhaul and time windows*, in Proceedings of the IEEE International Conference on Service Operations and Logistics and Informatics IEEE/SOLI 2008, 2008, pp. 1256–1261.

[150] X. WANG, J. NIU, X. HU, AND C. XU, *Rescue strategies of the VRPTW disruption*, Systems Engineering-Theory & Practice, 27 (2007), pp. 104–111.

[151] X. WANG, J. RUAN, H. SHANG, AND C. MA, *A combinational disruption recovery model for vehicle routing problem with time windows*, in Intelligent Decision Technologies, SIST 10, J. Watada, G. Phillips-Wren, L. C. Jain, and R. J. Howlett, eds., Springer-Verlag, Berlin, Heidelberg, 2011, pp. 3–12.

[152] X. WANG, X. WU, AND X. HU, *A study of urgency vehicle routing disruption management problem*, Journal of Networks, 5 (2010), pp. 1426–1433.

[153] X. WANG, C. XU, AND D. YANG, *Disruption management for vehicle routing problem with the request changes of customers*, International Journal of Innovative Computing, Information and Control, 5 (2009), pp. 2427–2438.

[154] X. WANG, K. ZHANG, AND D. YANG, *Disruption management of urgency vehicle routing problem with fuzzy time window*, ICIC Express Letters, 3 (2009), pp. 883–890.

[155] M. WEN, J.-Y. CORDEAU, G. LAPORTE, AND J. LARSEN, *The dynamic multiperiod vehicle routing problem*, Computers & Operations Research, 37 (2010), pp. 1615–1623.

[156] S. WOHLGEMUTH, R. OLORUNTOBA, AND U. CLAUSEN, *Dynamic vehicle routing with anticipation in disaster relief*, Socio-Economic Planning Services, 46 (2012), pp. 261–271.

[157] Z. XIANG, C. CHU, AND H. CHEN, *The study of a dynamic dial-a-ride problem under time-dependent and stochastic environments*, European Journal of Operational Research, 185 (2008), pp. 534–51.

[158] J. YANG, P. JAILLET, AND H. S. MAHMASSANI, *Real-time multivehicle truckload pickup and delivery problems*, Transportation Science, 38 (2004), pp. 135–148.

[159] S. YANG, M. HAMEDI, AND A. HAGHANI, *Online dispatching and routing model for emergency vehicles with area coverage constraints.*, Transportation Research Record: Journal of the Transportation Research Board, 1923 (2006), pp. 1–9.

[160] V. ZEIMPEKIS, C. D. TARANTILIS, G. M. GIAGLIS, AND I. MINIS, *Dynamic fleet management*, vol. 38 of Operations Research and Computer Science Interfaces Series, Springer-Verlag, New York, 2007.

[161] X. ZHANG AND L. TANG, *Disruption management for the vehicle routing problem with time windows*, in Advanced Intelligent Computing Theories and Applications: With Aspects of Contemporary Intelligent Computing Techniques, vol. 2 of Communications in Computer and Information Science, D. Huang, L. Heutte, and M. Loog, eds., Springer-Verlag, Berlin, Heidelberg, 2007, pp. 225–234.

Part III

Applications of the Vehicle Routing Problem

Chapter 12

Software Tools and Emerging Technologies for Vehicle Routing and Intermodal Transportation

Olli Bräysy
Geir Hasle

12.1 ▪ Introduction

Vehicle routing is a central task in a large number of private and public corporations (see, e.g., Golden, Raghavan, and Wasil [30]). Tours have to be planned in very diverse sectors of the economy, not only in the logistics and transport business but in virtually all industrial sectors producing physical goods. Variants of the vehicle routing problem manifest in a remarkably wide range of commercial and non-commercial enterprises: from waste/refuse collection to retail distribution; from construction material delivery planning to postal and express delivery routing; from inbound manufacturing component transportation to finished car distribution; from in-home primary health care service to hospital operations; from transportation network optimization to *third Party Logistics* (3PL) operation scheduling; and from bulk collection and delivery planning to passenger transportation routing. Hence, vehicle routing is key to logistics efficiency in industry, the public sector, and society in general. The high complexity of the vehicle routing problem renders purely human planning inadequate for most applications. Typically, human-made plans have a large potential for improvement. Also, non-assisted planning occupies valuable human resources. Therefore, high-quality software tools for decision support in vehicle routing are crucial to effective and efficient planning in many sectors of society.

Vehicle Routing Problems (VRPs) in their many guises have been the subject of intensive study for more than half a century now (see Laporte [41]). These research efforts have led to the development of a variety of models and solution algorithms and the publication of thousands of scientific papers. Our ability to solve VRPs has increased tremendously over the past half century due to a combination of better VRP methods and the general performance improvement of computers. These results have, together with the progress of key enabling technologies, led to the emergence of numerous software companies worldwide, selling commercial vehicle routing software tools and accompanying services (see Hallamäki et al. [32], Drexl [13], and Partyka and Hall [50]). Among the most important enabling technologies are *Geographic Information Systems* (GIS), positioning, tracking, and tracing technologies, mobile communication, web services, and cloud computing.

Despite the large number of scientific studies, most of the models and algorithms we find in the research literature are not directly applicable to industrial routing problems. Real-life applications usually incorporate many aspects that are not adequately modeled in standard VRP definitions: additional variables, idiosyncratic side constraints, and more or less exotic and incommensurable objective components. To give a simplified picture, VRP research has been reductionistic in its nature. One precisely defines somewhat stylized models that are studied in detail to reveal its structure. Highly optimized, effective, and efficient solution methods are developed, but they are typically not robust; often they cannot solve problem instances that have additional side constraints or optimization criteria, or their performance is based on some assumption that is often violated in practical applications. Problem generalizations have traditionally emerged in a piecewise and systematic way in the research community. Hence, a classification of various VRP variants exists, and it is gradually being extended (see, e.g., Eksioglu, Vural, and Reisman [17]).

Although some variant of the VRP in the research literature may be the core of the application problem at hand, more often than not the real-life problem is somehow richer: with more detail for some operational aspects, with a wider scope along the supply chain, with more focus on dynamics and uncertainty, or all of the above. Industrial problems tend to combine many aspects of richness in a way that renders *Operations Research* (OR) models and their accompanying solution methods inadequate.

Having said this, the substantial research efforts over the past 50 years have not only given deep insights into stylized variants of the VRP. The models and algorithms developed have also to a large degree been industrialized in the form of commercial software tools. In the past decade or so, focus in the research community has moved to more industrially relevant problem definitions, models, and corresponding solution algorithms. Computational experiments in the scientific community are now to a larger degree performed on instances with size and complexity that are comparable to industrial requirements. There is a tendency towards more cooperation between researchers in industry and academia. Furthermore, there is no strict segregation between the developers of industrial VRP tools and researchers in academia; some have positions in both camps, and some companies are tightly connected to groups in academia. However, there is still a considerable way to go to coordinate the efforts in these communities.

The trend towards more industrial relevance has led to a new category of VRP research. The term "Rich VRP" (Hasle and Kloster [33] and Drexl [13]) has been coined to describe a more holistic approach where one defines a more generic model that adequately captures all of the most important aspects of a selected segment of industrial applications. A main goal is to develop solution algorithms that produce high quality solutions in reasonable time for any instance of the generic model. Often, one tries to achieve this "instance robustness" with a uniform algorithmic approach.

VRPs are, in essence, highly complex optimization problems. Because of this complexity, software for supporting human planners and decision makers has been widely used for years. Computerized vehicle routing helps businesses to improve the utilization of their transportation resources. It can help to reduce journey times, vehicle mileage, and costs, but also increase revenues and improve customer service. This is achieved by rapidly processing the information concerning customer locations and quantities and types of goods to be transported and matching these to available vehicle capacity in order to make best use of all resources. In addition, users obtain substantial customer service benefits through improved reliability and environmental benefits through reduced mileage. It also reduces risk of error and helps to reduce customer lead times. The typical reported savings in mileage, total time, or vehicle fleet are 10–30%. Moreover, vendors typically claim that planning time and use of human resources is often reduced by 80–90%. A routing software tool can also modify driver behavior and improve driver safety.

Vehicle routing software companies often have customers in a wide range of sectors, including industry (raw materials and (semi-) finished goods transportation), wholesale and retail trade (consumer goods distribution), *Less-than-TruckLoad* (LTL) and *Full-TruckLoad* (FTL) forwarders, parcel delivery and letter mail services, reverse logistics and waste collection, service technicians, and salesmen.

The purpose of this chapter is to give an overview of the existing vehicle routing software solutions and their key features, and to discuss relevant new and emerging technologies. We will interchangeably use the terms (vehicle) routing software and (vehicle) routing tool to denote software systems that provide optimization-based decision support for planning the routes of a fleet of vehicles. By (vehicle) routing technology we mean the somewhat broader concept of vehicle routing software and the related technologies. By and large, we limit our discussion to routing software for land-based transportation, as they are the most widespread and typical applications of VRP solvers.

The survey is based on several different sources of information. Apart from searching the literature and the web pages of VRP tool vendors, we refer the reader to the OPT-LOG study conducted in 2007 by Hallamäki et al. [32]. We also conducted our own study in the form of questionnaires sent to 50 vendors worldwide (see Table 12.1 for a list of responding vendors). The personal experience of the authors on the development of several vehicle routing tools is another important source of information.

The rest of this chapter is structured as follows. In Section 12.2, we describe the basic functionalities of vehicle routing tools. Input and output are described in Section 12.3 and the various model properties in Section 12.4. The applied solution algorithms are discussed in Section 12.5, whereas Section 12.6 discusses implementation issues. Section 12.7 summarizes the results of our software survey. New and emerging technologies are discussed in Section 12.8. Section 12.9 contains a treatment of business aspects and prospects for the future. Conclusions are drawn in Section 12.10.

12.2 ▪ Basic Functionalities of Vehicle Routing Software

The main functionalities of vehicle routing software solutions include reading and displaying data on orders, vehicles, drivers, depots, as well as calculating distances and travel times between locations based on geocoded addresses. By calling a VRP algorithm, an optimized routing plan for the given data set (i.e., a problem instance) is generated automatically, possibly after entering a set of parameters to the solver. The resulting solution may be displayed several ways using the *Graphical User Interface* (GUI). After possible manipulations by the user, the solution may be exported to external systems.

The typical components of a tool are summarized in Figure 12.1 and in the following list:

- an interface to data files, database, or *Enterprise Resource Planning* (ERP) system,

- a GIS module for geocoding addresses, computing distance, cost, and travel time information, and visualizing data and solutions in digital maps,

- a planning module for automatic, manual, and interactive planning, which allows, e.g., to test the impact of various changes to costs and service levels,

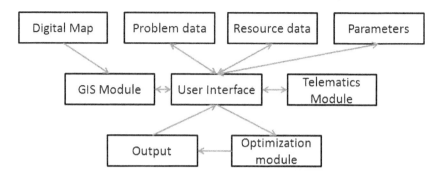

Figure 12.1. *Typical components of vehicle routing software.*

- a telematics module for data exchange between vehicles and the dispatching office, and for tracking and tracing of vehicles,

- a reporting module for presentation of plans, key performance indicators, and statistics in documents or via the GUI.

The GIS and automatic planning modules are the most crucial components. The differences between GIS modules lie in the algorithms, the accuracy, and types of supported maps. The map can directly affect the quality of the produced solutions and causes significant differences between systems. Most VRP tools calculate distances and times by running shortest path calculation (often with Dijkstra's algorithm [12]) on the electronic road network. Often, different types of vehicles may have different speed profiles. Here the term profile refers not only to varying vehicle dependent speeds, but to the fact that a set of different speeds must be defined for each vehicle type, depending on road categories and/or signed speed limits. The most sophisticated tools also consider speed profiles that account for rush hours. Another option supported by most systems is the use of Euclidean distances. Examples of accuracy issues include street level details, address geocoding, driving costs (e.g., uphill), and driving restrictions.

There are often significant differences between tools, depending on used data and modeling properties of the software. From the algorithmic viewpoint, the systems vary from the use of external software for distance and travel time calculation via simple Dijkstra implementations to very sophisticated and fast route and contraction hierarchies, filters, goal directed approaches, distance oracles, etc. (see, e.g., Geisberger et al. [20] and Puranen and Brigatti [53]). In the more sophisticated versions, the criterion for the shortest path may also vary and be based, e.g., on distance, time, cost, and safety. The input and output functionalities are discussed in more detail in Section 12.3 and the properties of automatic planning or optimization in Sections 12.4 and 12.5.

Vehicle routing software is usually used on three main planning levels. The first is operational planning, including both static and dynamic short-term planning in a period less than a week. In the operational context, software can also be used to validate manually planned routes and to test various what-if scenarios. Interactive planning is also supported in most cases. At the tactical level the key issue is typically the determination of resource use, such as vehicle fleet or drivers. One may also analyze different service policies and service networks as well as service areas and user-defined entities. For strategic purposes, the software can be used to plan the network, all resource requirements, sizing and locations, seasonal variations, and budget business based on forecasts and to evaluate alternative options, often in the form of scenarios. One can also use the software, e.g., on longer term service level studies or 3PL evaluations.

In addition to the above-mentioned planning entities, vehicle routing software can be used to

- review the distribution strategy regularly with changes in the business;

- consider larger entities by combining different elements of distribution, such as FTL routing and multi-depot operations;

- integrate with production planning, order processing, and warehouse activity;

- monitor key parts of the logistics chain using key cost drivers (such as visit density and volume);

- facilitate innovative changes within the business, for example, service level differentiation;

- adjust and design sales, service, delivery districts, and vehicle depots;

- to define fixed routes (geographic clustering);

- dispatch for police, fire, and emergency vehicles;

- optimize fuel purchase operations;

- automatically use real-time data in optimization; and

- optimize terminal and depot usage.

12.3 ▪ Input and Output

The input and output capabilities of vehicle routing software solutions are considerable. Often one has specific import and export wizard tools. The supported input data formats include text file, Excel, Access, ODBC, web services, SAP IDoc, and direct ERP integration and are variably available, depending on the software. The applied map data formats include Google maps, Navteq, TomTom, Microsoft, ShapeFile, MapInfo, and MID/MIF.

Often the imported data must be in a given format, though many systems allow adjusting the import format. Several software tools also include different error-correction mechanisms, such as an address matching and correction tool to automatically correct small errors in the input data using imported electronic address and map data. The data itself can vary from basic static single day data to dynamic, stochastic, or long-period data, and it can be based on single orders, or average or aggregate data.

In addition to actual data that includes order, terminal, vehicle fleet, and driver information, the user must specify a number of parameters. We distinguish between *operational parameters* and *planning parameters*.

Typical operational parameters include cost factors (e.g., used cost unit, vehicle and driver costs, revenue of order), units (e.g., distance and time unit, load unit), regulations (e.g., working regulations, turning costs, rules), and operational factors (e.g., vehicle speeds, loading rate). Correspondingly, the typical planning parameters include, for example, cost basis (time, cost, distance), length of planning period, use of depots (open, defined, to optimize), number of trips per vehicle, fleet type (fixed, defined by optimization), allowed maximum route length, and route departure times.

In addition one must define the actual optimization parameters used by the algorithms/optimization process. Often this process is facilitated by one or several alternative default settings or by automatic parameter setting.

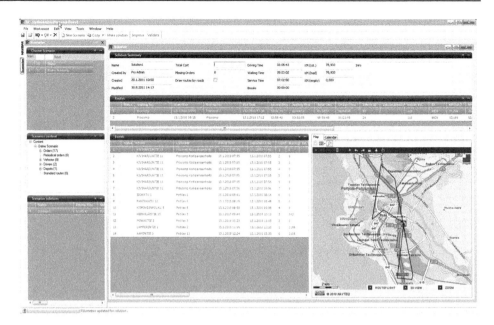

Figure 12.2. *An example of a vehicle routing software GUI. Software: R2, Procomp Solutions Oy (*www.procomp.fi*).*

Practically all systems allow manual adjusting of the routes, often with automatic suggestions. Apart from basic changes such as inserting or deleting orders, re-sequencing orders, exchanging depot, modifying used resources, or adding rest periods, most current systems include drag and drop functionality that also displays the effects of each change on feasibility and operating cost. An example of a GUI of the R^2 software with different types of plan visualization is given in Figure 12.2.

The available reports typically include printed color maps and tree view of data, resources, and results as well as route summaries (distance, time, points to visit, quantities, costs). Other reports include time utilization summaries showing the proportion of time spent on different activities and load manifests or daily traffic sheets showing the allocation of drivers and vehicles to routes. Often, a multi-language support, i.e., possibility to choose the language of the interface, is also provided. In addition, many systems have Gantt-charts (e.g., on time usage), performance monitoring reports (comparing actual results with planned), and status info on the execution of the plan and displaying of active resources. Some software solutions also include profit and loss reports (e.g., order or route basis), invoicing, interfacing to financial accounts, and a web interface.

A few vendors have recently also included carbon footprint display for each route, cost per order delivered, and automatic download of stops to a driver's handset. Other more rarely encountered features include management of diaries of driver shift patterns and holiday fleet maintenance diaries, storing and reporting of actual hours used and performance monitoring, as well as historical travel times and stop times. In some systems one may also utilize weather forecast information, display warnings for violating service levels and qualifications, and show animated load diagrams. Advanced vehicle tracking and monitoring, including GPS locations, vehicle maintenance information, and vehicle motion details is becoming more and more common. The same applies to integration with vehicle navigation and support for smartphone and tablet apps, SMS text alerts, email alerts, web portals, *Interactive Voice Response* (IVR), voice-assisted drive mode, and telematics platforms such as GTS, CSB-System, TomTom, Fleetboard, AVL/MDT systems, and RFID Scanner. Features of increasing importance also include multi-user or multi-site usage and a posteriori calculation of complex objective functions.

Normally, one may configure customized reports using external software applications; see, for instance, Goel [27] for details. Some software applications will display reasons for orders left unplanned (such as insufficient capacity, errors, or conflicts in data), allowing the user to quickly make adjustments to particular routes and display any constraints that would be broken as a result of the manual alterations. Reports can obviously be exported, for instance, to ERP systems, spreadsheets, and databases.

12.4 ▪ Model Properties

Solving any VRP requires at first determining the problem type and setting up key input data values. This definition is often done through parameter adjustment, but it is also possible via automatic analysis of the problem data. We discuss this setup phase in Section 12.4.1. In addition, one must always define the objective and the constraints for the problem at hand. In VRP software, the objective evaluations can be sophisticated, including numerous factors and even true multi-objective planning. This is discussed in Section 12.4.2. In Section 12.4.3, we have divided the constraints in three parts, depending on whether they are related to order, available resources, or the plan itself.

12.4.1 ▪ Basic Setup

The problem definition setup step includes defining the service/operation type, planning horizon, demand type and size, available resource set, start and end locations for the route, and service times at locations. These are discussed in detail below.

Order Type. One must distinguish between pickup, delivery, separate and simultaneous pickup and delivery, service, and node and arc routing–type operations (see, e.g., Gendreau et al. [23]). The problem may also consist of any combination of these. The supported order type has a strong impact on how to model and solve the problem, e.g., through precedence constraints, same tour constraints, and different location definitions. Some solutions support alternative pickup and delivery locations, referring to a combined problem where in addition to defining the pickup-and-delivery sequences, one must choose the most efficient pickup and/or delivery location from several alternatives.

A further complicating aspect is due to alternative ways of performing a request. In addition to the above listed basic order types, some tools also support more complex transshipment orders, linked orders, and optional orders (see, e.g., Drexl [14]). Transshipment of load as a term can refer not only to exchanging the complete load in a swap-body platform, but to partial exchange or one-way transfer of load from one vehicle to another, for example in multi-modal transportation. As an example, a load from A to B may be performed by a direct transportation, via a meet and turn operation, or via one or several hubs. All possibilities must be considered. According to Drexl [13] only a few systems can handle this. Optional requests refer to orders that need not be assigned to a route, but whose execution brings a bonus.

Demand Size. A key part of most practical VRPs is a vehicle capacity constraint. To deal with that, in the beginning one must define the actual demand size for each order and product type. Demand size can be measured, e.g., by weight, volume, loading meters, number of pallets, or a combination of these. Often, demand size is given directly in the data. However, there are several real-life complicating factors, such as time-dependent order size that may refer to production capacity at load site or consumption rate at de-livery location. In such applications of inventory routing, demand size depends on the timing of the visit, initial stock, and storage capacity (see, e.g., Andersson et al. [3]). The goal is to ensure that no customer runs out of stock within the planning period and to minimize transportation costs. Another issue is to determine realistic capacity usage. A practical example is school bus routing where pupils with wheelchair require more space. Such aspects are supported by some tools. In case order size exceeds vehicle capacity, one must split the order among several vehicles, potentially leading to a complex optimiza-tion problem, especially if the number of splitting options is large. Here, one may need to consider a minimum delivery amount.

Service Time.   In order to deal with time-related constraints such as customer time win-dows, maximum duration of routes, or drivers' working regulations, one must define the stop time at each visited location. In the basic case, this is simply given as a single service time value, but in real-life applications and software solutions this is rarely good enough. Possibly, a comprehensive model for service times may give the required accuracy. More often than not, stop times depend on several factors, such as the number of products to load/unload, the presence and type of loading dock, and the vehicle type. The use of a loading dock can reduce stop time considerably. On the other hand, if the loading docks are all in use, one must wait. A few tools have functionality to schedule visit times at loca-tions according to loading dock availability. Moreover, service times often depend heavily also on the vehicle type and the need for rearranging the current load. To illustrate, if the vehicle is too small to use the loading docks, or if it does not have necessary (un)loading equipment, the stop time often increases considerably.

To our knowledge, these properties are not fully supported by any tool on the market. One important issue is also vehicle maneuvering at the customer location. It must be considered separately to actual service time. Otherwise, it would be hard to deal with cases where multiple orders are serviced at same location. Some products have service time models that accommodate this aspect. Another practical problem is that, without setting a cost for parking a vehicle, optimization algorithms cannot avoid visiting centrally located customers (along multiple routes) several times. In addition to customer locations, one may also need to define stop times for the start and end locations for routes, for example due to servicing the vehicle or preparing it for the next day. Finally, the product type and used transport units may impact the stop time.

Route Start and End.   Many systems allow route start and end positions to be set freely. This means, e.g., that one may define different start and end positions for the route, op-timize the start and end depots, or generate open routes (Repoussis et al. [55]) without specifying the start and/or end depot.

Planning Horizon.   One of the most important problem definition issues is the plan-ning horizon. In basic delivery planning, the assumption is that the planning horizon is a single day and all data is known at the start of the day. This basic case is supported by all systems. In practice there are several variants. The most obvious one is a multi-day planning horizon, which again has several subvariants. It may consist of multiple single day plans such that the delivery day is fixed or such that the delivery day is deter-mined by the optimization. A further complication includes 24-hour planning horizons so that the routes do not have actual start or end positions or times but run in a contin-uous loop. This as well as overnight routes (considering, e.g., overnight cost, maximum overnights per journey) is supported by only a few tools. On the other hand, periodic, multi-period, or rolling horizon planning is supported by several systems. In periodic planning (Chao, Golden, and Wasil [9]) one typically has a longer planning horizon of several weeks or even months and given visit frequencies at each customer site, such as once a week. The goal is to define the optimal visit days for the whole period, given the allowed visit patterns. In rolling horizon planning (Rakke et al. [54]) the key idea is to repeatedly determine an optimized plan for a defined short-term planning horizon within a longer planning horizon. One may also distinguish important variants within the single day planning horizon, such as dynamic or real-time routing that is nowadays supported by most systems. The key idea in real-time routing (Psaraftis [52]) is allowing changes onto previously optimized routes and to reoptimize the routes quickly. The changes may include, e.g., new requests, cancelled requests, changed request size, delays, changed traffic conditions, or vehicle breakdowns.

Uncertainty. As most real-life route planning situations are full of uncertainties, stochastic planning (Gendreau, Laporte, and Séguin [22] and Flatberg et al. [19]) is an important issue. Given enough relevant information, parts of the input data should be given as probability distributions instead of crisp numbers. Of course, stochastic problems need more sophisticated algorithms that consume more computational resources. Only very few companies state that their algorithms are capable of handling stochastic customers or stochastic demand/supply. A few companies deal with the issue by using simulation to study the behavior (expected distance, arrival times, time window violations, and the robustness of solutions) of routing plans over a longer time horizon.

Resources. Apart from the constraints and objectives, there are a few resource-related issues that need to be set up in the beginning. Despite the very fast recent shortest path calculation engines, most of the systems are still based on the usage of distance and time matrices that are calculated in the beginning and stored before the actual route optimization can start. Here one must consider the used vehicle fleet. Even though practically all systems support multiple vehicle types, only a few practically allow calculating a separate distance and time matrix for each vehicle type, considering speeds, restrictions, impacts of different turns, etc. It is even rarer to find support for using bicycles and pedestrian routes and travel times, which are important in many applications (e.g., mail and newspaper delivery and school bus routing). Some systems provide support for multiple transport modes such as rail, barge, ship, and air, in addition to standard road-based transportation.

Another important aspect supported by some tools is the usage of trailers or semi-trailers (see, e.g., Villegas et al. [64]). This often requires defining separate tours for the tractor/vehicle and the vehicle-trailer combination and determining the optimal parking positions for trailers as well as trailer assignments. The vehicle fleet may be given, or it may be a part of the optimization objective, resulting in a fleet size and mix optimization problem (see, e.g., Bräysy et al. [7] and Hoff et al. [34]). Even in the case of a given vehicle fleet, assigning vehicles to routes is a complex optimization task if the fleet is heterogeneous. Unless this issue is not specifically supported, most heuristics fail to find the optimal vehicle allocation for practical problems. One may also need to consider the vehicle and driver rotations in planning the resource usage. This is, however, only rarely supported.

12.4.2 ▪ Objective Function

A large variety of objective functions is available in all or most systems. It is possible to minimize the number of vehicles used, the overall distance covered by all vehicles, and the total time or cost of all vehicles. The most common extensions include route balancing and penalty costs for violating soft constraints such as time windows, route total time, preferred vehicles, overnight charges, number of reloads or overloading (see, e.g., Jozefowiez, Semet, and Talbi [39] and Figliozzi [18]). There are a number of metrics used to balance the routes: duration, number of stops per week, stops per day, stops per driver, time on site per driver, driving distance per driver, and sales volume. A related issue is the goal of generating varying routes, e.g., in the context of transporting valuables or planning security patrol routes. Penalty costs are often used to consider undesired but not strictly infeasible properties of solutions or to allow infeasible solutions during the solution process. One may, e.g., penalize the waiting time of the route, which may also require the definition of optimal route start times. In the multi-depot context, the goal may also include defining the optimal depot. The most accurate cost models include even the daily allowances for drivers. Recently, some software vendors have added support to calculate CO_2 emissions.

Some systems allow the user to specify transport revenue details with order data, enabling a more realistic revenue maximization objective. Often this can be described in a variety of ways (e.g., per order, per pallet, or per kg). Similarly, some systems allow considering the inventory costs at depots, terminals, or hubs. Transportation tariffs are also supported by some tools. The tariffs are typically dependent on good type, order size, and distance and are still commonly used in several countries. The key idea here is that the cost is not dependent on the exact route but on the tariff. The goal is typically to design the optimal transportation network, given the tariffs.

The avoidance of toll roads and bridges is a quite common feature, but, for example, congestion charges, impact of vehicle weight along the route, or altitude related truck routing costs (e.g., trip routing to reduce hill climbing) are rarely supported.

Multi-objective optimization (Bãnos et al. [4]) is becoming more and more important. From this viewpoint, about half of all optimization tools support a weighted sum of one-dimensional objective functions that is a simple way to deal with multiple objectives. Real multi-criteria optimization (Steuer [61]) is supported by only very few systems, and even in these cases only for limited properties such as priorities, open orders, and total cost based on, e.g., rate, distance, time, and other input variables.

12.4.3 ▪ Constraints

In addition to the objective, a key issue is to specify a number of hard or soft constraints. We divide the constraints into three classes depending on their relation to order data, resource data, or plan generation. Some constraint types may also exist in several classes.

12.4.3.1 ▪ Order Specific Constraints

Time Windows. Some customers want to receive their deliveries within one or several alternative time windows. Note that the time windows can also be defined flexible or soft, meaning that violation is allowed against a penalty cost that is considered in the objective function (Figliozzi [18]). Time windows are supported by practically all routing tools, and soft and multiple time windows are also commonly available. A related, very rarely supported feature is determining multiple visits within a single time window to the same location. One may, for example, define that three visits are required within a working day within an 8–16 time window. One obviously cannot do the visits in a row—there must be approximately equal split between the visit times. An easy and often used solution is to generate additional orders with tighter non-overlapping time windows, but it does not work well in many practical cases. For proper solutions, an additional constraint on minimum time between visits is required. The time windows may also depend on the used vehicle type (for example, large delivery vehicles having more restrictive access to customers in inner-city zones than small ones). Again, this is only rarely supported (Drexl [13]).

Delivery Day. In some multi-day optimization cases a specific delivery day is defined for some customers. This can be used to force an order delivery on a specified day each week. In case the day is not specified, the systems will attempt to allocate the delivery in the most economic way. A related issue is the determination of visit patterns in the periodic optimization cases. One may, for instance, set the frequencies of visits per week or per month, or directly give allowed visit days or visit day combinations.

Priorities. Priorities can refer to two different issues. The most typical is to allocate priority sequences for orders so that higher-priority orders are visited first in time or they are preferred in case all orders cannot be served. Alternatively one may also define given delivery sequences for orders. The other possibility is to define for each order a priority order for vehicles and drivers servicing a customer so that the given worker, driver, or vehicle is preferred to serve that particular order or location.

Precedence Constraints. A precedence constraint is typically enforced by a routing tool to make sure that a pickup is performed before the corresponding delivery in pickup-and-delivery problems. In some tools, more general precedences may be defined among orders to model business aspects that are not directly supported.

Same Tour Constraint. To support pickup-and-delivery problems, tools will enforce same tour constraints to make sure that pickup-and-delivery tasks are performed on the same route. Again, this constraint can also be used to enforce user-defined combinations of requests on the same route, e.g., in case of pre-planned routes.

Time Limits. One may need to limit the maximum time that an order resides on a vehicle, e.g., in school bus routing, dial-a-ride operations, or distribution of perishable goods (Vidal et al. [63]). Related to this, some tools have support for real-time temperature control. In some applications one must consider the total time of multiple routes or even the whole supply chain.

Other Constraints. More rarely encountered constraints include interdependent requests where the key idea is to force two or more vehicles or workers to serve an order at the same location at the same time. In the literature this is often called synchronization constraint (Ioachim et al. [37] and Drexl [15]). Examples of applications include home care and street maintenance operations. Another restriction is time-based consistency, referring, for instance, to loading dock restrictions at the customer location. Sometimes one may also need to limit the maximum number of terminals per order.

12.4.3.2 ▪ Resource-Related Constraints

In this section we focus on constraints related to the common resources such as vehicles, drivers, depots, and terminals. Other important resources include vehicle equipment and trailers.

Capacity.   The most obvious and common resource-related constraint is capacity. Capacity is typically defined for vehicles, but in some software systems, terminal capacity can be considered as well. In the basic case there is only a single capacity constraint per vehicle, but often one needs to consider several dimensions simultaneously, such as volume, weight, and pallet places. Many systems support different capacities for trailers and for multiple vehicle compartments, with compatibility constraints between products and compartments. In multiple compartment cases there are often numerous alternative allocation options that should be optimized. In some cases one must also consider vehicle balance in the allocation, which further complicates the problem. In a few cases one can find support for special handling of container-type orders and for exact vehicle loading optimization, including vehicle loading patterns and loading restrictions, referring to 2D or 3D bin packing problems (see, e.g., Alvarez-Valdes, Parreno, and Tamarit [2]). A special case is 2.5D packing, where one has layers and towers based on can-stack-on rules, can-be-stacked-on rules, and product dimensions. Another very rarely supported real-life complication is nested containers, for example, packing of actual items in one container, platform or equivalent, and then the packing of these containers or platforms within the vehicle. In a few cases one can define the minimum load size for the vehicle and the impact of tail-lifts, side or rear loading.

Temporal Availability.   The temporal availability is similar to the time window defined above for order and can be set to all resource types. The availability can be used to schedule vehicle maintenance operations, drivers' working calendar and shifts, or terminal opening hours. These are supported by most systems, often via specific resource calendars.

Compatibility.   Compatibility is a very commonly encountered constraint supported by practically all tools. Compatibility comes in many forms and can refer to several resource or order combinations (Goel and Gruhn [29]). The most typical compatibility check is the vehicle-order check, i.e., can a given vehicle serve a given order? Other common examples are order-order (can the two orders be on the same tour?), driver-order (can a worker/driver serve the order?), driver-vehicle (does the driver have the necessary licenses?), and equipment-order (forces a special equipment to serve an order). From a driver compatibility viewpoint, in many cases in practice, drivers are assigned manually to vehicles or vehicle routes after the latter have been planned. A hard real-life example of equipment compatibility restrictions involves key management. More precisely, to access a customer location one often needs keys, and there may be only a single key per location available, resulting in complex key management and swapping issues, especially in real-time applications such as security services. Often one may also constrain the vehicle type servicing a location or driving a road, possibly due to height, length, weight, equipment, or emission restrictions. A common example of equipment constraint is snow plowing where the required plow type depends on the road type. According to Drexl [13] these are supported by about half of the systems that he surveyed. Related to this, a few systems support even axle weight restrictions and trailer kingpin settings. Another, more rarely supported check is sequential incompatibility, where the key idea is not to transport request B on a route that has transported request A before, at least unless a cleaning operation is performed in between. A related issue is a spatial loading constraint where, e.g., two reactive chemicals cannot be transported in neighboring compartments.

Driving Regulations.   Another matter of utmost importance in real-world transportation is driver rules. In the European Union (EU) and in other parts of the world, there is extensive safety legislation on driving, working, and break and rest times for drivers; see Goel [28] for an overview. From the viewpoint of vehicle routing software, several systems include some basic capabilities, such as driver compatibility or setting a maximum tour length, corresponding to a work day or shift length. Setting up the breaks is also supported by many tools, but rarely in an accurate way. For example, the tool often does not consider break locations (they can be anywhere), or it adjusts the break timings after changes to the solution have been made. It is also rare to find support for night and weekend rests and detailed shift planning. Moreover, only a few systems contain the rules for double-manned vehicles.

Zoning. Geographical limitations to planning are often encountered in practice (Ouyang [49]). The most obvious limitation is defining zones or areas for delivery. Some tools can define area preferences for vehicles or drivers, driver home terminals, zones where traffic is prohibited, and time-dependent regional driving bans for drivers. Another type of geographical limitation is allocation of orders to terminals and thus allowing service of a given order from a given terminal only.

12.4.3.3 ▪ Plan Related Constraints

The available systems also allow one to constrain the actual optimization or planning in multiple ways. The most common strategies supported by several tools include balanced routes, locked orders, and various maximum limits. As mentioned above, balancing refers to the soft constraint of having all tours be similar with respect to covered distance, duration, number of requests, or cost (Jozefowiez, Semet, and Talbi [39]). Often, in case of real-time routing or pre-planned routes, one has a given route allocation or already executed orders that cannot be changed. Some tools also allow one to create new separate routes for previously unscheduled tasks. This allows the planner to determine the use of additional (short-term) hired vehicles to handle a given volume of goods. Examples of maximum limits include maximum and minimum number of hours, stops, capacity, dump trips, and distance desired for each vehicle. These may differ from the values given in the actual data to analyze different scenarios. One may also set the maximum number of trips for each vehicle, maximum waiting time, or separately the maximum limit for additional distribution kilometers caused by a single order.

A very common in practice, and rather often supported feature, is multi-route planning, referring to the possibility that vehicles execute multiple routes within the planning horizon (Macedo et al. [44]). Even though multi-route is not a constraint as such, it impacts various other constraints such as time windows, maximum and minimum time limits, and compatibility. For example, from the time viewpoint, one must check the feasibility impact of each change not only to current route but also to routes preceding and following it. In a multi-route context one may in a few tools also limit the frequency and timing of "home visits".

More rarely supported issues are special geographic properties such as "visual beauty" constraints—a preference for routing plans that look nicely on the map. Such requirements have been observed with quite a few users. The beauty contest normally involves two subcriteria: compact routes and/or non-overlapping routes. The motivation for the preferred beauty may seem unclear. It is partly connected with a user belief that plans with compact routes are more efficient, which is not necessarily true. However, for instance in household newspaper delivery, it is not desirable that several carriers serve in the same area with the risk that neighboring subscribers receive their newspaper at very different times.

Some tools support detailed street routing constraints, like U-turns, turning penalties (Bräysy et al. [6]), street side management (e.g., meandering/zigzagging) (Irnich [38]), dynamic usability of roads (e.g., due to congestion or accidents), and hazmat (hazardous material) routing constraining the feasible path options (Akgün et al. [1]).

An important and common issue faced with transshipment loads is precedence constraints between different tours. That is, an order must have arrived to a terminal before it can be picked up by another route. This is, however, rarely supported explicitly, e.g., by unloading heavy orders first to save operating costs. On the other hand, according to Drexl [13], LIFO loading, i.e., delivering the last picked item first (pickup request A, pickup request B, deliver B, deliver A), is supported by half of the systems that he studied.

12.5 ▪ Algorithms

Information on the algorithms used by VRP tools is hard to acquire, as it is generally regarded as an important competitive factor. In practice, most of the tools are based on heuristics or metaheuristics, although exact methods are also utilized, the latter normally in a heuristic version. The solution methods are largely based on published research, although considerable adjustment and tailoring work has often been done in-house to accommodate a rich model. Some vendors have developed algorithms fully in-house.

The majority of the tools use a two-phase strategy that includes constructive and improvement or local search procedures (Bräysy and Gendreau [5]). The most commonly applied constructive heuristics include cheapest insertion (Solomon [60]), savings (Clarke and Wright [10]), nearest neighbor (Lin [42]), and cluster first-route second (Gillett and Miller [24]). These are implemented both in sequential and parallel versions; i.e., they construct the routes one at a time or all in parallel.

The most typical local search operators are relocate and exchange (Savelsbergh [57]), string exchange (Taillard et al. [62]), 2-opt and 3-opt (Lin [42]), Or-opt (Or [48]), Lin–Kernighan (Lin and Kernighan [43]), GENI (Gendreau, Hertz, and Laporte [21]), and large neighborhood search (Shaw [59]). In practice most systems are based on multiple operators.

The most common metaheuristic search strategies are tabu search (Glover [25, 26]), genetic algorithms (Holland [35]), and variants of simulated annealing (Kirkpatrick, Gelatt, and Vecchi [40]), such as threshold accepting (Dueck and Scheuer [16]). Several software solutions also apply dynamic programming (Powell [51]), guided local search (Voudouris and Tsang [65]), and variable neighborhood search (Mladenović and Hansen [45]). The applied exact solution methods are typically based on constraint programming (Rossi, Van Beek, and Walsh [56]), Branch-and-Cut (Gounaris et al. [31]), and column generation (Desaulniers, Desrosiers, and Solomon [11] and Oppen, Løkketangen, and Desrosiers [47]). In most software tools based on column generation, a heuristic variant is used.

Apart from the basic solution algorithms and their adaptations, real-life problem solving requires in some cases special tailored methods for specific tasks such as defining the visit days in periodic planning, splitting the orders, setting the delivery volume in inventory routing, and generating park-and-loop tours.

12.6 ▪ Implementation, Performance, and Price

Most of the vehicle routing tools are coded in C++. For the user interface, several vendors have used C# or Java. Some vendors have coded the whole system in Java. In rare cases also other languages, such as Delphi, have been used.

Mostly, the software is designed for Windows (Vista, 7, etc.), but in several cases also Linux and UNIX are supported. The standard implementation type is local installation, but a web version, Software as a Service (SaaS), components, and client-server variants are also commonly available.

Multi-threading or parallel computing is supported by about half of the systems. However, all approaches seem to be based on rather straightforward task parallelization with a limited number of cores used simultaneously. The significant computing power offered by *Graphic Processing Units* (GPUs) is not yet utilized in commercial systems, to our best knowledge. We return to parallel computing issues in Sections 12.7.3 and 12.8.

In the development, the vendors test their software most often with direct customer data or with synthetic problems based on customer data. Academic benchmarks are also commonly used, and some vendors have also created their own test problems.

Regarding computation time, a fair comparison between the systems is hard due to different hardware and the impact of problem type and solution quality. For example, according to Drexl [13] complex loading problems can cause a five-fold increase in run time. The computation times reported by the vendors may also not be reliable. The reported computation times to solve a routing problem with 100 delivery points varies from a few seconds up to a few minutes on an average desktop computer. When the number of stops is increased to 5000, the execution time range increases from 10 minutes to several hours. Some systems are not capable of solving problems of that large size. According to the survey of Partyka and Hall [50], the computation time to solve a 1000-point routing problem with hard time windows varies from seconds to 10 minutes. Such figures are difficult to assess, however, without detailed knowledge of the instance, the computer, and the quality of the solution.

The maximum problem size that is soluble by the systems is hard to define, as most of the vendors do not specify a maximum size. Another complicating factor is the usage of various aggregation mechanisms (Oppen and Løkketangen [46]) that often reduce the problem size considerably (10-fold is typical). Without aggregation mechanisms the reported maximum sizes vary from a few thousand up to 100.000.

Comparing the solution quality of the software is also very challenging. The first difficulty is that with real-life problems the output depends on the used map data and accuracy of the *Shortest Path Problem* (SPP) calculation. For example, according to the OPT-LOG investigation (Hallamäki et al. [32]), it was concluded that there are significant differences in address coding and street data accuracy between the tools. Some systems did not, e.g., consider turning restrictions or allowed driving directions and hence were able to report better results than the others. Comparing the different tools based on results reported by vendors is also unreliable, as the used computation time may not be comparable and possible manual intervention is not known. However, tests carried out by an independent academic group using software installed on university computers so that the differences in map data were corrected revealed significant differences already in small-scale problems of about 100 stops. The difference between the worst and best solutions in terms of total distance was more than 10%, and it increased in the case of larger problems (Hallamäki et al. [32]).

A basic, single-user license of routing software for commercial use costs 15,000 Euros on average. This does not include customizing and preparatory training of users (Drexl [13]). Analyzing the average price can be misleading also because systems differ a lot in their target market, special features, and integration capabilities. The vendors agree that from the viewpoint of basic static daily routing the markets are rather mature, but there are good business opportunities in the more strategic arena (ability to simulate strategic scenarios whereby all dimensions of the business are taken into account: fleet composition, network design, time windows redesign, etc.). It is also important to be able to combine IT and optimization modeling skills with business consulting. From the viewpoint of adapting the software according to the customer requirements, the vendors can be split in two. It is a common strategy to offer a software off the shelf without any tailoring. It is also very common that various tailoring services are used to create interfaces and to better adapt the system to the problem in question.

12.7 ▪ VRP Technology Survey

As mentioned in the introduction, we have conducted a survey by circulating a questionnaire to 50 vendors of VRP technology in the international market that we were able to identify. The questionnaire was circulated in February 2012, and we received the latest response in June 2012. We received 10 comprehensive responses from nine companies in the VRP tool market. The number of responses is admittedly low, but in our opinion, interesting, high quality information was received from what we regard as a good cross-section of vendors. It includes major players with a large customer base, as well as small enterprises established fairly recently. The companies vary in age between 5 and 33 years, but some have not offered VRP tools during the whole company lifetime. Some companies have not disclosed the number of customers or installations, but the concrete numbers we have received are between 2 and 1500. One vendor reports thousands. The reported number of employees involved in the development of vehicle routing technology is between 2 and 100, whereas the total number of employees varies from 2 to 650. The responding companies are all based in North America and Europe. Several have offices in different countries, and a couple have offices in more than one continent. The set of respondents is probably not representative of the full set of VRP tool vendors. There is reason to believe that the respondents will be more interested in academic research and have closer ties with academia than the average VRP technology vendor.

In this section, we expand and illustrate some of the general insights presented above with a more detailed treatment of questionnaire responses from tool vendors on a selection of aspects. For several reasons, the responses are kept anonymous. However, a full list of the vendors that responded is found in Table 12.1. The issues covered by the questionnaire are information on time dependency, uncertainty management, real-time planning, types of service interfaces, as well as applied parallel computing technology. In addition, the questionnaire covered critical success factors and key strengths, relation to VRP research, and important future research topics.

Also, responses are described in some detail for the treatment of new and emerging technologies and the future of the VRP business in Sections 12.8 and 12.9, respectively.

12.7.1 ▪ System Architectures and Service Interfaces

This issue is important, as the system architecture and supported service interfaces define how the vehicle routing functionality may be made available to the user, the hardware resources needed, the associated costs, the support for collaborative planning by multiple users, and the flexibility to configure the system. There are many system architecture options, and we can only introduce and describe the major ones here.

The simplest alternative is local, monolithic installation of the full system on a PC. All questionnaire respondents provide this option. It consists of a single user vehicle routing decision support system with a GUI, some form of database, and possibly interfaces to external systems such as ERP, mobile communication, and tracking systems. Such a

Table 12.1. *Responding VRP technology providers.*

Company name	Contact information	Email	Web site	Product name(s)
GIRO Inc	75 Port-Royal Street East, Suite 500 Montreal Quebec H3L 3T1, Canada	info@giro.ca	http://www.giro.ca	GeoRoute, GIRO/ACCES
GTS Systems and Consulting GmbH	Raiffeisenstr. 10 D-52134 Herzogenrath, Germany	info@gts-systems.com	http://www.gts-systems.de	TransIT
MJC2 Limited	33 Wellington Business Park Dukes Ride, Crowthorne Berkshire, RG45 6LS, UK	info@mjc2.com	http://www.mjc2.com	DISC, REACT
Optrak Distribution Software Ltd	Orland House, Mead Lane Hertford, SG13 7AT, UK	vrs-sales@optrak.com	http://optrak.co.uk	Optrak
ORTEC	Europe: Houtsingel 5, PO Box 75, 2700 AB, Zoetermeer, The Netherlands. USA: 3630 Peachtree Road NE, Suite 800, Atlanta, GA 30326, USA	info@ortec.com	http://www.ortec.com	ORTEC
Optit srl	Via Selice, 47, 40026 Imola (BO), Italy	info@optit.net	http://www.optit.net	EasyRoute, OptiRoute
Procomp Solutions Oy	Kiviharjuntie 11, 90220 Oulu, Finland	tuomas.kemppainen@procomp.fi	http://www.procomp.fi	R2
PTV AG	Haid-und-Neu-Strasse 15, 76131 Karlsruhe, Germany	karlsruhe@ptv.de	http://www.ptvgroup.com	SmarTour
Spider Solutions AS	P.O. Box 124, Blindern, 0314 Oslo, Norway	info@spidersolutions.no	http://www.spidersolutions.no	Spider 5

system works fine for most applications, at least if the operations have limited size and complexity or do not need support for multi-user planning.

It may be convenient to separate the GUI from the rest of the system, for instance to achieve centralized installation and management of the software for multiple, possibly decentralized users to limit investment and maintenance costs. This requires a client/server architecture, of which there are many types. The client/server model is composed of distributed applications that partition tasks between the providers of a resource or service, called servers, and service requesters, called clients. Clients and servers typically communicate over a computer network, for instance the Internet or a local area network, on separate hardware. Several VRP tool vendors offer client-server solutions for their product. A so-called thin client architecture, where basically only the GUI is run on the clients and the application processing (business logic) and data storage are executed on the server(s), is supported by a majority of our respondents. The hardware platform for thin clients may even be personal digital assistants, smartphones, or tablet computers. In contrast, so-called fat clients, where more of the application logic is executed on the client, are supported by two of our respondents. A three tier architecture, where there is separation between presentation, logic, and storage, is mentioned by one responding vendor.

In the past decade, SaaS has become a popular model for making software available. The definition of SaaS is arguably somewhat fuzzy, but it is most commonly used to denote a client/server like architecture where the GUI is provided by a web browser and the communication is based on the Internet. In this way, there is no need for client software installation if the users have access to a web browser, so the functionality is available virtually anywhere. The supported client hardware may include smartphones and tablet PCs. The software is installed and maintained centrally by the solution provider. Hence, the users do not need to invest in expensive hardware or worry about software maintenance or updates; these tasks are outsourced to the SaaS provider. Modern SaaS solutions have a so-called multi-tenant architecture, where one instance of the software is used to serve many users, potentially from different organizations. SaaS is one type of cloud computing, and multi-tenancy is regarded as a vital characteristic of cloud computing.

Centralized vehicle routing services offered through the web, i.e., VRP cloud computing, also pose challenges. As we know, VRP resolution is computationally demanding, and the demand pattern for a centralized VRP service will be unpredictable. The VRP SaaS provider will either have to dimension its computational resources according to an estimated peak demand, possibly with automatic load balancing, or be able to rapidly reconfigure the number of processes and the supporting hardware. Another form of cloud computing is utility computing, a.k.a. demand computing or elastic computing, in which computational resources and storage may be allocated dynamically according to need. As far as we know, no VRP technology vendor currently offers elastic computing. Data security is another general problematic issue in SaaS solutions. A VRP cloud computing provider will access and store company-specific data that may be critical to the user company, e.g., customer database, performance indicators, prices, etc., so care must be taken to protect sensitive information.

The SaaS model gives new opportunities for pricing. The common one-time software license model is typically replaced by a more fine-grained model where the user company pays for the actual use of the service. The price may depend not only on the number of VRPs solved but also on their complexity.

Five of our respondents answer that they provide a web GUI to their VRP tool, whereas two offer a SaaS solution, the latter probably with a multi-tenant architecture. Cloud computing is mentioned as a major emerging technology by one VRP technology provider.

We do not have enough information to estimate the distribution of installations over the various system architecture alternatives. We suspect that currently the majority of implementations utilize the basic local installation or thin client types of architecture. In our opinion, SaaS will become the most used approach.

All but one technology provider state that they tailor their solution to various types of customer demands, whereas the last one answers that tailoring is kept at a minimum. A different but related issue is the level of consultancy services offered. Tailoring is normally paid by the customer. Some VRP technology vendors receive major revenues, in some cases the lion's share, from pure consulting based on their portfolio of tools, for instance the development of static routes, optimizing fleet size and mix, or transportation network design for a transportation provider.

Finally, we note that there are vendors of VRP technology that offer a VRP solver as a software component, an optimization "engine" that must be somehow integrated with a larger system. For some vendors, this is their only product; they do not provide a GUI. Hence, their market consists predominantly of large organizations with in-house ICT departments, large ICT system (e.g., ERP) providers, and ICT consultant companies. The functionality of a VRP component may be exposed as a web service that constitutes the core of a SaaS VRP solution.

12.7.2 ▪ Time-Dependent Information, Uncertainty, and Dynamics

In the early years of VRP technology, VRP tools were almost exclusively used for static planning. The routing plans generated were supposed to be valid over a relatively long time horizon. Changes in the planning information would occur, and these could motivate a plan revision that was typically performed by replanning from scratch with the updated information. This "modus operandi" still works well for VRP applications with limited dynamics but poorly in highly dynamic applications such as urban courier service.

More recently, the introduction of tracking and tracing technology (positioning and tracking systems, barcoding, RFID) for vehicles and goods, on-board computers, and mobile communication has enabled ready access to updated information on fleet and order status in highly dynamic settings. Tight integration with external business systems such as ERP also enables rapid access to updated order information such as new order arrivals, cancellations, and changes in order size. Traffic conditions constitute an important source of dynamics for VRP planning. Historical speed profiles for roads are electronically accessible for large parts of Asia, Europe, and North America, e.g., through GIS and vehicle navigation systems, and can be used in predictive planning. Electronic access to real-time traffic information is available for many areas. The coverage of such information is proliferating and the use is increasing, both in real-time route finding and VRP planning systems. Some VRP technology vendors report that they have functionality for editing electronic road data, as GIS and vehicle navigation system providers may not offer the necessary quality or responsiveness to change. Moreover, methods for solving dynamic VRPs, i.e., VRPs where information changes rapidly and reactive replanning must be performed even while the plan is being executed, have matured. Another enabling factor for real-time planning is the general performance of VRP solvers. It has increased tremendously, due to both the rapid development of VRP algorithms and the general exponential performance increase of computers.

All these developments have enabled VRP planning based on more accurate information, as well as real-time (dynamic) replanning. Tool vendors including our respondents have exploited these opportunities in different ways. Most VRP tools we know of support speed profiles to make travel time predictions more accurate. A majority of our survey respondents somehow support reactive re-planning. The simplest way is through functionality for locking parts of the plan that have become history, and for replanning with updated information. Typically, these functionalities are not invoked automatically, but they are controlled by the human planner. In some tools, more sophisticated replanning algorithms are used that minimize plan disruption. The more advanced VRP tools have the ability to invoke automatic replanning when deviations from the plan become sufficiently large.

Another discriminatory aspect between VRP tools is the richness of sources of real-time information and their nature. They vary from driver and customer phone calls that need manual planning actions to a number of sources of electronic messages that may be handled automatically, often provided by third party software services. Among the external sources of real-time information are on-board computers, smartphones, business software, goods tracking systems, positioning systems, and traffic information systems. Barcoding, RFID tagging technologies, and proof of delivery systems support more or less the automatic status update of deliveries. Many VRP tools update the order status automatically, most often through integration with third party software services.

The issue of real-time planning is related to handling of uncertainty, an inherent aspect of real-life planning. We have seen no reports of VRP tools that utilize fully fledged stochastic VRP methods. Several vendors, however, report that they support the use of simulation techniques, e.g., for assessment of plan uncertainty and robustness. Some also report that they utilize prognoses, typically based on historical data, to predict demand that is included in the plan.

12.7.3 ▪ Parallel Computing

Given the computational hardness of industrial VRPs, the importance of quality vs. response time performance in the competition between VRP system vendors, and the embarrassingly parallel nature of many VRP solution methods, it is no wonder that providers have utilized parallel computing to boost tool performance.

Our VRP tool survey shows that the majority is designed for PC platforms. Modern PCs have a parallel and heterogeneous processor architecture, with several identical general computing cores that also support multi-threading and task parallelism, and one or more hardware accelerators, e.g., *Graphic Processing Units* (GPUs), each consisting of hundreds of simpler cores, that may be used for data parallelism, also known as stream processing. Software based on sequential algorithms can neither fully exploit current PC hardware nor profit from the general increase of computational performance that will be based on increasing levels of parallelism in a heterogeneous architecture.

Of the 10 tools in our survey, eight utilize task parallelism, one only for specific subtasks such as shortest path calculation. As far as we know, no commercial VRP tool utilizes GPU acceleration or other forms of stream processing, not to mention heterogeneous computing. One vendor states that GPU computing is now the most important technological trend and discloses current development efforts for a GPU implementation. We return to this trend in Section 12.8.

12.7.4 ▪ Relation to VRP Research

In some contexts, a dichotomy is portrayed between the VRP scientific community on the one hand and the industrial VRP community on the other. First, many VRP technology vendor companies are spin-offs from a university or research center. In many cases, there is still an umbilical cord, for some companies to several alma maters. Four out of nine companies in our survey say they have strong or substantial connections to researchers in academia or applied researchers. One states that connections with individual researches are the most rewarding. A couple of vendors have employees with part time positions in both camps. We must reiterate, however, that the respondents probably are more interested in academic VRP research than the average VRP software company.

Scientific research takes place also in the VRP tool industry, although publication of results may be selective, for obvious reasons. Two companies explicitly state that they contribute to scientific publication; several have paper referees on their payroll. With only one exception, all respondents state that they are keeping close tabs on the scientific VRP literature. Most participate or attend major VRP conferences, and several are active conference sponsors.

In our questionnaire we asked whether the VRP algorithm in use was taken from the literature or developed in-house. The typical answer is "both", which should not come as a surprise, with the specification that they have utilized published algorithms as a basis for in-house extensions. Typically, extensions and modifications are needed because the tool is based on a richer, more general VRP model than the variants studied in the literature. Several vendors, however, claim their VRP algorithm, or major parts of it, is developed completely in-house. Supposedly, some of these algorithms contain innovations that would be publishable in the scientific literature. Competition is a natural hindrance for publication.

The communication between the two communities is two-way. Unmet requirements from industry and key bottlenecks in VRP tools are communicated from users via tool vendors to academia and research institutes. Hence, industry participates in forming the VRP research agenda. Benchmarks with standard test cases for computational experiments are important to VRP research. There are a few examples of scientific benchmarks that are based on, or at least heavily inspired from, real-life cases.

The vendors in our survey seem to diverge on the role and importance of benchmarks to the VRP business. Some of them do not regard academic benchmarks as interesting, as they are too stylized. Instead, they use case data from customers. Most answer that they use a combination of academic benchmarks, test instances developed in-house, and customer data. Synthetic, in-house test instances are often generated from a set of customer data. One vendor emphasizes that their tool's performance on the Gehring and Homberger *VRP with Time Windows* (VRPTW) benchmark (Homberger and Gehring [36]) is a key strength that is probably also used in sales and marketing. Another describes the usefulness of academic benchmarks in the prototyping of new algorithms. Yet another says it would be nice if the run for the best new VRPTW algorithm would be replaced by the run to solve difficult rich VRPs with flexible algorithms and benchmarks would be made available for such problems, a view that seems to be common among vendors. One VRP tool provider is actively sponsoring work on producing benchmark real-life problems for academia and highlighting, for instance at scientific conferences, the need to use real-life problems in academic research. A few companies state that there is no need for additional standard benchmarks, but one would welcome large-scale and difficult instances for basic VRP versions.

In our view, it would be fruitful to both "camps" if the VRP industry could publish their benchmarks based on real-life cases, and the researchers in academia would focus their work more strongly on rich VRPs, i.e., comprehensive VRP definitions that include all important aspects of some industrial application. This view is supported by many questionnaire responses where the vendors would like to see a closer interaction and stronger link between their industry and academia. Yet, many companies actively participate in applied research projects with partners ranging from users via tool providers to academia.

12.7.5 ▪ Important Future Research Topics

The respondents mention a broad spectrum of important research topics. Several are related to rich VRPs, meaning VRP definitions that adequately represent a broad range of constraints and criteria that are found in real-life applications. Uncertainty, dynamics, and solution robustness are related aspects mentioned. Several companies advocate large-sized instances as a key topic for future research. The need to study integrated problems that cover larger parts of the value chain and extended problems with decisions that are beyond routing is also mentioned by several vendors. The main issues here include modeling and representation, as well as flexible solution algorithms. More specific topics related to richness include a general framework for handling complex constraints (such as the constraint programming methodology), the effect of real complex constraints on solution methodology, and the impact of problem richness and size on the performance of algorithms. Topics like multi-criteria problems, efficient handling of soft constraints, route compactness (see the discussion on visual beauty criteria in 12.4.3.3 above), driver issues, and loading issues are also often encountered. We also consider problems with time dependencies, complex trip structures, multi-modal planning, and modeling of real costs important.

The impact of real-time data and time-varying information such as rush hour speeds is mentioned by several respondents. One vendor describes GPU parallelization as an important research topic, not only as an implementation issue. A related issue mentioned is improvement of basic optimization performance.

12.7.6 ▪ Critical Success Factors and Key Strengths

It is interesting to note the respondents' views on critical success factors for the VRP tool industry and what they regard as key strengths of their own solution. Not surprisingly, these tend to have considerable overlap. The main success factors mentioned are marketing power, strength, and credibility in the market, customer focus and close relation to customers, and long-term relationships with customers. One must also consider expertise, reliability, cost, know-how of consultants, good demos, ability to deliver tailored solutions, and user-friendly, intuitive GUIs. Identified additional key success factors are coupling of technical skills with business consulting ability and experience, software integration possibilities, rich optimization model and accompanying high-performance solvers, ability to model all operational constraints, real-time features, and service-oriented models.

The key strengths mentioned include very fast return on investment, intuitive and customizable GUI and workflow, modern system architecture, flexible import/export interfaces, and integration of real-time information. Furthermore, the vendors consider full multi-user support, configurable solvers based on rich models, emission calculation, fast and reliable trip planning algorithms, integrated algorithms for loading, and sophisticated rush-hour modeling as important strengths. It is also important to have configurable objectives, a rich set of cost components, strong solution algorithms that yield high quality solutions even for complex, large-sized instances, fast and well-scalable systems, high-quality results on benchmarks, and fast shortest path calculation.

12.8 ▪ New and Emerging Technologies

VRP tools are based not only on methods for modeling and solving VRPs inspired from the OR literature and general software engineering methods but also on several underpinning methods and technologies that are utilized to provide the necessary information and computational resources. The main technologies described above include GIS, positioning, tracking and tracing, and mobile communication. Although there are still challenges, these technologies are fairly mature. There are commercial, external providers of software (services) that VRP technology providers may utilize, often with application interfaces or message protocols that have become industry standards.

On this background, and based on our questionnaire, the most important new and emerging technologies are quantum processors, cloud computing and SaaS, parallel computing and GPUs, and smartphones and tablet PCs. It is also important to utilize speed profile databases, forecasting models and data mining, and technologies for real-time information.

As was expected, the responses vary widely from "blue skies" basic research topics that are far from commercialization, such as quantum computers, via technologies that are emerging and have not been utilized in commercial VRP tools yet, to mature technologies for which there are already commercial service providers that VRP technology vendors may use, such as transportation telematics. Even for mature technologies that offer high value to VRP technology providers, the cover and quality of information are major bottlenecks.

Several VRP tools support *smartphones and tablet PCs* as client platforms. They are typically used as terminals for drivers but may also be used for general purpose clients. *Technologies for real-time information* include vehicle tracking and goods tracing technologies, and traffic information systems, as briefly discussed in Section 12.7.2 above. Reasonably accurate information on travel distance, travel time, and travel cost, as well as a good reactive planning capability, are prerequisites for a successful VRP tool in dynamic applications.

The coverage, level of detail, and quality of available GIS road network and address data are improving rapidly but still vary considerably between regions and countries. Likewise, road instrumentation and other technologies for real-time monitoring of traffic have become mature. There are well-known live traffic services by major providers that may be accessed through the web and via smartphones and car navigation computers. However, there are still many uncharted areas, and there are methodological challenges regarding the use of real-time traffic information in a VRP context.

In predictive planning, *speed profile databases* based on historical data are important for increasing the accuracy of travel time and cost predictions. Some electronic road information vendors provide good speed profiles, but again the cover and quality vary widely between regions. It is clear that the quality and cover of both static and dynamic information from GIS and transportation telematics systems are improving rapidly. This development will enable a more widespread use of VRP technology.

Another key technology and business model that supports a wider exploitation of VRP tools is SaaS. It lowers the investments needed for implementation and maintenance of a VRP tool, and enables a more fine-grained pricing model that may attract new users, small- and medium-sized companies in particular. In Section 12.7.1 above we discuss *cloud computing and SaaS*.

One of the more recent, less mature, and less understood underpinning technologies mentioned above is *heterogeneous computing based on new PC architectures*. For routing applications, there is still a large gap between the requirements and the performance of today's optimization-based decision support systems. The ability to provide better solutions in a shorter time will give substantial savings through better optimization performance of existing tools. Moreover, applications that are too complex to be effectively solved by the technology of today may become within reach of the optimization technology of tomorrow. More integrated, larger, and richer routing problems may be solved.

Further performance increase will result from a combination of better optimization algorithms that are implemented in more efficient ways on more powerful computers.

More than that, the new PC architecture, and the prospects for general PC performance that will partly be based on stream processing accelerators such as the GPU, call for novel VRP algorithms. Early investigations by the research community indicate a large potential for performance increase of VRP solvers. GPU-parallelization is already used in real scientific computing applications, as well as in other applications of discrete optimization, for instance bioinformatics.

It should be clear that heterogeneous computing including utilization of modern GPUs may significantly increase the performance of commercial routing tools and hence boost their improvement effects in industry. VRP technology vendors can hardly ignore the opportunities for performance increase offered by recent and future developments in PC hardware. While the development efforts involved in task parallelization of sequential VRP solver software may be low or moderate, the cost of developing software that also exploits data parallel accelerators such as the GPU is considerably higher. For a tutorial and literature survey focused on routing problems, we refer the reader to Brodtkorb et al. [8] and Schulz et al. [58].

12.9 ▪ The Future of the Vehicle Routing Business

Software tools for vehicle routing have been around for more than 30 years. This fact is often taken as an indication of success for VRP science. Despite the huge improvement potential in transportation and logistics, it is our opinion that the growth of the VRP technology business has been moderate, even if we consider the general slumps in the economy that the world has seen over the past decades. In fact, economical setbacks may sometimes motivate the use of optimization technology. Lack of awareness among potential users is one explanation for the limited growth; another is the lack of ready access to the necessary basic information; a third is the investments needed in terms of money and human capital. The expertise needed to operate VRP tools is often considerable. Finally, model richness and planning performance may be inadequate for important applications. All these hindrances seem to be diminishing.

We asked the software vendors about their views on the future of the VRP business. In summary, they emphasize that there are still a lot of challenges, but the market is growing rapidly and that there will be a continued evolution of the business. The VRP technology still has a low adoption percentage, and many large distribution companies do not use it. The pure routing applications are already relatively mature, but the vendors also think that there are a lot of market opportunities and numerous problem types that have not been considered yet, e.g., identifying new business models for transportation and logistics. In the future, customers will become more sophisticated and demanding, expecting more powerful and flexible solutions rather than just being satisfied with the "industry standard off-the-shelf" approach, and there will be stronger requirements on customer service and timeliness. There shall also be stronger demands for strategic planning functionality, and sustainability will grow in importance. We also see stronger requirements on solving more integrated problems in the value chain, and intermodal transport planning, E-mobility, and social aspects are growing in importance. Clearly, there are differing opinions on the future prospects for the business, but there seems to be a good alignment of views between vendors on important requirements for the future.

12.10 ▪ Summary and Conclusions

The first vehicle routing decision support system was available on the market more than 30 years ago. The huge research efforts devoted to the VRP since its first definition in 1959, the accompanying scientific results, the industrial relevance of this family of problems, and the general improvement of underpinning technologies and computational power, are all factors that have enabled the creation of a successful international VRP technology business. Today, we can identify more than 50 VRP technology vendor companies. Despite the still existing large potential for improvement in transportation and logistics through better coordination, the growth of the VRP business has been rather slow, and most of the companies are small.

In this chapter, we have first given a general description of vehicle routing software with basic functionalities, input and output, model properties, solution algorithms, implementation, and performance. Second, we have given a more detailed account of responses to a questionnaire survey. Nine VRP technology companies have provided product information regarding time-dependent information, uncertainty, real-time information and real-time planning, types of service interface, and parallel computing. Moreover, the survey covers the responding companies' views on critical success factors and key strengths, the relation to VRP research, future research topics, new and emerging technologies, and the future of the VRP business. We cannot claim that the cross-section of responding vendors is representative, but in our opinion their views on these important aspects are interesting.

According to our survey, the software vendors consider the integration possibilities, real-time features, rich optimization models with excellent solvers, and ease of use of the system important from the business viewpoint. There is a growing interest in service oriented models, such as the Software as a Service (SaaS) model, whereby the software vendor generates solutions and manages data from their own servers (Partyka and Hall [50]). Real-time optimization is also becoming more and more important due to consolidation of field equipment and progress in computational power and technology. Customers will become more sophisticated and demanding, expecting more powerful and flexible solutions rather than just being satisfied with the "industry standard off-the-shelf" approach. This requires an easily extensible general framework for handling various constraints and objectives as well as the efficient implementation of basic elements.

A majority of the responding vendors seem to have a positive outlook on the future of their business, despite conservative customers and the current economical crisis in large parts of the Western world. On our own account we would like to mention the business opportunities that should exist in the developing countries with rapid economic growth, although commercial dissemination and exploitation of VRP software requires a high-quality information and communication infrastructure.

Many VRP technology provider companies have their origins in academia, and many still have good connections with individual VRP researchers or research groups. It is interesting to observe that several of our respondents would like to see a tighter interaction between industry and academia, primarily for discussion of recent scientific results, unmet user requirements, and technology bottlenecks, as well as concrete exchange of case information and test data.

The authors have observed clear signs of increasing interaction between industry and academia in recent conferences. We believe that both VRP science and VRP business have ample, hard, and exciting work ahead and a bright future that can be made even brighter through tighter collaboration.

Acknowledgments

We extend our sincere thanks to the companies listed in Table 12.1 for their valuable response to our questionnaire. This work has been supported by the eVita and SMARTRANS programs of the Research Council of Norway under contracts 205298/

V30 (DOMinant II), 192905/I40 (Collab), and 217108 (Respons) and the Future Optimization Models project, partially financed by TEKES.

Bibliography

[1] V. AKGÜN, A. PAREKH, R. BATTA, AND C. M. RUMP, *Routing of a hazmat truck in the presence of weather systems*, Computers & Operations Research, 34 (2007), pp. 1351–1373.

[2] R. ALVAREZ-VALDES, F. PARRENO, AND J. M. TAMARIT, *A grasp/path relinking algorithm for two- and three-dimensional multiple bin-size bin packing problems*, Computers & Operations Research, 40 (2013), pp. 3081–3090.

[3] H. ANDERSSON, A. HOFF, M. CHRISTIANSEN, G. HASLE, AND A. LØKKETANGEN, *Industrial aspects and literature survey: Combined inventory management and routing*, Computers & Operations Research, 37 (2010), pp. 1515–1536.

[4] R. BÃNOS, J. ORTEGA, C. GIL, A. L. MARQUEZ, AND F. DE TORO, *A hybrid metaheuristic for multi-objective vehicle routing problems with time windows*, Computers & Industrial Engineering, 65 (2013), pp. 286–296.

[5] O. BRÄYSY AND M. GENDREAU, *Vehicle routing problem with time windows, Part I: Route construction and local search algorithms*, Transportation Science, 39 (2005), pp. 104–118.

[6] O. BRÄYSY, E. MARTÍNEZ, Y. NAGATA, AND D. SOLER, *The mixed capacitated general routing problem with turn penalties*, Expert Systems with Applications, 38 (2011), pp. 12954–12966.

[7] O. BRÄYSY, P. P. PORKKA, W. DULLAERT, P. P. REPOUSSIS, AND C. D. TARANTILIS, *A well-scalable metaheuristic for the fleet size and mix vehicle routing problem with time windows*, Expert Systems with Applications, 36 (2009), pp. 8460–8475.

[8] A. BRODTKORB, T. R. HAGEN, C. SCHULZ, AND G. HASLE, *GPU computing in discrete optimization. Part I: Introduction to the GPU*, EURO Journal on Transportation and Logistics, 2 (2013), pp. 129–157.

[9] I.-M. CHAO, B. L. GOLDEN, AND E. A. WASIL, *An improved heuristic for the period vehicle routing problem*, Networks, 26 (1995), pp. 25–44.

[10] G. CLARKE AND J. W. WRIGHT, *Scheduling of vehicles from a central depot to a number of delivery points*, Operations Research, 12 (1964), pp. 568–581.

[11] G. DESAULNIERS, J. DESROSIERS AND M. M. SOLOMON, *Column Generation*, Springer, New York, 2005.

[12] E. W. DIJKSTRA, *A note on two problems in connexion with graphs*, Numerische Mathematik, 1 (1959), pp. 269–271.

[13] M. DREXL, *Rich vehicle routing in theory and practice*, Logistics Research, 5 (2012), pp. 47–63.

[14] ———, *Applications of the vehicle routing problem with trailers and transshipments*, European Journal of Operational Research, 227 (2013), pp. 275–283.

[15] ——, *Synchronization in vehicle routing—A survey of VRPs with multiple synchronization constraints*, Transportation Science, 46 (2012), pp. 297–316.

[16] G. DUECK AND T. SCHEUER, *Threshold accepting: A general purpose optimization algorithm appearing superior to simulated annealing*, Journal of Computational Physics, 90 (1990), pp. 161–175.

[17] B. EKSIOGLU, A. V. VURAL, AND A. REISMAN, *The vehicle routing problem: A taxonomic review*, Computers & Industrial Engineering, 57 (2009), pp. 1472–1483.

[18] M. A. FIGLIOZZI, *An iterative route construction and improvement algorithm for the vehicle routing problem with soft time windows*, Transportation Research Part C: Emerging Technologies, 18 (2010), pp. 668–679.

[19] T. FLATBERG, G. HASLE, O. KLOSTER, E. J. NILSSEN, AND A. RIISE, *Dynamic and stochastic vehicle routing in practice*, in Dynamic Fleet Management: Concepts, Systems, Algorithms and Case Studies, V. S. Zeimpekis, C. D. Tarantilis, G. M. Giaglis, and I. Minis, eds., Springer, Berlin, 2007, pp. 41–63.

[20] R. GEISBERGER, P. SANDERS, D. SCHULTES, AND D. DELLING, *Contraction hierarchies: Faster and simpler hierarchical routing in road networks*, in Proceedings of the 7th International Conference on Experimental Algorithms (WEA'08), Springer-Verlag, Berlin, Heidelberg, 2008, pp. 319–333.

[21] M. GENDREAU, A. HERTZ, AND G. LAPORTE, *New insertion and postoptimization procedures for the traveling salesman problem*, Operations Research, 40 (1992), pp. 1086–1094.

[22] M. GENDREAU, G. LAPORTE, AND R. SÉGUIN, *Stochastic vehicle routing*, European Journal of Operational Research, 88 (1996), pp. 3–12.

[23] M. GENDREAU, J.-Y. POTVIN, O. BRÄYSY, G. HASLE, AND A. LØKKETANGEN, *Metaheuristics for the vehicle routing problem and its extensions: A categorized bibliography*, in The Vehicle Routing Problem: Latest Advances and New Challenges, B. L. Golden, S. Raghavan, and E. A. Wasil, eds., vol. 43 of Operations Research/Computer Science Interfaces, Springer, New York, 2008, pp. 143–169.

[24] B. GILLETT AND R. MILLER, *A heuristic for the vehicle dispatching problem*, Operations Research, 22 (1974), pp. 340–349.

[25] F. GLOVER, *Tabu search—Part I*, ORSA Journal on Computing, 1 (1989), pp. 190–206.

[26] ——, *Tabu search—Part II*, ORSA Journal on Computing, 2 (1990), pp. 4–32.

[27] A. GOEL, *Fleet Telematics: Real-Time Management and Planning of Commercial Vehicle Operations*, Springer, New York, 1990.

[28] ——, *Vehicle scheduling and routing with drivers' working hours*, Transportation Science, 43 (2009), pp. 17–26.

[29] A. GOEL AND V. GRUHN, *A general vehicle routing problem*, European Journal of Operational Research, 191 (2008), pp. 650–660.

[30] B. L. GOLDEN, S. RAGHAVAN, AND E. A. WASIL, eds., *The Vehicle Routing Problem: Latest Advances and New Challenges*, vol. 43 of Operations Research/Computer Science Interfaces Series, Springer, New York, 2008.

[31] C. E. GOUNARIS, P. P. REPOUSSIS, C. D. TARANTILIS, AND C. A. FLOUDAS, *A hybrid branch-and-cut approach for the capacitated vehicle routing problem*, Computer Aided Chemical Engineering, 29 (2011), pp. 507–511.

[32] A. HALLAMÄKI, P. HOTOKKA, J. BRIGATTI, P. NAKARI, O. BRÄYSY, AND T. RUOHONEN, *Vehicle Routing Software: A Survey and Case Studies with Finnish Data*, Technical Report, University of Jyväskylä, Finland, 2007.

[33] G. HASLE AND O. KLOSTER, *Industrial vehicle routing problems*, in Geometric Modelling, Numerical Simulation, and Optimization—Applied Mathematics at SINTEF, G. Hasle, K.-A. Lie, and E. Quak, eds., Springer, Berlin, Heidelberg, 2007, pp. 397–435.

[34] A. HOFF, H. ANDERSSON, M. CHRISTIANSEN, G. HASLE, AND A. LØKKETANGEN, *Industrial aspects and literature survey: Fleet composition and routing*, Computers & Operations Research, 37 (2010), pp. 2041–2061.

[35] J. H. HOLLAND, *Adaptation in Natural and Artificial Systems*, The University of Michigan Press, Ann Arbor, MI, 1975.

[36] J. HOMBERGER AND H. GEHRING, *A two-phase hybrid metaheuristic for the vehicle routing problem with time windows*, European Journal of Operational Research, 162 (2005), pp. 220–238.

[37] I. IOACHIM, J. DESROSIERS, F. SOUMIS, AND N. BÉLANGER, *Fleet assignment and routing with schedule synchronization constraints*, European Journal of Operational Research, 119 (1999), pp. 75–90.

[38] S. IRNICH, *Undirected postman problems with zigzagging option: A cutting-plane approach*, Computers & Operations Research, 35 (2008), pp. 3998–4009.

[39] N. JOZEFOWIEZ, F. SEMET, AND E.-G. TALBI, *An evolutionary algorithm for the vehicle routing problem with route balancing*, European Journal of Operational Research, 195 (2009), pp. 761–769.

[40] S. KIRKPATRICK, C. D. GELATT, AND M. P. VECCHI, *Optimization by simulated annealing*, Science, 220 (1983), pp. 671–680.

[41] G. LAPORTE, *Fifty years of vehicle routing*, Transportation Science, 43 (2009), pp. 408–416.

[42] S. LIN, *Computer solutions of the traveling salesman problem*, Bell Systems Technical Journal, 44 (1965), pp. 2245–2269.

[43] S. LIN AND B. W. KERNIGHAN, *An effective heuristic algorithm for the travelling-salesman problem*, Operations Research, 21 (1973), pp. 498–516.

[44] R. MACEDO, C. ALVES, J. M. V. DE CARVALHO, F. CLAUTIAUX, AND S. HANAFI, *Solving the vehicle routing problem with time windows and multiple routes exactly using a pseudo-polynomial model*, European Journal of Operational Research, 214 (2011), pp. 536–545.

[45] N. MLADENOVIĆ AND P. HANSEN, *Variable neighborhood search*, Computers & Operations Research, 24 (1997), pp. 1097–1100.

[46] J. OPPEN AND A. LØKKETANGEN, *Arc routing in a node routing environment*, Computers & Operations Research, 33 (2006), pp. 1033–1055.

[47] J. OPPEN, A. LØKKETANGEN, AND J. DESROSIERS, *Solving a rich vehicle routing and inventory problem using column generation*, Computers & Operations Research, 37 (2010), pp. 1308–1317.

[48] I. OR, *Traveling salesman-type combinatorial problems and their relation to the logistics of regional blood banking*, UMI order 77-10076, Xerox University Microfilms, Ann Arbor, MI, 1976.

[49] Y. OUYANG, *Design of vehicle routing zones for large-scale distribution systems*, Transportation Research Part B: Methodological, 41 (2007), pp. 1079–1093.

[50] J. PARTYKA AND R. HALL, *Vehicle routing software survey–On the road to innovation*, OR/MS Today, 39 (2012), pp. 38–45.

[51] W. POWELL, *Approximate Dynamic Programming*, John Wiley and Sons, New York, 2007.

[52] H. N. PSARAFTIS, *Dynamic vehicle routing: Status and prospects*, Annals of Operations Research, 61 (1995), pp. 143–164.

[53] T. PURANEN AND J. BRIGATTI, *Topology-based optimal road network reduction—a method for speeding up shortest path computation*, in Evolutionary Methods for Design, Optimization and Control, P. Neittaanmäki, J. Periaux and T. Tuovinen, eds., CIMNE, Barcelona, Spain, 2007, pp. 409–415.

[54] J. G. RAKKE, M. STÅLHANE, C. R. MOE, M. CHRISTIANSEN, H. ANDERSSON, K. FAGERHOLT, AND I. NORSTAD, *A rolling horizon heuristic for creating a liquefied natural gas annual delivery program*, Transportation Research Part C: Emerging Technologies, 19 (2011), pp. 896–911.

[55] P. P. REPOUSSIS, C. D. TARANTILIS, O. BRÄYSY, AND G. IOANNOU, *A hybrid evolution strategy for the open vehicle routing problem*, Computers & Operations Research, 37 (2010), pp. 443–455.

[56] F. ROSSI, P. VAN BEEK, AND T. WALSH, *Handbook of Constraint Programming*, Elsevier, Amsterdam, 2006.

[57] M. W. P. SAVELSBERGH, *The Vehicle Routing Problem with Time Windows: Minimizing Route Duration*, Memorandum COSOR, Eindhoven University of Technology, Department of Mathematics and Computing Science, Eindhoven, The Netherlands, 1991.

[58] C. SCHULZ, G. HASLE, A. R. BRODTKORB, AND T. R. HAGEN, *GPU computing in discrete optimization - Part II: Survey focused on routing problems*, EURO Journal on Transportation and Logistics, 2 (2013), pp. 159–186.

[59] P. SHAW, *A new local search algorithm providing high quality solutions to vehicle routing problems*, Technical Report, Department of Computer Science, University of Strathclyde, Glasgow, Scotland, 1997.

[60] M. M. SOLOMON, *Algorithms for the vehicle routing and scheduling problems with time window constraints*, Operations Research, 35 (1987), pp. 254–265.

[61] R. E. STEUER, *Multiple Criteria Optimization: Theory, Computations, and Application*, Wiley, New York, 1986.

[62] E. D. TAILLARD, P. BADEAU, M. GENDREAU, F. GUERTIN, AND J.-Y. POTVIN, *A tabu search heuristic for the vehicle routing problem with soft time windows*, Transportation Science, 31 (1997), pp. 170–186.

[63] T. VIDAL, T. G. CRAINIC, M. GENDREAU, AND C. PRINS, *A hybrid genetic algorithm with adaptive diversity management for a large class of vehicle routing problems with time-windows*, Computers & Operations Research, 40 (2013), pp. 475–489.

[64] J. G. VILLEGAS, C. PRINS, C. PRODHON, A. L. MEDAGLIA, AND N. VELASCO, *A matheuristic for the truck and trailer routing problem*, European Journal of Operational Research, 230 (2013), pp. 231–244.

[65] C. VOUDOURIS AND E. TSANG, *Guided local search*, European Journal of Operational Research, 113 (1998), pp. 80–119.

Chapter 13

Ship Routing and Scheduling in Industrial and Tramp Shipping

Marielle Christiansen
Kjetil Fagerholt

13.1 ▪ Introduction

International trade heavily depends on maritime transportation with more than eight billion tons of goods carried at sea annually (UNCTAD [73]). It is common to distinguish between the following three modes of operation in maritime transportation (Lawrence [52]): liner, industrial, and tramp shipping. Liner vessels follow a fixed route according to a published schedule, trying to maximize profit, similar to a public bus service. An industrial operator owns the cargoes and controls the ships, trying to minimize the cost of transporting its own cargoes, similar to a private fleet. In a tramp operation the vessels follow the available cargoes, with a mix of mandatory contract cargoes and optional spot ones. Since some cargoes are optional, the tramp operator must in addition to the routing and scheduling also determine which spot cargoes to accept/reject, trying to maximize profit. Industrial and tramp shipping usually have several similarities when it comes to operational characteristics.

In this chapter we present models and discuss solution methods for some important routing and scheduling problems in industrial and tramp shipping, which include most of what is referred to as bulk shipping, i.e., transportation of wet (oil, chemicals, and oil products) and dry bulk products (iron ore, grain, coal, bauxite/alumina, and phosphate). These product types constitute more than 60% of the weight transported at sea (UNCTAD [73]). The world fleet consists of more than 7.000 tankers with more than 10.000 deadweight tons for carrying wet bulk products and a similar number of dry bulk carriers.

Given the large amounts of goods transported and the high revenues and costs related to maritime transportation it is easy to imagine that the impact of *Operations Research* (OR) models and methods in planning of ship routes and schedules can be huge, especially given the large concentration of players in this billion dollar industry. As an example, a *Very Large Crude Carrier* (VLCC) with a cargo carrying capacity of about 200.000 tons has a building cost of approximately 100 million *United States Dollars* (USD), a daily time-charter rate of 70.000 USD, a daily fuel cost of 50.000 USD, while the value of one full shipload is 80–100 million USD. (All values depend heavily, of course, on the market and are only intended to give an impression of the order of magnitude of the values related to a typical segment in bulk shipping). Therefore, efficient operation of this fleet is in the best interest of the world economy and vessel operators, and may reduce bunker fuel consumption and environmental impact. In the short to intermediate run, routing and scheduling of this fleet determines the efficiency of its operation. Despite this, there are still relatively few applications of OR in maritime transportation compared to land-based transportation (Christiansen, Fagerholt, and Ronen [24] and Christiansen et al. [25]).

In this chapter, we use the definition from Al-Khayyal and Hwang [3] and distinguish between cargo routing and inventory routing. We use *cargo routing* to denote the planning problem of routing a fleet of ships to service a number of specified cargoes that are given as input to the planning process, in contrast to *inventory routing*, where the cargoes are determined through the planning process itself. In Section 13.2 we present mathematical models for some important cargo routing and scheduling problems in tramp shipping. The industrial shipping counterparts can easily be derived from the corresponding tramp shipping versions of the problems and will be explained. Section 13.3 deals with maritime inventory routing. Illustrative examples are provided in both Sections 13.2 and 13.3. In Section 13.4 we discuss dynamic and stochastic routing in industrial and tramp shipping before we summarize in Section 13.5.

13.2 ▪ Cargo Routing and Scheduling

In this section we present mathematical models and discuss solution methods for some important cargo routing and scheduling problems in tramp shipping. We start in Section 13.2.1 by presenting a basic version of the problem, which can be modeled as a maritime pickup-and-delivery problem with time windows. Then in Section 13.2.2 we proceed by presenting and discussing some important extensions of the basic version of the problem, such as flexible cargo sizes, split loads, and variable sailing speeds, amongst others.

13.2.1 ▪ Maritime Pickup-and-Delivery Problem with Time Windows

Cargo routing and scheduling is an important problem arising in industrial and tramp shipping. A cargo consists of a specified amount of product(s) to be picked up at a specified port, transported, and unloaded at a specified delivery port. There is usually a time window during which the loading (pickup) of the cargo must start, and there may also be an unloading (delivery) time window. The operator controls a heterogeneous fleet of ships that are available to transport the cargoes. For various reasons some cargoes may not be compatible with certain ships (e.g., due to loading and/or unloading ports' draft limitations). Generally, the ship capacities and the cargo quantities are such that ships can carry multiple cargoes simultaneously. Whereas for major bulk commodities a cargo is usually a full shipload, for minor bulk commodities and chemicals where the shipments are smaller, the ship capacity may accommodate several cargoes simultaneously. The model that follows reflects this more general case.

A ship operator in industrial shipping must transport all cargoes while minimizing costs, whereas a tramp operator focuses on profit maximization. The tramp operator usually has a set of mandatory contracted cargoes and will try to increase its revenue by transporting optional spot cargoes. The mandatory cargoes come from long-term agreements between the shipping company and the cargo owners. The challenge for the tramp shipping company is to select spot cargoes and construct routes and schedules that maximize profit. Here, the profit is defined as the revenue from all transported cargoes minus the variable sailing costs, which mainly consist of fuel and port/canal costs, and sometimes also costs for spot charters (i.e., chartering in ships from the market to service given cargoes). We focus here on the tramp routing and scheduling problem, as this is the more general case and includes most of the characteristics of the corresponding industrial shipping problem. The problem that we discuss here in detail has similarities with the *multi-vehicle pickup and delivery problem with time windows* described by Desrosiers et al. [28] (see also Chapter 6).

Figure 13.1(a) illustrates a small example with two ships with capacities of 45 and 60 units, respectively. We assume that the ships are empty in the beginning of the planning period, and have an initial position somewhere at sea, as shown in the figure. There are three cargoes with quantities of 30, 55, and 20 units, respectively, where, for example, (1):30 in the figure denotes the pickup port node of cargo 1 of 30 units, while $(n+1)$ denotes the corresponding delivery port node. Cargo 3 is an optional spot cargo. The example reflects a situation from deep sea shipping where the loading and unloading of each cargo take place in different continents (e.g., North America and Europe with the Atlantic Ocean in between). To make the example simple, we do not include time windows.

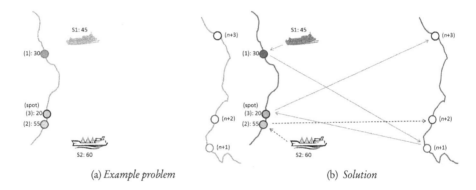

(a) *Example problem*   (b) *Solution*

Figure 13.1. *Example of multi-vehicle pickup-and-delivery problem and its solution.*

Figure 13.1(b) shows a possible solution to the problem, where ship 1 first transports cargo 1 and then sails back (e.g., over the Atlantic Ocean) to service spot cargo 3, while ship 2 carries cargo 2. If we had included time windows, it is easy to imagine that ship 1 would have arrived at the pickup port of the optional spot cargo too late, so that one would have had to reject that cargo, thus missing the revenue from transporting it.

An excerpt from a real ship routing and scheduling working sheet used by a shipping company in the planning of ship routes and schedules is provided in Figure 13.2 (ship names and other information have been modified to preserve confidentiality). The pickup ports are divided into four different regions: United Kingdom and Continental Europe (UK/CONT), Mediterranean Sea (MEDIT), Caribbean Sea (CARIB), and United States and Gulf of Mexico (US/GULF). The cargoes are given in the upper row of the table, based on the locations of their pickup ports. Each cargo is specified by its pickup and delivery ports, its weight, its product type (i.e., MTBE or methanol), and the time window during which pickup can start. In the lower row, the available ships based in each region are specified with a time and a port. When planning ship routes and schedules the shipping company uses such a sheet as an aid to assign cargoes to the available ships.

In the following we present a mathematical model for the maritime pickup-and-delivery problem with time windows in tramp shipping. Let each cargo be represented by an

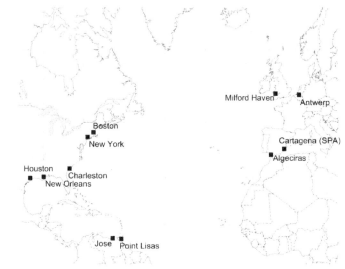

UK/CONT	MEDIT	CARIB	US/GULF
Milford Haven – New York 40.000 ton MTBE 3/11-07 – 9/11-07	Cartagena (SPA) – Houston 28.000 ton Methanol 15/11-07 – 19/11-07	Point Lisas – Houston 17.000 ton Methanol 4/11-07 – 10/11-07	Houston – New York 18.000 ton MTBE 6/11-07 – 11/11-07
Antwerp – Boston 37.000 ton MTBE 5/11-07 – 10/11-07	Algeciras – Charleston 16.000 ton Methanol 15/11-07 – 20/11-07	Jose – New Orleans 30.000 ton Methanol 4/11-07 – 10/11-07	
		Jose – New York 18.000 ton Methanol 10/11-07 – 15/11-07	
Cecilia Antwerp 2/11-07 Catharina Antwerp 7/11-07		Suzanna Jose 15/10-07	Alberta Boston 4/11-07 Maria Charleston 23/10-07

Figure 13.2. *Excerpt from a real ship routing and scheduling working sheet.*

index i. Associated with cargo i is a loading port node i and a unloading port node $n+i$, where n is the number of cargoes that might be transported during the planning horizon. Note that different nodes may correspond to the same physical port. Let $N^P = 1, 2, \ldots, n$ be the set of pickup nodes and $N^D = n+1, n+2, \ldots, 2n$ be the set of delivery nodes. The set of pickup nodes is partitioned into two subsets, N^C and N^O, where N^C is the set of pickup nodes for the mandatory contracted cargoes and N^O is the set of pickup nodes for the optional spot cargoes.

Let K be the set of ships. A network (N_k, A_k) is associated with each ship k. Here, N_k is the set of nodes that can be visited by ship k, including the origin and an artificial destination for ship k, denoted by $o(k)$ and $d(k)$, respectively. Geographically, the origin can be either a port or a point at sea, while the artificial destination is the last planned unloading port for ship k. If the ship is not used, $d(k)$ will represent the same location as $o(k)$. From this, we can extract the sets $N_k^P = N^P \cap N_k$ and $N_k^D = N^D \cap N_k$ consisting of the pickup and delivery nodes that ship k may visit, respectively. The set A_k contains all feasible arcs for ship k, which is a subset of $N_k \times N_k$.

For each ship $k \in K$ and each arc $(i, j) \in A_k$, let T_{ijk}^S be the sailing time from node i to node j, while T_{ik}^P represents the service time in port at node i with ship k. The variable transportation costs C_{ijk} consist of the sum of the sailing costs from node i to node j and the port costs of node i for ship k. In this model, we assume that a (contract) cargo i can be serviced by a ship chartered from the spot market at a given cost, C_i^S. Further, let $[\underline{T}_{ik}, \overline{T}_{ik}]$ denote the time window for ship k associated with node i, where $\underline{T}_{ik}$ is the earliest time for start of service and $\overline{T}_{ik}$ is the latest. Each cargo i has a quantity Q_i and generates a revenue R_i per unit if it is transported. Let Q_k^C be the capacity of ship k.

The binary variable x_{ijk} is assigned the value 1 if ship k sails directly from node i to node j, and 0 otherwise. The variable t_{ik} represents the time for start of service for ship k at node i. The variable l_{ik} is the load (weight) on board ship k when leaving node i. To ease the reading of the model, we assume that each ship is empty when leaving the origin and when arriving at the artificial destination, i.e., $l_{o(k)k} = l_{d(k)k} = 0$. Let z_i be a binary variable that is equal to 1 if cargo i is serviced by a ship from the spot market, and 0 otherwise. Finally, the binary variable y_i is equal to 1 if the optional spot cargo i is transported, and 0 otherwise.

The basic tramp ship routing and scheduling problem can now be formulated as follows:

$$(13.1) \qquad \text{maximize} \sum_{i \in N^C} R_i Q_i + \sum_{i \in N^O} R_i Q_i y_i$$
$$- \sum_{k \in K} \sum_{(i,j) \in A_k} C_{ijk} x_{ijk} - \sum_{i \in N^C} C_i^S z_i$$

$$(13.2) \qquad \text{s.t.} \quad \sum_{k \in K} \sum_{j \in N_k} x_{ijk} + z_i = 1 \qquad \forall\, i \in N^C,$$

$$(13.3) \qquad \sum_{k \in K} \sum_{j \in N_k} x_{ijk} - y_i = 0 \qquad \forall\, i \in N^O,$$

$$(13.4) \qquad \sum_{j \in N_k} x_{o(k)jk} = 1 \qquad \forall\, k \in K,$$

$$(13.5) \qquad \sum_{j \in N_k} x_{ijk} - \sum_{j \in N_k} x_{jik} = 0 \qquad \forall\, k \in K,$$
$$i \in N_k \setminus \{o(k), d(k)\},$$

$$(13.6) \qquad \sum_{i \in N_k} x_{id(k)k} = 1 \qquad \forall\, k \in K,$$

$$(13.7) \qquad l_{ik} + Q_j - l_{jk} - Q_k^C(1 - x_{ijk}) \le 0 \qquad \forall\, k \in K,\, (i,j) \in A_k$$
$$| j \in N_k^P,$$

$$(13.8) \qquad l_{ik} - Q_j - l_{n+j,k} - Q_k^C(1 - x_{i,n+j,k}) \le 0 \qquad \forall\, k \in K,\, (i, n+j) \in A_k$$
$$| j \in N_k^P,$$

$$(13.9) \qquad \sum_{j \in N_k} Q_i x_{ijk} \le l_{ik} \le \sum_{j \in N_k} Q_k^C x_{ijk} \qquad \forall\, k \in K,\, i \in N_k^P,$$

$$(13.10) \qquad 0 \le l_{n+i,k} \le \sum_{j \in N_k} (Q_k^C - Q_i) x_{n+i,jk} \qquad \forall\, k \in K,\, i \in N_k^P,$$

$$(13.11) \qquad t_{ik} + T_{ik}^P + T_{ijk}^S - t_{jk} - M_{ijk}(1 - x_{ijk}) \le 0 \qquad \forall\, k \in K,\, (i,j) \in A_k,$$

$$(13.12) \qquad \sum_{j \in N_k} x_{ijk} - \sum_{j \in N_k} x_{n+i,jk} = 0 \qquad \forall\, k \in K,\, i \in N_k^P,$$

(13.13) $t_{ik} + T_{ik}^P + T_{i,n+i,k}^S - t_{n+i,k} \leq 0$ $\forall\, k \in K,\, i \in N_k^P,$

(13.14) $\underline{T}_{ik} \leq t_{ik} \leq \overline{T}_{ik}$ $\forall\, k \in K,\, i \in N_k,$

(13.15) $l_{ik} \geq 0$ $\forall\, k \in K,\, i \in N_k,$

(13.16) $x_{ijk} \in \{0,1\}$ $\forall\, k \in K,\, (i,j) \in A_k,$

(13.17) $y_i \in \{0,1\}$ $\forall\, i \in N^O,$

(13.18) $z_i \in \{0,1\}$ $\forall\, i \in N^C.$

The objective function (13.1) maximizes the profit from operating the fleet. The four terms are the revenue gained by transporting the mandatory contracted cargoes, the revenue from transporting the optional spot cargoes, the variable sailing costs, and the cost of using spot charters. The fixed revenue for the contracted cargoes can be omitted, but is included here to obtain a more complete picture of the profit. Constraints (13.2) state that all mandatory contract cargoes are transported, either by a ship in the fleet or by a spot charter. The corresponding requirements for the optional spot cargoes are given by constraints (13.3). Constraints (13.4)–(13.6) describe the flow along the sailing route used by ship k. Constraints (13.7) and (13.8) keep track of the load on board at the pickup and delivery nodes, respectively. Constraints (13.9) and (13.10) represent the ship capacity constraints at the loading and discharging nodes, respectively. Constraints (13.11) ensure that the time of starting service at node j must be greater than or equal to the departure time from the previous node i, plus the sailing time between the nodes. The big M coefficient in constraints (13.11) can be calculated as $M_{ijv} = \max(0, \overline{T}_{iv} + T_{iv}^P + T_{ijv}^S - \underline{T}_{jv})$. Constraints (13.12) ensure that the same ship k visits both loading node i and the corresponding discharging node $n+i$. Constraints (13.13) force node i to be visited before node $n+i$, while constraints (13.14) define the time window within which service must start. If ship k is not visiting node i, we will get an artificial starting time within the time windows for that (i,k)-combination. The non-negativity requirements for the load on board the ship are given by constraints (13.15). Constraints (13.16), (13.17), and (13.18) impose the binary requirements on the flow, spot cargo, and spot charter variables, respectively.

In the industrial shipping case, the objective will be to minimize the variable sailing costs, which correspond to the third and fourth terms in objective function (13.1), while constraints (13.3) and variable y_i are no longer required since in industrial shipping all cargoes are mandatory.

Brown, Graves, and Ronen [18], Fisher and Rosenwein [33], Kim and Lee [48], and Bausch, Brown, and Ronen [8] were among the first to study different versions of the maritime pickup-and-delivery problem with time windows. Their solution methods are based on set-covering path flow formulations where all feasible ship routes are generated a priori. Later, the problem modeled by (13.1) - (13.18) was studied by Brønmo et al. [14], who suggest a multi-start local search heuristic to solve the problem. The study shows that optimal or near-optimal solutions are obtained within reasonable time on eight real-life planning problems from four different shipping companies. Korsvik, Fagerholt, and Laporte [50] propose a unified tabu search heuristic, which is shown to perform better than the heuristic by Brønmo et al. [14], especially for large and tightly constrained problems. In a recent paper, Malliappi, Bennell, and Potts [56] present a variable neighborhood search heuristic for the same problem. Their method produced good results on modified benchmark data originating from benchmark instances for the pickup-and-delivery problem with time windows.

Jetlund and Karimi [45] present a different formulation for a real-life tramp routing and scheduling problem for a shipping company engaged in shipping bulk liquid chemicals in the Asia Pacific region. Lin and Liu [54] also consider a real tramp ship routing and scheduling problem for a shipping company operating seven handy-max dry bulk vessels for transportation of various types of dry cargoes in simple packaging (e.g., steel coils, wood pulp, or stone). They suggest a genetic algorithm to solve the problem.

13.2.2 ▪ Application-Driven Modeling Extensions

Real-life problems often impose additional complexities than considered by the mathematical model from Section 13.2.1. These additional complexities may sometimes be viewed as opportunities. In this section we discuss and present the modeling extensions for some important opportunities that arise in several real-life problems.

13.2.2.1 ▪ Flexible Cargo Sizes

Bulk cargoes are frequently shipped on a recurrent basis (e.g., under contracts of affreightment). In such cases the exact cargo size is not that important and the ship operator has some flexibility in the size of the cargo. Normally there is a target cargo size with allowed variability around it (e.g., 20.000 tons $+/-$ 10%). This is also known as a *More Or Less Owner's Option* (MOLOO) contract. Under such a contract the ship operator is paid per unit delivered. Such a contract provides the operator with additional flexibility in assigning cargoes to vessels and in utilizing the vessel capacity. Then, the tramp ship routing and scheduling problem also includes determining the optimal size of each cargo to transport (within its interval).

Figure 13.3 illustrates the potential benefit of introducing flexibility in cargo sizes. Here, the same example as in Section 13.2.1 is used, except that flexibility in the cargo sizes is introduced. For example, in Figure 13.3(b), (1):[27,33] denotes that the size of cargo 1 is flexible within the interval 27 and 33 units. In the solution with flexible sizes, ship 1 carries the minimum sizes of cargoes 1 and 3, while ship 2 can carry 60 units of cargo 2. By utilizing this flexibility, it can be noted that the total sailing distance of the ships, and hence the fuel costs, are significantly reduced.

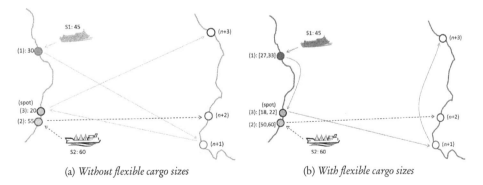

(a) *Without flexible cargo sizes*   (b) *With flexible cargo sizes*

Figure 13.3. *Example solution without and with flexibility in cargo sizes.*

In order to represent flexible cargo sizes the mathematical formulation from Section 13.2.1 has to be modified as follows. Instead of specifying the size of cargo i as Q_i, the quantity that can be transported is now flexible within the interval $[\underline{Q}_i, \overline{Q}_i]$, where $\underline{Q}_i$ and $\overline{Q}_i$ are the minimum and maximum quantities that must be transported (if serviced at all), respectively. Let T_{ik}^Q be the time required to load or discharge one unit of cargo i

with ship k. We now need an additional set of variables, q_{ik}, representing the quantity of cargo i that is transported by ship k.

To include the flexible cargo sizes, we need the following adjustments of the model for the basic tramp ship routing and scheduling problem:

$$(13.19) \qquad \text{maximize} \sum_{k \in K} \sum_{i \in N^C \cup N^O} R_i q_{ik} - \sum_{k \in K} \sum_{(i,j) \in A_k} C_{ijk} x_{ijk} - \sum_{i \in N^C} C_i^S z_i$$

s.t. $(13.2)-(13.6)$, (13.12), $(13.14)-(13.18)$, and

$$(13.20) \qquad l_{ik} + q_{jk} - l_{jk} - Q_k^C(1 - x_{ijk}) \le 0 \qquad \forall\, k \in K,\ (i,j) \in A_k$$
$$| j \in N_k^P,$$

$$(13.21) \qquad l_{ik} - q_{jk} - l_{n+j,k} - Q_k^C(1 - x_{i,n+j,k}) \le 0 \qquad \forall\, k \in K,\ (i, n+j) \in A_k$$
$$| j \in N_k^P,$$

$$(13.22) \qquad q_{ik} \le l_{ik} \le \sum_{j \in N_k} Q_k^C x_{ijk} \qquad \forall\, k \in K,\ i \in N_k^P,$$

$$(13.23) \qquad 0 \le l_{n+i,k} \le \sum_{j \in N_k} Q_k^C x_{n+i,jk} - q_{ik} \qquad \forall\, k \in K,\ i \in N_k^P,$$

$$(13.24) \qquad t_{ik} + T_{ik}^Q q_{ik} + T_{ijk}^S - t_{jk} - M_{ijk}(1 - x_{ijk}) \le 0 \qquad \forall\, k \in K,\ (i,j) \in A_k,$$

$$(13.25) \qquad t_{ik} + T_{ik}^Q q_{ik} + T_{i,n+i,k}^S - t_{n+i,k} \le 0 \qquad \forall\, k \in K,\ i \in N_k^P,$$

$$(13.26) \qquad \sum_{j \in N_k} \underline{Q}_i x_{ijk} \le q_{ik} \le \sum_{j \in N_k} \overline{Q}_i x_{ijk} \qquad \forall\, k \in K,\ i \in N_k^P.$$

The change in the objective function (13.19) compared with the original one (13.1) is that the revenue now depends on the quantity transported of each cargo. The only difference between the new constraints (13.20)–(13.23) and the original constraints (13.7)–(13.10) is that the cargo size parameter Q_i has been replaced by the variable q_{iv}. Time constraints (13.24) and (13.25) are similar to the original constraints (13.11) and (13.13), except for that the service time in port now depends on the quantity loaded or discharged. Constraints (13.26) are new compared to the original formulation and specify the upper and lower bounds for the new quantity variables.

Tramp ship routing and scheduling problems with flexible cargo sizes have been studied by Brønmo, Christiansen, and Nygreen [15], Brønmo, Nygreen, and Lysgard [16], and Korsvik and Fagerholt [49]. These studies show that this flexibility can be utilized to significantly improve the profit (e.g., reducing the size of some cargoes in order to free enough ship capacity to carry additional spot cargoes by the controlled fleet). Brønmo, Christiansen, and Nygreen [15] solve a path flow formulation where all feasible ship routes are generated a priori and optimized with respect to the cargo sizes. Eight small- to medium-sized problem instances from two different shipping companies are solved. For larger instances this method becomes intractable due to the exponential increase in the number of feasible ship routes and hence variables. Therefore, Brønmo, Nygreen, and Lysgard [16] suggest a dynamic column generation scheme where ship routes are generated as needed. However, they discretize the cargo quantities, which turns the solution method into a heuristic column generation approach. Korsvik and Fagerholt [49] develop a tabu search algorithm for the same problem and show that very good solutions are obtained with their method within reasonable time.

13.2.2.2 ▪ Split Loads

In the previous models a cargo cannot be transported by more than one ship. By introducing split loads this restriction is relaxed and a cargo may be split among several ships.

Figure 13.4 shows the example considered in Figure 13.1, now illustrating the potential benefit from load splitting. Figure 13.4(a) illustrates the solution without load splitting, while Figure 13.4(b) shows a solution where load splitting is utilized (though without flexible cargo sizes). Here, ship 1 services cargo 1 and uses the remaining capacity to take 15 units of cargo 2, while ship 2 transports the remaining 40 units of cargo 2 together with spot cargo 3. It is easy to imagine the solution in Figure 13.4(b) as better than the no-split solution in Figure 13.4(a), as this solution requires only two sailing legs between the continents. Furthermore, as previously mentioned, time windows could have restricted cargo 3 to be picked up after delivering cargo 1, like shown in Figure 13.4(a). Note that the spot cargo was not split due to the location at the unloading side.

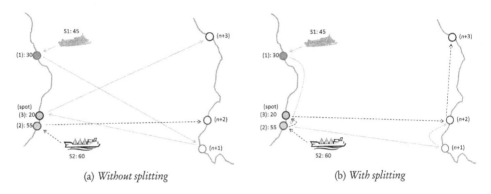

(a) *Without splitting* (b) *With splitting*

Figure 13.4. *Example solution without and with load splitting.*

Also here (same as for the former extension for flexible cargo sizes) we need quantity variables, q_{ik}, specifying how much of cargo i ship k is carrying. In addition we need the following adjustments to the original model (13.1)–(13.18):

$$\text{maximize (13.19)}$$

s.t. $(13.4) - (13.6), (13.11) - (13.18), (13.20) - (13.23),$ and

$$(13.27) \quad \sum_{k \in K} \sum_{j \in N_k} x_{ijk} + z_i \geq 1 \qquad \forall\, i \in N^C,$$

$$(13.28) \quad \sum_{k \in K} \sum_{j \in N_k} x_{ijk} - y_i \geq 0 \qquad \forall\, i \in N^O,$$

$$(13.29) \quad l_{ik} + q_{jk} - l_{jk} - Q_k^C(1 - x_{ijk}) \leq 0 \qquad \begin{aligned} &\forall\, k \in K,\, (i,j) \in A_k, \\ &|\, j \in N_k^P, \end{aligned}$$

$$(13.30) \quad l_{ik} - q_{jk} - l_{n+j,k} - Q_k^C(1 - x_{i,n+j,k}) \leq 0 \qquad \begin{aligned} &\forall\, k \in K,\, (i, n+j) \in A_k, \\ &|\, j \in N_k^P, \end{aligned}$$

$$(13.31) \quad q_{ik} \leq l_{ik} \leq \sum_{j \in N_k} Q_k^C x_{ijk} \qquad \forall\, k \in K,\, i \in N_k^P,$$

$$(13.32) \quad 0 \leq l_{n+i,k} \leq \sum_{j \in N_k} Q_k^C x_{n+i,jk} - q_{ik} \qquad \forall\, k \in K,\, i \in N_k^P,$$

$$(13.33) \quad \sum_{k \in K} q_{ik} + Q_i z_i = Q_i \qquad \forall\, i \in N^C,$$

$$(13.34) \quad \sum_{k \in K} q_{ik} - Q_i y_i = 0 \qquad \forall\, i \in N^O,$$

$$(13.35) \quad q_{ik} \geq 0 \qquad \forall\, k \in K,\, i \in N_k^P.$$

Constraints (13.27) replace the original constraints (13.2) and state that all mandatory contract cargoes are transported. Each of these cargoes can be transported by one or several ships since it is now possible to split each cargo. The corresponding requirements for the optional cargoes are given by constraints (13.28). Constraints (13.29)–(13.32) are similar to the original constraints (13.7)–(13.10), except that the cargo quantity parameter has been replaced with the quantity variable q_{ik}. It can be noted that constraints (13.29)–(13.32) are similar to constraints (13.20)–(13.23) for the flexible cargo case. Constraints (13.33)–(13.35) are new and specific for the split problem. Constraints (13.33) ensure that the total quantity of contracted cargo i is lifted by one or several ships in the fleet, or by a spot charter. A similar requirement for optional cargoes is given by constraints (13.34), while constraints (13.35) impose non-negativity requirements on the load variables.

This tramp ship routing and scheduling problem with split loads was recently studied by Andersson, Christiansen, and Fagerholt [5], Korsvik, Fagerholt, and Laporte [51], and Stålhane et al. [71]. Andersson, Christiansen, and Fagerholt [5] suggest a solution method based on a priori generation of single ship routes and two alternative path flow models that deal with the selection of ship schedules and assignment of cargo quantities to the schedules. Computational results show that the solution method can provide optimal solutions only to realistic problems of small sizes. In order to overcome this limitation Stålhane et al. [71] propose a Branch-and-Price approach, while Korsvik, Fagerholt, and Laporte [51] suggest a large neighborhood search heuristic for the same problem. Both are able to find optimal solutions to the same instances as Andersson, Christiansen, and Fagerholt [5] within a short time, as well as solving larger problems. All three papers show that utilizing split loads can result in improved utilization of the fleet and hence significantly increased profit.

A segment in maritime transportation where splitting cargoes can be important is crude oil tanker routing and scheduling. Hennig et al. [41] present an extensive mathematical formulation to illustrate various aspects of that problem. One characteristic is that in contrast to the problem modeled above there are no predefined cargoes. There is rather a quantity requirement for each crude grade that must be picked up at each loading port within some (rather wide) time windows. This quantity may be picked up by several tankers. Similarly, there are quantity requirements also for the delivery ports, which can also be satisfied by deliveries from more than one tanker (or more than one source). These pickup-and-delivery requirements are not paired into cargoes beforehand, which results in a problem with split pickups and split deliveries. Logically, this problem lies between inventory routing (see Section 13.3) and cargo routing and scheduling. Hennig et al. [40] propose a path flow model with a priori route generation for the problem in Hennig et al. [41], though omitting some of the complicating aspects. They introduce continuous variables to distribute the cargo among the different routes. It is demonstrated that small realistic instances can be solved to optimality.

It should be mentioned that even though the advantages of load splitting can be significant, as demonstrated by the small example in Figure 13.1, there may also be some drawbacks. Load splitting gives an increased number of port calls. In the above example both the pickup-and-delivery ports of cargo 2 will be visited twice. However, the resulting increase in port costs can be more than outweighed by reduced sailing cost, especially in deep sea shipping with long sailing distances and high fuel costs compared to port costs. It may still be considered as reduced service quality for customers that may have to manage the service of more than one ship. It may also lead to increased operational complexity for the planners.

13.2.2.3 ▪ Variable Sailing Speed

Oil price increases during recent years have brought the issue of optimizing sailing speed to the forefront. Bunker fuel cost is a major component of the variable operating cost of a ship, and when fuel prices are high it may amount to the majority of the operating costs. The bunker fuel consumption of a cargo ship per time unit is often estimated to be proportional to the third power of its speed (within normal operating speeds) (Ronen [64]). Bunker fuel consumption per unit of distance is thus proportional to the second power of the speed. Thus, reducing the sailing speed by 10% will reduce the bunker fuel consumption for a given sailing leg by close to 20%. Figure 13.5 shows an example of the fuel consumption per nautical mile as a function of sailing speed over the speed interval that is feasible in practice for a *Liquefied Natural Gas* (LNG) carrier. It should be evident that reducing sailing speed provides a significant potential for cost savings. On the contrary, speeding up can sometimes result in that a vessel is able to arrive at a port in time to service an additional spot cargo. Determining optimal speeds within the planning of ship routes and schedules can therefore obviously influence both the fuel costs and revenues.

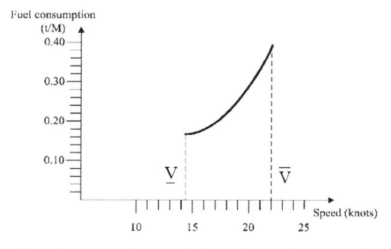

Figure 13.5. *Fuel consumption [ton/nautical mile] as a function of speed for an LNG carrier.*

The following amendments to the basic model presented in Section 13.2.1 are necessary in order to incorporate sailing speed optimization. Let D_{ij} be the sailing distance from node i to node j. The variable s_{ijk} defines the speed of travel from node i to node j with ship k. The time it takes to sail along arc (i, j) is D_{ij}/s_{ijk}. The non-linear function $C_k(s)$, defined on the speed interval $[\underline{V}_k, \overline{V}_k]$, represents the sailing costs per unit of distance for ship k sailing at speed s. The cost of sailing an arc (i, j) with ship k at speed s_{ijk} is then $D_{ij} C_k(s_{ijk})$.

The model for the basic tramp ship routing and scheduling problem (13.1)–(13.18) can now be adjusted as follows:

$$(13.36) \quad \text{maximize} \sum_{i \in N^C} R_i Q_i + \sum_{i \in N^O} R_i Q_i y_i - \sum_{k \in K} \sum_{(i,j) \in A_k} D_{ij} C_k(s_{ijk}) x_{ijk} - \sum_{i \in N^C} C_i^S z_i$$

$$\text{s.t.} \quad (13.2)-(13.10), \ (13.12), \ (13.14)-(13.18), \ \text{and}$$

$$(13.37) \quad t_{ik} + T_{ik}^P + \frac{D_{ij}}{s_{ijk}} - t_{jk} - M_{ijk}(1 - x_{ijk}) \leq 0 \qquad \forall \, k \in K, \ (i,j) \in A_k,$$

$$(13.38) \quad t_{ik} + T_{ik}^P + D_{i,n+i}/s_{i,n+i,k} - t_{n+i,k} \leq 0 \qquad \forall \, k \in K, \ i \in N_k^P,$$

$$(13.39) \quad \underline{V}_k \leq s_{ijk} \leq \overline{V}_k \qquad \forall \, k \in K, \ (i,j) \in A_k.$$

The objective function (13.36) has now become a non-linear function because of the non-linear relationships between fuel consumption and speed. Constraints (13.37) and (13.38) correspond to constraints (13.11) and (13.13) in the original formulation. These constraints are also non-linear because the sailing time depends on the speed variable. The new constraints (13.39) define the lower and upper bounds for the speed variables.

Psaraftis and Kontovas [61] provide an excellent taxonomy and survey of speed models in maritime transportation. Fagerholt, Laporte, and Norstad [31] present some alternative mathematical models for the speed optimization problem for a given route. Norstad, Fagerholt, and Laporte [57] develop a local search heuristic, including a specialized algorithm for determining optimal speeds for given ship routes, to solve the combined tramp ship routing and speed optimization problem discussed above. It is demonstrated that incorporating sailing speeds as decision variables when planning vessel routes significantly improves fleet utilization and profit. Gatica and Miranda [36] deal with minimization of the cost of serving a set of mandatory single trip cargoes while determining the speed for each trip. The port time windows were discretized, which facilitated the use of a network model.

13.2.2.4 ▪ Other Extensions

In many industrial and tramp shipping problems, the time window for when a cargo must be picked up or delivered is somewhat negotiable, especially when planning some time in advance. This gives rise to soft time windows, which is yet another such model extension that increases solution space and can give improved solutions. Fagerholt [30] shows that huge cost reductions were achieved on real-life industrial shipping problems by introducing soft time windows with a maximum allowed violation of the target time windows.

In contrast to the above extensions, which can all be viewed as opportunities since the solution space increases, there may also exist important additional limiting constraints that must be considered in real-life ship routing and scheduling problems. One such example is stowage on board the ships. In many maritime bulk shipping operations the ships can have tens of tanks and carry several cargoes of different products simultaneously, and one must decide which tanks should be used for each cargo. When allocating loads to the different tanks on board the ship, numerous constraints must be satisfied, such as capacity, stability, and hazardous material regulation constraints. Stowage problems in bulk shipping have been studied by Vouros, Panayiotopoulos, and Spyropoulos [76] and Hvattum, Fagerholt, and Armentano [42].

Sometimes one must also consider bunkering/refueling decisions when planning shipping routes. Besbes and Savin [10] and Kim et al. [47] study the handling of refueling decisions for a single vessel taking into account varying fuel prices between ports. This problem is also closely related to the speed optimization problem.

Other examples of additional practical constraints that sometimes must be considered are draft limits in ports (Rakke et al. [62]), time slot assignments in berths to avoid clashes (Pang, Xu, and Li [58]), and restricted port opening hours (Christiansen and Fagerholt [21]).

13.3 ▪ Maritime Inventory Routing

In this section we present mathematical models and refer to solution methods for the *Maritime Inventory Routing Problem* (MIRP). This problem can be defined as a planning problem where an actor has the responsibility for both the inventory management at one or both ends of the maritime transportation legs and for the ships' routing and scheduling. This actor is most often operating in the industrial bulk shipping mode, and the problem appears both in deep and short sea shipping. We start in Section 13.3.1 by presenting a basic version of the problem including a simple example. Then, in Section 13.3.2, we describe a mathematical model for the basic MIRP, and related research on models and solution methods is briefly reviewed. Finally, application-driven extensions are discussed in Section 13.3.3.

13.3.1 ▪ Problem Description and Example

Vital management issues in maritime supply chains are often inventory control at processing facilities and consumption at the customer sites as well as the routing and scheduling of ships. MIRPs often occur in maritime supply chain management when the inventory management and the routing and scheduling of the ship fleet have to be coordinated simultaneously. By coordinating these planning challenges it is possible to achieve monetary benefits, flexibility in services, and improved robustness. This has resulted in increased attention towards MIRP both in the OR community and the industry. These activities have formed the basis of the following surveys in the last decade: Christiansen and Fagerholt [22] and Andersson et al. [6].

Much of the research described in the literature on MIRPs is developed based on real planning problems, particularly from the chemical, petroleum, and pulp industries. In recent years, we have seen the beginning of successful implementation stories of *Decision Support Systems* (DSSs) for MIRPs. Dauzère-Pérès et al. [27] report the development of a DSS for Omya Hustadmarmor transporting calcium carbonate sloppy, which is a product used in paper manufacturing. Furman et al. [35] describe a DSS for feedstock routing in the ExxonMobil downstream sector. A LNG inventory routing model developed by Fodstad et al. [34] is used by Statoil and GDF SUEZ.

According to Christiansen and Fagerholt [22], the basic MIRP concerns the transportation of a single product. The product is stored in inventories at or close to loading (production) and unloading (consumption) ports. Inventory storage capacities are defined in all ports. Further, we assume that the production and consumption rates are given in all ports and are constant during the planning horizon. A heterogeneous fleet of ships is used to transport the product. Each ship has a given capacity and sailing speed. The ships can wait outside a port before entering for (un)loading. A ship can both load and unload at multiple ports in succession. The initial position and load on board each ship are known at the beginning of the planning horizon. The sailing and port costs are all ship-dependent. Inventory carrying costs are not considered because the inventories at both ends of the shipping legs usually belong to the same company. The objective of the MIRP is to design routes and schedules for the fleet that minimize the total transportation cost and determine the (un)loading quantity at each port visit without violating the inventory limits. A typical planning horizon may span from one week up to several months depending on the shipping segment.

In contrast to the cargo routing problem, the MIRP has no predetermined number of calls at a given port during the planning horizon and no predetermined quantity to be (un)loaded at each port call. There are also no predefined pickup and delivery pairs in the basic MIRP, as opposed to cargo routing, where each cargo has a specified pickup and delivery port. Normally, no time windows are given.

Figure 13.6(a) illustrates a small example with two ships with capacities of 30 and 40 units, respectively. For simplicity, we assume that the ships are empty in the beginning of the planning period. Ship 1 is positioned in port 5, while ship 2 has an initial position at sea. The time horizon is 36 days. There are five ports with storages placed close to the ports. At ports 1 and 3, the product is consumed and the ports are defined as consumption (delivery or unloading) ports. Likewise, ports 2, 4, and 5 are production ports and the product is picked up or loaded in these ports. Each of the storages at the ports has defined lower and upper inventory limits and a specific production or consumption rate. For instance, at port 4 the production rate is one unit per day and the lower and upper storage limits are 1 and 32 units, respectively. The corresponding numbers for consumption port 3 are −2, 1, and 28. See Figures 13.7(a) and 13.7(b).

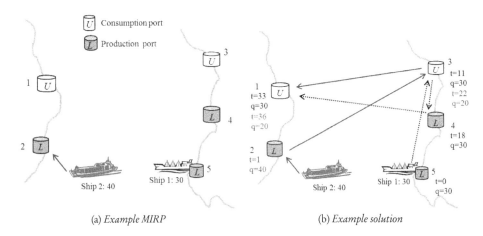

(a) *Example MIRP*   (b) *Example solution*

Figure 13.6. *Example MIRP and its solution with routes, start times (t), and (un)load quantities (q).*

In this example, the berth capacity in all ports is one, which means that just one ship can load or unload at the same time in a port. The (un)loading rate of both ships is 10 units per day. Figure 13.6(b) indicates a solution to the problem where the routes and schedules satisfy the inventory and berth constraints. In port 5, ship 1 loads up to the capacity of the ship before sailing to port 3, where it unloads this quantity. Ship 1 starts loading at port 5 in the beginning of the planning period, and the loading operation lasts for 3 days, while the sailing takes 8 days. Without any waiting in port 3, the unloading starts at day 12 and lasts for 3 days. The inventory limits are tight and the consumption rate big, so the port needs to be visited often. The unloaded quantity is 30 in total, and this quantity is larger than the storage capacity. However, the consumption rate is two units per day, so 30 units can be unloaded during the 3 days. The ship continues on a

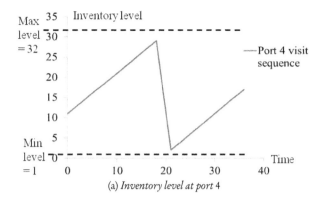

(a) *Inventory level at port 4*

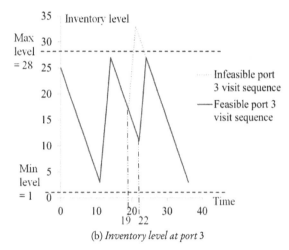

(b) *Inventory level at port 3*

Figure 13.7. *Inventory levels at port 4 and port 3 during the planning horizon.*

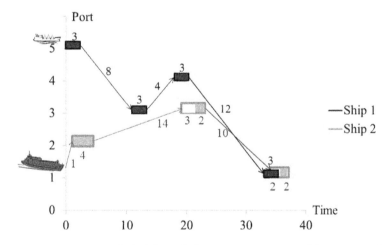

Figure 13.8. *Routes and schedules including time consumption of the various tasks.*

4-day sailing leg to port 4 before loading to its capacity for 3 days. Finally, the ship leaves port 4 at day 22 and sails toward port 1. Here it starts unloading at day 34 and unloads for 3 days. See Figure 13.8 for an illustration of the ship routes and schedules including the time consumption of each task.

Ship 2 starts at sea and reaches port 2 after 1 day. Here ship 2 loads to its capacity before sailing toward port 3. The loading at port 2 lasts for 4 days, while the sailing leg takes 14 days. This means that the unloading can start after 19 days. To unload half the ship's capacity, the ship has to wait until day 23 (after 22 days) before unloading because of the tight inventory constraints; see Figure 13.7(b). Ship 2 continues to port 1 and unloads the rest of the quantity on board. Also here, the ship has to wait because ship 1 is unloading when ship 2 arrives and the berth capacity is one. Ship 2 calls two consumption ports in succession.

13.3.2 ▪ Mathematical Model and Solution Methods

In order to present a mathematical model for the MIRP, most of the mathematical notations from Section 13.2 have to be modified. In addition, some new notation will be defined. Each port is represented by an index i, and the set of ports is given by N. Still let K be the set of available ships and N_k the set of ports (including $o(k)$ and $d(k)$) that can be visited by ship k. Each port can be visited several times during the planning horizon, and M_i is the set of possible calls at port i, while M_{ik} is the set of calls at i that can be made by ship k. The port call number is represented by an index m, and $|M_i|$ is the last possible call at port i.

The set of nodes in the flow network represents the set of port calls, and each port call is specified by (i, m), $i \in N$, $m \in M_i$. In addition, we specify flow networks for each ship k with nodes (i, m), $i \in N_k$, $m \in M_{ik}$. Finally, A_k contains all feasible arcs for ship k, which is a subset of $\{i \in N_k, m \in M_{ik}\} \times \{i \in N_k, m \in M_{ik}\}$.

Some of the parameters remain the same as defined in Section 13.2, such as Q_k^C for the ship capacity, T_{ik}^Q representing the unit loading time, the sailing time T_{ijk}^S, and, finally, the sailing cost C_{ijk}. The berth capacities in MIRPs are often limited, and just one ship can call a port at a time. Due to small port areas or narrow channels from the port to the pilot station a minimum time, T_i^B, between a departure of one ship and the arrival of the next ship might be given. Let T denote the length of the planning horizon.

The levels of the inventory (or stock) have to be within a given interval at each port $[\underline{S}_i, \overline{S}_i]$. The production rate R_i is positive if port i is producing the product and negative if port i is consuming the product. Further, constant I_i is equal to 1 if i is a loading port, -1 if i is an unloading port, and 0 if i is $o(k)$ or $d(k)$.

In the mathematical formulation many of the variables are similar to the ones defined in Section 13.2 but are modified. We use the following types of variables: the binary flow variable x_{imjnk} equals 1 if ship k sails from node (i, m) directly to node (j, n), and 0 otherwise, and the slack variable w_{im} is equal to 1 if no ship takes port call (i, m), and 0 otherwise. The time variable t_{im} represents the time at which service begins at node (i, m). Variable l_{imk} gives the total load on board ship k just after the service is completed at node (i, m), while variable q_{imk} represents the quantity loaded or unloaded at port call (i, m) when ship k calls (i, m). Finally, s_{im} represents the inventory (or stock) level when service starts at port call (i, m). It is assumed that, for each ship k, nothing is loaded or unloaded at the artificial origin $o(k)$, $q_{o(k)1k} = 0$ and that the ship starts at $o(k)$ in the beginning of the planning horizon, $t_{o(k)1}$. At this time, the ships may have cargo on board, L_k^0; $l_{o(k)1k} = L_k^0$, and the stock level at each port i is S_i^0. These fixed variables are omitted in the presentation of the model.

The arc flow formulation of the basic MIRP is as follows:

$$(13.40) \qquad \text{maximize} \sum_{k \in K} \sum_{(i,m,j,n) \in A_k} C_{ijk} x_{imjnk}$$

$$\text{(13.41)} \quad \text{s.t.} \quad \sum_{k \in K} \sum_{j \in N_k} \sum_{n \in M_{jk}} x_{imjnk} + w_{im} = 1 \qquad \forall\, i \in N,\ m \in M_i,$$

$$\text{(13.42)} \quad \sum_{j \in N_k} \sum_{n \in M_{jk}} x_{o(k)1jnk} = 1 \qquad \forall\, k \in K,$$

$$\text{(13.43)} \quad \sum_{i \in N_k} \sum_{m \in M_{ik}} x_{imjnk} - \sum_{i \in N_k} \sum_{m \in M_{ik}} x_{jnimk} = 0 \qquad \forall k \in K,\ j \in N_k \setminus \{o(k), d(k)\},$$
$$n \in M_{jk},$$

$$\text{(13.44)} \quad \sum_{i \in N_k} \sum_{m \in M_{ik}} x_{imd(k)1k} = 1 \qquad \forall\, k \in K,$$

$$\text{(13.45)} \quad w_{im} - w_{i(m-1)} \geq 0 \qquad \forall\, i \in N,\ m \in M_i,$$

$$\text{(13.46)} \quad l_{imk} + I_j q_{jnk} - l_{jnk} - Q_k^C (1 - x_{imjnk}) \leq 0 \qquad \forall\, k \in K,$$
$$(i, m, j, n) \in A_k \mid j \neq d(k),$$

$$\text{(13.47)} \quad l_{imk} + I_j q_{jnk} - l_{jnk} + Q_k^C (1 - x_{imjnk}) \geq 0 \qquad \forall\, k \in K,$$
$$(i, m, j, n) \in A_k \mid j \neq d(k),$$

$$\text{(13.48)} \quad q_{imk} \leq l_{imk} \leq \sum_{j \in N_k} \sum_{n \in M_{jk}} Q_k^C x_{imjnk} \qquad \forall\, k \in K,\ i \in N_k,$$
$$m \in M_{ik} \mid I_i = 1,$$

$$\text{(13.49)} \quad 0 \leq l_{imk} \leq \sum_{j \in N_k} \sum_{n \in M_{jk}} Q_k^C x_{imjnk} - q_{imk} \qquad \forall\, k \in K,\ i \in N_k,$$
$$m \in M_{ik} \mid I_i = -1,$$

$$\text{(13.50)} \quad t_{im} + T_{ik}^Q q_{imk} + T_{ijk}^S - t_{jn} - T(1 - x_{imjnk}) \leq 0 \qquad \forall\, k \in K,$$
$$(i, m, j, n) \in A_k \mid j \neq d(k),$$

$$\text{(13.51)} \quad t_{im} - t_{i(m-1)} - \sum_{k \in K} T_{ik}^Q q_{i(m-1)k} + T_i^B w_{im} \geq T_i^B$$
$$\forall\, i \in N,\ m \in M_i \setminus \{1\},$$

$$\text{(13.52)} \quad s_{i1} - R_i t_{i1} = S_i^0 \qquad \forall\, i \in N,$$

$$\text{(13.53)} \quad s_{i(m-1)} - \sum_{k \in K} I_i q_{i(m-1)k} + R_i (t_{im} - t_{i(m-1)}) - s_{im} = 0$$
$$\forall\, i \in N,\ m \in M_i \setminus \{1\},$$

$$\text{(13.54)} \quad \underline{S}_i \leq s_{im} \leq \overline{S}_i \qquad \forall\, i \in N,\ m \in M_i,$$

$$\text{(13.55)} \quad \underline{S}_i \leq s_{im} - \sum_{k \in K} I_i q_{imk} + R_i (T - t_{im}) \leq \overline{S}_i \qquad \forall\, i \in N,\ m = |M_i|,$$

$$\text{(13.56)} \quad t_{im} \geq 0 \qquad \forall\, (i \in N, m \in M_i) \cup (i \in o(k),\ k, m = 1),$$

$$\text{(13.57)} \quad q_{imk} \geq 0 \qquad \forall\, k \in K,\ i \in N_k \setminus \{d(k)\}, m \in M_{ik},$$

$$\text{(13.58)} \quad x_{imjnk} \in \{0, 1\} \qquad \forall\, k \in K,\ (i, m, j, n) \in A_k,$$

$$\text{(13.59)} \quad w_{im} \in \{0, 1\} \qquad \forall\, i \in N,\ m \in M_i.$$

The objective function (13.40) minimizes the total costs. Constraints (13.41) ensure that each port call is visited at most once. Constraints (13.42)–(13.44) describe the flow on the sailing route used by ship k. One or several of the calls in a specified port can be made by a dummy ship, and the highest call numbers will be assigned to dummy ships in constraints (13.45). For the calls made by a dummy ship, we get artificial starting times and artificial inventory levels within their limits. Constraints (13.46)–(13.49) keep track of the load on board. The relationship between the binary flow variables and the ship load at each port call is given by constraints (13.46)–(13.47). We need both (13.46) and (13.47) to ensure the balance for the load on board after each port visit. This is in contrast to the problems presented in Section 13.2 where just the first type of constraints is necessary. There the quantity loaded is the same as the quantity unloaded for a particular cargo with known port pair. For the MIRP the quantity loaded in one port might be unloaded in several ports. Constraints (13.48) and (13.49) give the ship capacity intervals at the port calls for loading and unloading ports, respectively. The scheduling of the route is taken into account in constraints (13.50). Constraints (13.51) prevent service overlap in the ports and ensure the order of real calls in the same port. A ship must complete its service before the next ship starts its service in the same port. Most of the constraints (13.41)–(13.50) are similar to the constraints for the tramp ship routing and scheduling problem with flexible cargo sizes given in Section 13.2. In addition, we have some inventory constraints for this problem. The inventory level at the first call in each port is calculated in constraints (13.52). From constraints (13.53), we find the inventory level at any port call (i, m) from the inventory level upon arrival at the port in the previous call $(i, m-1)$, adjusted for the loaded/unloaded quantity at the port call and the production/consumption between the two visits. The general inventory limit constraints at each port call are given in (13.54). Constraints (13.55) ensure that the inventory level at the end of the planning horizon is within its limits. The non-negativity requirements for the time for start of service and the quantity loaded or unloaded are given by (13.56) and (13.57), respectively. Finally, the formulation involves binary requirements (13.58) and (13.59) on the flow variables and port call slack variables, respectively.

The model could easily be extended by quantity intervals for minimum and maximum quantities to be (un)loaded at a port call given that the port is called. It is also possible to derive time windows for start of service based on data in the model, such as the inventory conditions. Finally, it is possible to calculate the minimum number of visits to each port based on the data, and the dummy ship variable should then only be defined for the optional visits.

A real planning problem similar to this basic MIRP is studied by Christiansen [20]. There, a company produces and consumes ammonia in its factories worldwide. The planners at the company are responsible for keeping the inventory levels within their limits at all its factories. The factories are placed close to ports, and the ammonia is transported by ships from production ports to consumption ports. In addition to the inventory management, the planners have to design routes and schedules for their fleet of ships. Many of the single product MIRPs described in the literature are studied for transportation of petroleum products. For example, Furman et al. [35] present an industrial MIRP for the transportation of vacuum gas oil. LNG is usually considered a single product for distribution planning; see, for instance, Grønhaug and Christiansen [37]. However, the structure of the particular MIRPs deviates in some respects from the model (13.40)–(13.59).

The combination of inventory management and ship routing and scheduling makes the MIRP a very complex problem to solve. Many of the studies in the literature formulate a MIP model and use this model in various solution approaches. We can divide the models into two main classes of formulations: arc-flow models and path-flow models.

Model (13.40)–(13.59) is an arc-flow model where each sailing leg for each ship is defined as a binary variable. Both Christiansen [20] and Al-Khayyal and Hwang [3] present MIRP arc-flow models. These models (like (13.40)–(13.59)) have a weak linear programming relaxation, so often different types of valid inequalities are developed to strengthen the formulation. Agra, Christiansen, and Delgado [1] consider a short sea fuel oil MIRP where an arc-flow model is improved by tightening bounds, using extended formulations and including valid inequalities. Song and Furman [70] solve subproblems that are restricted versions of their MIRP problem by Branch-and-Cut. These subproblems are solved iteratively in a large neighborhood search heuristic. Arc-flow models are also solved by metaheuristic-based algorithms as in Dauzère-Pérès et al. [27]. Siswanto, Essam, and Sarker [69] combine an arc-flow model with a heuristic. Bredström, Carlsson, and Rönnqvist [13] develop a hybrid algorithm based on a genetic algorithm and linear programming. Finally, Sherali and Al-Yakoob [68] use a rolling horizon heuristic based on an arc-flow model. Path-flow models make the other class of MIRP formulations, and here each path is most often defined as a binary variable. The path is described by the route and/or schedule, and sometimes also the quantities loaded or unloaded on the route are included. Path-flow models are used in Branch-and-Price methods (see, e.g., Christiansen [20], Persson and Göthe-Lundgren [59], Grønhaug et al. [38], Andersson [4], and Engineer et al. [29]), rolling horizon heuristics (Rakke et al. [63]), and various fix and relax heuristics (Gunnarsson, Rönnqvist, and Carlsson [39] and Bilgen [11]). Pure heuristics are also used to solve complicated MIRPs; see, for example, Christiansen et al. [23], who develop a construction heuristic that was embedded in a genetic algorithmic framework.

13.3.3 ▪ Application-Driven Extensions for the MIRP

Many real applications of MIRPs have a more complex structure than the basic MIRP presented in Sections 13.3.1 and 13.3.2. In this section we discuss some of the extensions that are described in the literature. In many real-life applications, several of the extensions are combined.

13.3.3.1 ▪ Multiple Products and Allocation of Products to Compartments

In the basic MIRP several cargoes may be transported simultaneously in one ship, but the product is assumed to be the same. This means that the product does not need to be transported in separate compartments on board the ship or stored in separate storages at the ports. When we move from the single-product MIRP to a multi-product MIRP, we need to keep track of the storages of each product in each port and the amount of each product on board each ship during the entire planning horizon.

Most often the allocation of products to compartments is not considered, and it is assumed that the allocation can be solved by the people responsible for stowage, or as a separate planning problem. Christiansen et al. [23], Persson and Göthe-Lundgren [59], and Siswanto, Essam, and Sarker [69] present multi-product MIRPs without taking the allocation of products to compartments into account. In other studies it is assumed that the ship compartments are dedicated to specific products. Both Al-Khayyal and Hwang [3] and Li, Karimi, and Srinivasan [53] assume that compartments are dedicated to specific products. Recently, Agra, Christiansen, and Delgado [1] considered both the case without any allocation of different fuel products into different cargo tanks as well as the case where there are dedicated tanks for families of products. See Section 13.2.2.4 for additional comments regarding stowage on board the ships.

Recently we have witnessed an increasing attention to multi-product MIRPs relative to single product MIRPs, and these problems are frequently encountered by chemical and petroleum transport companies. Al-Khayyal and Hwang [3] consider the transportation of petrochemicals, Persson and Göthe-Lundgren [59] deal with bitumen products, and Ronen [65] considers refinery products, while Li, Karimi, and Srinivasan [53] study different types of chemicals. However, we can also find studies of multi-product MIRPs in the cement (Christiansen et al. [23]), wheat (Bilgen and Ozkarahan [12]), pulp (Andersson [4]), and calcium carbonate slurry (Dauzère-Pérès et al. [27]) industries.

13.3.3.2 ▪ Particular Network Structures

In model (13.40)–(13.59) we describe the basic MIRP including a network structure consisting of many production and consumption ports. However, some real MIRPs concern inventory constraints at just one of the port types, either in the production or consumption ports. In addition, there might be just one central producer or one central customer. Such networks correspond to a classical vehicle routing structure with one depot and a set of customers. Sherali and Al-Yakoob [67] handle the problem of determining the optimal ship fleet mix and schedules for a problem with a single source and destination, while in Sherali and Al-Yakoob [68] they extend the problem to multiple sources and destinations.

Several real studies with a particular network structure can be found in the literature. Rakke et al. [63] present a real LNG MIRP for one central producer with inventory considerations and many customers with contract requirements instead of inventory. Similarly, the transportation of calcium carbonate slurry products by Dauzère-Pérès et al. [27] starts at a central producer, but in contrast to Rakke et al. [63] the inventories are managed at the unloading ports.

In real life MIRPs we find many unique network structures and constraints regarding an acceptable route structure for each ship. For instance, several studies present problems with a limited number of loadings or unloadings in succession. Bilgen and Ozkarahan [12] define paths with at most two loading ports and one unloading port. On the contrary, in the LNG business considered by Grønhaug and Christiansen [37] it is assumed that an LNG ship is always loaded to its capacity but may unload at two regasification terminals in succession. At each unloading port the unloaded quantity is equal to the capacity of a number of cargo tanks, such that a tank is either empty or fully loaded. This loading and unloading policy is a result of avoiding active movements within the cargo tanks (sloshing) on board the LNG ships.

13.3.3.3 ▪ Time Varying Production and Consumption Rates

In the basic MIRP, the production and consumption rates, R_i, are assumed fixed and constant during the planning horizon for all port storages. This property of the production and consumption rates allows for a mathematical formulation based on continuous time variables; see Christiansen [20], Al-Khayyal and Hwang [3], and Siswanto, Essam, and Sarker [69]. For many real planning problems this assumption is too coarse, and varying production and consumption must be taken into account in the modeling. When the production and/or consumption rate is varying during the planning horizon, the production or consumption rate can, for instance, be given for each day t, by R_{it}, and a discrete time model is applied (see, e.g., Persson and Göthe-Lundgren [59], Sherali and Al-Yakoob [67, 68], Bilgen and Ozkarahan [12], Song and Furman [70], Andersson [4], and Furman et al. [35]).

Finally, the production or consumption in a port might not be known and given to the model. In Grønhaug et al. [38] decision variables are defined for the production and consumption of LNG in each port and each time period.

13.3.3.4 ▪ Other Extensions

Often, the companies facing a MIRP trade cargoes with other operators in order to better utilize the fleet and to ensure the product balance at their own factories. These traded volumes are determined by negotiations. The transporter undertakes to load or unload cargoes with defined pickup and delivery ports and determined quantity intervals and to arrive at a particular port within a given time window. For these cargoes, no inventory management problem exists. This is an example of a shipping company moving from industrial shipping towards tramp shipping. In the real problem described by Christiansen [20], the transporter trades ammonia with other operators. On the contrary, some tramp shipping companies have for a while considered the possibility of introducing a *Vendor Managed Inventory* (VMI) service for some of their customers. This service may replace the more traditional *Contracts of Affreightment* (COAs), which have been the standard agreement between a tramp shipping company and a charterer.

The basic MIRP concerns the sea transportation and the inventory management at both ends of the sailing leg. In many real planning situations, it is sensible to include supply chain activities beyond the MIRP. For instance, Persson and Göthe-Lundgren [59] include the process scheduling at the oil refineries (production ports), while Bilgen and Ozkarahan [12] address bulk grain blending. Bredström, Carlsson, and Rönnqvist [13], Fodstad et al. [34], and Andersson [4] extend the supply chain at the customer side.

13.4 ▪ Dynamic and Stochastic Ship Routing

As explained by Psaraftis [60], a problem is dynamic if some of the problem inputs are not known beforehand but are revealed as time goes by. When probabilistic information concerning the unknown inputs is available, one also faces a stochastic optimization problem. Recent contributions on land-based routing and scheduling problems that are treated as both dynamic and stochastic exist; see, for example, the work of Hvattum, Løkketangen, and Laporte [43], Van Hentenryck, Bent, and Upfal [74], and Schilde, Doerner, and Hartl [66]. We also refer the reader to Chapters 8 and 11 in this book for further references.

Even though most planning of ship routes and schedules in maritime transportation is still performed manually, it is to an increasing degree performed with assistance from optimization-based decision support systems; see, e.g., Fagerholt and Lindstad [32] and Kang et al. [46]. However, most algorithms for ship routing and scheduling solve static and deterministic versions of the problem. In land-based transportation, previous research on dynamic and stochastic routing has indicated that the inclusion of stochastic information within a dynamic planning process is valuable; see, for example, the work of Bent and Van Hentenryck [9] and Hvattum, Løkketangen, and Laporte [43].

Industrial and tramp ship routing and scheduling problems are most often dynamic and stochastic in nature. Sailing times represent one type of problem input that is highly stochastic due to varying environmental conditions. Lo and McCord [55] minimized the expected fuel consumption for a given sailing leg by exploiting uncertain ocean currents, while Azaron and Kianfar [7] represented weather conditions in a stochastic dynamic network. Considering decisions on the tactical level, Cheng and Duran [19] developed a decision support system for a crude oil transportation and inventory problem that takes into account uncertainty in demand. The problem is formulated as a discrete time Markov decision process and solved heuristically.

Recently, Agra et al. [2] presented a robust solution approach to a full load (meaning that only one cargo could be on board a ship at any time) ship routing and scheduling problem where the sailing times are uncertain. The aim is to find robust solutions, where routes are feasible for all travel times defined by a predetermined uncertainty set.

Hwang, Visoldilokpun, and Rosenberg [44] considered the risk associated with fluctuating spot rates and sought to maximize the profit while constraining the variance. The problem is relevant in both tramp and industrial shipping, as optional cargoes are included. Two different formulations were presented, and one of them was solved using a Branch-and-Cut-and-Price method. Only full load instances were considered, and the problem was considered as a static problem. Yet another problem input that can be considered as stochastic in maritime routing problems is the occurrence of future cargoes to be transported. Tirado et al. [72] consider such a problem in industrial shipping. They present three different heuristics for the problem, two of which use stochastic information represented by scenarios. These two heuristics were adapted from the algorithms for land-based routing and scheduling by Bent and Van Hentenryck [9] and Hvattum, Løkketangen, and Laporte [43], respectively. Computational experiments show that the use of stochastic information within the proposed solution methods yields an average cost saving of 2.5% on a set of realistic test instances.

Almost no studies are concerned with the uncertainty in the parameters of the MIRP. However, Rakke et al. [63] and Sherali and Al-Yaakob [67, 68] introduce penalty functions for deviating from the customer contracts and the storage limits, respectively. Christiansen and Nygreen [26] introduce soft inventory levels to handle uncertainties in sailing time and time in port, and these levels are transformed into soft time windows.

Even though there are few research studies on stochastic and dynamic routing and scheduling in the maritime sector, as shown above, we expect an increased interest for this topic in the future. This research area is still in its infancy, and we believe significant improvements in planning, as suggested by Tirado et al. [72], can be made by treating maritime routing and scheduling problems as both dynamic and stochastic.

13.5 ▪ Conclusions and Future Research Directions

The size and importance of the shipping market, which is tightly connected to the world's financial health, create a competitive environment where mainly the skillful actors succeed in the long run. A key to success partly relies on the shipping companies' ability to optimize fleet utilization. Furthermore, considering the focus on environmental emissions and the fact that maritime transportation accounts for 5% of total global CO_2 emissions, which is twice as much as the emissions from airlines (Vidal [75]), it is also easy to recognize the environmental benefits from good planning.

This chapter presents mathematical models and briefly discusses solution methods for some important routing and scheduling problems in industrial and tramp shipping. We hope this chapter will stimulate researchers and students to an increased interest in maritime routing problems. Even though research on ship routing and scheduling has blossomed during the last decade, there are still a number of topics which remain to be properly addressed. As an example, we have observed that some of the recent research is addressing problems that are less grounded in real operations but rather focuses more on theoretical contributions. This trend creates a need for benchmark data sets for the different types of industrial and tramp shipping problems, like those available in land-based transportation as well as the recent ones for network design in liner shipping (Brouer et al. [17]). Such datasets accommodate comparisons among competing solution approaches and would probably attract even more research interest to maritime transportation problems. Furthermore, in maritime transportation there is significant uncertainty in sailing and port times as well as in demand, freight rates, and other inputs. So far just a few studies have explicitly considered uncertainty.

Acknowledgments

This research was carried out with financial support from the MARFLIX and DOMinant II projects, partly funded by the Research Council of Norway. The authors also thank PhD student Giovanni Pantuso for his valuable comments as well as the involvement in the presentation of the content.

Bibliography

[1] A. AGRA, M. CHRISTIANSEN, AND A. DELGADO, *Mixed integer formulations for a short sea fuel oil distribution problem*, Transportation Science, 47 (2013), pp. 108–124.

[2] A. AGRA, M. CHRISTIANSEN, R. FIGUEIREDO, L. M. HVATTUM, M. POSS, AND C. REQUEJO, *The robust vehicle routing problem with time windows*, Computers & Operations Research, 40 (2013), pp. 856–866.

[3] F. AL-KHAYYAL AND S. J. HWANG, *Inventory constrained maritime routing and scheduling for multi-commodity liquid bulk, part I: Applications and model*, European Journal of Operational Research, 176 (2007), pp. 106–130.

[4] H. ANDERSSON, *A maritime pulp distribution problem*, INFOR: Information Systems and Operational Research, 49 (2011), pp. 125–138.

[5] H. ANDERSSON, M. CHRISTIANSEN, AND K. FAGERHOLT, *The maritime pickup and delivery problem with time windows and split loads*, INFOR: Information Systems and Operational Research, 49 (2011), pp. 79–91.

[6] H. ANDERSSON, A. HOFF, M. CHRISTIANSEN, G. HASLE, AND A. LØKKETAN- GEN, *Industrial aspects and literature survey: Combined inventory management and routing*, Computers & Operations Research, 37 (2010), pp. 1515–1536.

[7] A. AZARON AND F. KIANFAR, *Dynamic shortest path in stochastic dynamic networks: Ship routing problem*, European Journal of Operational Research, 144 (2003), pp. 138–156.

[8] D. O. BAUSCH, G. G. BROWN, AND D. RONEN, *Scheduling short-term marine transport of bulk products*, Maritime Policy & Management, 25 (1998), pp. 335–348.

[9] R. W. BENT AND P. VAN HENTENRYCK, *Scenario-based planning for partially dynamic vehicle routing with stochastic customers*, Operations Research, 52 (2004), pp. 977–987.

[10] O. BESBES AND S. SAVIN, *Going bunkers: The joint route selection and refueling problem*, Manufacturing & Service Operations Management, 11 (2009), pp. 694–711.

[11] B. BILGEN, *An iterative fixing variable heuristic for solving a combined blending and distribution planning problem*, in Numerical Methods and Applications, T. Boyanov, S. Dimova, K. Georgiev, and G. Nikolov, eds., vol. 4310 of Lecture Notes in Computer Science, Springer, Berlin, Heidelberg, 2007, pp. 231–238.

[12] B. BILGEN AND I. OZKARAHAN, *A mixed-integer linear programming model for bulk grain blending and shipping*, International Journal of Production Economics, 107 (2007), pp. 555–571.

[13] D. BREDSTRÖM, D. CARLSSON, AND M. RÖNNQVIST, *A hybrid algorithm for distribution problems*, Intelligent Systems, IEEE, 20 (2005), pp. 19–25.

[14] G. BRØNMO, M. CHRISTIANSEN, K. FAGERHOLT, AND B. NYGREEN, *A multistart local search heuristic for ship scheduling: A computational study*, Computers & Operations Research, 34 (2007), pp. 900–917.

[15] G. BRØNMO, M. CHRISTIANSEN, AND B. NYGREEN, *Ship routing and scheduling with flexible cargo sizes*, Journal of the Operational Research Society, 58 (2007), pp. 1167–1177.

[16] G. BRØNMO, B. NYGREEN, AND J. LYSGAARD, *Column generation approaches to ship scheduling with flexible cargo sizes*, European Journal of Operational Research, 200 (2010), pp. 139–150.

[17] B. D. BROUER, J. F. ALVAREZ, C. E. M. PLUM, D. PISINGER, AND M. M. SIGURD, *A base integer programming model and benchmark suite for liner-shipping network design*, Transportation Science, 48 (2014), pp. 281–312.

[18] G. G. BROWN, G. W. GRAVES, AND D. RONEN, *Scheduling ocean transportation of crude oil*, Management Science, 33 (1987), pp. 335–346.

[19] L. CHENG AND M. A. DURAN, *Logistics for world-wide crude oil transportation using discrete event simulation and optimal control*, Computers & Chemical Engineering, 28 (2004), pp. 897–911.

[20] M. CHRISTIANSEN, *Decomposition of a combined inventory and time constrained ship routing problem*, Transportation Science, 33 (1999), pp. 3–16.

[21] M. CHRISTIANSEN AND K. FAGERHOLT, *Robust ship scheduling with multiple time windows*, Naval Research Logistics, 49 (2002), pp. 611–625.

[22] ——, *Maritime inventory routing problems*, in Encyclopedia of Optimization, C. Floudas and P. Pardalos, eds., Springer, Berlin, Heidelberg, 2009, pp. 1947–1955.

[23] M. CHRISTIANSEN, K. FAGERHOLT, T. FLATBERG, Ø. HAUGEN, O. KLOSTER, AND E. H. LUND, *Maritime inventory routing with multiple products: A case study from the cement industry*, European Journal of Operational Research, 208 (2011), pp. 86–94.

[24] M. CHRISTIANSEN, K. FAGERHOLT, AND D. RONEN, *Ship routing and scheduling: Status and perspectives*, Transportation Science, 38 (2004), pp. 1–18.

[25] M. CHRISTIANSEN, K. FAGERHOLT, D. RONEN, AND B. NYGREEN, *Maritime transportation*, in Handbook in Operations Research and Management Science, C. Barnhart and G. Laporte, eds., vol. 14, North–Holland, Amsterdam, 2007, pp. 189–284.

[26] M. CHRISTIANSEN AND B. NYGREEN, *Robust inventory ship routing by column generation*, in Column Generation, G. Desaulniers, J. Desrosiers, and M. Solomon, eds., Springer, New York, 2005, pp. 197–224.

[27] S. DAUZÈRE-PÉRÈS, A. NORDLI, A. OLSTAD, K. HAUGEN, U. KOESTER, M. P. OLAV, G. TEISTKLUB, AND A. REISTAD, *Omya hustadmarmor optimizes its supply chain for delivering calcium carbonate slurry to European paper manufacturers*, Interfaces, 37 (2007), pp. 39–51.

[28] J. DESROSIERS, Y. DUMAS, M. M. SOLOMON, AND F. SOUMIS, *Time constrained routing and scheduling*, in Handbooks in Operations Research and Management Science, C. L. Monma, M. O. Ball, T. L. Magnanti, and G. Nemhauser, eds., vol. 8, North–Holland, Amsterdam, 1995, pp. 35–139.

[29] F. G. ENGINEER, K. C. FURMAN, G. L. NEMHAUSER, M. W. P. SAVELSBERGH, AND J. H. SONG, *A branch-price-and-cut algorithm for single-product maritime inventory routing*, Operations Research, 60 (2012), pp. 106–122.

[30] K. FAGERHOLT, *Ship scheduling with soft time windows: An optimisation based approach*, European Journal of Operational Research, 131 (2001), pp. 559–571.

[31] K. FAGERHOLT, G. LAPORTE, AND I. NORSTAD, *Reducing fuel emissions by optimizing speed on shipping routes*, Journal of the Operational Research Society, 61 (2010), pp. 523–529.

[32] K. FAGERHOLT AND H. LINDSTAD, *Turborouter: An interactive optimisation-based decision support system for ship routing and scheduling*, Maritime Economics & Logistics, 9 (2007), pp. 214–233.

[33] M. L. FISHER AND M. B. ROSENWEIN, *An interactive optimization system for bulk-cargo ship scheduling*, Naval Research Logistics, 36 (1989), pp. 27–42.

[34] M. FODSTAD, K. T. UGGEN, F. RØMO, A. LIUM, G. STREMERSCH, AND S. HECQ, *LNGscheduler: A rich model for coordinating vessel routing, inventories and trade in the liquefied natural gas supply chain*, The Journal of Energy Markets, 3 (2010), pp. 31–64.

[35] K. C. FURMAN, J. H. SONG, G. R. KOCIS, M. K. MCDONALD, AND P. H. WARRICK, *Feedstock routing in the ExxonMobil downstream sector*, Interfaces, 41 (2011), pp. 149–163.

[36] R. A. GATICA AND P. A. MIRANDA, *A time based discretization approach for ship routing and scheduling with variable speed*, Networks and Spatial Economics, 11 (2010), pp. 465–485.

[37] R. GRØNHAUG AND M. CHRISTIANSEN, *Supply chain optimization for the liquefied natural gas business*, in Innovations in Distribution Logistics, J. Nunen, M. Speranza, and L. Bertazzi, eds., vol. 619 of Lecture Notes in Economics and Mathematical Systems, Springer, Berlin, Heidelberg, 2009, pp. 195–218.

[38] R. GRØNHAUG, M. CHRISTIANSEN, G. DESAULNIERS, AND J. DESROSIERS, *A branch-and-price method for a liquefied natural gas inventory routing problem*, Transportation Science, 44 (2010), pp. 400–415.

[39] H. GUNNARSSON, M. RÖNNQVIST, AND D. CARLSSON, *A combined terminal location and ship routing problem*, Journal of the Operational Research Society, 57 (2006), pp. 928–938.

[40] F. HENNIG, B. NYGREEN, M. CHRISTIANSEN, K. FAGERHOLT, K. C. FURMAN, J. SONG, G. R. KOCIS, AND P. H. WARRICK, *Maritime crude oil transportation—a split pickup and split delivery problem*, European Journal of Operational Research, 218 (2012), pp. 764–774.

[41] F. HENNIG, B. NYGREEN, K. C. FURMAN, J. SONG, AND G. R. KOCIS, *Crude oil tanker routing and scheduling*, INFOR: Information Systems and Operational Research, 49 (2011), pp. 153–170.

[42] L. M. HVATTUM, K. FAGERHOLT, AND V. A. ARMENTANO, *Tank allocation problems in maritime bulk shipping*, Computers & Operations Research, 36 (2009), pp. 3051–3060.

[43] L. M. HVATTUM, A. LØKKETANGEN, AND G. LAPORTE, *A branch-and-regret heuristic for stochastic and dynamic vehicle routing problems*, Networks, 49 (2007), pp. 330–340.

[44] H. S. HWANG, S. VISOLDILOKPUN, AND J. M. ROSENBERG, *A branch-and-price-and-cut method for ship scheduling with limited risk*, Transportation Science, 42 (2008), pp. 336–351.

[45] A. S. JETLUND AND I. A. KARIMI, *Improving the logistics of multi-compartment chemical tankers*, Computers & Chemical Engineering, 28 (2004), pp. 1267–1283.

[46] M. H. KANG, H. R. CHOI, H. S. KIM, AND B. J. PARK, *Development of a maritime transportation planning support system for car carriers based on genetic algorithm*, Applied Intelligence, 36 (2012), pp. 585–604.

[47] H. J. KIM, Y. T. CHANG, K. T. KIM, AND H. J. KIM, *An epsilon-optimal algorithm considering greenhouse gas emissions for the management of a ship's bunker fuel*, Transportation Research Part D: Transport and Environment, 17 (2012), pp. 97–103.

[48] S. H. KIM AND K. K. LEE, *An optimization-based decision support system for ship scheduling*, Computers & Industrial Engineering, 33 (1997), pp. 689–692.

[49] J. E. KORSVIK AND K. FAGERHOLT, *A tabu search heuristic for ship routing and scheduling with flexible cargo quantities*, Journal of Heuristics, 16 (2010), pp. 117–137.

[50] J. E. KORSVIK, K. FAGERHOLT, AND G. LAPORTE, *A tabu search heuristic for ship routing and scheduling*, Journal of the Operational Research Society, 61 (2010), pp. 594–603.

[51] ——, *A large neighbourhood search heuristic for ship routing and scheduling with split loads*, Computers & Operations Research, 38 (2011), pp. 474–483.

[52] S. A. LAWRENCE, *International Sea Transport: The Years Ahead*, Lexington Books, Lexington, MA, 1972.

[53] J. LI, I. A. KARIMI, AND R. SRINIVASAN, *Efficient bulk maritime logistics for the supply and delivery of multiple chemicals*, Computers & Chemical Engineering, 34 (2010), pp. 2118–2128.

[54] D. Y. LIN AND H. Y. LIU, *Combined ship allocation, routing and freight assignment in tramp shipping*, Transportation Research Part E: Logistics and Transportation Review, 47 (2011), pp. 414–431.

[55] H. K. LO AND M. R. MCCORD, *Adaptive ship routing through stochastic ocean currents: General formulations and empirical results*, Transportation Research Part A: Policy and Practice, 32 (1998), pp. 547–561.

[56] F. MALLIAPPI, J. A. BENNELL, AND C. N. POTTS, *A variable neighborhood search heuristic for tramp ship scheduling*, in Computational Logistics, J. W. Böse, H. Hu, C. Jahn, X. Shi, R. Stahlbock, and S. Voß, eds., vol. 6971 of Lecture Notes in Computer Science, Springer, Berlin, Heidelberg, 2011, pp. 273–285.

[57] I. NORSTAD, K. FAGERHOLT, AND G. LAPORTE, *Tramp ship routing and scheduling with speed optimization*, Transportation Research Part C: Emerging Technologies, 19 (2011), pp. 853–865.

[58] K. W. PANG, Z. XU, AND C. L. LI, *Ship routing problem with berthing time clash avoidance constraints*, International Journal of Production Economics, 131 (2011), pp. 752–762.

[59] J. A. PERSSON AND M. GÖTHE-LUNDGREN, *Shipment planning at oil refineries using column generation and valid inequalities*, European Journal of Operational Research, 163 (2005), pp. 631–652.

[60] H. N. PSARAFTIS, *Dynamic vehicle routing problems*, in Vehicle Routing: Methods and Studies, B. L. Golden and A. A. Assad, eds., vol. 16, North–Holland, Amsterdam, 1988, pp. 223–248.

[61] H. N. PSARAFTIS AND C. A. KONTOVAS, *Speed models for energy-efficient maritime transportation: A taxonomy and survey*, Transportation Research Part C: Emerging Technologies, 26 (2013), pp. 331–351.

[62] J. G. RAKKE, M. CHRISTIANSEN, K. FAGERHOLT, AND G. LAPORTE, *The traveling salesman problem with draft limits*, Computers & Operations Research, 39 (2012), pp. 2161–2167.

[63] J. G. RAKKE, M. STÅLHANE, C. R. MOE, M. CHRISTIANSEN, H. ANDERSSON, K. FAGERHOLT, AND I. NORSTAD, *A rolling horizon heuristic for creating a liquefied natural gas annual delivery program*, Transportation Research Part C: Emerging Technologies, 19 (2011), pp. 896–911.

[64] D. RONEN, *The effect of oil price on the optimal speed of ships*, Journal of the Operational Research Society, 33 (1982), pp. 1035–1040.

[65] ——, *Marine inventory routing: Shipments planning*, Journal of the Operational Research Society, 53 (2002), pp. 108–114.

[66] M. SCHILDE, K. F. DOERNER, AND R. F. HARTL, *Metaheuristics for the dynamic stochastic dial-a-ride problem with expected return transports*, Computers & Operations Research, 38 (2011), pp. 1719–1730.

[67] H. D. SHERALI AND S. M. AL-YAKOOB, *Determining an optimal fleet mix and schedules: Part I—single source and destination*, in Integer Programming: Theory and Practice, CRC Press, Boca Raton, FL, 2006, pp. 137–166.

[68] ——, *Determining an optimal fleet mix and schedules: Part II—multiple sources and destinations, and the option of leasing transshipment depots*, in Integer Programming: Theory and Practice, CRC Press, Boca Raton, FL, 2006, pp. 167–194.

[69] N. SISWANTO, D. ESSAM, AND R. SARKER, *Solving the ship inventory routing and scheduling problem with undedicated compartments*, Computers & Industrial Engineering, 61 (2011), pp. 289–299.

[70] J. H. SONG AND K. C. FURMAN, *A maritime inventory routing problem: Practical approach*, Computers & Operations Research, 40 (2013), pp. 657–665.

[71] M. STÅLHANE, H. ANDERSSON, M. CHRISTIANSEN, J.-F. CORDEAU, AND G. DESAULNIERS, *A branch-price-and-cut method for a ship routing and scheduling problem with split loads*, Computers & Operations Research, 39 (2012), pp. 3361–3375.

[72] G. TIRADO, L. M. HVATTUM, K. FAGERHOLT, AND J.-F. CORDEAU, *Heuristics for dynamic and stochastic routing in industrial shipping*, Computers & Operations Research, 40 (2013), pp. 253–263.

[73] UNCTAD, *Review of maritime transport*, 2011, United Nations, New York and Geneva, 2011.

[74] P. VAN HENTENRYCK, R. BENT, AND E. UPFAL, *Online stochastic optimization under time constraints*, Annals of Operations Research, 177 (2010), pp. 151–183.

[75] J. VIDAL, *CO2 output from shipping twice as much as airlines*, 2007. Available at: http://www.guardian.co.uk/environment/2007/mar/03/travelsenvironmentalimpact.transportintheuk (accessed December 11, 2012).

[76] G. A. VOUROS, T. PANAYIOTOPOULOS, AND C. D. SPYROPOULOS, *A framework for developing expert loading systems for product carriers*, Expert Systems with Applications, 10 (1996), pp. 113–126.

Chapter 14

Vehicle Routing Applications in Disaster Relief

Bruce L. Golden
Attila A. Kovacs
Edward A. Wasil

14.1 ▪ Introduction

The working assumption in most *Vehicle Routing Problems* (VRPs) is that profitable customer requests have to be processed at minimal cost. However, there are VRP applications where servicing a customer has no monetary benefit, but all available resources must be mobilized to provide the best possible service. These applications appear in the context of humanitarian relief, where the key principle is to prevent or alleviate human suffering. This chapter examines a class of problems that obey this principle, namely, VRPs in disaster relief operations.

We define a disaster as an extraordinary event that can occur with or without limited forewarning and has devastating effects on the population. Disasters might have different causes: natural (e.g., earthquakes and hurricanes) or man made (e.g., terrorist attacks and industry disasters). Furthermore, one disaster may be the trigger for another, such as an earthquake leading to a tsunami and a tsunami causing an industrial accident. Regardless of the cause, disasters often bring destruction, suffering, and loss of lives at a level that cannot be managed by local emergency units. The severity of the impact on the victims depends on the event's magnitude and on two additional factors (International Federation of Red Cross and Red Crescent Societies [23]): the affected population's vulnerability, expressed as the inability to resist the hazard, and the available capacities to cope with the effects of the event. As local capacities are overwhelmed after a disaster, national and international aid agencies play an important role in providing the support that is required to alleviate the suffering. Even though local authorities are responsible for the relief operations, the aid agencies are heavily involved in the disaster management process. Their functions include deploying rescue teams and providing essential supplies such as water, food, and medicine to the affected regions.

Research in the field of disaster relief routing can make important contributions to support aid agencies in their operational activities and eventually to help save lives.

409

14.1.1 ▪ Logistics in Disaster Management

One of the most important tools to make the best use of existing resources and to enable fast relief, even under aggravated circumstances, is logistics. After the damage has been assessed and the requirements have been determined, supplies are shipped from all over the world to the disaster area. At this stage, effective logistics strategies are needed to enable a smooth flow of commodities from the point of origin to the victims. As a consequence, more and more logistics experts are involved in the establishment of distribution networks. In fact, each disaster has different effects, yet the basic way in which a network is established is similar in all relief operations (Stephenson [57]). Typically, international supplies enter the affected area through a large port or airport. Next, these supplies are transferred to a primary warehouse that is nearby. Commodities (supplies) then flow to final storage locations and are eventually delivered to the victims. The first stage of the transportation process is usually carried out by long-haul trucks or trains. Smaller vehicles or vans cover the last miles to remote victims. This procedure is depicted in Figure 14.1. Clearly, humanitarian logistics considers several difficult optimization problems, including warehouse location and vehicle routing. In this chapter, we focus on the last-mile VRPs that involve direct victim assistance (e.g., damage assessment, evacuations, moving of rescue teams, and the delivery of essential supplies).

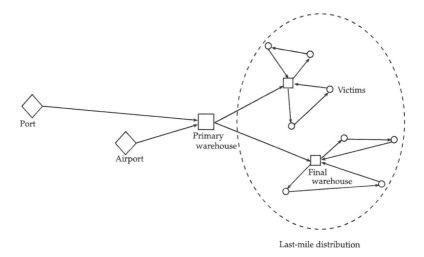

Figure 14.1. *Disaster relief supply chain* [57].

The humanitarian objective to use all possible means to alleviate human suffering clearly deviates from the classical VRP objective to minimize cost. In addition, the dangerous environment in which vehicles operate after a disaster poses new challenges in modeling and solving these problems.

The road network on which vehicles travel might be in poor condition (e.g., roads with damaged bridges might not be passable, rendering the entire planning useless). Gathering sufficient means of transportation to handle the incoming supply may be difficult due to shortages of vehicles, competition among the aid agencies for vehicles, or the refusal of local carriers to operate in areas exposed to the risk of looting. Even if the fleet capacity is sufficient despite these difficulties, fuel shortages might force some vehicles to remain idle. In addition to the extreme circumstances that have to be captured in meaningful models, decisions have to be made quickly. Typically, international rescue teams are able to arrive at the disaster site within 24 hours. They cannot wait a day or more for the results from an optimization algorithm. In such a case, computational efficiency is not a question of cost or user convenience, but rather a question of saving lives.

The consideration of a number of alternative objectives further complicates the modeling effort. Although it is clear that the ultimate objective is to reduce suffering, this is too vague to be used as a quantitative measure to help make good decisions. Therefore, in accordance with the literature, we define three specific objectives that directly originate from the nature of a disaster.

First, the situation in the direct aftermath of a disaster requires a quick response. People may be injured, buried under rubble, traumatized, and left without shelter and food. Every second counts at this stage, and time becomes the most important resource to prevent the further loss of lives. Hence, accelerating the response is an important aspect in disaster relief. Second, the number of affected people is very high after a disaster. As a result, there may be insufficient supplies and limited means of transportation. If transportation is the bottleneck, then vehicle routing models can be used to maximize the amount delivered.

All people have the right to assistance. The third objective is to devote impartial and fair attention to everyone. The significance of this objective increases with the scarcity of the resources. Finally, though not a humanitarian objective, we generally need to take costs into account. In fact, quick-onset disasters are often overfinanced by donations (Van Wassenhove [66]). However, in later phases of a disaster relief operation, the budget is likely to become an important concern.

In this chapter, we review the relevant literature on disaster relief routing and highlight those phases of the disaster management process in which VRP variants are involved. Given the complexity and the scope of disaster management, we examine three consecutive phases: preparedness, response, and recovery.

In addition, we describe how the humanitarian perspective is implemented in various routing models and show how the different focus and objectives affect the solution approaches and the resulting solutions. Finally, we close the chapter with suggestions for future research.

14.1.2 ▪ Complementary Literature

There have been several recent survey articles on humanitarian logistics. A survey with a special focus on disaster relief routing problems is presented in de la Torre, Dolinskaya, and Smilowitz [13]. The authors review the challenges in this field and the way in which these are captured in state-of-the-art articles. Directions for future research are given from a researcher's and a practitioner's point of view. A description of the health care, emergency, and disaster services offered by the Austrian Red Cross is given in Doerner and Hartl [16]. Here, the organization's basic optimization problem in relief operations is classified as a warehouse location-routing problem.

Several papers examine the field of humanitarian logistics in more general terms. In Kovacs and Spens [30], a deeper understanding of humanitarian logistics is promoted by examining the different actors (e.g., donors, aid agencies, and governments), phases, and processes involved in disaster relief operations and the way these relate to each other. Additionally, parallels between humanitarian and commercial logistics are drawn. A comparison of humanitarian and commercial logistics is also given in Beamon and Balcik [7] and Van Wassenhove [66]. In Van Wassenhove [66], the complexities involved in managing humanitarian supply chains and the fundamental differences between humanitarian and commercial supply chains are illustrated. The author points out that the two fields share many similarities that should be exploited. Humanitarian logistics can learn from new and successful concepts developed for commercial applications. The private sector can learn from the agility and adaptability of humanitarian supply chains. The lack of effective performance measures in humanitarian applications is addressed in Beamon and Balcik [7]. The authors compare humanitarian and commercial supply chains and adapt existing commercial measures to evaluate relief operations. This is done with special attention to the requirements of relief operations after quick-onset disasters. Pedraza-Martinez and Van Wassenhove [48] highlight challenges in humanitarian logistics that have received limited attention in the operations research literature. Challenges are posed by different objectives of short-term relief programs and long-term development programs, by decentralized decision making and conflicting objectives at different levels, by fundings that are earmarked for specific programs, and by difficult operating conditions in the field. The current state of humanitarian logistics and the gaps in terms of practice, research, and education are presented in Kovacs and Spens [31]. The authors find that quantitative models are still too immature to be applicable in the case of emergency situations. Early work in the field of humanitarian logistics is presented in Knott [28, 29].

A detailed survey on general operations research and management science approaches in disaster management is presented in Altay and Green [1]. They classify the literature in terms of research methodology, disaster type, and the disaster life cycle. The review in Simpson and Hancock [56] is not limited to large-scale disaster management. The review includes urban services (e.g., police, ambulance, and fire fighters) and general emergency response.

The practitioner's point of view is illustrated by publications of different aid organizations. The Sphere Project was initiated by a group of humanitarian organizations with the aim of improving the quality of disaster management. The project's handbook is a recognized summary of common principles and universal minimum standards for humanitarian response (The Sphere Project [58]). Insights into disaster relief operations of the United Nations are given by a logistics training module provided by the United Nations Development Programme (Stephenson [57]). The importance of logistics in humanitarian operations is pointed out in Thomas and Kopczak [59]. After describing the difficulties encountered in the field of humanitarian logistics, the authors present different strategies for overcoming them.

14.2 ▪ Phases in Disaster Management

Disaster management is a complex and comprehensive process that extends over a long period of time with changing emphases and constraints. The process can be divided into four phases: mitigation, preparedness, response, and recovery.

No community can rule out the possibility of a disaster. Therefore, the management of a disaster starts long before a disaster occurs with the mitigation and the preparedness phases. The mitigation phase is concerned with activities that lessen the potential impact of a disaster. Typical activities include the reinforcement of buildings to withstand disasters or the prevention of populating dangerous areas. This phase will not be discussed further here. In the preparedness phase, arrangements are made that allow a quick and appropriate response after a disaster. Logistics-related activities include the prepositioning of important supplies and the rehearsal of distribution plans.

The response phase focuses on operations that satisfy the urgent needs of a population immediately after a disaster, including assessment of the destruction, evacuation and rescue of the disaster victims, and the supply of essential commodities. These activities are carried out under severe time constraints. The recovery phase includes long-term activities that support the population's self-sustainability and return to normalcy, including the removal of debris and providing health care and food. Typically, the four phases are carried out in a cyclic manner (i.e., affected communities start with the mitigation phase right after they have recovered from a disaster). Next, we describe the logistics-related problems that emerge in each phase.

14.2.1 ▪ Preparedness Phase

In the preparedness phase, strategic decisions are made before a disaster occurs. When a disaster strikes, there is a tremendous demand for essential supplies in a short period of time. It is crucial to have emergency supplies such as tents, first-aid kits, and food at hand in order to be prepared for the worst-case scenario. The better the preparation, the faster the response time after a disaster, and the more lives that can be saved. However, this phase is mostly characterized by the lack of information about the location and the scope of a disaster which makes preplanning difficult.

Several papers deal with the prepositioning of supplies close to disaster-prone areas to enable a fast response. In Dessouky et al. [14] and Jia, Ordóñez, and Dessouky [25], facility location models are presented that deal with the characteristics of disasters. The objective is to locate warehouses that provide the best possible coverage to meet the demand of an impacted area. In contrast to standard location problems, the proposed models consider the quantity and the quality of the coverage. The quantity of coverage refers to the number of facilities that must cover a location. Redundant facilities can serve as backup sites if the demand is larger than expected. The quality of coverage depends on the distance between a demand location and the facilities that cover it. In addition, Balcik and Beamon [2] and Jia, Ordóñez, and Dessouky [24] consider stochastic demand by optimizing the expected objective value over a set of discrete scenarios. Balcik and Beamon [2] maximize the total expected demand covered by the established warehouses. Jia, Ordóñez, and Dessouky [24] formulate several objective functions that represent the quality of covering the uncertain demand.

Facility location and routing decisions are made in Ukkusuri and Yushimito [63] and Van Hentenryck, Bent, and Coffrin [64]. In Ukkusuri and Yushimito [63], each node and edge in the transportation network have a chance to be destroyed by a disaster. The objective is to place facilities at positions in the network such that the probability that the stored supplies survive and can reach the demand locations is maximized. The objective in Van Hentenryck, Bent, and Coffrin [64] is to allocate sufficient supplies before a disaster to enable fast distribution after the disaster strikes. The problem is solved in multiple, consecutive stages. The first stage is a stochastic location problem with uncertainty in the demands, travel times, and surviving supplies. The routing aspect is approximated. The stages that follow the disaster deal with the routing problem when the real-world outcomes are revealed.

Two-stage stochastic programming models with recourse capture uncertainty in the supply, demand, and condition of the road network. Decisions that are implemented before a disaster strikes are made in the first stage. These decisions include prepositioning the facilities and determining the volume of supplies to be stored at each facility. The second-stage decisions provide a response to the random outcome of a disaster. At this stage, the post-event transportation is planned based on the decisions from the first stage. The objective is to place resources (e.g., warehouses, shelters, and supplies) that allow an appropriate response under various disaster scenarios. Problems and solution approaches that adopt two-stage stochastic modeling are presented in Mete and Zabinsky [35], Rawls and Turnquist [50, 51], Salmerón and Apte [53], and Zhu et al. [72]. The model presented by Döyen, Aras, and Barbarosoğlu [17] extends the second stage by considering location and routing decisions simultaneously. Regional facilities must be located before a disaster strikes (first stage). Local facilities are established after a disaster strikes to serve the demand locations (second stage). In Barbarosoğlu and Arda [5], both stages are transportation problems. In the first stage, decisions are made about the allocation of goods just before a disaster strikes. A recourse is made in the second stage, based on the information about the actual scope of the damage.

In order to reduce the complexity of the models, a simplified formulation of the transportation problems is considered in several papers. Barbarosoğlu and Arda [5], Döyen, Aras, and Barbarosoğlu [17], and Rawls and Turnquist [50, 51] model the transportation problem as a network flow problem with fractional commodity flows. Mete and Zabinsky [35] and Zhu et al. [72] consider only direct flows of supplies between warehouses and demand locations. In Mete and Zabinsky [35], the flows are converted into vehicle routes in a post-processing phase. Salmerón and Apte [53] have only one location per vehicle route.

Shen, Dessouky, and Ordóñez [54] give an example of a VRP in the preparedness phase. Vehicle routes are planned well in advance of a potential terrorist attack based on stochastic information about the demands and the travel times. The pre-planned routes are used to provide training opportunities for the responsible authorities. When the actual attack occurs, the pre-planned and rehearsed routes are modified and used to distribute supplies quickly.

14.2.2 ▪ Response Phase

The response phase starts immediately after a disaster and can continue for several weeks. In this phase, all available resources are mobilized to minimize the immediate threat to the population's well-being. Tomasini and Van Wassenhove [60] point out that a successful humanitarian operation is one that satisfies the urgent needs of the population in the shortest amount of time. The cost associated with the operations plays a minor role. Instead, time is the most critical factor. VRPs that explicitly aim to reduce the response time are discussed in Section 14.3.

Facility location problems that emerge in the response phase are presented in Horner and Downs [20] and Murali, Ordóñez, and Dessouky [36]. Given a set of eligible facilities, Murali, Ordóñez, and Dessouky [36] consider the problem of selecting a subset of facilities to be opened. The objective is to open facilities that maximize the coverage of the affected population. The demand is uncertain, and the number of people that can be served decreases with increasing distance between facility and demand location. (Tricoire, Graf, and Gutjahr [61] assume a similar demand characteristic in a covering tour problem.) Horner and Downs [20] try to establish intermediate depots between the demand locations and the prepositioned facilities. As soon as the disaster location is known, the population's accessibility to supplies can be improved by establishing depots that are nearby.

Van Hentenryck, Coffrin, and Bent [65] investigate recovering from power outages after a hurricane. The problem involves the selection of damaged components in the power network and the routing of repair crews to the selected sites. The objective is to restore the power infrastructure as quickly as possible. Rath and Gutjahr [49] consider the problem of placing intermediate depots between supply sources and demand locations combined with routing the vehicles. Intermediate depots are delivered in full truck loads. The supply is then reloaded to smaller vehicles and distributed to the demand locations. The strategic part of the problem is concerned with location decisions, and the operational part deals with routing decisions. Multiple objective functions are defined that maximize satisfied demand and minimize cost.

One stream of research focuses on the first stage of the supply chain. This stage is characterized by the large number of supplies that enter the affected area through the ports of entry and have to be distributed as quickly as possible. The number of locations is usually small because only larger cities and warehouses are considered. Nevertheless, the demand for essentials can be quite large and can exceed the total capacity of the vehicle fleet. To meet the demands, a single location can be visited by several vehicles or by the same vehicle several times.

This problem is modeled as a network flow problem in the literature. Here, supplies and vehicles may be dispersed over the entire network and there are no special rules on the delivery strategy; e.g., demand can be satisfied completely or partially, by one or more vehicles, from one or more depots. Accordingly, the definition of a route is rather broad. In general, any sequence of nodes is called a feasible route as long as the associated vehicle has an uninterrupted flow through the network and its capacity is not exceeded. The flow may contain cycles if vehicles need to return to a depot for replenishment. The focus is on matching commodity flows from the sources to the destinations to the corresponding vehicle flows that carry the commodities. We point out that similar single commodity flow models have been developed for the VRP (see Laporte and Nobert [32] and Magnanti [34] for a discussion of these models).

In Figure 14.2, we show an example with two depots (squares D_1 and D_2) with enough supply to serve seven locations (circles 1–7). There are four identical vehicles that have a capacity of three units each. Two vehicles are located at each depot. The left side of

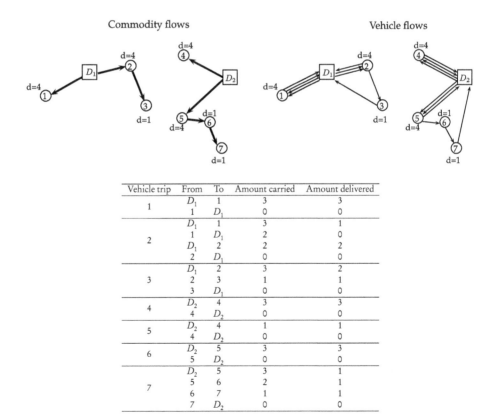

Vehicle trip	From	To	Amount carried	Amount delivered
1	D_1	1	3	3
	1	D_1	0	0
2	D_1	1	3	1
	1	D_1	2	0
	D_1	2	2	2
	2	D_1	0	0
3	D_1	2	3	2
	2	3	1	1
	3	D_1	0	0
4	D_2	4	3	3
	4	D_2	0	0
5	D_2	4	1	1
	4	D_2	0	0
6	D_2	5	3	3
	5	D_2	0	0
7	D_2	5	3	1
	5	6	2	1
	6	7	1	1
	7	D_2	0	0

Figure 14.2. *Network flow model: A square denotes a depot and a circle denotes a demand location. Demand* (d) *is given next to a location.*

the figure shows the commodity flows that are required to meet the entire demand. The right side shows the vehicle flows that are needed for a feasible solution. Since the total demand at the locations (19 units) exceeds the total vehicle capacity (12 units), vehicles have to reload at the two depots. Furthermore, locations 1, 2, 4, and 5 are visited twice to deliver their entire demand. In this example, the vehicle flows have to end at one of the depots. The solution of the network flow problem does not provide a unique routing plan. A possible routing plan is given in the accompanying table. The second and the third columns show the traversed arcs of the vehicle trips. Amount carried refers to the amount loaded on the vehicle when traversing the respective arc, and amount delivered is the amount unloaded at a visited location.

To model this problem using binary vehicle flow variables, which is common in VRP formulations, a fourfold indexing is used. Two indices capture the respective arc, one index for the vehicle and one for the trip in which the arc is traversed. Variable x_{ijkr} takes the value 1 if vehicle k traverses arc (i, j) in its rth trip, and 0 otherwise. By using a network flow model, we do not have to explicitly track each vehicle on each trip. This is possible because separate commodity flow variables guarantee that the demand will be met. Therefore, it does not matter which vehicle is involved in transporting supplies to locations. It is only important that the available vehicle capacity on an arc is sufficient to carry the assigned supplies. The binary flow variables can be replaced by integers that aggregate the total number of vehicles that flow through an arc ($x_{ij} \in \mathbb{N}$). This approach makes the model more compact and easier to solve with MIP solvers because the number of integer variables is independent of the number of vehicles. However, it is not possible to directly determine the vehicle routes from the solution or to include constraints that limit route travel time.

Transportation problems that are modeled as network flow problems are presented in Özdamar [42] and Özdamar and Demir [43]. In Vitoriano, Ortuño, and Tirado [68] and Vitoriano et al. [67], a vehicle may visit a node only once; i.e., cycles to replenish the vehicles are not allowed. The authors consider alternative objectives such as the minimization of looting risk by consolidating vehicles via a convoy or the minimization of the maximum difference in the ratio of satisfied demand to requested demand of the locations.

Several papers account for the time perspective by dividing the planning period into discrete time intervals (Clark and Culkin [11], Haghani and Oh [18], Oh and Haghani [41], Özdamar and Yi [45], Özdamar, Ekinci, and Küçükyazici [44], Yi and Kumar [70], and Yi and Özdamar [71]). Using these time intervals, urgency and time-dependent supply and demand can be modeled by adding an index to the decision variables that gives the respective time interval in which the flow between two nodes takes place. In this way, the model is more precise over shorter intervals, but the computational burden increases.

In Tzeng, Cheng, and Huang [62], only continuous commodity flows are considered and the flows have to go through candidate transshipment depots. There is a fixed cost to locate a depot. Fairness is considered by maximizing the minimum satisfaction rate among all customers.

14.2.3 ▪ Recovery Phase

The main goal in the recovery phase is to restore the self-sufficiency of the affected population (Thomas and Kopczak [59]). The recovery starts when the immediate threat is brought under control and the seriously injured people are being treated. However, it is almost always difficult to determine a specific point in time for the transition from the response phase to the recovery phase. We can associate an optimization problem with the recovery phase if it accounts for the cost at which relief operations are executed and promotes the self-help initiatives of the population (e.g., by asking people in unvisited locations to walk to their nearest tour stop where supply is provided). Considering the cost is important to enable long-term assistance. One goal should be to get the affected population to support the disaster relief effort (The Sphere Project [58]). In Section 14.3, we present VRPs for the recovery phase.

Another characteristic of the recovery phase is the large demand for bulk commodity transportation. The removal of debris, the transportation of construction material, and the long-term supply of essentials such as water, food, and medicine are major planning problems. These operations are extremely expensive, so cost is a major concern (Stephenson [57]). The network flow approach described above can be applied to deal with these problems on a strategic level.

14.3 ▪ Performance Metrics in Disaster Operations

The most common performance metric for evaluating commercial vehicle routing solutions is the total travel cost associated with the routes. If the cost of a route is not directly available, it can be approximated by the travel time or the travel distance. Therefore, the cost measure is convenient to differentiate between bad, good, and very good solutions and, moreover, to guide the decision-making process.

However, routing plans in disaster operations cannot be evaluated on the basis of monetary considerations. Financial aspects play a role, but the crucial question is how well the service alleviates human suffering. Accordingly, money issues fade into the background and are treated as constraints, rather than an objective (Kaplan [27]). On the other hand, the qualitative, and often abstract, goals stated by aid agencies cannot be used readily to evaluate the success of an operation. Instead, these goals have to be transformed into quantitative metrics that can be used as objective functions in mathematical models. In Beamon and Balcik [7], different performance measures are proposed to quantify the efficiency and effectiveness of humanitarian operations. The most suitable measures for disaster relief routing problems are response time (the time that elapses until urgent needs are met), service equity (the differences in the assistance provided to the people at various locations), demand satisfaction (the volume of goods distributed to the population), and transportation cost. The importance of each measure depends on the phase of disaster management with which the specific problem is associated. For example, response time is the main concern in the response phase, while cost is more important in the recovery phase.

In this section, we examine alternative performance metrics in more detail and show how they have been implemented in the literature. Recent papers that consider the four metrics are listed in Table 14.1. In Rath and Gutjahr [49] and Van Hentenryck, Bent, and Coffrin [64], decisions about locating a depot and routing the vehicles from each depot are made in a hierarchical fashion. In the first phase, locations for the depots are selected. In the second phase, vehicle routes are generated from each depot. Only the routing part is considered in this section.

14.3.1 ▪ Response Time

Response time is the interval from the occurrence of a need to its satisfaction. Due to the high degree of urgency in disaster relief operations, a fast response is crucial. Any delay in assistance could lead to further suffering of the affected population. Correspondingly, time is the most precious resource in the immediate aftermath of a disaster. If necessary, goods are flown in from abroad to speed up the relief operations regardless of the resulting

Table 14.1. *Performance metrics considered in 24 recent papers.*

	Response time	Service equity	Demand satisfaction	Transportation cost
Balcik, Beamon, and Smilowitz [3]	✓	✓	✓	✓
Barbarosoğlu, Özdamar, and Çevik [6]	✓	✓		
Bish [8]	✓	✓		✓
Campbell, Vandenbussche, and Hermann [9]	✓	✓		
De Angelis et al. [12]			✓	
Doerner, Focke, and Gutjahr [15]			✓	✓
Hodgson, Laporte, and Semet [19]		✓		✓
Huang, Smilowitz, and Balcik [21]	✓	✓		✓
Huang, Smilowitz, and Balcik [22]	✓			
Jozefowiez, Semet, and Talbi [26]		✓		✓
Lin et al. [33]	✓	✓	✓	✓
Ngueveu, Prins, and Wolfler Calvo [37]	✓			
Nolz et al. [38]	✓	✓	✓	✓
Nolz, Doerner, and Hartl [39]			✓	✓
Nolz, Semet, and Doerner [40]	✓		✓	✓
Panchamgam et al. [47]	✓	✓		✓
Rath and Gutjahr [49]			✓	
Rekik, Renaud, and Berkoune [52]	✓	✓		
Shen, Dessouky, and Ordóñez [54]	✓		✓	
Shen, Ordóñez, and Dessouky [55]	✓		✓	
Tricoire, Graf, and Gutjahr [61]			✓	✓
Van Hentenryck, Bent, and Coffrin [64]	✓	✓		
Van Hentenryck, Coffrin, and Bent [65]	✓			
Wohlgemuth, Oloruntoba, and Clausen [69]	✓			✓

costs (Van Wassenhove [66]). Vehicle routing is an important part of the supply chain and can significantly accelerate the delivery if urgency is incorporated properly in the optimization problem.

The most commonly used method for generating solutions with low response time is the minimization of the maximum arrival time. This objective function is often referred to as the min-max objective. The definition of the maximum arrival time varies in the literature. In Campbell, Vandenbussche, and Hermann [9], Nolz et al. [38], and Rekik, Renaud, and Berkoune [52], the maximum arrival time is the time when the last location is serviced. In Barbarosoğlu, Özdamar, and Çevik [6], Bish [8], and Van Hentenryck, Bent, and Coffrin [64], the maximum arrival time reflects the time when the last vehicle returns to its depot. The first definition is useful when each destination has to be visited once to satisfy the urgent needs of the population. Vehicles can then return to the depot without any further time pressure. The second approach makes more sense in certain pickup operations, e.g., evacuations, where the safe arrival of the vehicle back at the depot is important. Campbell, Vandenbussche, and Hermann [9], Huang, Smilowitz, and Balcik [22], Ngueveu, Prins, and Wolfler Calvo [37], and Van Hentenryck, Coffrin, and Bent [65] propose an alternative to the min-max objective that minimizes the average arrival time among all locations (a min-avg objective). This objective is equivalent to the minimization of the sum of arrival times (min-sum objective). The main difference between these two objectives is that the min-max objective considers only the bottleneck route with the latest arrival time and ignores all shorter routes. Accordingly, several optimal solutions may have the same objective value. In contrast, the min-avg and the min-sum objectives consider each location's arrival time and, therefore, take into account all routes.

Huang, Smilowitz, and Balcik [21] extend the min-sum objective function by considering the arrival time at each location and the amount delivered. Their objective minimizes the supply weighted arrival times and produces a solution in which large amounts of supply are distributed quickly (see Figure 14.3). In Figure 14.3, the locations (circles) are serviced from one depot (square). The demand (d) is given for each location. The figure shows how the preference to visit larger demand locations first leads to crossings of the route segments. The edges back to the depot, depicted as dashed lines, do not contribute to the objective value and might also cross other edges.

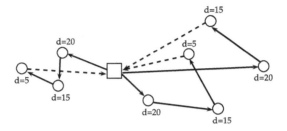

Figure 14.3. *Minimizing the supply weighted arrival times* [21].

The approaches that we have just described do not consider different degrees of urgencies at the locations. Panchamgam et al. [47] introduce the *Hierarchical Traveling Salesman Problem* (HTSP), where the sequence in which locations are visited depends on the urgency of their needs. Destinations with similar degrees of urgency are grouped together into priority classes. A location can only be visited after all locations in a higher priority class have been serviced. The authors also address a relaxed version of the HTSP. Panchamgam [46] extends the HTSP by considering multiple capacitated vehicles and presents two approaches to enforce the priority rule. The local timing priority rule restricts the visit sequence only within a single route; i.e., priorities must be obeyed on each individual route. The global timing priority rule requires that priorities be taken into account with respect to the entire solution; i.e., no location may be serviced later than any lower priority location. The two approaches are illustrated in Figure 14.4. The locations (circles) are grouped together into priority classes according to their distances from an earthquake's epicenter. The locations have to be serviced by two vehicles, located at the depot (square) and with sufficient capacity to make deliveries to three locations each. The left figure shows the solution when each route is considered separately. The local rule allows vehicles to visit low priority locations before high priority locations as long as they are not on the same route. The solution that results from applying the global priority rule is shown on the right. The global rule forces vehicles to visit all high priority locations before a lower priority location is visited. The decision on which rule to apply mainly depends on the availability of the resources. The global rule enables a fast response according to the urgency levels, but it also requires more travel time. The local rule is less binding and generates shorter routes. This can be important when the vehicle's travel time is restricted. A more flexible variant of the described priority rules is the d-relaxed priority rule (Panchamgam [46] and Panchamgam et al. [47]). If p is the highest priority class among all unvisited locations, the relaxed rule allows the vehicle to visit locations with priority $p, p + 1, \ldots, p + d$ before visiting all locations in class p.

Balcik, Beamon, and Smilowitz [3] and Lin et al. [33] describe routing problems where commodities with different priorities have to be delivered over a planning horizon of several days. Urgency is considered by minimizing artificial penalty costs for unsatisfied demand. In Balcik, Beamon, and Smilowitz [3], the amount of the penalty cost at each location is weighted by the location's priority and the importance of the commodity type being demanded by the population. The more important a commodity or the

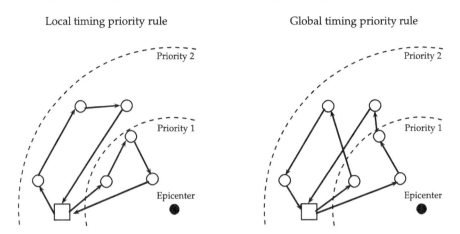

Figure 14.4. *Local vs. global timing priority rules (adapted from Panchamgam [46]).*

higher the priority of a location, the larger is the penalty cost. In Lin et al. [33], location priority is disregarded and only commodity priorities are considered. Goods may be delivered within a soft time window. Penalty costs are incurred when demand is not satisfied within the specified time window.

In Shen, Dessouky, and Ordóñez [54] and Shen, Ordóñez, and Dessouky [55], an appropriate response time is modeled with hard constraints. This is motivated by bioterrorism attacks where the exposed population must be treated within a specified time limit. A delivery after a deadline leads to a large decline in the population's health status or the loss of lives and is, therefore, not feasible.

The risk of encountering an impassable route during the execution of a plan directly affects the response time. Nevertheless, this concern has received only limited attention in the vehicle routing literature. Given the risk values for each single path to become impassable, Nolz, Semet, and Doerner [40] examine five different measures to quantify the overall attractiveness of a solution. The measures, defined for each pair of locations, are the number of alternative paths, the minimal travel time (gives the risk of the fastest path), the minimal risk (gives the risk of the safest path), unreachability (gives the accumulated risk over all paths), and the threshold risk (gives the number of arcs along a path with a risk larger than a specified threshold). For each measure, the attractiveness of a solution is defined by the least attractive link among all pairs of visited locations. An example is presented in Figure 14.5. The figure shows a graph with one depot (square) and five locations (circles). The table next to the graph gives the attractiveness for each pair of nodes. Attractiveness can be defined according to one of the five measures described above. The higher the score, the greater the attractiveness. The table at the bottom shows two solutions. Solution 1 has an attractiveness of one because it contains the unattractive link between location 2 and 3. The least attractive link in the second solution is between location 2 and the depot, and so the attractiveness of the solution is two. Therefore, solution 2 would be preferred.

Shen, Dessouky, and Ordóñez [54] and Shen, Ordóñez, and Dessouky [55] treat the travel times as random variables and formulate the delivery problem as a stochastic programming model.

Route planning in a dynamic environment with varying travel times and unknown demand locations is investigated in Wohlgemuth, Oloruntoba, and Clausen [69]. Dynamic optimization is especially appropriate when reliable information about the scale of a disaster becomes available only during the relief operations.

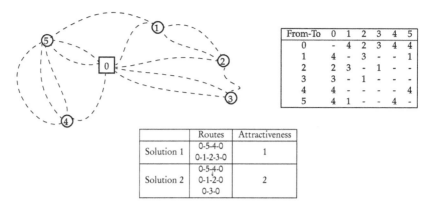

From-To	0	1	2	3	4	5
0	-	4	2	3	4	4
1	4	-	3	-	-	1
2	2	3	-	1	-	-
3	3	-	1	-	-	-
4	4	-	-	-	-	4
5	4	1	-	-	4	-

	Routes	Attractiveness
Solution 1	0-5-4-0 0-1-2-3-0	1
Solution 2	0-5-4-0 0-1-2-0 0-3-0	2

Figure 14.5. *Example of solution attractiveness in terms of risk.*

14.3.2 ▪ Demand Satisfaction

In disaster relief operations, aid agencies are faced with a large demand that must be managed with scarce resources. They need to find routes for each available vehicle such that the distributed supply is maximized, i.e., the number of lives saved is increased.

This problem resembles the *Team Orienteering Problem* (TOP) (Chao, Golden, and Wasil [10]), where as many destinations as possible must be visited within a given time limit. In the TOP, each destination has a score that is earned during a visit. The objective is to maximize the total score. In order to apply TOP concepts to disaster relief routing, the destinations are the affected regions and the score is each region's demand. An example of a TOP solution is given in Figure 14.6. Eight locations (circles) with different demands (d) have to be served from one depot (square). Due to the constraints on the number of vehicles, vehicle capacity, and travel time, only a subset of the locations can be visited. Locations with higher demand are visited with higher priority in order to maximize the distributed supply.

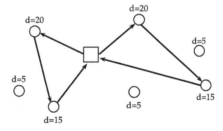

Figure 14.6. *Team orienteering problem* [54].

Rath and Gutjahr [49] extend the TOP by considering multiple depots to service the locations. In addition, a vehicle may reload at its depot as long as supply is available or the total travel time for the fleet has not been exceeded.

Several papers in the literature consider periodic delivery problems. In Balcik, Beamon, and Smilowitz [3] and Lin et al. [33], demand satisfaction is increased by minimizing a penalty cost that is incurred for unsatisfied demand. Lin et al. [33] consider demand that occurs regularly during the planning period. Balcik, Beamon, and Smilowitz [3] consider commodities with recurrent and one-time demand. In De Angelis et al. [12], supplies are delivered to locations from different depots according to the available supply. The demand at each location occurs once at the beginning of the planning period and is satisfied by full-vehicle loads. A tour contains only one location that has to be visited several times. The objective is to satisfy as much demand as possible within a specified planning period. Figure 14.7 shows an example of a daily schedule for one vehicle. The vehicle starts at depot D_1 and visits demand locations and depots alternately until the allowed travel time is exceeded. The vehicle is parked at depot D_3 for the night and resumes making deliveries from D_3 the next day.

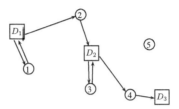

Figure 14.7. *Daily schedule for one vehicle of the problem presented in De Angelis et al.* [12].

Shen, Dessouky, and Ordóñez [54] and Shen, Ordóñez, and Dessouky [55] present a variant of the stochastic vehicle routing problem. The objective is to minimize the unsatisfied demand when demand and travel times are uncertain. Here the demand can only be satisfied if the delivery is made before a specified deadline. Because of the uncertainty in the demand and the travel times, there is a tradeoff between long-and-safe routes and short-and-risky routes. The difference between long-and-safe routes and short-and-risky routes is illustrated in Figure 14.8. The left side of the figure shows a routing plan that favors long arcs with a lower risk of long travel times (e.g., congestions) over short arcs with higher risk. Not all locations can be visited before the deadline in the routing plan, but the planned routes can be executed with a high probability. The routing plan on the right side ignores risk and has short arcs. All locations can be visited before the deadline in the routing plan. However, some locations may not be reached within the time limit if the real travel times are longer on the actual route.

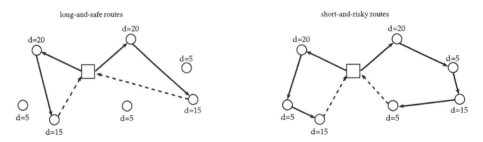

Figure 14.8. *Difference between long-and-safe routes and short-and-risky routes.*

To avoid underutilization of the vehicles due to conservative planning, all locations are scheduled on vehicle routes even if they cannot be delivered before the deadline. Locations that cannot be served in the planning phase still have the chance to be served in the execution phase. Figure 14.9 shows an example of this approach. The scheduled locations are the same as those on the left side of the example above (Figure 14.8), but here all remaining locations are appended at the end of the routes (depicted by dashed lines). These locations may also be served if the actual travel times or demands are smaller than the estimates.

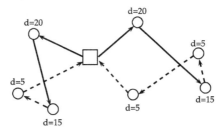

Figure 14.9. *Avoid underutilization of the vehicles due to conservative planning* [54].

Several papers focus on routing and allocation problems in which only a subset of the locations must be visited. However, the routes must be generated in such a way that people in an unvisited location have the opportunity to reach a location on a vehicle's tour without difficulty. The underlying assumption is that the people from the unvisited locations will be able to collect the supply on their own. This assumption restricts the usability of the models in the immediate aftermath of a disaster. Typically, people require direct assistance in the early phase of disaster relief. It may be difficult to inform the people about their nearest tour stop when the local environment is chaotic. However, involving the entire affected population in this way enables a larger distribution of supplies when transportation resources are scarce. This is shown in Figure 14.10. Again, only four locations are delivered directly (solid lines). The unvisited locations are assigned to their closest tour stop to collect supplies that are demanded (dashed lines).

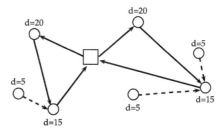

Figure 14.10. *Routing and allocation problems.*

In Nolz et al. [38], Nolz, Semet, and Doerner [40], and Nolz, Doerner, and Hartl [39], the objective is to minimize the average walking distance and maximize the proportion of the population that is covered within a certain distance. Both objectives aim for a large distribution of supplies. Tricoire, Graf, and Gutjahr [61] assume that the demand at each location is stochastic. In addition, the proportion of the population that collects supply at stops on a tour depends on the walking distance. With longer walking distances, fewer people are able to satisfy their demands.

14.3.3 ▪ Service Equity

Disaster relief operations are conducted with limited resources in terms of time, supply, and transportation. Therefore, it may not be possible to meet all urgent requests to the same extent. The concept of sequential visits on vehicle routes implies that not all tour stops are reached at the same time; some are visited before others. The short time frame in which operations have to be performed may be better utilized if vehicles visit easy access locations with larger demands first instead of trying to visit remote locations. Each person has the same right to receive assistance, so operations should be planned to avoid the unfair treatment of any group. Furthermore, unfair supply distribution may cause tensions between close-by communities and increase the risk of looting (The Sphere Project [58]). Therefore, appropriate equity metrics must be considered during operations planning in order to balance the service levels between different locations.

A review of equity in VRP applications is given in Balcik, Iravani, and Smilowitz [4]. The authors state that an operation is equitable when it has equal effects on the beneficiaries. This does not mean that all parties should be treated equally. Rather, differences in the initial needs have to be taken into account when estimating the effect of a service (e.g., severely injured people benefit more from quick assistance than uninjured people, who have lost only property). The goal is to balance the initial requirements and to ensure that all persons have similar living conditions and health after the relief operation. Equity metrics do not indicate how well the requirements have been satisfied. Instead, they consider the service level at each location and reflect the differences in the effects caused by the service. The smaller the differences, the more equitable is the service and the more balanced are the requirements after the operation. Clearly, equity is difficult to measure quantitatively.

In Figure 14.11, we show two different solutions to a distribution problem with scarce supply. The left figure shows the solution when the objective is to service all locations equally. The figure and the accompanying table show that each location obtains half the requested demand (second and third columns in the table). Even though there is no difference between the fractions of unsatisfied demand (fourth column), this solution is only considered equitable if all locations have similar initial needs. In the right solution, different need levels are considered based on the different distances from the origin of a disaster (e.g., the epicenter of an earthquake). The closer that a location is to the epicenter, the greater the level of destruction and the more important it is to satisfy the demand. To capture these differences, the unsatisfied demand rates are weighted by a penalty term (sixth

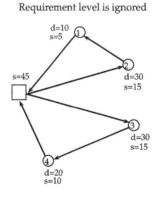
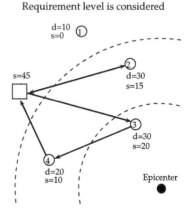

Location	Demand (d)	Equal Distribution		Equitable Distribution			
		Supply (s)	Unsatisfied Demand (1-s/d)	Supply (s)	Penalty (p)	Unsatisfied Demand (1-s/d)	Weighted Unsatisfied Demand ((1-s/d)p)
1	10	5	0.50	0	1	1.00	1
2	30	15	0.50	15	2	0.50	1
3	30	15	0.50	20	3	0.33	1
4	20	10	0.50	10	2	0.50	1

Figure 14.11. *Solutions with equal distribution and equitable distribution when supply is scarce.*

column). The right figure shows the solution obtained by minimizing the differences in the priority weighted unsatisfied demands (last column). The corresponding satisfaction rates (seventh column) indicate that this approach assigns supplies in proportion to the need. Therefore, the solution is considered equitable.

In the literature, different approaches have been presented to consider equity in terms of delivery quantity and response time. In Lin et al. [33], supply equity is achieved by minimizing the maximum difference in the demand satisfaction rates among all locations. The satisfaction rate is defined as the ratio of the satisfied demand to the requested demand. The model formulated by Lin et al. [33] incorporates several commodities that are required with different urgencies. To account for differences in the initial requirements, the satisfaction rates are weighted by the priority with which supplies are needed.

Response time and supply equity are considered in Balcik, Beamon, and Smilowitz [3] and Huang, Smilowitz, and Balcik [21]. Balcik, Beamon, and Smilowitz [3] consider a periodic routing problem where two types of commodities with different priorities have to be delivered. For each location, a penalty cost for unsatisfied demand is incurred for each day for each commodity type. The penalties are weighted by the urgency level of each location to account for different initial needs. Equity is obtained by minimizing the sum of the maximum penalties over all days and over both commodity types. Given that the unsatisfied demand is backordered, this objective also balances the time at which deliveries are made in order to avoid accumulating large penalties. In Huang, Smilowitz, and Balcik [21], different performance metrics are examined that measure the equity of routing plans when demand satisfaction may be split among several deliveries. The authors point out that not all equity metrics are suitable to guide an optimization process. For example, the minimization of the maximum difference in the supply weighted arrival times might lead to solutions with equal, but equally bad, service levels. This type of solution is shown in Figure 14.12. Here, two vehicles traverse the same route in opposite directions and deliver one unit of supply. The total demand of each customer (two units) is satisfied after time tt, where tt is the travel time of each vehicle (we assume a symmetric distance matrix). This solution minimizes the objective function at the cost of the response time.

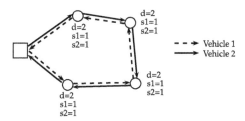

Figure 14.12. *Equal but inefficient delivery when demand satisfaction may be split. Demand (d) is given for each location. The supply s1 is delivered by vehicle 1 and supply s2 is delivered by vehicle 2 [21].*

The authors propose an objective function that promotes partial deliveries to locations in order to save goods for delivery to other destinations. However, in contrast to the previously described objective, this one favors a rapid delivery through the use of a convex penalty term that depends on the fraction of unsatisfied demand at each location. The lower the satisfaction rate, the higher the penalty; i.e., the penalty decreases with each unit delivered. The objective is to minimize the sum of the penalty costs that are incurred over time. Initial differences in the needs are not considered.

In terms of equity in response time, there are many similarities with the performance measurements presented in Section 14.3.1. The minimization of the maximum arrival time bounds the difference between the earliest arrival time and the latest arrival time (Barbarosoğlu, Özdamar, and Çevik [6], Bish [8], Campbell, Vandenbussche, and Hermann [9], Nolz et al. [38], Rekik, Renaud, and Berkoune [52], Van Hentenryck, Bent, and Coffrin [64]). Therefore, this objective also bounds the differences between the effects on the locations. This approach does not consider differences in the initial requirements and leads to equitable tour schedules when all locations have similar initial requirements.

Setting urgency-dependent precedence constraints on the sequence in which locations are visited, as done in the HTSP (Panchamgam et al. [47]), favors equitable solutions. In this case, equity is enforced by hard constraints in the model rather than the objective function.

Routing and allocation problems emerge when the urgent needs have been satisfied, but the population depends on further assistance in health care or in the supply of essential goods. In these problems, equity is taken into account by bounding the longest distance a person has to walk to the nearest tour stop (Hodgson, Laporte, and Semet [19]). Instead of bounding the maximum walking distance, it is minimized in Jozefowiez, Semet, and Talbi [26]. Both approaches yield solutions with balanced walking distances because the differences are bounded either by a constraint or by the objective function.

14.3.4 ▪ Transportation Costs

Minimizing costs is not the primary objective of aid organizations, even though disaster relief can require lots of money. Immediately after a disaster, urgent needs are satisfied regardless of the expenses. However, after the situation has stabilized, aid agencies have to operate economically in order to be able to provide assistance until self-sustainability of the population is recovered.

Commercial and relief routing models are very similar in terms of cost as a performance metric. Relief routing models consider travel costs with lower priority or in conjunction with humanitarian constraints. The first priority in Bish [8] and Huang, Smilowitz, and Balcik [21] is the optimization of humanitarian oriented metrics. Routing costs are only considered to break ties if the primary objective has multiple optimal solutions. In Balcik, Beamon, and Smilowitz [3] and Lin et al. [33], humanitarian objectives and routing costs are aggregated into one objective function. The respective weights are set to favor the humanitarian metrics. Minimizing the travel distance is the only objective in Panchamgam et al. [47]. The urgency of need at locations is taken into account by precedence constraints in the sequence of visits. The objective function in Wohlgemuth, Oloruntoba, and Clausen [69] minimizes the number of vehicles used and the total travel time. The applied dynamic solution approach incorporates two features of disaster relief operations: changing road conditions (i.e., varying travel times) and the arrival of new orders during the execution of the tours.

In the routing and allocation problem presented in Hodgson, Laporte, and Semet [19], transportation costs are minimized while the walking distance from the unvisited locations to their nearest tour stop is constrained. In Doerner, Focke, and Gutjahr [15], Jozefowiez, Semet, and Talbi [26], Nolz et al. [38], Nolz, Semet, and Doerner [40], Nolz, Doerner, and Hartl [39], and Tricoire, Graf, and Gutjahr [61], a multi-objective optimization approach is applied where costs are optimized along with the walking distance or the share of the population that is covered within a specified distance.

14.4 ▪ Commercial VRPs vs. Disaster Relief VRPs

The different requirements in commercial vehicle routing and disaster relief routing raise the question of how much the corresponding solutions are different from each other. Examining the differences in the performance metrics is important since routing models, algorithms, and software products have been developed mainly for commercial applications. In order to justify adapting these routing tools, the service provided to the disaster victims by humanitarian-oriented approaches must be significantly better after applying the tools to the problem. In this section, we illustrate the improvement in solutions that can be achieved by using humanitarian-focused models instead of models that only minimize cost. At the same time, we highlight the increase in transportation cost that results when we focus on optimizing humanitarian objectives.

14.4.1 ▪ Low Cost vs. High-Quality Relief

In Campbell, Vandenbussche, and Hermann [9], classical TSP and VRP models are used to investigate the impact of two objective functions on the service quality and on the total travel cost. The first objective is the minimization of the latest arrival time at a location (min-max). The second objective is the minimization of the sum of arrival times (min-sum). The return trip from the last node visited by a vehicle to the depot is ignored in both cases. The authors derive theoretical worst-case bounds on the relationship between the cost-based objective and the min-max and min-sum objectives. The results show that the optimal TSP tour could double the latest arrival time compared with the optimal min-max solution. The optimal min-max tour increases the TSP travel cost by at most 50%. The TSP solution might also be significantly worse in terms of optimal min-sum objective. However, in this case, the increase in the travel cost caused by minimizing the sum of arrival times cannot be bounded theoretically. So, the min-sum objective might not be applicable when the budget for relief operations is scarce.

The differences in the solutions, as a result of using the two alternative objectives, increase with the size of the vehicle fleet. If the triangle inequality holds, there is no incentive in the classical VRP to use more vehicles than required to serve all locations. However, the min-max and the min-sum objectives can be improved by visiting locations simultaneously with multiple vehicles. In addition to the worst-case bounds, Campbell, Vandenbussche, and Hermann [9] use empirical experiments to demonstrate that solutions for commercial routing problems differ significantly from those found with the two alternative (min-max and min-sum) objectives. In fact, the numerical results show modest differences in the solutions when compared to the worst-case bounds. The results confirm that by minimizing the travel cost, the increase in the latest arrival time is much larger than the increase in total cost when the maximum arrival time is minimized. The same is true in the context of the sum of arrival times.

An extension of the min-sum objective is examined in Huang, Smilowitz, and Balcik [21]. In their paper, a location's demand can be satisfied by more than one vehicle. Thus, each location may be associated with several arrival times. The objective is to minimize the sum of the arrival times weighted by the supply delivered to a location. The authors use empirical experiments to identify the impact of the weighted min-sum objective on the solution. A comparison with a cost-based objective reveals that the alternative objective causes the travel cost to increase by 17% to 18% on average. When cost minimization is the primary objective, the value of the weighted min-sum objective deteriorates by 29% on average.

These large differences occur because the min-max and the (weighted) min-sum objective implicitly consider the cost by favoring shorter routes. The ignored trip back to the depot drives up the cost to a minor extent. The cost-based objective, on the other hand, does not consider any humanitarian-related requirement.

Bish [8] presents a bus evacuation problem where bus routes have to be planned to transport transit-dependent people to shelters in the case of an emergency. The vehicle capacity is constrained, but buses may perform several trips to manage the enormous demand. The objective is to minimize the duration of the evacuation, i.e., the maximum route length. In this problem, the evacuation duration includes the final trip back to the shelter. Therefore, this objective has an impact on the solution different from the min-max objective, which ignores the final trip. If only one vehicle is involved, the minimization of the evacuation duration and the minimization of the routing cost are equivalent and generate the same solution. When the fleet consists of multiple vehicles, the minimization of the travel cost can lead to evacuation durations that are arbitrarily larger than the optimal duration. Again, this is a worst-case bound that has not been observed in empirical experiments.

In general, the min-max objective leads to shorter maximum route lengths with a larger fleet size. Bish [8] shows that the min-max objective value does not always decrease in a convex manner with more vehicles. Adding one more vehicle to a fleet might have larger effects in some cases and smaller effects in others. A small effect can be expected when several vehicles have the same maximum route length and not all routes can be shortened by adding a single vehicle. Furthermore, increasing the fleet size above a certain threshold does not affect the optimal solution. In Figure 14.13, we illustrate these observations. Four routing plans are presented with one, two, three, and four vehicles. Vehicles start from the depot (square), visit the assigned locations (circles), and return to the depot. All unmarked arcs have a length of 1. The graph at the bottom of the figure shows the relation between the min-max objective and the number of vehicles.

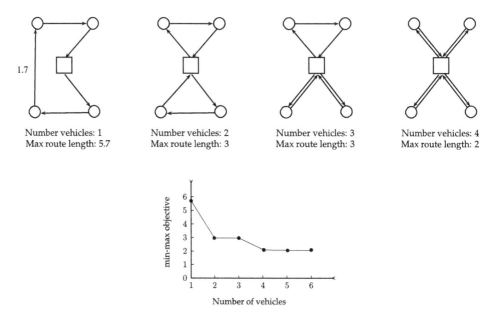

| Number vehicles: 1 | Number vehicles: 2 | Number vehicles: 3 | Number vehicles: 4 |
| Max route length: 5.7 | Max route length: 3 | Max route length: 3 | Max route length: 2 |

Figure 14.13. *Non-convex relation between number of vehicles and the maximum route length.*

The results suggest that aid agencies should not be overeager to acquire vehicles. Instead, the available vehicles should be shared among different agencies and regions such that the overall service quality is improved.

In the HTSP, locations are assigned to different priority classes. The additional cost associated with serving the locations in the strict order of their priorities is examined in Panchamgam et al. [47]. The comparison between the classical TSP and the HTSP shows that, in the worst case, the ratio of the optimal HTSP length to the optimal TSP length is equal to the number of priority classes.

All four papers (Bish [8], Campbell, Vandenbussche, and Hermann [9], Huang, Smilowitz, and Balcik [21], and Panchamgam et al. [47]) indicate that the approaches developed for commercial VRPs are not readily applicable to solve humanitarian-focused problems. Much better service quality can be provided for those in need if the models and solution approaches are adapted to comply with the requirements in real-world relief operations.

14.4.2 ▪ Solving Disaster Relief Problems

The significant differences in the optimal solutions between commercial and relief routing problems require special attention in the design of solution algorithms. Changing the objective function of an algorithm that performs well on commercial routing problems may not be a viable option to solve humanitarian routing problems. This point is made in Campbell, Vandenbussche, and Hermann [9]. The authors state that modifying existing heuristics in order to achieve good results for the min-max and the min-sum objective functions is nontrivial.

An interesting characteristic of optimal min-max and min-sum solutions can be observed with regard to crossings of route segments. Given the triangle inequality, crossings are always associated with additional travel cost. Even though crossings might not be avoidable because of certain constraints, several approaches have been developed to avoid them. In the min-max and min-sum solutions, however, crossings might not influence the objective value at all or, in some cases, they might improve it. For example, consider the case where the objective function ignores the trip from the last location back to the depot. An example is shown in Figure 14.3. The last edge (depicted by a dashed horizontal line) crosses another edge without having any impact on the latest arrival time or the sum of arrival times. The same figure shows the positive effect of a crossing in terms of supply weighted arrival times. Vehicles are primarily routed to high-demand locations in order to deliver large amounts of supply. The benefit of serving large amounts quickly outweighs the loss from crossings at the end of the routes. Campbell, Vandenbussche, and Hermann [9] show that this effect is observable even if the supplied amounts are neglected. This is illustrated on the graph in Figure 14.14. The optimal route in terms of cost is the sequence 0-1-2-3-4-5-0 (called tour A) that does not contain crossings. The sequence 0-1-4-3-2-5-0 (called tour B) has a smaller sum of arrival times. This is because the longest edge with cost $3\sqrt{7}$ is traversed later and only contributes to the arrival times of location 2 and 5. The savings that result outweighs the loss from the crossing.

Computation time is a major concern in disaster relief. Stephenson [57] points out that locally organized relief actions are the most effective and appropriate. Usually, only laptops are available for solving hard optimization problems in the field. When every second counts, decision makers cannot wait long to generate reasonable solutions. Furthermore, the continuously changing environment requires frequent revisions and modifications to the generated solutions. Researchers need to develop robust, low-complexity

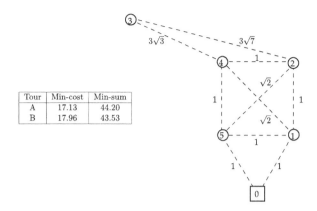

Tour	Min-cost	Min-sum
A	17.13	44.20
B	17.96	43.53

Figure 14.14. *Positive influence of edge crossing in min-sum objective* [9].

algorithms that provide good solutions quickly for a variety of input parameters. Solving problems to proven optimality or struggling to close the last few percent in the optimality gap is neither important nor possible when time is short.

14.5 ▪ Conclusions and Future Research Directions

Disaster management is a complex process that includes several difficult optimization problems. In this chapter, we illustrated how vehicle routing is integrated into the disaster management process and showed how VRPs are different from other transportation-related problems in this area. We argued that disaster management can be divided into three phases that depend on the timing of the decisions. In the preparedness phase, decisions are made under uncertainty, before a disaster occurs. In the response phase, the main objective is to provide assistance to the victims immediately after a disaster. In the recovery phase, both humanitarian-focused objectives and cost-based objectives are taken into account to enable long-term assistance. Routing problems occur in all three phases.

Each phase has specific objectives and constraints with the overriding goal to alleviate human suffering and provide equitable assistance to all victims. Cost-based performance metrics, such as transportation cost, are not suitable to evaluate the success of humanitarian operations. We examined three performance metrics that reflect how much an affected population will benefit from a planned activity. These metrics are response time, demand satisfaction, and equity.

The most sensitive phase in disaster relief is the response phase. Here, decisions are made under extreme time pressure. Fundamental differences arise in the solutions by optimizing either cost or response time. We provided two important insights. First, a much higher service quality occurs if humanitarian requirements are considered cautiously in the models. Second, solution approaches that have been developed for commercial applications cannot be used directly to solve humanitarian-focused problems.

Research in the field of disaster relief routing can make an important contribution to provide the best possible service to those in need. Work in this field may be attractive to researchers because of its unique and interesting characteristics. The humanitarian objectives and the associated environment in which routes are generated pose new challenges in modeling and solving these problems. Many aspects of disaster relief have already been addressed in the literature. Nevertheless, there is need for a deeper understanding of the effect of humanitarian objectives on different kinds of disaster relief scenarios. One aspect that requires further consideration is vehicle-location compatibility. This extension is important when the transportation network is damaged and roads are impassable so that only off-road vehicles or aircraft can be used to reach some locations. Furthermore, evacuation operations modeled as pickup-and-delivery problems have received little attention in the disaster routing literature. Another aspect that requires further research is the lack of information about travel time, demand, and urgency of demand after a disaster. The lack of information could be taken into account by using stochastic models and perhaps by deterministic models that optimize artificial risk measures for route interruptions. In situations where time is short, the assessment of the destruction and the delivery of supplies might occur at the same time. In this case, the information gathered during the execution could be incorporated into dynamic solution approaches to route the vehicles to locations with larger needs.

From an algorithmic point of view, the development of solution approaches that account for the humanitarian requirements generates many possibilities for future research. Due to the complexity of disaster relief routing problems and the fact that solutions have to be generated in the shortest amount of time, the focus should be on heuristic approaches. The development of exact algorithms and examining their performance on test instances could be used to identify good solution features. Knowing the characteristics of optimal solutions enables good solutions to be implemented even without sophisticated computational support.

Bibliography

[1] N. ALTAY AND W. GREEN, *OR/MS research in disaster operations management*, European Journal of Operational Research, 175 (2006), pp. 475–493.

[2] B. BALCIK AND B. BEAMON, *Facility location in humanitarian relief*, International Journal of Logistics, 11 (2008), pp. 101–121.

[3] B. BALCIK, B. BEAMON, AND K. SMILOWITZ, *Last mile distribution in humanitarian relief*, Journal of Intelligent Transportation Systems, 12 (2008), pp. 51–63.

[4] B. BALCIK, S. IRAVANI, AND K. SMILOWITZ, *A review of equity in nonprofit and public sector: A vehicle routing perspective*, in Wiley Encyclopedia of Operations Research and Management Science, J. Cochran, ed., John Wiley & Sons, New York, 2010.

[5] G. BARBAROSOĞLU AND Y. ARDA, *A two-stage stochastic programming framework for transportation planning in disaster response*, Journal of the Operational Research Society, 55 (2004), pp. 43–53.

[6] G. BARBAROSOĞLU, L. ÖZDAMAR, AND A. ÇEVIK, *An interactive approach for hierarchical analysis of helicopter logistics in disaster relief operations*, European Journal of Operational Research, 140 (2002), pp. 118–133.

[7] B. BEAMON AND B. BALCIK, *Performance measurement in humanitarian relief chains*, International Journal of Public Sector Management, 21 (2008), pp. 4–25.

[8] D. BISH, *Planning for a bus-based evacuation*, OR Spectrum, 33 (2011), pp. 629–654.

[9] A. CAMPBELL, D. VANDENBUSSCHE, AND W. HERMANN, *Routing for relief efforts*, Transportation Science, 42 (2008), pp. 127–145.

[10] I.-M. CHAO, B. L. GOLDEN, AND E. A. WASIL, *The team orienteering problem*, European Journal of Operational Research, 88 (1996), pp. 464–474.

[11] A. CLARK AND B. CULKIN, *A network transshipment model for planning humanitarian relief operations after a natural disaster*, Technical Report, Department of Mathematical Sciences, University of the West of England, Bristol, 2007. Available at: http://www.cems.uwe.ac.uk/~arclark/.

[12] V. DE ANGELIS, M. MECOLI, C. NIKOI, AND G. STORCHI, *Multiperiod integrated routing and scheduling of world food programme cargo planes in Angola*, Computers & Operations Research, 34 (2007), pp. 1601–1615.

[13] L. DE LA TORRE, I. DOLINSKAYA, AND K. SMILOWITZ, *Disaster relief routing: Integrating research and practice*, Socio-Economic Planning Sciences, 46 (2012), pp. 88–97.

[14] M. DESSOUKY, F. ORDÓÑEZ, H. JIA, AND Z. SHEN, *Rapid distribution of medical supplies*, in Patient Flow: Reducing Delay in Healthcare Delivery, R. Hall, ed., vol. 91 of International Series in Operations Research & Management Science, Springer, New York, 2006, pp. 309–338.

[15] K. DOERNER, A. FOCKE, AND W. GUTJAHR, *Multicriteria tour planning for mobile healthcare facilities in a developing country*, European Journal of Operational Research, 179 (2007), pp. 1078–1096.

[16] K. DOERNER AND R. HARTL, *Health care logistics, emergency preparedness, and disaster relief: New challenges for routing problems with a focus on the Austrian situation*, in The Vehicle Routing Problem: Latest Advances and New Challenges, B. L. Golden, S. Raghavan, and E. A. Wasil, eds., vol. 43 of Operations Research/Computer Science Interfaces Series, Springer, New York, 2008, pp. 527–550.

[17] A. DÖYEN, N. ARAS, AND G. BARBAROSOĞLU, *A two-echelon stochastic facility location model for humanitarian relief logistics*, Optimization Letters, 6 (2012), pp. 1123–1145.

[18] A. HAGHANI AND S.-C. OH, *Formulation and solution of a multi-commodity, multi-modal network flow model for disaster relief operations*, Transportation Research Part A: Policy and Practice, 30 (1996), pp. 231–250.

[19] M. HODGSON, G. LAPORTE, AND F. SEMET, *A covering tour model for planning mobile health care facilities in Suhum District, Ghana*, Journal of Regional Science, 38 (1998), pp. 621–638.

[20] M. HORNER AND J. DOWNS, *Testing a flexible geographic information system-based network flow model for routing hurricane disaster relief goods*, Transportation Research Record: Journal of the Transportation Research Board, 2022 (2007), pp. 47–54.

[21] M. HUANG, K. SMILOWITZ, AND B. BALCIK, *Models for relief routing: Equity, efficiency and efficacy*, Transportation Research Part E: Logistics and Transportation Review, 48 (2012), pp. 2–18.

[22] ——, *A continuous approximation approach for assessment routing in disaster relief*, Transportation Research Part B: Methodological, 50 (2013), pp. 20–41.

[23] INTERNATIONAL FEDERATION OF RED CROSS AND RED CRESCENT SOCIETIES, *What is a disaster?*, 2011. Available at: http://www.ifrc.org/en/what-we-do/disaster-management/about-disasters/what-is-a-disaster/ (accessed on May 8, 2012).

[24] H. JIA, F. ORDÓÑEZ, AND M. DESSOUKY, *A modeling framework for facility location of medical services for large-scale emergencies*, IIE Transactions, 39 (2007), pp. 41–55.

[25] ———, *Solution approaches for facility location of medical supplies for large-scale emergencies*, Computers & Industrial Engineering, 52 (2007), pp. 257–276.

[26] N. JOZEFOWIEZ, F. SEMET, AND E.-G. TALBI, *The bi-objective covering tour problem*, Computers & Operations Research, 34 (2007), pp. 1929–1942.

[27] R. KAPLAN, *Strategic performance measurement and management in nonprofit organizations*, Nonprofit Management and Leadership, 11 (2001), pp. 353–370.

[28] R. KNOTT, *The logistics of bulk relief supplies*, Disasters, 11 (1987), pp. 113–115.

[29] ———, *Vehicle scheduling for emergency relief management: A knowledge-based approach*, Disasters, 12 (1988), pp. 285–293.

[30] G. KOVACS AND K. SPENS, *Humanitarian logistics in disaster relief operations*, International Journal of Physical Distribution & Logistics Management, 37 (2007), pp. 99–114.

[31] ———, *Trends and developments in humanitarian logistics—a gap analysis*, International Journal of Physical Distribution & Logistics Management, 41 (2011), pp. 32–45.

[32] G. LAPORTE AND Y. NOBERT, *Exact algorithms for the vehicle routing problem*, Annals of Discrete Mathematics, 31 (1987), pp. 147–184.

[33] Y.-H. LIN, R. BATTA, P. ROGERSON, A. BLATT, AND M. FLANIGAN, *A logistics model for emergency supply of critical items in the aftermath of a disaster*, Socio-Economic Planning Sciences, 45 (2011), pp. 132–145.

[34] T. MAGNANTI, *Combinatorial optimization and vehicle fleet planning: Perspectives and prospects*, Networks, 11 (1981), pp. 179–213.

[35] H. METE AND Z. ZABINSKY, *Stochastic optimization of medical supply location and distribution in disaster management*, International Journal of Production Economics, 126 (2010), pp. 76–84.

[36] P. MURALI, F. ORDÓÑEZ, AND M. DESSOUKY, *Facility location under demand uncertainty: Response to a large-scale bio-terror attack*, Socio-Economic Planning Sciences, 46 (2012), pp. 78–87.

[37] S. NGUEVEU, C. PRINS, AND R. WOLFLER CALVO, *An effective memetic algorithm for the cumulative capacitated vehicle routing problem*, Computers & Operations Research, 37 (2010), pp. 1877–1885.

[38] P. NOLZ, K. DOERNER, W. GUTJAHR, AND R. HARTL, *A bi-objective metaheuristic for disaster relief operation planning*, in Advances in Multi-Objective Nature Inspired Computing, C. Coello Coello, C. Dhaenens and L. Jourdan, eds., vol. 272 of Studies in Computational Intelligence, Springer, Berlin, Heidelberg, 2010, pp. 167–187.

[39] P. NOLZ, K. DOERNER, AND R. HARTL, *Water distribution in disaster relief*, International Journal of Physical Distribution & Logistics Management, 40 (2010), pp. 693–708.

[40] P. NOLZ, F. SEMET, AND K. DOERNER, *Risk approaches for delivering disaster relief supplies*, OR Spectrum, 33 (2011), pp. 543–569.

[41] S.-C. OH AND A. HAGHANI, *Testing and evaluation of a multi-commodity multimodal network flow model for disaster relief management*, Journal of Advanced Transportation, 31 (1997), pp. 249–282.

[42] L. ÖZDAMAR, *Planning helicopter logistics in disaster relief*, OR Spectrum, 33 (2011), pp. 655–672.

[43] L. ÖZDAMAR AND O. DEMIR, *A hierarchical clustering and routing procedure for large scale disaster relief logistics planning*, Transportation Research Part E: Logistics and Transportation Review, 48 (2012), pp. 591–602.

[44] L. ÖZDAMAR, E. EKINCI, AND B. KÜÇÜKYAZICI, *Emergency logistics planning in natural disasters*, Annals of Operations Research, 129 (2004), pp. 217–245.

[45] L. ÖZDAMAR AND W. YI, *Greedy neighborhood search for disaster relief and evacuation logistics*, Intelligent Systems, IEEE, 23 (2008), pp. 14 –23.

[46] K. PANCHAMGAM, *Essays in retail operations and humanitarian logistics*, PhD thesis, Robert H. Smith School of Business, University of Maryland, College Park, MD, 2011.

[47] K. PANCHAMGAM, Y. XIONG, B. L. GOLDEN, B. DUSSAULT, AND E. A. WASIL, *The hierarchical traveling salesman problem*, Optimization Letters, 7 (2013), pp. 1517–1524.

[48] A. PEDRAZA-MARTINEZ AND L. VAN WASSENHOVE, *Transportation and vehicle fleet management in humanitarian logistics: Challenges for future research*, EURO Journal on Transportation and Logistics, 1 (2012), pp. 185–196.

[49] S. RATH AND W. GUTJAHR, *A math-heuristic for the warehouse location-routing problem in disaster relief*, Computers & Operations Research, 42 (2014), pp. 25–39.

[50] C. RAWLS AND M. TURNQUIST, *Pre-positioning of emergency supplies for disaster response*, Transportation Research Part B: Methodological, 44 (2010), pp. 521–534.

[51] ———, *Pre-positioning planning for emergency response with service quality constraints*, OR Spectrum, 33 (2011), pp. 481–498.

[52] M. REKIK, J. RENAUD, AND D. BERKOUNE, *Two-stage solution methods for vehicle routing in disaster area*, Technical Report, CIRRELT-2012-31, Montréal, Canada, 2012.

[53] J. SALMERÓN AND A. APTE, *Stochastic optimization for natural disaster asset prepositioning*, Production and Operations Management, 19 (2010), pp. 561–574.

[54] Z. SHEN, M. DESSOUKY, AND F. ORDÓÑEZ, *A two-stage vehicle routing model for large-scale bioterrorism emergencies*, Networks, 54 (2009), pp. 255–269.

[55] Z. SHEN, F. ORDÓÑEZ, AND M. DESSOUKY, *The stochastic vehicle routing problem for minimum unmet demand*, in Optimization and Logistics Challenges in the Enterprise, W. Chaovalitwongse, K. Furman, and P. Pardalos, eds., vol. 30 of Springer Optimization and Its Applications, Springer, New York, 2009, pp. 349–371.

[56] N. SIMPSON AND P. HANCOCK, *Fifty years of operational research and emergency response*, Journal of the Operational Research Society, 60 (2009), pp. 126–139.

[57] R. STEPHENSON, *Logistics*, UNDP/UNDRO Disaster Management Training Programme, 1st ed., 1993. Available at: http://www.pacificdisaster.net/pdnadmin/data/documents/633.html.

[58] THE SPHERE PROJECT, *The Sphere Project: Humanitarian Charter and Minimum Standards in Humanitarian Response*, The Sphere Project, 2011. Available at: http://www.sphereproject.org/handbook/.

[59] A. THOMAS AND L. KOPCZAK, *From logistics to supply chain management: The path forward in the humanitarian sector*, Fritz Institute, 2005. Available at: www.fritzinstitute.org/pdfs/whitepaper/fromlogisticsto.pdf.

[60] R. TOMASINI AND L. VAN WASSENHOVE, *A framework to unravel, prioritize and coordinate vulnerability and complexity factors affecting a humanitarian response operation*, Technical Report, 2004/41/TM. INSEAD, Fontainebleau, France, 2004. Available at: http://www.insead.edu/facultyresearch/centres/isic/humanitarian/PapersHumanitarianResearch.cfm.

[61] F. TRICOIRE, A. GRAF, AND W. GUTJAHR, *The bi-objective stochastic covering tour problem*, Computers & Operations Research, 39 (2012), pp. 1582–1592.

[62] G.-H. TZENG, H.-J. CHENG, AND T. HUANG, *Multi-objective optimal planning for designing relief delivery systems*, Transportation Research Part E: Logistics and Transportation Review, 43 (2007), pp. 673–686.

[63] S. UKKUSURI AND W. YUSHIMITO, *Location routing approach for the humanitarian prepositioning problem*, Transportation Research Record: Journal of the Transportation Research Board, 2089 (2008), pp. 18–25.

[64] P. VAN HENTENRYCK, R. BENT, AND C. COFFRIN, *Strategic planning for disaster recovery with stochastic last mile distribution*, in Integration of AI and OR Techniques in Constraint Programming for Combinatorial Optimization Problems, A. Lodi, M. Milano, and P. Toth, eds., vol. 6140 of Lecture Notes in Computer Science, Springer, Berlin, Heidelberg, 2010, pp. 318–333.

[65] P. VAN HENTENRYCK, C. COFFRIN, AND R. BENT, *Vehicle routing for the last mile of power system restoration*, in Proceedings of the 17th Power Systems Computation Conference, Stockholm, Sweden, 2011.

[66] L. VAN WASSENHOVE, *Humanitarian aid logistics: Supply chain management in high gear*, Journal of the Operational Research Society, 57 (2005), pp. 475–489.

[67] B. VITORIANO, M. ORTUÑO, G. TIRADO, AND J. MONTERO, *A multi-criteria optimization model for humanitarian aid distribution*, Journal of Global Optimization, 51 (2011), pp. 189–208.

[68] B. VITORIANO, T. ORTUÑO, AND G. TIRADO, *HADS, a goal programming-based humanitarian aid distribution system*, Journal of Multi-Criteria Decision Analysis, 16 (2009), pp. 55–64.

[69] S. WOHLGEMUTH, R. OLORUNTOBA, AND U. CLAUSEN, *Dynamic vehicle routing with anticipation in disaster relief*, Socio-Economic Planning Sciences, 46 (2012), pp. 261–271.

[70] W. YI AND A. KUMAR, *Ant colony optimization for disaster relief operations*, Transportation Research Part E: Logistics and Transportation Review, 43 (2007), pp. 660–672.

[71] W. YI AND L. ÖZDAMAR, *A dynamic logistics coordination model for evacuation and support in disaster response activities*, European Journal of Operational Research, 179 (2007), pp. 1177–1193.

[72] J. ZHU, J. HUANG, D. LIU, AND J. HAN, *Resources allocation problem for local reserve depots in disaster management based on scenario analysis*, in the 7th International Symposium on Operations Research and its Applications. Lijiang, China, 2008, pp. 395–407.

Chapter 15

Green Vehicle Routing

Richard Eglese
Tolga Bektaş

15.1 ▪ Environmentally Sustainable Routing

Transport services enable economic growth but at the same time have negative environmental impacts including land use, *Greenhouse Gas* (GHG) emissions, pollution, noise, summer fog, and toxic effects on the ecosystem such as acid rain. According to the TERM 2011 Report published by the European Environment Agency, transport (including international maritime) contributed 24% of the overall GHG emissions in the EU-27 countries in 2009, with road transport accounting for 17% of the total GHG emissions (see Vicente [58]).

The focus of this chapter is to review and explore a new and a growing line of research, namely "green" logistics, and in particular "green" vehicle routing, which aims to minimize the harmful effects of transportation on the environment. The main concern of this chapter is to look at vehicle routing models where environmental issues are taken into account on top of the normal economic issues, primarily at an operational level of decision making. Emphasis will be placed on modeling rather than solution methods, which are often modifications of existing methods covered elsewhere in this book. Two related areas of research which fall under the broad area of green logistics, namely routing of hazardous materials and waste collection, will be excluded from this chapter (see, e.g., Ghiani et al. [29]). Although problems arising in such contexts are related to the environment, the hazards or recycling issues are associated with the commodities being transported rather than the environmental effects of transportation per se. We refer interested readers to the general surveys by Sbihi and Eglese [50] and Dekker, Bloemhof, and Mallidis [12] for other types of problems not covered here. Similarly, we do not cover the broad field of green supply chain management, which encompasses a wide variety of activities from product design and material sourcing to manufacturing and remanufacturing but is clearly beyond the scope of this chapter. Interested readers can refer to Srivastava [52] for a review of the literature on green supply chain management.

15.1.1 ▪ External Costs of Transportation

The environmental damage caused by transportation activities has long been recognized since the 1950s (see McKinnon [40]), although no significant work addressing this issue was done until the early 2000s. The most prominent environmental impacts of freight transportation are as follows:

- *Atmospheric emissions.* This is due to combustion engines used in goods vehicles which, while converting fuel into energy, emit pollutants such as CO, CO_2, NO_x, and particulate matter. These pollutants have harmful effects on humans (e.g., respiratory problems and asthma) and the environment (e.g., acid rain and summer fog) (see Cullinane and Edwards [11]).

- *Noise pollution.* The three sources generating noise are propulsion, tyre/road contact, and acceleration (see Cullinane and Edwards [11]). Noise from a vehicle's power unit comprising the engine, air intake, and exhaust becomes dominant at low speeds of 15–20 mph and at high acceleration rates of 2 m/sec^2 (see Knight, ed.) [37]). Annoyance, communication difficulties, and loss of sleep are some of the negative consequences on human life, although road vibrations caused by very heavy vehicles might also damage the neighboring buildings over time.

- *Accidents.* Road traffic accidents are responsible for personal injuries and death.

McKinnon [40] identifies nine factors as being critical for reducing the externalities of logistics activities. These are modal split, average handling factor, average length of haul, vehicle utilization (average payload and empty running), energy efficiency, emissions, other externalities which cannot be measured through energy use (e.g., noise and accidents), and monetary valuation of externalities. Although it is, in general, difficult to calculate the exact external costs of transportation, estimations exist. For intercity truck freight transportation, Forkenbrock [26] proposes four general types of costs: accidents, emissions, noise, and those associated with the provision, operation, and maintenance of public facilities. Forkenbrock [27] presents a similar analysis for freight trains and compares this to external costs of trucking. The general conclusion is that the external costs of trucking are over three times that of freight trains.

Of the different types of externalities mentioned above, a review of the emerging literature on "green" vehicle routing shows an increasing trend in looking at emissions and fuel consumption and their minimization in operational route planning (see Eglese and Black [18]). This is not surprising given the detrimental consequences of GHG emissions, a by-product of fuel usage, not to mention the implications of fuel on the economy. We now turn our attention to this body of research and first show how fuel consumption and emissions can be estimated, and then discuss how they can be accounted for within the traditional approaches for the VRP.

15.2 ▪ Fuel Consumption and Emission Models for Road Transportation

The amount of CO_2 emitted by a vehicle is directly proportional to fuel consumption. In the literature, two ways to estimate fuel consumption for vehicles have been suggested: *on-road measurements*, which are based on real-time collection of emissions data on a running vehicle, and *analytical fuel consumption* (or *emission*) models, which estimate fuel consumption based on a variety of vehicle, environment, and traffic-related parameters, such as vehicle speed, load, and acceleration. In this section, we briefly review some of the analytical models described in the literature. The reader is referred to Ardekani, Hauer, and Jamei [1], Esteves-Booth et al. [21], and Boulter, McCrae, and Barlow [6] for general reviews of vehicle emission models.

Analytical models for fuel consumption can be broadly classified into three classes, namely (i) emission factor models, (ii) average speed models, and (iii) modal (including instantaneous) models (see Esteves-Booth et al. [21]). Emission factor models are the simplest in form and are used at a macroscale level (regional or national emission estimations), particularly when data related to a vehicle's journey are limited. These models use an emission factor often expressed per unit of distance. Average speed models are speed-related functions to estimate emissions at a road network scale and do not include detailed enough parameters for an analysis at a microscale level. Finally, modal models operate at a higher level of complexity, specific enough for use at a microscale level and use detailed inputs, such as acceleration and road gradient, which are drawn from a running vehicle engine even on a second-by-second basis. As emission factor models are rather simplistic with major disadvantages such as their inability to represent driving cycles with good accuracy, we will restrict our review below only to the last two classes of models but refer the interested readers to Esteves-Booth et al. [21] for a detailed exposition of emission factor models.

15.2.1 ▪ Average Speed Models

In this section, we will present two models for estimating emissions which are primarily based on speed and are obtained using regression techniques. The first of these is due to the MEET report published by the European Commission (see Hickman et al. [31]), where the authors present the following general expression to calculate the rate of emissions $E(v)$ (g/km) for an unloaded goods vehicle on a road with a zero gradient as a function of the average speed v of the vehicle (km/h):

$$(15.1) \qquad E(v) = \zeta_0 + \zeta_1 v + \zeta_2 v^2 + \zeta_3 v^3 + \zeta_4/v + \zeta_5/v^2 + \zeta_6/v^3,$$

where ζ_0–ζ_6 are predefined coefficients for different types of vehicles classified according to their weight. The MEET report describes similar functions for correction factors for additional aspects, such as gradient and load, to be applied to $E(v)$. As an example, the rate of emissions according to the MEET report for a vehicle of less than 3.5 tonnes weight is given by $E(v) = 0.0617v^2 - 7.8227v + 429.51$. Another average speed model, namely COPERT described by Ntziachristos and Samaras [42], is similar to the MEET report in that it also describes an emissions model based on regression with speed as the primary determinant. The COPERT model describes two models for estimating emissions, to be used for vehicles of different classes and speed ranges. In particular, if the vehicle speed $v \in [0, v^*]$ for a given v^*, then $E(v) = Kv^{-\delta}$; otherwise $E(v) = K + av + bv^2$. As an example, for a vehicle with weight between 3.5 and 7.5 tonnes, $E(v) = 1425.2v^{-0.7593}$ if $v \leq v^* = 47$ km/h, and $E(v) = 60.12 - 0.0430v + 0.0082v^2$ otherwise.

15.2.2 ▪ Instantaneous Models

The models in this category only deal with "hot" emissions, i.e., exhaust emissions of a running engine, and aim to estimate emission rates of an operating vehicle for short time intervals of its driving cycle, e.g., on a second-by-second basis. As these models require detailed and precise measurements for an operating vehicle, which are often difficult and costly to collect, these models have been claimed to be of restricted use to the research community (see Boulter, McCrae, and Barlow [6]). Below, we review some of these models.

An energy-related emissions estimation model, called the *instantaneous fuel consumption model*, or *instantaneous model* in short, is described by Bowyer, Biggs, and Akçelik [7]. The model uses vehicle characteristics such as mass, energy, efficiency parameters, drag force, and fuel consumption components associated with aerodynamic drag and rolling resistance and approximates the fuel consumption per second. The model assumes that changes in acceleration and deceleration levels occur within a one second time interval. The instantaneous model is

$$(15.2) \qquad f_t = \begin{cases} \alpha + \beta_1 R_t v + \beta_2 M a^2 v / 1000 & \text{for } R_t > 0, \\ \alpha & \text{for } R_t \leq 0, \end{cases}$$

where f_t is the fuel consumption per unit time (mL/s), α is the constant fuel consumption rate of an idle running engine (mL/s), and R_t is the total tractive force (kN = kilonewtons) required to move the vehicle and calculated as the sum of force induced by drag, inertia, and road grade. In this function, β_1 is the fuel consumption per unit of energy (mL/kJ) and β_2 is the fuel consumption per unit of energy-acceleration mL/(kJ m/s^2). The grade force R_t is further calculated as $R_t = b_1 + b_2 v^2 + M a / 1000 + g M \theta / 100000$, where b_1 is the rolling drag force (kN), b_2 is the rolling aerodynamic force kN/(m/s^2), a is instantaneous acceleration (m/s^2), M is the total vehicle weight (kg), v is the speed (m/s), θ is the percent grade, and $g = 9.91$ (m/s^2) is the acceleration due to gravity. The model operates at a microscale level and is better suited for short trip emission estimations. Although this version of the model presented above does not use macro-level (aggregated) data such as the number of stops, an aggregated version suitable for longer journeys, as well as a more detailed version for more accurate estimations also appear in Bowyer, Biggs, and Akçelik [7]. The three versions differ with respect to the number of parameters required for estimation.

A more *Comprehensive Modal Emissions Modeling* (CMEM) is described by Scora and Barth [51], and details for heavy-duty vehicles (vehicles with a maximum operating mass of 11,794kg or above) are given in Barth, Younglove, and Scora [3]. According to this model, the fuel rate (g/s) is estimated by the following expression:

$$(15.3) \qquad FR = \phi (\bar{k} \bar{N} \bar{V} + P / \eta) / x,$$

where ϕ is fuel-to-air mass ratio, $\bar{k}$ is the engine friction factor, $\bar{N}$ is the engine speed, $\bar{V}$ is the engine displacement, P is the second-by-second engine power output (kW), η is an efficiency parameter for diesel engines, and x is a constant. Engine speed $\bar{N}$ is approximated by the vehicle speed v (m/s) as $\bar{N} = S(R(L)/R(L_g))v$, where S is the engine-speed/vehicle-speed ratio in top gear L_g, and $R(L)$ is the gear ratio in gear $L = 1, \ldots, L_g$. The engine power output P, on the other hand, is calculated as $P = P_t / \eta_t + P_a$, where η_t is the vehicle drive train efficiency, and P_a is the engine power demand associated with running losses of the engine and the operation of vehicle accessories such as usage of air conditioning. The total tractive power requirement placed on the vehicle at the wheels is shown by P_t (kW), which is further calculated as follows:

$$(15.4) \qquad P_t = (M a + M g \sin \theta + 0.5 C_d \rho \bar{A} v^2 + M g C_r \cos \theta) v / 1000.$$

As seen from (15.4), P_t depends on a variety of parameters, including air density ρ (kg/m^3), frontal surface area of the vehicle $\bar{A}$ (m^2), coefficients of aerodynamic drag C_d, and rolling resistance C_r, in addition to those already defined above.

To give the reader an idea of the way in which some of these emission models behave, we present Figure 15.1. This figure shows the total amount of fuel consumption for a 6350kg vehicle traveling on a distance of 10 km in the vertical axis estimated by MEET and CMEM, with values of speed varying from 20–110 km/h, and assuming zero acceleration and road gradient.

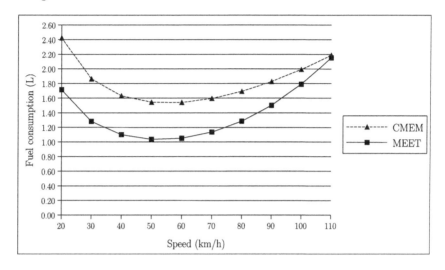

Figure 15.1. *Change of fuel consumption with speed.*

The two curves shown in Figure 15.1 are typical of the behavior of fuel consumption vs. speed and show an "optimal" speed which minimizes the total consumption. The shape of these curves changes with factors such as vehicle type, weight, acceleration, and road gradient. The reader is referred to Demir [13] and Demir, Bektaş, and Laporte [14] for a numerical comparison of a number of such models.

15.3 ▪ Minimizing Emissions in Vehicle Routing

A first line of research in accounting for emissions or fuel consumption in the VRP initially started out by looking primarily at vehicle weight as the sole determinant of emissions and the optimization thereof. Later work focused on speed optimization, mostly in combination with vehicle weight, although a number of papers have taken into account other factors as well. Some work also looked at the effect of time dependency. This section will review the work along these lines and will use the classification presented in Table 15.1 as the basis of the structure of the review.

The models presented in the remainder of the chapter will use the following notation. We are given a complete undirected graph $G = (V, A)$, where $V = \{0, \ldots, n\}$ is the set of nodes, 0 is a depot, and $A = \{(i, j) : i, j \in V \text{ and } i \neq j\}$ is the set of arcs. The distance from i to j is denoted by d_{ij}. A fixed-size fleet of vehicles denoted by the set K is available, and each vehicle has capacity Q. The set $N = V \setminus \{0\}$ is a customer set, and each customer $i \in N$ has a non-negative demand q_i as well as a time interval $[a_i, b_i]$ in which service of s_i time units long must commence.

15.3.1 ▪ Time-Independent VRP

Time independency in the context of the VRP implies an assumption that the problem data, in particular travel times between pairs of nodes, does not change with time. In this

Table 15.1. *Classification of "green" vehicle routing*

	Time-independent	*Time-dependent*
Speed fixed	Kara, Kara, and Yetiş [36]	Kuo [38]
	Palmer [43]	Eglese, Maden, and Slater [19]
	Tavares et al. [54]	Maden, Eglese, and Black [39]
	Suzuki [53]	
	Ubeda, Arcelus, and Faulin [56]	Conrad and Figliozzi [9]
	Xiao et al. [59]	Figliozzi [24]
		Figliozzi [25]
Speed variable	Bektaş and Laporte [5]	Franceschetti et al. [28]
	Demir, Bektaş, and Laporte [15]	Jabali, Van Woensel, and de Kok [34]
		Qian [46]

section, we review studies on green vehicle routing where this assumption is maintained. In accordance with the classification presented in Table 15.1, we first review approaches assuming constant speeds in the next section and then look at those with variable speed in the subsequent section.

15.3.1.1 ▪ Constant Speeds

Kara, Kara, and Yetiş [36] introduce the *Energy-Minimizing Vehicle Routing Problem* (EMVRP) as an extension of the *Capacitated Vehicle Routing Problem* (CVRP), where a weighted load function (load multiplied by distance), rather than just the distance, is minimized. The authors present a model for this problem that is based on a flow formulation of the VRP with a load-based objective function derived from simple physics. Xiao et al. [59] present a similar but slightly extended version of the EMVRP by factoring the fuel consumption rate of a vehicle, defined primarily with respect to the weight of the vehicle, into a standard flow-based CVRP formulation. The authors also propose a solution algorithm based on simulated annealing for this problem and report computational results on benchmark instances. Neither Kara, Kara, and Yetiş [36] nor Xiao et al. [59] consider speed as a factor in the development of their models and do not assume any time-window constraints. Palmer [43] extends this line of study by presenting an integrated routing and emissions model for freight vehicles and investigates the role of speed in reducing CO_2 emissions. The model uses a known *Vehicle Routing Problem with Time Windows* (VRPTW) heuristic as a black-box solver to produce the routing plans within the model, where speeds, as well as acceleration and deceleration, are inputs to the model rather than optimized outputs. Testing the approach on a case study of grocery stores in the UK concerning home deliveries, Palmer [43] finds that an average saving of 4.8% in

CO_2 emissions can be achieved over routes that only minimize time, but at the expense of an average increase in time by 3.8%. A smaller saving of 1.2% in CO_2 emissions can also be obtained on average compared to distance-minimizing routes, but at the expense of a 2.4% increase in distance. The work by Palmer [43] is the first to look at the effect of speeds in vehicle routing planning, although his approach does not optimize speeds but rather uses them as fixed inputs for calculating a matrix of "least cost" routes where cost might correspond to emissions. His approach does not account for vehicle loads either.

Suzuki [53] presents three formulations for a single-truck routing problem as variants of an assignment-based formulation of the TSP with time windows. The first of these formulations minimizes the total distance traveled. The other two both aim to minimize fuel consumption, but differ in the way in which this quantity is estimated. In particular, the second formulation calculates fuel consumption as a function of a vehicle's speed, whereas in the third formulation it is measured by payload and waiting time at customers. In this approach, the vehicle's fuel consumption rate (mpg) has been modeled through the regression function in the form of (15.1) with $\zeta_2, \ldots, \zeta_6 = 0$ and ζ_0, ζ_1 estimated as 2.82 and 0.07, respectively, based on data provided by US Department of Energy in 2009 on heavy-duty trucks. The vehicle speed v is provided as an exogenous parameter. The second formulation uses a similar function to estimate fuel consumption, but has payload instead of speed. Simulation results provided by Suzuki [53] indicate that the third formulation yields the highest savings in fuel consumption, suggesting that delivering heavy items earlier on in a tour is worthwhile in reducing fuel requirements.

Pradenas, Oportus, and Parada [44] extend the modeling approach put forward by Bektaş and Laporte [5] to the VRP with backhauls, in which customers are either of type linehaul or backhaul, and, in each route, the latter type should be served only after the former have been visited. A mixed integer programming formulation of the problem is presented, although the problem itself has been solved using a scatter search algorithm where speeds are treated as constant. Computational results obtained on standard benchmark instances suggest that savings of around 2% in GHG emissions can be obtained at the expense of an increase of 2–8% in operational costs. An interesting feature of this study is to incorporate cooperation among transport companies as a means to reduce emissions and use Shapley's value to calculate the value of a coalition.

Case studies using time-independent approaches for fuel and emission minimization in the context of vehicle routing have been presented by Tavares et al. [54] and Ubeda, Arcelus, and Faulin [56]. The former study is based on optimization of municipal solid waste in Cape Verde with an aim to minimize fuel consumption, the estimation of which is done through the use of the emissions function described in Ntziachristos and Samaras [42]. For short-distance waste collection, the authors report fuel savings of 9%, which implies only a small increase of 0.2% in the distance traveled. As for long-distance waste collection, savings of 52% can be achieved in fuel but at an increase of 34% in the distance traveled. The latter study looks at the use of emission factors in planning routes of a fleet of trucks for food delivery, using models for the Capacitated VRP or VRP with backhauls, and reports savings of around 25.5% in CO_2 emissions and around 30% in the distance covered using the proposed approach, but this is primarily due to cutting down the number of routes by about 43%. A similar problem concerning waste collection is described by Ramos, Gomes, and Barbosa-Póvoa [47] and is modeled as a multi-depot, multiple product VRP and solved through a multi-stage method. In a scenario where only vehicle routes are optimized, the authors report savings both in CO_2 emissions and distance traveled, reduced by 20% and 23%, respectively, over the currently employed solution.

15.3.1.2 · Variable Speeds

Assuming each road segment is a server to which vehicles arrive at a certain rate, Van Woensel, Creten, and Vandaele [57] show how queueing models could be used to describe traffic flows and to calculate emissions using the MEET emissions function. Their results show that constant speed, an assumption commonly made in the VRP literature, can be misleading and leads to an underestimation of emissions, particularly under congestion when speeds are lower. These results also suggest that speed plays a fundamental role in reducing emissions and hence should be optimized, along with load, within the route planning.

This is precisely the approach taken in Bektaş and Laporte [5], who present the *Pollution-Routing Problem* (PRP) as an extension of the classical VRPTW. The PRP consists of routing a number of vehicles to serve a set of customers within preset time windows, and determining their speed on each route segment, so as to minimize a function comprising fuel, emission, and driver costs. The emission function used within the PRP is the CMEM described by Barth, Younglove, and Scora [3], and differs from previous work in that it allows one to optimize *both* load and speed as well as to account for the effect of, among other parameters, acceleration and road gradient. The modeling approach proposed in [5] assumes a vehicle of empty weight w and carrying a load f, traveling at a constant speed v on a given arc of length d (see Figure 15.2) and proposes to estimate emissions using the following formula:

$$(15.5) \qquad P \approx \quad P_t d/v$$
$$(15.6) \qquad \approx \quad (a + g\sin\theta + gC_r\cos\theta)(w+f)d$$
$$(15.7) \qquad + (0.5C_dA\rho)v^2 d.$$

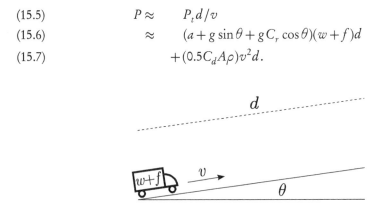

Figure 15.2. *Estimating emissions for a vehicle traveling at constant speed.*

The expression (15.5) is divided into two terms: (15.6), which shows the part of emissions induced primarily by total vehicle weight $w+f$, and (15.7), which shows the other part fundamentally induced by speed v. This divide will later be reflected in the integer linear programming formulation.

An integer programming formulation for the PRP works with a discretized speed function defined by a set R of non-decreasing speed levels $\bar{v}^r$ ($r = 1, \ldots, |R|$) (Bektaş and Laporte [5] and Demir, Bektaş, and Laporte [15]). Binary variables x_{ij} are equal to 1 if and only if arc (i,j) appears in solution. Continuous variables f_{ij} represent the total amount of flow on each arc $(i,j) \in A$. Continuous variables T_j represent the time at which service starts at node $j \in N$. Moreover, σ_j represents the total time spent on a route that has a node $j \in N$ as last visited before returning to the depot. Finally, binary variables z_{ij}^r indicate whether or not arc $(i,j) \in A$ is traversed at a speed level r. Bektaş and Laporte [5] provide illustrative examples as to the difference load and speed make in reducing the emissions, with respect to time-window restrictions and customer demand distribution, propose a non-linear mixed integer mathematical model for the problem, and show how it could be linearized. An integer linear programming formulation of the PRP is shown below:

$$(15.8) \qquad \text{minimize} \qquad \sum_{(i,j)\in A} c_f \bar{k} \bar{N} \bar{V} \lambda d_{ij} \sum_{r=1}^{|R|} z_{ij}^r / \bar{v}^r$$

$$(15.9) \qquad\qquad + \qquad \sum_{(i,j)\in A} c_f w \gamma \lambda \alpha_{ij} d_{ij} x_{ij}$$

$$(15.10) \qquad\qquad + \qquad \sum_{(i,j)\in A} c_f \gamma \lambda \alpha_{ij} d_{ij} f_{ij}$$

$$(15.11) \qquad\qquad + \qquad \sum_{(i,j)\in A} c_f \beta \gamma \lambda d_{ij} \sum_{r=1}^{|R|} z_{ij}^r (\bar{v}^r)^2$$

$$(15.12) \qquad\qquad + \qquad \sum_{j\in N} c_w \sigma_j$$

$$(15.13) \qquad \text{s.t.} \quad \sum_{j\in V} x_{0j} = |K|,$$

$$(15.14) \qquad\qquad \sum_{j\in V} x_{ij} = 1 \qquad\qquad \forall\, i \in N,$$

$$(15.15) \qquad\qquad \sum_{i\in V} x_{ij} = 1 \qquad\qquad \forall\, j \in N,$$

$$(15.16) \qquad\qquad \sum_{j\in V} f_{ji} - \sum_{j\in V} f_{ij} = q_i \qquad\qquad \forall\, i \in N,$$

$$(15.17) \qquad q_j x_{ij} \leq f_{ij} \leq (Q - q_i) x_{ij} \qquad \forall\, (i,j) \in A,$$

$$(15.18) \quad T_i - T_j + s_i + \sum_{r\in R} d_{ij} z_{ij}^r / \bar{v}^r \leq K_{ij}(1 - x_{ij}) \qquad \forall\, i \in V, j \in N, i \neq j,$$

$$(15.19) \qquad\qquad a_i \leq T_i \leq b_i \qquad\qquad \forall\, i \in N,$$

$$(15.20) \quad T_j + s_j - \sigma_j + \sum_{r\in R} d_{j0} z_{j0}^r / \bar{v}^r \leq L(1 - x_{j0}) \qquad \forall\, j \in N,$$

$$(15.21) \qquad\qquad \sum_{r=1}^{|R|} z_{ij}^r = x_{ij} \qquad\qquad \forall\, (i,j) \in A,$$

$$(15.22) \qquad\qquad x_{ij} \in \{0,1\} \qquad\qquad \forall\, (i,j) \in A,$$

$$(15.23) \qquad\qquad f_{ij} \geq 0 \qquad\qquad \forall\, (i,j) \in A,$$

$$(15.24) \qquad\qquad T_i \geq 0 \qquad\qquad \forall\, i \in N,$$

$$(15.25) \qquad\qquad z_{ij}^r \in \{0,1\} \qquad\qquad \forall\, (i,j) \in A,\ r = 1,\dots,|R|.$$

This mathematical formulation of the PRP presented here is an extension of the one presented in Bektaş and Laporte [5] to account for speeds 40 km/h or lower through the term (15.8) of the objective function. The objective function (15.8)–(15.11) is derived from (15.3), where $\lambda = \phi/\varkappa\psi$ and $\gamma = 1/1000\eta_t \eta$ are constants and ψ is the conversion factor of fuel from gram/second to liter/second. Furthermore, $\alpha = a + g\sin\theta + gC_r\cos\theta$ is a vehicle-arc-specific constant and $\beta = 0.5C_d\rho\bar{A}$ is a vehicle-specific constant. The parameters c_f and c_w are used to denote the per liter cost of fuel and hourly driver wage, respectively. The terms (15.9) and (15.10) calculate the cost incurred by the vehicle curb weight and payload and correspond to expression (15.6). The term (15.9) captures the effect of the vehicle speed on total cost, as indicated earlier by the expression (15.7). Finally, the term (15.12) measures the total driver costs. Constraints (15.13) state that each vehicle must leave the depot. Constraints (15.14) and (15.15) are the degree constraints which ensure that each customer is visited exactly once. Constraints (15.16) and (15.17) define the arc flows. Constraints (15.18)–(15.20), where $K_{ij} = \max\{0, b_i + \sigma_i + d_{ij}/l_{ij} - a_j\}$, and L is a large number, enforce the time-window restrictions. Constraints (15.21) ensure that only one speed level is selected for each arc and $z_{ij}^r = 1$ if $x_{ij} = 1$. This formulation can only solve small-sized PRP instances. To tackle larger instances, an adaptive large neighborhood search algorithm is described by Demir, Bektaş, and Laporte [15], where the authors present computational results for PRP instances of up to 200 nodes.

15.3.2 ▪ Time-Dependent VRP and Congestion

In contrast to the time-independent VRP, allowing for problem data to change with time gives rise to the time-dependent VRP. The changes in the time to travel along a road section can be attributed to foreseen events, such as peak-hour congestion, or unexpected changes in the traffic conditions such as those caused by accidents or changes in weather conditions. Time dependency is particularly suited to model traffic congestion, a phenomenon which occurs when the capacity of a particular transportation link is insufficient to accommodate an incoming flow at a particular point in time. Congestion has a number of adverse consequences: it increases average times of trips and increases variability in trip time which results in decreased transport reliability. Heavy congestion results in low speeds with fluctuations, e.g., 16 km/h ($\approx$ 10 mph), often accompanied by frequent acceleration and deceleration, and greatly contributes to CO_2 emissions (Barth and Boriboonsomsin [2]). According to the *International Road Transport Union* (IRU), around 100 billion liters of wasted fuel or 250 billion tonnes of CO_2 were attributed to traffic congestion in the United States in 2004 (International Road Transport Union [33]). In this section, we review the existing work on green vehicle routing where time dependency is explicitly taken into account to allow for introducing congestion into the problem setting. The next two sections look at studies with constant and variable speeds, respectively.

15.3.2.1 ▪ Constant Speeds

Kuo [38] considers a time-dependent VRP with an aim to minimize fuel consumption. The time-dependent VRP breaks away from the assumption that travel speeds or times are constant and models travel time as a function of the time of departure from a given node. The approach described by Kuo [38] assumes that time is divided into a number of intervals and v_{ij}^k represents the (known) speed between nodes i and j in each interval k. Using fuel consumption rates (e.g., miles per gallon), the total fuel consumption on a given route can be calculated. The approach uses simulated annealing to solve the underlying VRP initially using shortest distances, for which total fuel consumption is calculated. Computational results presented on randomly generated instances of 100 customers indicate that routing plans that are fuel-optimized increase both total transportation time and the total distance traveled.

Eglese, Maden, and Slater [19] show how traffic speed data collected at different times on individual road sections of a network can be used to create a Road Timetable that provides the quickest times and paths between nodes starting at different times of the day. This information takes into account the effect of regular traffic congestion that has been observed in the past. It is assumed that the vehicles being scheduled will travel at a preferred speed unless congestion on a road section makes that impossible. In that case the vehicles travel according to the average speed of the traffic on that road section. Maden, Eglese, and Black [39] present a tabu-search algorithm, called LANTIME, which uses the Road Timetable information and aims to minimize the total travel time. The approach allows more reliable routes and schedules to be planned. In addition, by minimizing the total travel time, LANTIME tends to produce routes where congestion is avoided and so provides environmental benefits. Results from a case study based on the distribution plans for an electrical goods wholesaler show that CO_2 emissions, total distance, and time can be reduced by around 7%, 7%, and 6%, respectively. This is due to avoiding routes where congestion is high, speeds are low, and emissions are high.

Figliozzi [24] presents an *Emission Minimizing VRP* (EVRP), a variant of the *Time-Dependent VRP* (TDVRP) with time windows, which takes into account congestion so as to minimize a speed-dependent CO_2 emissions function described in Hickman et al. [31]. The EVRP is modeled on a partition $T^1, T^2, \ldots, T^M$ of the total working time where in each interval a constant travel speed s^m is assumed. A set of speeds $\{v_{ij}^m(\Delta_i), v_{ij}^{m+1}(\Delta_i), \ldots, v_{ij}^{m+p}(\Delta_i)\}$ for each arc (i, j) of the network is defined as a function of the departure time Δ_i from node i, where $v_{ij}^m(\Delta_i)$ is the speed at the time of departure and $v_{ij}^{m+p}(\Delta_i)$ is the speed at the time of arrival and $p+1$ is the number of time intervals required to traverse arc (i, j). The amount of emissions on a particular link (i, j) of length d_{ij} is then calculated based on the departure time, using a function suggested in Hickman et al. [31], as follows:

$$(15.26) \quad F_{ij}(\Delta_i) = f_e \sum_{l=0}^{p} \left(\xi_0 + \xi_1 v_{ij}^{m+l}(\Delta_i) + \xi_3 (v_{ij}^{m+l}(\Delta_i))^3 + \xi_5 / (v_{ij}^{m+l}(\Delta_i))^2 \right) d_{ij}^l,$$

where d_{ij}^l is the distance traveled in each period $l = m, \ldots, m+p$ and f_c is the cost of emissions. The travel time on arc (i, j) is also modeled as a function of the departure time Δ_i from node i and is shown as $\tau_{ij}(\Delta_i)$. An integer programming formulation for the EVRP given by Figliozzi [24] is modeled on an extended network with $V' = V \cup \{n+1\}$, where node $n+1$ is a copy of the depot, $A' = \{(i, j) : i, j \in V'\}$, and $N' = V' \setminus \{0, n+1\}$. The model uses a binary variable x_{ijk} to denote whether vehicle k travels on arc (i, j) and T_{ik} to indicate the start of service at customer i served by vehicle k. The model also allows for pre-service idle waits in that service might start later than time of arrival, even if the arrival time is within the time window $[a_i, b_i]$ of customer i. If c_k denotes the unit cost for vehicle $k \in K$ and c_t is the cost of travel per unit time, a model of the EVRP as described by Figliozzi [24] is given below:

$$(15.27) \qquad \text{minimize} \quad \sum_{k \in K} \sum_{j \in N'} c_k x_{0jk}$$

$$+ \sum_{k \in K} \sum_{(i,j) \in A} d_{ij} x_{ijk}$$

$$+ \sum_{k \in K} \sum_{j \in N'} c_k (T_{n+1,k} - T_{0k}) x_{ijk}$$

$$+ \sum_{k \in K} \sum_{(i,j) \in A'} c_e F_{ij}(T_{ik} + s_i)$$

$$\text{(15.28)} \qquad \text{s.t.} \quad \sum_{j \in V'} x_{0jk} = 1 \qquad \forall\, k \in K,$$

$$\text{(15.29)} \qquad \sum_{j \in V'} x_{j,n+1,k} = 1 \qquad \forall\, k \in K,$$

$$\text{(15.30)} \qquad \sum_{k \in K} \sum_{j \in V'} x_{ijk} = 1 \qquad \forall\, i \in N',$$

$$\text{(15.31)} \qquad \sum_{i \in V'} x_{ilk} - \sum_{i \in V'} x_{lik} = 0 \qquad \forall\, l \in N',\, k \in K,$$

$$\text{(15.32)} \qquad \sum_{i \in N'} q_i \sum_{j \in V'} x_{ijk} \le Q \qquad \forall\, k \in K,$$

$$\text{(15.33)} \qquad a_i \sum_{j \in V'} x_{ijk} \le T_{i,k} \qquad \forall\, i \in V',\, k \in K,$$

$$\text{(15.34)} \qquad b_i \sum_{j \in V'} x_{ijk} \ge T_{i,k} \qquad \forall\, i \in V',\, k \in K,$$

$$\text{(15.35)} \qquad x_{ijk}(T_{ik} + s_i + \tau_{ij}(T_{ik} + s_i)) \le T_{jk} \qquad \forall\, (i,j) \in A',\, k \in K,$$

$$\text{(15.36)} \qquad x_{i0k} = 0 \qquad \forall\, i \in V',\, \forall\, k \in K,$$

$$\text{(15.37)} \qquad x_{n+1,ik} = 0 \qquad \forall\, i \in V',\, \forall\, k \in K,$$

$$\text{(15.38)} \qquad x_{ijk} \in \{0,1\} \qquad \forall\, (i,j) \in A',\, k \in K$$

$$\text{(15.39)} \qquad T_{ik} \ge 0 \qquad \forall\, i \in V',\, k \in K.$$

Figliozzi [24] presents a version of the EVRP with three objective functions, one to minimize the number of vehicles, one to minimize the emissions, and the last to minimize the total distance and duration of the resulting routes. The author describes a solution algorithm to solve this version of the EVRP which initially solves a TDVRP to minimize the number of vehicles, denoted by $|K|$, and then minimizes emissions by modifying both departure times Δ_i on a given route and assignments x_{ijk} subject to an additional constraint limiting the fleet size to $|K|$. Using the algorithm, computational tests are conducted on the Solomon benchmark problems of 100 nodes under three settings of congestion, namely uncongested, somewhat congested, and congested. These results indicate that small increases in fleet size imply reductions in emissions, and that uncongested travel speeds tend to reduce emissions on average but some instances exhibit the opposite trend. These instances are when the uncongested travel speed is higher than the speed which minimizes the emissions per unit distance. The author concludes that changes in emissions are related to certain characteristics of the problem, e.g., randomly distributed customers vs. clustered, tight vs. loose time windows, and average number of customers.

A continuous approximation model for solving the EVRP is described by Saberi and Verbas [49]. The authors present results of computational experimentation on benchmark instances, to look at the spatial and temporal effects of congestion due to variations in speed. The results suggest that the optimal number of dispatches during the peak period is smaller than that of off-peak. An application of the EVRP and the algorithm just described on a case study in Portland, Oregon is presented in Figliozzi [25], where scenarios with and without congestion are considered. Of the implications of the results presented in Figliozzi [25], we mention the importance of optimizing travel times and the location of the depot on reducing emission levels.

15.3.2.2 ▪ Variable Speeds

Jabali, Van Woensel, and de Kok [34] take a similar approach to Figliozzi [25] by using the same emissions function in a formulation of the time-dependent VRP (but without time windows) with speed as an additional decision variable. Travel times are modeled by partitioning the planning horizon into two parts, where one partition corresponds to a peak period in which there is congestion and assumes the vehicle speed as fixed, whereas the other partition assumes free-flow speeds which could be optimized. The primary difference between the work of Jabali, Van Woensel, and de Kok [34] and Figliozzi [24] is that the latter explicitly models a "transient" zone in which a vehicle changes from congestion to free-flow speeds. In other words, for a vehicle traveling on a given link, part of the link would be traveled at lower levels of speed due to congestion and the rest would be free-flow for when the congestion dissipates. If v_c denotes the congestion speed and v_f denotes free-flow speed on a road segment of length d, and the vehicle is subject to congestion speed for Ψ units of time from when it starts to traverse the link, then the total travel time is described by

$$(15.40) \quad T(t) = \begin{cases} d/v_c & \text{if } t \leq \Psi - d/v_c, \\ t(v_c - v_f)/v_f + \Psi(v_f - v_c)/v_f + d/v_f & \text{if } \Psi - d/v_c < t < \Psi, \\ d/v_f & \text{if } t \geq \Psi \end{cases}$$

for a given start time t. To estimate emissions, (15.1) with parameters $(\zeta_0, \ldots, \zeta_6) = (1576, -17.6, 0, 0.00117, 0, 36067, 0)$ is used, which prescribes an optimal free-flow speed of 71 km/h, minimizing emissions. Jabali, Van Woensel, and de Kok [34] describe a tabu search heuristic for the solution of the problem, which is tested on standard benchmark instances. Their results suggest that achieving reducing CO_2 emissions of about 11.4% implies a 17.1% increase in travel times, on average, using a congestion speed of 50 km/h and an upper speed limit of 90 km/h.

The PhD dissertation by Qian [46] studies a VRP with time-dependent travel data with an aim to minimize fuel consumption that is based on speed. The author first develops a dynamic programming algorithm and a heuristic solution procedure to optimize speeds on each segment of a route. The sequence of customers to be visited is fixed, but the route between the customers and the speeds on the road segments are determined by the algorithm. Qian [46] uses a real road network, where each road junction is modeled as a node. For instance, one of the data sets considered in this work is a Bristol data set of a single depot and 14 customers, which, when transformed into a full network, results in 9,256 nodes (customer nodes and road junctions) and 6,643 arcs. Results on this network indicate that average savings of 6–7% in fuel emissions can be achieved, though at the cost of increasing the trip time by around 9–10%. The speed optimization procedure is then incorporated into a column-generation framework to optimize the routes as well as speeds. Tests of this method on a London data set with 208,488 nodes and 477,411 arcs and time-varying speeds show the importance of looking at route optimization in addition to speed optimization. Furthermore, allowing drivers to wait at customer nodes is shown to reduce fuel emissions by avoiding congestion. Qian [46] suggests that if idle waiting at customers is to be used as an option, driver costs should be considered.

Following a similar line of research, Franceschetti et al. [28] look into a similar problem where time dependency is modeled as in Jabali, Van Woensel, and de Kok [34]. In this work, the costs of fuel, emissions, and drivers are explicitly taken into account by extending the modeling approach of Bektaş and Laporte [5]. The authors describe a mathematical formulation of the problem and a complete characterization of optimal solutions on a single-link version, which offers insights as to when it is profitable to wait idle at customer nodes in order to minimize the total cost. The authors also present a procedure on fixed routes, which not only optimizes the speed on each segment of the route, but also the departure times from each node which might be *later* than the service completion time, in order to avoid congestion.

15.4 ▪ Speed Optimization on Fixed Routes

Most approaches described above for modeling or solving "green" VRPs are either variations or extensions of the existing body of work on the VRP, e.g., incorporating emissions functions within VRP models or algorithms. However, there is one other interesting problem type, namely schedule optimization for a fixed route with (linear or convex) inconvenience costs (see, e.g., Dumas, Soumis, and Desrosiers [17] and Fagerholt [22]). This problem has now been applied to the Speed Optimization Problem (SOP) on a given route. The SOP is defined on a given route of customers indexed by $k = 1, 2, \ldots, n$, who have (hard) time windows $[a_k, b_k]$ in which the service needs to start, and consists of determining the speed on each leg of the route so as to minimize the total fuel consumption. Fagerholt, Laporte, and Norstad [23] present, to our knowledge, the first description of SOP in the context of shipping and describe mathematical models for the problem along with a solution algorithm. One of these models uses speed as a primary decision variable, defined as $v_{k,k+1}$ on each arc $(k, k+1)$ of length $d_{k,k+1}$ for $k = 1, 2, \ldots, n-1$, using which one can calculate the travel time on an arc $(k, k+1)$ as $t_{k,k+1} = d_{k,k+1}/v_{k,k+1}$. The model is presented below:

$$(15.41) \qquad \text{minimize} \qquad \sum_{k=1,\ldots,n-1} d_{k,k+1} f(v_{k,k+1})$$

$$(15.42) \qquad \text{s.t.} \quad T_{k+1} - T_k - d_{k,k+1}/v_{k,k+1} \geq 0 \qquad \forall\, k = 1, \ldots, n-1,$$

$$(15.43) \qquad\qquad a_k \leq T_k \leq b_k \qquad \forall\, k = 1, \ldots, n,$$

$$(15.44) \qquad\qquad v_{\min} \leq v_{k,k+1} \leq v_{\max} \qquad \forall\, k = 1, \ldots, n-1.$$

The objective function (15.41) minimizes the total fuel consumption on the given route, where $f(v)$ is the fuel consumption as a function of speed v. Constraints (15.42) are used to restrict the start time of service at a given node only after the ship has arrived to that node. Time-window restrictions are modeled through constraints (15.43), whereas the last set of constraints (15.44) enforce minimum and maximum speed limits on each arc as $v_{\min}$ and $v_{\max}$, respectively. As $f(v)$ is, in general, a quadratic convex function for ships, this model is non-linear in the objective function and the constraints. Fagerholt, Laporte, and Norstad [23] show that, by using traveling time as a primary decision variable, the nonlinearity in the constraints can be avoided. Furthermore, by discretizing the arrival times at each node and replicating each node as many times as the number of discretizations, the SOP can be modeled and solved as a shortest path problem on a directed acyclic graph. Norstad, Fagerholt, and Laporte [41] later on present a recursive smoothing algorithm for the SOP that is based on the convexity of the fuel consumption function which runs very fast, and later on was shown to be optimal by Hvattum et al. [32].

This algorithm was later on adapted to the PRP by Demir, Bektaş, and Laporte [15] and reported to result in an average improvement of 3%, with respect to fuel consumption, on fixed vehicle routes. A similar algorithm appears in Figliozzi [25] which optimizes departure times at each node on a fixed route by taking into account time-window constraints and emissions.

15.5 ▪ Multicriteria Analysis

Minimizing emissions in VRPs might sometimes conflict with the traditional objective of minimizing total travel distance or time. One way of dealing with multiobjective VRPs in which some objectives are environmental (e.g., CO_2 emissions) and others are operational (e.g., financial) is to use a weighted combination of the multiple objectives to convert the problem into a single objective problem. However, the solutions produced through such an approach heavily depend on the weights assigned to the objectives. An alternative way to overcome this difficulty, as well as to capture the tradeoffs between several objectives, is to use multiobjective programming techniques. One example of the use of such an approach is described by Jemai, Zekri, and Mellouli [35], who describe an evolutionary algorithm to solve a bi-objective VRP where one objective minimizes the total distance traveled whereas the other minimizes CO_2 emissions. In a similar vein, Demir, Bektaş, and Laporte [16] study the bi-objective PRP in which the two objective functions pertaining to minimization of fuel consumption and driving time are conflicting and are thus considered separately. Four a posteriori methods, namely the weighting method, the weighting method with normalization, the epsilon-constraint method, and a new *Hybrid Method* (HM), are evaluated and compared with one another.

15.6 ▪ Routing in Other Modes of Transport

This chapter has so far focused on environmental hazards of routing vehicles for road transportation. Similar studies exist in other domains of transportation, and shipping is one which has been given particular attention in the literature given the size of ships and the significant amount of freight transported on this mode of transport. Fuel consumption for vessels can be described by a function which, among other technical factors, is linear in the distance of travel and cubic in the speed (see Corbett, Wang, and Winebrake [10]). For container ships running on diesel engines, reducing speeds from design speed to slow steaming is shown to have reduced emissions by around 11% between 2008 and 2010 (see Cariou [8]). Furthermore, the approach described by Fagerholt, Laporte, and Norstad [23] to optimize speeds for ships has been shown to yield savings in fuel consumption by 21% on average. However, reduction in speed may have undesirable consequences, including increased fleet size and inventory costs (see Psaraftis, Kontovas, and Kakalis [45]).

Few studies combine routing decisions with environmental considerations in rail transportation planning. The study by Bauer, Bektaş, and Crainic [4] looks at an intermodal rail transportation system operating over several countries where intermodality arises due to the need to transfer containers between rail services of different characteristics at borders. In this study, rail fuel consumption is modeled using the emissions function introduced by Ross [48] and used later by Barth, Younglove, and Scora [3] assuming constant speed. The authors present two variants of an integer programming formulation for the service network design problem, one to minimize service time and the other to minimize emissions. Using the model on case study data, the authors present results of computational experiments that capture the tradeoff between minimizing time and emissions.

Environmental concerns are gradually being introduced into other variants of the VRP. One particular study is due to Treitl, Nolz, and Jammernegg [55], who look at a multi-period inventory-routing problem arising in a case study in the petrochemical industry. The particular problem the authors consider has deterministic demands and no inventory costs, as the retailers and the suppliers belong to the same company. Instead they model emissions resulting from electricity usage for inventory transfer at customer nodes, as well as emissions due to the routing itself modeled in the spirit of Bektaş and

Laporte [5], and minimize emission costs along with the usual routing costs. The authors also introduce a carbon cap for the total amount of emissions. An integer linear programming formulation of the problem is tested on case study data of a single depot and 45 customer nodes, and results indicate that not only can the total routing distance (loaded and empty) be significantly reduced, but also the average load on the vehicles can be increased. The authors also mention that, given the current low costs of carbon in the EU ETS, the solution is not affected by consideration of CO_2 emissions. This conclusion parallels those of Bektaş and Laporte [5] and Jabali, Van Woensel, and de Kok [34].

15.7 ▪ Alternative Fuel-Powered Vehicles

For many years, vehicles have mainly been powered by petroleum-based fuels. Environmental concerns have encouraged the development and use of alternative fuels that have reduced environmental impact. Examples include electricity and ethanol. In many cases, the use of an alternative fuel will not affect the structure of the associated VRP, but in some cases an alternative fuel can have a significant effect.

A good example is given by Erdoğan and Miller-Hooks [20], who describe a VRP that applies to alternative fuel vehicles that have limited range and where there are limited locations for the alternative fuel stations where the vehicles may refuel. They define a "Green Vehicle Routing Problem" where the objective is to minimize the total distance covered by a vehicle fleet, with the additional constraint that no vehicle must exceed its maximum range without refuelling and vehicles may only refuel at alternative fuel stations located in specific places. The problem is formulated as a mixed integer program, and heuristics are developed for its solution. Tests are carried out based on locations for publicly available fuel stations for biodiesel in an area of the USA showing how the total distance for a fleet of vehicles may depend on the number of fuel stations and the range of the vehicles.

15.8 ▪ Conclusions and Future Research Directions

This chapter has presented a review of the state of the art in green vehicle routing, an emerging area of research within vehicle routing attracting growing interest from the scientific community. The review presented the available fuel consumption and emissions models in the literature, and ways in which these models could be integrated into the existing formulations or approaches for the VRP. In particular, we have presented a classification whereby problems are differentiated with respect to time dependency and whether speed is considered as a decision variable. The paper has also briefly touched on the "green" literature for other modes of transport, as well as relevant areas such as alternative fuel-powered vehicles.

The review has revealed several interesting characteristics of the relevant body of green vehicle routing literature, listed below, along with suggestions for further research:

- Of the various externalities of transportation including noise and accidents, the VRP-related literature mainly focuses on minimizing GHG emissions and fuel consumption. This is not surprising for several reasons. First, there is rich literature on analytical expressions for emissions, but this is not so much the case for other externalities. Second, externalities such as accidents are difficult to model analytically and are usually the subject of other areas of research, such as accident prevention and policy making.

- The emphasis of the existing research is predominantly on new problem definitions, such as the EVRP, rather than new solution techniques. In fact, most new problems of green vehicle routing can be solved by variations of the existing techniques available for the traditional VRP. However, one novel problem that has been identified in this context is the speed optimization problem on fixed routes, which has previously been studied primarily for theoretical interest, but has now found applications in a more relevant domain of research.

- An accurate estimation of fuel consumption (and emissions) requires significantly more parameters than what has previously been used in the VRP literature. Travel speed, elevation and inclination, and time-dependent data are but a few examples. This might sound too difficult a task, but the recent developments in GIS and traffic system management software (see, e.g., Hansen et al. [30]) might facilitate gathering such data and in sufficient detail to be used as input to some of the methods described in this chapter.

- Whereas some studies report that environmental objectives (e.g., minimizing emissions) are in conflict with those that aim to reduce operational costs, this is not always the case. As some of the results mentioned in this chapter suggest, there are cases where minimizing fuel consumption also leads to minimizing the total distance. In some cases, this is due to using rural roads as opposed to highways, which require slower speeds but often provide direct routes. In other cases, this might be due to avoiding congestion or using fewer vehicles. Whether the two objectives are in conflict or not depends on the particular application and the decisions involved, and our view is that this will become more evident as more studies are conducted in this area of research and will provide further evidence into the tradeoffs.

- Apart from road-based transportation, one other area of green logistics that has received particular attention is in maritime transportation. To the best of our knowledge, little work has been done on how vehicle routing aspects in rail or air transportation are affected by environmental considerations.

- Of the problems hitherto addressed, those assuming time independency seem to have been sufficiently well studied. The green vehicle routing literature is gradually moving towards the time-dependent setting, and this is an area that needs further investigation. Phenomena such as congestion are one of the primary causes of pollution, particularly for urban transportation. Although one type of congestion can be predicted (e.g., peak hours), allowing for vehicle routes to be planned in advance, other types of congestion arising due to unforeseen events warrant research into stochastic, dynamic, or real-time vehicle route planning to minimize the impact on the environment. This is an open research direction.

Acknowledgments

Thanks are due to Emrah Demir for his help with drawing Figure 15.1 and to an anonymous reviewer for the valuable comments and suggestions on an earlier version of this chapter.

Bibliography

[1] S. ARDEKANI, E. HAUER, AND B. JAMEI, *Traffic impact models*, in Traffic Flow Theory, US Federal Highway Administration, Washington D.C., 1996, pp. 1–7.

[2] M. BARTH AND K. BORIBOONSOMSIN, *Energy and emissions impacts of a freeway-based dynamic eco-driving system.*, Transportation Research D, 14 (2009), pp. 400–410.

[3] M. BARTH, T. YOUNGLOVE, AND G. SCORA, *Development of a heavy-duty diesel modal emissions and fuel consumption model*, Technical Report, UC Berkeley: California Partners for Advanced Transit and Highways (PATH), Berkeley, CA, 2005.

[4] J. BAUER, T. BEKTAŞ AND T. G. CRAINIC, *Minimizing greenhouse gas emissions in intermodal freight transport: An application to rail service design*, Journal of the Operational Research Society, 61 (2010), pp. 530–542.

[5] T. BEKTAŞ AND G. LAPORTE, *The pollution-routing problem*, Transportation Research B, 45 (2011), pp. 1232–1250.

[6] P. G. BOULTER, I. S. MCCRAE, AND T. J. BARLOW, *A review of instantaneous emission models for road vehicles*, Technical Report, Transport Research Laboratory, Berks, UK, 2007.

[7] D. P. BOWYER, D. C. BIGGS, AND R. AKÇELIK, *Guide to fuel consumption analysis for urban traffic management*, Technical Report 32, Australian Road Research Board Transport Research Ltd., Vermont South, Australia, 1985. Available at: http://www.sidrasolutions.com/Cms_Data/Contents/SIDRA/Folders/Resources/Articles/Articles/~contents/XCMJQVHWJAJ6ALYW/BowyerAkcelikBiggs_SR32_Fuel.pdf (accessed on August 5, 2013).

[8] P. CARIOU, *Is slow steaming a sustainable means of reducing CO_2 emissions from container shipping?*, Transportation Research D, 16 (2011), pp. 260–264.

[9] R. G. CONRAD AND M. A. FIGLIOZZI, *Algorithms to quantify impact of congestion on time-dependent real-world urban freight distribution networks*, Transportation Research Record: Journal of the Transportation Research Board, 2168 (2010), pp. 104–113.

[10] J. J. CORBETT, H. WANG, AND J. J. WINEBRAKE, *The effectiveness and costs of speed reductions on emissions from international shipping*, Transportation Research D, 14 (2009), pp. 593–598.

[11] S. CULLINANE AND J. EDWARDS, *Assessing the environmental impacts of freight transport*, in Green Logistics: Improving the Environmental Sustainability of Logistics, A. McKinnon, S. Cullinane, M. Browne, and A. Whiteing, eds., Kogan Page, London, 2010, pp. 31–48.

[12] R. DEKKER, J. BLOEMHOF, AND I. MALLIDIS, *Operations research for green logistics—an overview of aspects, issues, contributions and challenges*, European Journal of Operational Research, 219 (2012), pp. 671–679.

[13] E. DEMIR, *Models and Algorithms for the Pollution-Routing Problem and Its Variations*, PhD thesis, Southampton Management School, Southampton, UK, 2012.

[14] E. DEMIR, T. BEKTAŞ, AND G. LAPORTE, *A comparative analysis of several vehicle emission models for road freight transportation*, Transportation Research D, 16 (2011), pp. 347–357.

[15] ———, *An adaptive large neighborhood search heuristic for the pollution-routing problem*, European Journal of Operational Research, 223 (2012), pp. 346–359.

[16] ———, *The bi-objective pollution-routing problem*, European Journal of Operational Research, 232 (2014), pp. 464–478.

[17] Y. DUMAS, F. SOUMIS, AND J. DESROSIERS, *Optimising the schedule for a fixed vehicle path with convex inconvenience costs*, Transportation Science, 24 (1990), pp. 145–152.

[18] R. W. EGLESE AND D. BLACK, *Optimizing the routeing of vehicles*, in Green Logistics: Improving the Environmental Sustainability of Logistics (2nd edition), M. Browne, A. Whiteing, and A. McKinnon, eds., Kogan Page, London, 2012, pp. 223–235.

[19] R. W. EGLESE, W. MADEN, AND A. SLATER, *A road timetable^{TM} to aid vehicle routing and scheduling*, Computers & Operations Research, 33 (2006), pp. 3508–3519.

[20] S. ERDOĞAN AND E. MILLER-HOOKS, *A green vehicle routing problem*, Transportation Research Part E: Logistics and Transportation Review, 48 (2012), pp. 100–114.

[21] A. ESTEVES-BOOTH, T. MUNEER, J. KUBIE, AND H. KIRBY, *A review of vehicular emission models and driving cycles*, Proceedings of the Institution of Mechanical Engineers, Part C: Journal of Mechanical Engineering Science, 216 (2002), pp. 777–797.

[22] K. FAGERHOLT, *Ship scheduling with soft time windows: An optimisation based approach*, European Journal of Operational Research, 131 (2001), pp. 559–571.

[23] K. FAGERHOLT, G. LAPORTE, AND I. NORSTAD, *Reducing fuel emissions by optimizing speed on shipping routes*, Journal of the Operational Research Society, 61 (2010), pp. 523–529.

[24] M. A. FIGLIOZZI, *Vehicle routing problem for emissions minimization*, Transportation Research Record: Journal of the Transportation Research Board, 2197 (2010), pp. 1–7.

[25] ———, *The impacts of congestion on time-definitive urban freight distribution networks CO_2 emission levels: Results from a case study in Portland, Oregon*, Transportation Research Part C: Emerging Technologies, 19 (2011), pp. 766–778.

[26] D. J. FORKENBROCK, *External costs of intercity truck freight transportation*, Transportation Research A, 33 (1999), pp. 505–526.

[27] ———, *Comparison of external costs of rail and truck freight transportation*, Transportation Research A, 35 (2001), pp. 321–337.

[28] A. FRANCESCHETTI, D. HONHON, T. VAN WOENSEL, T. BEKTAŞ, AND G. LAPORTE, *The time-dependent pollution routing problem*, Transportation Research B, 56 (2013), pp. 265–293.

[29] G. GHIANI, C. MOURÃO, L. PINTO, AND D. VIGO, *Routing in waste collection applications*, in Arc Routing: Problems, Methods, and Applications, Á. Corberán and G. Laporte, eds., vol. 20 of MOS-SIAM Series on Optimization, SIAM, Philadelphia, 2014, ch. 15.

[30] S. HANSEN, A. BYRD, A. DELCAMBRE, A. RODRIGUEZ, S. MATTHEWS, AND R. L. BERTINI, *PORTAL: An On-Line Regional Transportation Data Archive with Transportation System Management Applications*, Technical Report, Department of Civil and Environmental Engineering, Nohad A. Toulan School of Urban Studies and Planning, Portland State University, Portland, OR, 2005.

[31] J. HICKMAN, D. HASSEL, R. JOUMARD, Z. SAMARAS, AND S. SORENSON, *MEET methodology for calculating transport emissions and energy consumption*, Technical Report, European Commission, DG VII, Luxembourg, 1999. Available at: http://www.transport-research.info/Upload/Documents/200310/meet.pdf (accessed on August 5, 2013).

[32] L. M. HVATTUM, I. NORSTAD, K. FAGERHOLT, AND G. LAPORTE, *Analysis of an exact algorithm for the vessel speed optimization problem*, Networks, 62 (2013), pp. 132–135.

[33] INTERNATIONAL ROAD TRANSPORT UNION, *Congestion is responsible for wasted fuel*, 2012. Available at: http://www.iru.org/en_policy_co2_response_wasted (accessed on August 5, 2013).

[34] O. JABALI, T. VAN WOENSEL, AND A. G. DE KOK, *Analysis of travel times and CO_2 emissions in time-dependent vehicle routing*, Production and Operations Management, 21 (2012), pp. 1060–1074.

[35] J. JEMAI, M. ZEKRI, AND K. MELLOULI, *An NSGA-II algorithm for the green vehicle routing problem*, in Evolutionary Computation in Combinatorial Optimization, J.-K. Hao and N. Middendorf, eds., vol. 7245 of Lecture Notes in Computer Science, Springer-Verlag, Berlin, 2012, pp. 37–48.

[36] I. KARA, B. Y. KARA, AND M. K. YETIŞ, *Energy minimizing vehicle routing problem*, in Combinatorial Optimization and Applications, Y. X. A. Dress and B. Zhu, eds., vol. 4616 of Lecture Notes in Computer Science, Springer, Berlin, Heidelberg, 2007, pp. 62–71.

[37] R. KNIGHT ed., *Mobility 2030: Meeting the challenges to sustainability*, Technical Report, World Business Council for Sustainable Development, Hertfordshire, England, 2004. Available at: http://www.wbcsd.org/web/publications/mobility/mobility-full.pdf (accessed on October 17, 2014).

[38] Y. KUO, *Using simulated annealing to minimize fuel consumption for the time-dependent vehicle routing problem.*, Computers & Industrial Engineering, 59 (2010), pp. 157–165.

[39] W. MADEN, R. W. EGLESE, AND D. BLACK, *Vehicle routing and scheduling with time-varying data: A case study*, Journal of the Operational Research Society, 61 (2010), pp. 515–522.

[40] A. McKinnon, *Environmental sustainability: A new priority for logistics managers*, in Green Logistics: Improving the Environmental Sustainability of Logistics, A. McKinnon, S. Cullinane, M. Browne, and A. Whiteing, eds., Kogan Page, London, 2010, pp. 3–30.

[41] I. Norstad, K. Fagerholt, and G. Laporte, *Tramp ship routing and scheduling with speed optimization*, Transportation Research C, 19 (2011), pp. 853–865.

[42] L. Ntziachristos and Z. Samaras, *COPERT III computer programme to calculate emissions from road transport: Methodology and emission factors (version 2.1)*, Technical Report, European Environment Agency, Copenhagen, Denmark, 2000.

[43] A. Palmer, *The Development of an Integrated Routing and Carbon Dioxide Emissions Model for Goods Vehicles*, PhD thesis, Cranfield University, School of Management, Bedford, UK, 2007.

[44] L. Pradenas, B. Oportus, and V. Parada, *Mitigation of greenhouse gas emissions in vehicle routing problems with backhauling*, Expert Systems with Applications, 40 (2013), pp. 2985–2991.

[45] H. N. Psaraftis, C. A. Kontovas, and N. M. P. Kakalis, *Speed reduction as an emissions reduction measure for fast ships*, in Proceedings of the 10th International Conference on Fast Sea Transportation FAST 2009, Athens, Greece, 2009. Available at: http://www.martrans.org/documents/2009/air/fast2009-psaraftiskontovas.pdf (accessed on August 5, 2013).

[46] J. Qian, *Fuel emission optimization in vehicle routing problems with time-varying speeds*, PhD thesis, Lancaster University, School of Management, Lancaster, UK, 2012.

[47] T. R. P. Ramos, M. I. Gomes, and A. P. Barbosa-Póvoa, *Minimizing CO$_2$ emissions in a recyclable waste collection system with multiple depots*, in Proceedings of the EUROMA/POMS Joint Conference, Amsterdam, 2012. Available at: http://docentes.fct.unl.pt/sites/default/files/mirg/files/euroma_2012_fullpaper_final.pdf (accessed on June 15, 2013).

[48] M. Ross, *Fuel efficiency and the physics of automobiles*, Contemporary Physics, 38 (1997), pp. 381–394.

[49] M. Saberi and I. O. Verbas, *Continuous approximation model for the vehicle routing problem for emissions minimisation at the strategic level*, Journal of Transportation Engineering, 138 (2012), pp. 1368–1376.

[50] A. Sbihi and R. W. Eglese, *Combinatorial optimization and green logistics*, 4OR, 5 (2007), pp. 99–116.

[51] G. Scora and M. Barth, *Comprehensive modal emission model (CMEM), version 3.01, user's guide*, Technical Report, Center for Environmental Research and Technology, University of California, Riverside, 2006. Available at: http://www.cert.ucr.edu/cmem/docs/CMEM_User_Guide_v3.01d.pdf (accessed on August 5, 2013).

[52] S. K. Srivastava, *Green supply-chain management: A state-of-the-art literature review*, International Journal of Management Reviews, 9 (2007), pp. 53–80.

[53] Y. SUZUKI, *A new truck-routing approach for reducing fuel consumption and pollutants emission*, Transportation Research D, 16 (2011), pp. 73–77.

[54] G. TAVARES, Z. ZSIGRAIOVA, V. SEMIAO, AND M. G. CARVALHO, *A case study of fuel savings through optimisation of MSW transportation routes*, Management of Environmental Quality: An International Journal, 19 (2008), pp. 444–454.

[55] S. TREITL, P. C. NOLZ, AND W. JAMMERNEGG, *Incorporating environmental aspects in an inventory routing problem. A case study from the petrochemical industry*, Flexible Services and Manufacturing Journal, 26 (2014), pp. 143–169.

[56] S. UBEDA, F. J. ARCELUS, AND J. FAULIN, *Green logistics at Eroski: A case study*, International Journal of Production Economics, 131 (2011), pp. 44–51.

[57] T. VAN WOENSEL, R. CRETEN, AND N. VANDAELE, *Managing the environmental externalities of traffic logistics: The issue of emissions*, Production and Operations Management, 10 (2001), pp. 207–223.

[58] A. S. VICENTE, *Laying the foundations for greener transport. TERM 2011: Transport indicators tracking progress towards environmental targets in Europe*, EEA Report 7/2011, European Environment Agency, Copenhagen, Denmark, 2011.

[59] Y. XIAO, Q. ZHAO, I. KAKU, AND Y. XU, *Development of a fuel consumption optimization model for the capacitated vehicle routing problem*, Computers & Operations Research, 39 (2012), pp. 1419–1431.

Index